Quantitative Genetics in the Wild

Quantitative Genetics in the Wild

EDITED BY

Anne Charmantier

Centre National de la Recherche Scientifique, Montpellier, France

Dany Garant

Université de Sherbrooke, Canada

Loeske E. B. Kruuk

University of Edinburgh, United Kingdom

and

The Australian National University, Australia

OXFORD
UNIVERSITY PRESS

Great Clarendon Street, Oxford, OX2 6DP,
United Kingdom

Oxford University Press is a department of the University of Oxford.
It furthers the University's objective of excellence in research, scholarship,
and education by publishing worldwide. Oxford is a registered trade mark of
Oxford University Press in the UK and in certain other countries

First Edition published in 2014

Published in the United States of America by Oxford University Press
198 Madison Avenue, New York, NY 10016, United States of America

British Library Cataloguing in Publication Data
Data available

Library of Congress Control Number: 2013955895

ISBN 978–0–19–967424–4

Foreword

Why do individuals differ from each other? How can we tease apart the complex effects of environments, parents, family and genes? What do the size of these effects, and the way that they contribute to differences between populations, mean in terms of the way that evolution has shaped biodiversity? And what do they mean when we think about the rapidly changing world in which we live? Questions of this sort lie at the heart of a vigorous and vibrant field: the application of quantitative genetics to wild populations. This book is both a summary of the state of the art as well as a mission statement for the future.

That this research field has a future, and indeed, one that is brighter and more relevant than ever, is abundantly clear from the work described in this book. The vitality of this field is notable, because it is emerging unscathed, indeed, strengthened, from what might have been considered a major threat derived from other ways of understanding the genetic basis of characters. Quantitative genetics had its origin among the biometricians at the turn of the 19th century, at about the time that 'classical' (i.e. Mendelian) genetics was rediscovered. Although Fisher showed how the two approaches could be combined almost a century ago, they have, in recent times, with the explosion of molecular genetic and genomic approaches, often been seen as offering competing frameworks. On reflection, 'competing' is perhaps the wrong term; at least, for competition to be perceived, both approaches must acknowledge the other! I have lost count of the number of times, in conversation with colleagues working on lab model organisms, that I have been told that, unless I understood the molecular genetic basis of a character, I didn't know anything about the genetic basis of a trait. On the other hand, I do remember, quite vividly, how hard I had to argue with a representative of one of the UK research councils that work on quantitative genetics of wild great tits did fall legitimately within their remit of fundamental research into genetics.

Perhaps field ecologists interested in quantitative genetics have been too reticent in the face of such dogmatism. Whilst the past few years have seen several high profile papers in the weekly 'tabloid' journals dissecting the single-locus genetic basis of functional traits in wild populations, there is a growing realisation that these may be relatively rare examples. The huge effort expended, for relatively meagre return, in studies of genetics of human quantitative characters and disease is a salutary lesson that even with enormous sample sizes, and genetic marker density at levels that are only just within reach of studies of wild organisms, relatively little variance in quantitative traits may be attributable to the effects of specific identifiable loci. Aulchenko *et al.* (2009) illustrated this with the case of human height, showing that genotyping the (at the time) 54 SNPs with largest effect, in a sample of 5748 people, explained only about a tenth as much variance as did the 'Victorian' method developed by Galton, which simply used the parental mid-point as a prediction. There is something simultaneously remarkable and encouraging about the fact that a centuries-old method requiring no more than a ruler, a pencil and (I suppose) a slide rule, outperformed, by an order of magnitude, the fruits of the genomic revolution. This gap will continue to narrow, of course, but this example, many others like it, and emerging evidence from wild populations of the highly polygenic nature of many quantitative characters, serves to legitimise the quantitative genetic approach.

As the chapters in this book demonstrate, the great strength of a quantitative genetic approach is that it is a flexible way to ask questions about the causes of variation and their effects: I share the editors' enthusiasm for viewing the application of quantitative genetics to wild populations as providing a broader analytical framework to think about all sorts of causes of variation, including environmental, genetic and developmental processes. As a consequence, we can fit questions about the adaptive influence of mothers on offspring, about epigenetics, about the developmental processes associated with ageing, about mechanisms of sexual selection, about effects of climate change, and about the influence of social processes within one coherent framework, and doing so provides a much richer understanding of the role of genetics in evolution and ecology.

This book is also forward-looking, and there are two particular aspects of this, among many, that I wish to highlight. First, it is clear that there are huge opportunities to be gained by combining classical 'phenotype-based' quantitative genetics with molecular genomics. These range from the ability to determine relatedness in systems where this has been impossible, or impractical, via deriving true measures of pairwise relatedness, rather than expected ones, to combining pedigrees with markers to test models of genetic architecture. Ironically, because it gets ever easier and cheaper to derive genetic information, the limiting step in such combined studies, and in quantitative genetic studies of wild populations generally, may be the quality and extent of phenotypic data. In many cases, long-term studies are limited by the decisions made by previous generations about which phenotypes to study. Digital techniques and remote- or automated-tracking of organisms offer the scope to collect very rich phenotypic data, including that relating to social and behavioural traits, and we should perhaps be thinking harder about how we can lay the foundation, in terms of phenotypic data, for the (academic) generations that may follow us.

Second, the current foundations of quantitative genetics in the wild are based almost entirely on vertebrate populations, with a disproportionate number of estimates derived from a very limited sample of species. It is very encouraging to see active consideration being given here as to how these taxonomic blinkers can be lifted, and very stimulating to think about how the experimental approach to quantitative genetics that typifies work in invertebrates and plants can inform these more ecologically framed studies.

Whilst the history of application of quantitative genetics in the wild is almost four decades old at the time of writing, the explosion of interest is more recent. There are probably many reasons for that, some of which are outlined in the following chapters, but one that I think should not be neglected is the series of meetings held among a group of practitioners of this approach, at 2–3 year intervals since the first in 2004. These meetings (known as the Wild Animal Model Biennial Meeting) have always been held in quite remote locations (Rum, Scotland 2004; Gotland, Sweden 2007; Dejioz, Italy 2009; Corsica 2011), far from the usual big-city hotel milieu of conference centres, and always close to a study site that hosted a population that was an active model for quantitative genetics in the wild. Informal, and with a timetable that was sufficiently elastic to incorporate extended, sometimes very extended, discussions of the points made by speakers, these have been among the most intellectually satisfying and invigorating of meetings, with the feeling that, after each one, genuine progress had been made in the field. Many, but by no means all, of the authors of the chapters that follow have also been key participants in these meetings (indeed, two of the editors organised a meeting each), and the feeling on reading the chapters here is not unlike that of attending one of those meeting: real progress has been made, and there are tremendous opportunities for more work in the future.

Ben Sheldon
Oxford & Uppsala
October 2013

Reference Cited

Aulchenko, Y.S., Struchalin, M.V., Belonogova, N.M., Axenovich, T.I., Weedon, M.N., Hofman, A., Uitterlinden, A.G., Kayser, M., Oostra, B.A., van Duijn, C.M., Janssens, A.C. & Borodin, P.M. (2009) Predicting height by Victorian and genomic methods. *European Journal of Human Genetics*, **17**, 1070–1075.

Acknowledgements

We have many people to thank for their involvement in the production of this book.

Most importantly, we are very grateful to all the authors for their excellent contributions, for (mostly!) meeting the deadlines, for taking on board the editorial and reviewers' comments, for reviewing other chapters and overall for making the editorial process an enjoyable experience for us. Many thanks also to the external reviewers of all the chapters for their time.

Many contributors to this book have gathered over the years in informal biennial meetings to discuss methodological, theoretical and empirical advancements in quantitative genetics applied to wild populations. These meetings, held in remote study sites across Europe (the Isle of Rum, Scotland; the island of Gotland, Sweden; Gran Paradiso National Park, Italy; and Fango Valley in Corsica, France) have been incredibly inspiring for the exchange of ideas and techniques, and for triggering new collaborations. We thank all the past organisers and attendees of these meetings.

We would all like to thank our respective institutes for stimulating environments. Anne Charmantier thanks the Centre d'Ecologie Fonctionnelle et Evolutive in Montpellier, as well as the Large Animal Research Group (LARG) at the University of Cambridge for a very welcoming sabbatical stay. Dany Garant thanks the Université de Sherbrooke and the Molecular Ecology and Evolution Laboratory (MEEL) at Lund University for offering ideal conditions to oversee the editing stage for this book. Loeske Kruuk has been based at the Institute of Evolutionary Biology, University of Edinburgh, and particularly thanks the Wild Evolution Group there, and at the Division of Evolution, Ecology and Genetics at the Australian National University, Canberra, and is very grateful for the supportive and stimulating environments of both. Finally, there are many colleagues whom we would like to thank for useful discussions and inspiration over the years. In addition to the other authors in this book, these include: Ben Sheldon, Josh Auld, Nick Barton, Louis Bernatchez, Jacques Blondel, Sue Brotherstone, Luc Bussière, Luis-Miguel Chevin, Andrew Cockburn, Tim Coulson, Etienne Danchin, Patrice David, Marco Festa-Bianchet, Juan Fornoni, Peter Grant, Jarrod Hadfield, Adam Hayward, Thomas Hansen, Bengt Hansson, Bill Hill, Elise Huchard, Philippe Jarne, Mark Kirkpatrick, Jeff Lane, Fanie Pelletier, Daniel Promislow, Benoit Pujol, Katja Räsänen, Denis Réale, Derek Roff, Marcel Visser and Craig Walling.

AC was funded by the French Agence Nationale pour la Recherche (grant ANR-12-ADAP-0006) and by an Overseas Fellowship from the Service pour la Science et la Technologie de l'Ambassade de France au Royaume-Uni. DG was supported by funding from the Natural Sciences and Engineering Research Council of Canada (NSERC). LK was funded by an Australian Research Council Future Fellowship. We are grateful to the OUP team, and especially Lucy Nash, for all their work. Thanks to them, the whole process was in fact smoother than expected.

Finally, we thank our families for their love and support, and dedicate this book to our children, Vincent, Laetitia, Alexanne, Émilien, Saskia, Lyndon and Edward.

Anne Charmantier
Dany Garant
Loeske Kruuk

Contents

List of Authors

Alexander V. Badyaev Department of Ecology and Evolutionary Biology, University of Arizona, Tucson, AZ 85721, United States
abadyaev@email.arizona.edu

Russell Bonduriansky Evolution & Ecology Research Centre and School of Biological, Earth and Environmental Sciences, University of New South Wales, Sydney NSW 2052, Australia
r.bonduriansky@unsw.edu.au

Jon E. Brommer Department of Biology, University Hill, 20014 University of Turku, Finland
jon.brommer@utu.fi

Anne Charmantier Centre d'Ecologie Fonctionnelle et Evolutive, UMR 5175, Campus CNRS, F34293 Montpellier Cedex 5, France
anne.charmantier@cefe.cnrs.fr

Tim Clutton-Brock Department of Zoology, University of Cambridge, Cambridge CB2 3EJ, United Kingdom
t.h.clutton-brock@zoo.cam.ac.uk

Niels J. Dingemanse Behavioural Ecology, Department Biology II, Ludwig-Maximilians University of Munich, Planegg-Martinsried, Germany
and
Evolutionary Ecology of Variation Group, Max Planck Institute for Ornithology, Seewiesen, Germany
ndingemanse@orn.mpg.de

Ned A. Dochtermann Department of Biological Sciences, North Dakota State University, United States
ned.dochtermann@gmail.com

Blandine Doligez Université Lyon 1, CNRS, UMR 5558, Laboratoire de Biométrie et Biologie Evolutive, 69622 Villeurbanne, France
blandine.doligez@univ-lyon1.fr

Dany Garant Département de Biologie, Université de Sherbrooke, Sherbrooke, QC, J1K 2R1, Canada
Dany.Garant@usherbrooke.ca

Phillip Gienapp Netherlands Institute of Ecology (NIOO-KNAW), Department of Animal Ecology, P.O. Box 50, 6700 AB Wageningen, The Netherlands
p.gienapp@nioo.knaw.nl

Olivier Gimenez Centre d'Ecologie Fonctionnelle et Evolutive, UMR 5175, Campus CNRS, F34293 Montpellier Cedex 5, France
olivier.gimenez@cefe.cnrs.fr

Henrik Jensen Centre for Biodiversity Dynamics, Dept. of Biology, Norwegian University of Science and Technology, NO-7491 Trondheim, Norway
Henrik.Jensen@ntnu.no

Lukas F. Keller Institute of Evolutionary Biology and Environmental Studies, University of Zurich, Winterthurerstrasse 190, CH-8057 Zurich, Switzerland
lukas.keller@ieu.uzh.ch

Loeske E. B. Kruuk Institute of Evolutionary Biology, School of Biological Sciences, University of Edinburgh, Edinburgh EH9 3JT, United Kingdom
and
Division of Evolution, Ecology & Genetics, Research School of Biology, The Australian National University, Canberra, ACT 0200, Australia
Loeske.Kruuk@ed.ac.uk

Andrew G. McAdam Department of Integrative Biology, University of Guelph, Guelph, N1G 2W1, ON, Canada
amcadam@uoguelph.ca

Juha Merilä Ecological Genetics Research Unit, Department of Biosciences, PO Box 65 (Biocenter 3, Viikinkaari 1), FIN-00014 University of Helsinki, Finland
Juha.Merila@helsinki.fi

Michael B. Morrissey School of Biology, University of St. Andrews, St Andrews, KY16 9TH, United Kingdom
michael.morrissey@st-andrews.ac.uk

Daniel H. Nussey Institute of Evolutionary Biology, School of Biological Sciences, University of Edinburgh, Edinburgh EH9 3JT, United Kingdom
dan.nussey@ed.ac.uk

Josephine M. Pemberton Institute of Evolutionary Biology, School of Biological Sciences, University of Edinburgh, Edinburgh EH9 3JT, United Kingdom
j.pemberton@ed.ac.uk

Erik Postma Institute of Evolutionary Biology and Environmental Studies, University of Zurich, CH-8057 Zurich, Switzerland
erik.postma@ieu.uzh.ch

Jane M. Reid Institute of Biological and Environmental Sciences, School of Biological Sciences, University of Aberdeen, Aberdeen, AB24 2TZ, United Kingdom
jane.reid@abdn.ac.uk

Matthew R. Robinson Department of Animal and Plant Science, University of Sheffield, Sheffield, S10 2TN, United Kingdom
and
Queensland Brain Institute (QBI), The University of Queensland, St Lucia, QLD 4072, Australia
m.robinson11@uq.edu.au

Jon Slate Department of Animal and Plant Sciences, University of Sheffield, Sheffield, S10 2TN, United Kingdom
j.slate@sheffield.ac.uk

John R. Stinchcombe Department of Ecology and Evolutionary Biology & Koffler Scientific Reserve at Joker's Hill, University of Toronto, Toronto, ON, M5S 3B2, Canada
John.Stinchcombe@utoronto.ca

Marta Szulkin Centre d'Ecologie Fonctionnelle et Evolutive, UMR 5175, Campus CNRS, F34293 Montpellier Cedex 5, France
and
Department of Zoology, University of Oxford, Oxford, OX1 3PS, United Kingdom
marta.szulkin@zoo.ox.ac.uk

Céline Teplitsky Centre des Sciences de la Conservation, UMR 7204 CNRS - MNHN - UPMC, CP 51, 75005 Paris, France
teplitsky@mnhn.fr

Pierre de Villemereuil Université Joseph Fourrier, Laboratoire d'Ecologie Alpine, 38400 Saint-Martin d'Hères, France
bonamy@magbio.ens.fr

J. Bruce Walsh Department of Ecology and Evolutionary Biology, University of Arizona, Tucson, AZ 85721, United States
jbwalsh@u.arizona.edu

Alastair J. Wilson Daphne du Maurier, Centre for Ecology and Conservation, College of Life and Environmental Sciences, University of Exeter, Cornwall Campus, TR10 9EZ, United Kingdom
A.Wilson@exeter.ac.uk

Matthew E. Wolak Department of Biology and Graduate Program in Evolution, Ecology, and Organismal Biology, University of California, Riverside, United States
matthewwolak@gmail.com

Felix Zajitschek Department of Animal Ecology, Evolutionary Biology Centre, Uppsala University, 752 36 Uppsala, Sweden
felix.zajitschek@biodemography.net

CHAPTER 1

The study of quantitative genetics in wild populations

Loeske E. B. Kruuk, Anne Charmantier and Dany Garant

1.1 Why study quantitative genetics?

A core aim of evolutionary biology is to explain the biological diversity of natural populations. This diversity occurs at multiple levels: between species or higher taxonomic groups, between populations of the same species, between individuals of the same population, or between different time points in an individual's life. Quantitative genetics, the study of the genetic basis of complex (or 'quantitative') traits, is concerned with these lower levels, and in particular with the diversity between individuals in a population, and the extent to which it is determined by genetic vs non-genetic causes (Fisher 1918; Wright 1921). In addition to addressing the fundamental question of the relative contribution of 'nature' vs 'nurture' to variation, knowledge of levels of genetic variance is critical for assessing the extent to which changes in phenotypic traits due to selection are passed on from one generation to the next—i.e. the microevolutionary dynamics of traits. Plant or animal breeders therefore use quantitative genetics to determine how artificial selection can change the distribution of phenotypes within a population (Lush 1937; Falconer & Mackay 1996). Evolutionary biologists also want to understand and even predict the effects of selection, but with a focus on natural or sexual selection: quantitative genetic analyses provide information about the raw material on which selection can work (Roff 1997; Lynch & Walsh 1998). The application of quantitative genetics to evolutionary biology has generated a large and rapidly changing field (for an excellent history of the subject, see Lynch & Walsh 1998). In this book, we aim to provide an overview of one particular area of this wide field: quantitative genetic studies of wild populations inhabiting natural environments, motivated by evolutionary ecologists wishing to address core evolutionary questions in realistic ecological settings.

The last decade has seen a rapid expansion in quantitative genetic studies in natural environments (see Chapter 2, Postma), fuelled by methodological advances in molecular genetics and statistical techniques (Kruuk *et al.* 2008), and by increasing availability of suitable long-term datasets, especially in animals (Clutton-Brock & Sheldon 2010). As a result, studies of 'wild quantitative genetics' have provided insights into a range of important questions in evolutionary ecology, some in well-established fields such as life-history theory, behavioural ecology and sexual selection, others addressing relatively new issues such as the response of populations to climate change, or the process of senescence. This work is motivated in part by the increasing appreciation of the need to quantify the genetic—rather than just phenotypic—diversity in key traits, and the genetic basis of the associations between traits (Roff 2007): phenotypic associations may not be accurate representations of the underlying genetic associations that will ultimately determine evolutionary dynamics, especially in studies of populations experiencing natural environmental heterogeneity (Kruuk *et al.* 2008).

We use the term 'quantitative genetics' somewhat loosely, to cover a range of aspects of the

Quantitative Genetics in the Wild. Edited by Anne Charmantier, Dany Garant, and Loeske E. B. Kruuk

evolutionary ecology of populations—in fact a more accurate (if even less appealing) term might have been 'variance component analysis'. Thus, whilst a core aim is often to estimate levels of genetic variance and heritability for particular traits, as well as the structure of the multivariate genetic relationships between them, we are also interested in the other sources of variation that may be important for a wild population: for example, effects of environmental variation due to phenotypic plasticity, maternal effects (genetic or environmental), genotype–environment interactions, dominance variance, or the effects of ageing. The statistical tools of quantitative genetics, and the pedigree data required to estimate levels of genetic variance, fortunately provide efficient ways of exploring these fascinating questions.

1.1.1 Ten big questions

A research field is driven by the central questions or hypotheses it aims to address. Below is a (non-exhaustive) list of what we see as core questions in current evolutionary quantitative genetics.

1. What is the genetic basis of variation in phenotypic traits, and of covariation between traits?
2. Is there heritable genetic variance for fitness? Across traits, how is genetic variance maintained in the face of erosion by selection?
3. To what extent do genetic trade-offs shape the evolution of life histories? More generally, how widespread are evolutionary genetic constraints?
4. Can we predict evolutionary responses to selection pressures? Or, why does artificial selection generate predictable evolutionary responses, but natural selection does not?
5. To what extent is the phenotype of an individual shaped by the genotypes of other individuals in the population—for example by maternal effects?
6. Do individuals vary in their response to environmental conditions, and is this variation genetically based: how prevalent are genotype–environment interactions? Do other components of variance change with environmental conditions?
7. Why does senescence occur?
8. Why does sexual dimorphism occur?
9. How much inbreeding and inbreeding depression are there in a population?
10. How will climate change affect the evolutionary dynamics of natural populations?

These questions can be addressed with many types of study populations, but as we discuss below—and as we hope this book illustrates—they address issues into which studies in natural environments can provide valuable insights.

1.1.2 Why in the wild?

Quantifying genetic effects in artificial (domestic or laboratory) populations under controlled conditions is undoubtedly easier than in wild populations experiencing natural environments, and obviously provides invaluable insights into evolutionary processes (Roff 1997). However, the importance of genetic variation is arguably best assessed relative to other causes of variation, requiring an understanding of both genetic and environmental variation, and by extension a need for relevant environmental conditions. Furthermore, there is increasing evidence for the impact of environmental conditions both on the selection processes in which we are interested (Endler 1986; Wade & Kalisz 1990) and on the expression of genetic variance (Hoffmann & Merilä 1999; Charmantier & Garant 2005)[1]. This suggests that extrapolation of estimates from artificial conditions to more realistic ecological contexts may be difficult. Third, simple theoretical

[1] One point to note here is that in referring to estimates in 'wild' populations, we mean exactly that: phenotypes are typically measured in individuals inhabiting natural environments. Previous comparisons of 'lab' vs 'field' heritabilities (e.g. Simons & Roff 1994; Roff 1997; Hoffmann 2000) have involved 'field' populations in which individuals have been collected in the field and brought into and bred in the lab, so that the 'field' vs 'lab' contrast lies in the source of the population, not in the location in which phenotypic variation is expressed. Although some of these comparisons suggest that lab heritabilities provide good surrogates for field heritabilities (Roff 1997), we believe it is worth bearing this distinction in mind. Comparison of lab with true field heritabilities is inevitably difficult given that lab studies tend to involve shorter-lived and smaller organisms, predominately invertebrates, whereas field studies tend to involve relatively longer-lived species in which individuals are easily monitored in the field.

predictions for the expected cross-generational responses to selection (the 'breeder's equation', see Box 1.1), which work for artificial selection on single traits in controlled conditions (Roff 2007), do not seem to hold when considering natural selection in wild populations (Merilä *et al.* 2001). Multiple explanations for this mismatch have been proposed, but most centre on the fact that real-world natural selection, involving multiple traits, is likely to be much more complex than artificial selection (Rausher 1992; Kruuk *et al.* 2008; Walsh & Blows 2009; Morrissey *et al.* 2010). Fourth, there are arguably many important traits, for example life-history or behavioural traits, which will not be expressed properly in artificial conditions, but which are critical components of a species' biology. In particular, estimates of natural variation in individual fitness, comprising natural variation in survival and fecundity, may only be feasible in field studies. Fifth, increased appreciation of the potential feedbacks between the ecological and evolutionary dynamics of a population underlines the value of investigating evolutionary parameters in a relevant ecological setting (Pelletier *et al.* 2009). In relation to this, it is worth noting that almost all of the quantitative genetic analyses of field data discussed in this book have arisen as extensions of ecological or behavioural studies (see below), reflecting a rapid expansion of activity at the interface between evolutionary biology and ecology.

However, despite these arguments, we do not wish to imply any artificial distinction between evolutionary quantitative genetic studies under artificial or natural conditions. Clearly some of the interesting aspects of the latter, such as natural (i.e. uncontrollable) environmental heterogeneity, can also constitute serious drawbacks, and opportunities for experimental manipulation are greatly reduced. As the following chapters illustrate, research in the field is motivated by general questions such as those above, and in evaluating empirical evidence we need to consider results drawn from both artificial and 'wild' populations.

In this chapter, we first outline briefly the basic principles of a quantitative approach and of the most commonly used statistical tools, by way of introduction to the subject for readers with less familiarity with the concepts (Section 1.2). Box 1.1 contains a glossary of important terminology which appears repeatedly throughout the book. We then provide an overview of the different chapters in the book (Section 1.3), and finally we discuss some recurrent challenges (Section 1.4) and then consider some emerging topics in the field (Section 1.5).

1.2 How? The basic tools of quantitative genetics

1.2.1 Estimating similarity between relatives

At the core of a quantitative genetic analysis is estimation of the extent of genetic control of traits and of the associations between different traits, i.e. levels of genetic variance, its magnitude relative to the overall phenotypic variance or the trait's heritability, and the genetic determinants of correlations between traits (Falconer & Mackay 1996, see Box 1.1 for definitions). This estimation relies on the concept that if a complex (or continuous, or 'quantitative') trait is genetically determined, then individuals who share the same genes should have similar phenotypes: in other words, the degree of phenotypic similarity between relatives should reflect the genetic control of that trait. A trait can be any measure on an individual, for example body size, number of babies, antibody levels, aggression score, plumage colouration, or when it breeds. The approach relies on an assumption that quantitative traits are likely determined by very large numbers of genes spread across the genome, an assumption that (reassuringly) appears to be upheld by both the results of selection experiments and recent molecular data (Hill & Kirkpatrick 2010; Hill 2012). The degree to which two individuals share the same genes depends on their relatedness, which can be quantified either via knowledge of a pedigree (or family tree, constructed from knowledge of each individual's parents), or with appropriate genomic marker data (Lynch & Walsh 1998).

The similarity (covariance) between pairs of individuals for a given phenotypic trait is therefore determined by i) the relatedness of the pair and ii) the degree of genetic variance underlying the trait. The phenotypic covariance can be observed and the relatedness can be estimated, so we can solve statistically for an estimate of the additive genetic variance (see Box 1.1). These calculations can be done in different ways, the simplest being to use only relatives of a certain type and consider, for example, the similarity of offspring to their parents (a 'parent–offspring regression'), or among groups of full or half siblings. In practice, if we have phenotypic information on individuals in a population, it is most efficient to consider the covariance between as many pairs as possible, which is feasible using a form of mixed-effect model known as an 'animal model' (Henderson 1975; Lynch & Walsh 1998). An animal model partitions the phenotypic trait of an individual into contributions from predictable effects (e.g. sex, age, climate), termed 'fixed effects', and other effects for which we wish only to estimate the overall variance in individual effects, known as 'random effects'. For the latter, given pedigree or relatedness information, we can fit an additive genetic term which exploits the fact that the effect of an individual's genotype (or specifically, the additive genetic value of its genotype; see Box 1.1) will be similar to that of its relatives, and the degree of similarity will scale with the degree of relatedness. Box 1.2 contains a brief overview of animal models; for more details, see Lynch & Walsh (1998).

For no particularly clear reason, other than possibly computational demands, application of the animal model to evolutionary quantitative genetic studies outside plant and animal breeding is surprisingly recent (for a brief history, see Kruuk 2004): the earliest applications to data from free-ranging populations being for rhesus macaques (*Macaca mulatta*; Konigsberg & Cheverud 1992), and three populations of ungulates: bighorn sheep (*Ovis canadensis*; Réale *et al.* 1999), Soay sheep (*Ovis aries*; Milner *et al.* 2000) and red deer (*Cervus elaphus*; Kruuk *et al.* 2000). The late arrival of the animal model in studies of the quantitative genetics of wild species, relative to its ubiquity in applied research, is especially surprising given that some of its strongest advantages are in dealing with the problems posed by data from natural populations: it is relatively tolerant of unbalanced designs, missing trait data and pedigree links, and the complexities of heterogeneous environmental conditions (Kruuk 2004; Wilson *et al.* 2010). However, despite the late start, it has now been applied to dozens of different populations (see Chapter 2, Postma), and this application has provided great impetus to the current interest in wild (and also non-wild) evolutionary quantitative genetics. For a practical guide to application of the animal model for ecologists, see Wilson *et al.* (2010).

1.2.2 Role of long-term studies

The vast majority of quantitative genetics in the wild has to date been conducted on populations that have been the subject of long-term study, in many cases over several decades (Clutton-Brock & Sheldon 2010). Clearly such studies offer many advantages, one of which is that most were set up by ecologists and have been used for extensive investigations into the effects of natural environmental variation as well as the mating systems and behavioural ecology of the study species: quantitative genetic analyses are therefore generally founded in a thorough understanding of a population's ecology. However, reliance on long-term studies has obvious drawbacks: a new study on a new species cannot obviously be created and used immediately, funding bodies do not work to delivery points several decades away, and continuous maintenance of ongoing studies in a harsh funding environment can be difficult. The ability to estimate relevant genetic parameters from genomic data will change this dependence on historical information to some extent, but even if it removes the need for a multigenerational pedigree, it still cannot generate estimates of the impact of temporal environmental variation, nor of any interaction of environmental and genetic variance. Use of historical data from long-term studies also generally relies on correlational associations between traits, despite the value of experimental manipulations such as cross-fostering for separating genetic and non-genetic causes of similarity between relatives (Merilä & Sheldon 2001).

The timeline of analyses in one study population of our cover species, the great tit (*Parus major*), illustrates the development of quantitative genetic studies of wild populations. The long-term study of the great tit population in Wytham Woods, Oxford, UK (running since 1947, Lack 1964) generated possibly the earliest field heritability estimate from a wild population, the inheritance of clutch size (Perrins & Jones 1974). Subsequent analyses have progressed from single-trait models using either parent–offspring regressions (van der Jeugd & McCleery 2002) or the animal model (McCleery *et al.* 2004), to bivariate models (Garant *et al.* 2008), random regressions to test for genotype-by-environment interactions (Charmantier *et al.* 2008; Husby *et al.* 2010), tests for environmentally induced variation in inbreeding depression (Szulkin & Sheldon 2007) and senescence (Bouwhuis *et al.* 2010), analysis of trends in breeding values (Garant *et al.* 2008) and subsequent reanalysis with more appropriate methods (Hadfield *et al.* 2010), and most recently, genomic marker-based partitioning of variances and covariances (Santure *et al.* 2013). Studies have therefore progressed from simple estimates of heritability to much more sophisticated tests of some of the key hypotheses at the heart of quantitative genetics.

1.3 Overview of chapters

In this book we invited a range of researchers in the field to illustrate how quantitative genetics research in the wild has developed over the years and to provide an up-to-date resource covering the most important topics addressed by this area of research.

Defining the heritable basis of a trait was the main goal of most early studies of quantitative genetics in the wild (see the great tit examples above; Boag & Grant 1978; and reviews in Mousseau & Roff 1987; Merilä & Sheldon 2001; Visscher *et al.* 2008). The book thus starts with an in-depth analysis of the variation in heritability estimates published over the last four decades from wild populations (Chapter 2, Postma). Postma analysed 1600 heritability estimates from over 50 species and traits, showing that heritabilities have become more precise and less biased over time. This seems to result from both the application of the animal model, and the inevitable strengthening of datasets over time, with resultant improvements in the quality of pedigree information. Postma also assesses the relationships between the estimates of heritability and the coefficient of additive genetic variance, and shows that it is weak at best (and even negative) and thus that there is little concordance between the two metrics (see also Houle 1992; Hansen *et al.* 2011), re-emphasising the need to report and compare both in future studies.

Chapters 3 to 5 illustrate that the methods associated with quantitative genetic analyses have now been successfully applied in natural populations for the study of a variety of fundamental ecological and evolutionary processes. In Chapter 3, Reid shows how quantitative genetics can be applied to deriving and testing pertinent sexual selection theory in wild populations experiencing natural genetic and environmental variation. She uses two case studies in birds to illustrate how quantitative genetics can bring new insights in the evolutionary causes and consequences of mate choice and sexual selection, as well as trait and mating system evolution. In Chapter 4, Dingemanse and Dochtermann show how the theory and tools already adopted by quantitative geneticists can be used by behavioural ecologists interested in the adaptive nature of between-individual variation in behaviour. They further show that theory and empirical research in behavioural ecology might inform quantitative geneticists as to how and why trait variation is distributed, thus illustrating how these fields would gain from a more integrative approach and sustained exchange of ideas (Owens 2006). Finally, the authors suggest how we can bridge the gap between the two disciplines by presenting theoretical and empirical demonstrations of the statistical language familiar to quantitative geneticists in order to explain behavioural patterns of current interest. Charmantier, Brommer and Nussey (Chapter 5) follow with a review of the concepts and analyses related to senescence in the wild. They start by discussing the main classical evolutionary theories of ageing, emphasising the importance of estimating age-dependent patterns of genetic (co)variance (G × Age interactions). They then outline a detailed statistical framework with which to quantify G × Age, and review the literature supporting evidence for individual differences in senescence rates in wild vertebrates. They conclude their chapter by identifying important statistical issues, forthcoming challenges, and recommendations for future work in this field of research. In particular they call for higher standards of analysis and reporting to facilitate generalisation about senescence patterns across populations and species.

Besides the assessment of additive genetic variance and heritabilities, the importance of quantifying other variance components that are relevant for evolution has been increasingly recognised (see Crnokrak & Roff 1995; Mousseau & Fox 1998; Keller & Waller 2002; Räsänen & Kruuk 2007). As a result, there is a growing interest in estimating these components in the wild, as emphasised by the following two chapters. In Chapter 6, McAdam, Garant and Wilson emphasise the importance of considering 'indirect genetic effects' (Box 1.1) for studies of evolutionary dynamics. In particular, they provide conceptual and analytical background to the importance of maternal effects, the best studied type of indirect effects. They point out important emerging questions in the field such as the need to explore the evolutionary implications of social interactions across a wider range of contexts and scenarios. In the next chapter, Wolak and Keller (Chapter 7) review the main issues related to the estimation of non-additive variance, especially dominance variance. They present an overview of empirical estimates obtained in laboratory and agricultural populations, and conclude that dominance variance is a major contributor to phenotypic variation, and may even rival additive genetic variance. As estimates of dominance variance in the wild are still lacking, the authors explore the practical considerations for quantifying these effects in wild populations. They conclude their chapter by discussing how inbreeding affects estimates of non-additive genetic variance.

It is evident from the literature content of the field that, despite several years of research in quantitative genetics in the wild, most studies are still based on a rather limited number of species/populations (see below, and also Chapter 2, Postma). Yet, several systems offer promising perspectives for future developments in order to reach a broader taxonomic coverage in this field. For example, Stinchcombe (Chapter 8) provides an original and constructive approach comparing studies published on long-lived mobile animals in the wild with those focussing on short-lived plants mainly performed on a single generation and/or under common garden conditions. In particular, he explores the conceptual, analytical, and biological insights that might be obtained from applying lessons and techniques of experimental studies in plant evolutionary ecology to studies of wild vertebrate populations, and vice versa. This chapter reviews important findings in plant evolutionary ecology and their potential implications for wild animals, and also assesses the main challenges that have so far prevented the potential application of wild quantitative genetic approaches in free-living plant populations. In Chapter 9, Zajitschek and Bonduriansky consider recent developments in assessing genetic variation in fitness-related traits in wild populations of arthropods. The life-history characteristics of insects—which made them typical model species for many laboratory studies—have resulted in a near complete absence of genetic parameter estimates from wild populations. They suggest potential ways to fill this gap, and discuss some examples of suitable systems for doing so. They emphasise that much will be gained from studies of quantitative genetic parameters for natural populations of invertebrates as they will allow for comparison with the enormous literature on captive invertebrate populations, as well as extend our knowledge of quantitative genetics in the wild to a broader array of taxonomic coverage.

Development of research in quantitative genetics in the wild has resulted in a transition from studies conducted on single traits to applications of multivariate analyses (Arnold *et al.* 2008; Walsh & Blows 2009). As such, both theoretical and empirical considerations of the ***G***-matrix in nature are presented in the next three chapters. In Chapter 10, Kruuk, Clutton-Brock and Pemberton present an empirical case study to illustrate recent developments in applications of quantitative genetic analyses, using 40 years of data to apply a multivariate quantitative genetic approach to sexually selected antler traits in a red deer population from the Isle of Rum, Scotland. Despite computational constraints due to the demanding nature of multivariate analyses, they find significant positive covariances between antler traits, positive phenotypic selection, and genetic variance for annual breeding success. However, their results also reveal that environmentally driven associations between traits and components of fitness can generate the appearance of selection which has no evolutionary relevance because of the lack of appropriate genetic covariance between trait and fitness component. In Chapter 11, Badyaev and Walsh consider the contribution of epigenetic developmental dynamics to the maintenance of multivariate genetic variation in complex traits that are subject to strong natural selection. They combine geometric and developmental perspectives to the understanding of the evolution of genetic architecture that reconciles precise adaptation, evolutionary diversification, and environmentally contingent developmental variation. As a case study, they assess the importance of forces that shape the current **G**-matrix of beak traits for a population of house finches (*Carpodacus mexicanus*) studied over several generations. In doing so, they show that the dimensionality estimated at the genetic level of a structure is often far smaller than is expected from the dimensionality of its phenotype. Finally, Teplitsky, Robinson and Merilä (Chapter 12) provide an overview of our current knowledge and limitations in the study of evolutionary potential and constraints in wild populations. They then examine available data regarding the stability of genetic architecture across different ecological timescales. They focus especially on the current state of the field in dealing with the assessment of multivariate evolutionary potential, the evaluation of genetic constraints and the effect of evolutionary forces on the structure of **G**-matrices. Finally, they use a simulation-based approach to compare several matrix comparison statistics with respect to their capacity to detect differences in **G**-matrices.

Quantitative genetics in the wild is still expanding as a field of research, and the final three chapters suggest promising avenues for future developments. First, Jensen, Szulkin and Slate (Chapter 13) tackle important aspects related to the newly emerging field of molecular quantitative genetics by showing how high-throughput genomic approaches are increasingly being applied to evolutionary quantitative genetics research. They first describe how newly available molecular approaches promise to enhance our understanding of the genetic architecture and evolutionary dynamics of fitness-related traits in non-model species in the wild. They then examine how the integration of genomic data is allowing detailed population genetic analyses of natural populations and emphasise how these approaches are highly complementary to quantitative genetics; for instance, they allow identification of genes and/or genomic regions that are under selection. Morrissey, de Villemereuil, Doligez and Gimenez (Chapter 14) then provide an overview of Bayesian statistics and their applications to quantitative genetic analyses of empirical data in the wild. They focus primarily on how Bayesian Markov Chain Monte Carlo (MCMC) algorithms are particularly suitable for such analyses. They provide examples of models in the BUGS statistical programming language which aim to demystify some aspects of these methods. They then discuss ways in which Bayesian tools can be used to make quantitative genetic inferences of complex data from natural populations and outline a range of benefits afforded by such applications. In particular, they emphasise that more direct inference of key evolutionary parameters and their associated error can be achieved than is often possible in frequentist frameworks. Finally in Chapter 15, Gienapp and Brommer emphasise the importance of improving our understanding of how climate change affects selection and the genetic variation in important traits in wild populations. To do so, they explore evidence for selection on phenological traits driven by climate change and then review quantitative genetic studies of these traits. They emphasise that very few studies reporting presumed evolutionary changes in response to climate change also considered phenotypic plasticity as a possible mechanism for such change, despite the need to assess whether observed changes related to climate are plastic and/or genetic. Their overview of the field suggests that evidence for genetic changes in response to climate change is scarce, yet it is still unclear if such absence also stems from a lack of statistical power and/or appropriate methods in previous studies.

1.4. Challenges

This book demonstrates that the field of quantitative genetics applied to populations studied in natural environments has extended substantially in the last two decades, providing fundamental insights into a wide range of topics in evolutionary ecology. Nevertheless, almost every chapter of this book contains discussion of problems inherent in applying quantitative genetic tools to data collected in natural populations; here we briefly outline some recurrent issues, before considering some promising emerging topics in Section 1.5.

1.4.1 Evaluating the potential for microevolution

There is increasing incentive to integrate knowledge of the heritable (co)variance displayed by key adaptive traits into policies for the conservation and management of wild populations (e.g. Hendry *et al.* 2003). Adaptive evolutionary change is predicted to be proportional to the force of selection combined with the level of heritability in the focal trait (Falconer & Mackay 1996). However, previous attempts to estimate the net intensity of selection based on observed evolutionary changes between generations (e.g. Hendry & Kinnison 2001) yielded estimates of selection forces that were far smaller than those estimated using direct regression-based approaches to quantifying selection (the Lande-Arnold approach: Lande & Arnold 1983; see examples in Kingsolver *et al.* 2001). In accordance with this observation, evolutionary stasis is commonly reported, i.e. we commonly observe no change over generations in traits that are apparently heritable and under directional selection (Merilä *et al.* 2001). Explaining this mismatch between natural world observations and theoretical expectations—which work under controlled conditions—has been a major motivation for much of the recent work on the quantitative genetics of wild populations (see above), and has generated many valuable insights into the microevolutionary dynamics of populations. However it has also clearly highlighted the challenges inherent in measuring both selection and genetic (co)variance accurately in natural populations. Thus, recent work has raised awareness of problems such as how to separate genetic from non-genetic causes of similarity between relatives (Kruuk & Hadfield 2007), estimate the form and force of natural selection (e.g. Morrissey & Sakrejda 2013), account for the 'invisible fraction' (the fact that selection may remove a proportion of individuals before they express a trait, Hadfield 2008), as well as the need to consider alternative predictions of evolutionary responses (Morrissey *et al.* 2010). Such discussions will hopefully lead to changes in analytical practice that should improve our ability to evaluate the evolutionary potential of natural populations. Furthermore, a fundamental question with regard to our assessment of evolutionary potential is whether the rate of evolution that we can measure on contemporary time scales (typically over 5–20 generations) is relevant for the adaptation of populations to environmental changes. Whilst there is clear evidence for microevolutionary change on such time scales (Hendry & Kinnison 2001), studies on the stability of the **G**-matrix over time (Arnold *et al.* 2008; see also Chapter 12, Teplitsky *et al.*) and on fluctuations of selection (Siepielski *et al.* 2009; Morrissey & Hadfield 2011) need to be extensively developed before we can answer this question.

1.4.2 Biostatistical issues

A recurrent limitation mentioned in many chapters of this book is the statistical power that pedigrees from wild populations offer to allow unbiased decomposition of the phenotypic variance into several genetic and non-genetic influences. Although sample sizes typically obtained from long-term monitoring in the wild are not small compared to standard behavioural ecology or life-history evolution studies, they are not comparable to the much larger sample sizes used in animal breeding. The limited sample sizes are partly explained by the logistical limitations of fieldwork, and because the organisms studied in natural conditions typically have longer lifespans than those most commonly used in laboratory studies. Thus, even studies based on decades of work can have limited statistical power to estimate quantitative genetic parameters, as witnessed by often large standard errors. The increasing awareness of the importance of considering multivariate associations (Blows 2007; Walsh & Blows 2009) heightens the need for large sample sizes given the additional demands on statistical power of higher-dimension multivariate models that include covariance components. The issue of statistical power is also particularly pertinent when one aims to disentangle dominance genetic variance estimates from other variance components using wild pedigrees (see Chapter 7, Wolak & Keller).

Another recurrent problem is the fact that pedigree data may not be entirely accurate. For example, most studies on birds use pedigrees constructed from the observed social bonds rather than based on neutral genetic markers, and may therefore contain errors due to extra-pair paternities. Genetic assignment of parentage may also involve low levels of error (for discussion, see Kruuk 2004). Attempts have been made to provide rules of thumb on how to avoid substantial biases in heritability estimations due to pedigree errors, with simulation studies concluding that the levels of error typical of most pedigrees may not have substantial impacts on estimates (Charmantier & Réale 2005). However, power and sensitivity analyses using simulation-based frameworks calibrated for specific datasets and pedigree structures are also now available (e.g. the R-package pedantics, Morrissey & Wilson 2010) and should become common practice.

Application of statistical techniques initially developed for data of a different type has had its problems, and the rapid expansion of the field has not been without pitfalls. Using estimates of breeding values (BLUPs) derived from pedigrees without sufficiently accounting for potential contributions of environmental variation (Postma 2006) and without considering their associated error (Hadfield *et al.* 2010) can generate misleading, anticonservative results, which can be avoided with application of suitable statistical approaches (Hadfield *et al.* 2010). Looking back over the development of the analytical tools over the last two decades, it is obvious that we are facing a challenge of increasing computational complexity. Whilst we need, as evolutionary biologists, to keep up with this progress and to create a positive feedback loop with methodological advances, it is also desirable to keep it simple where possible. Parent–offspring regressions are not always wrong, and over-specified animal models are often unnecessary and misinterpreted (Kruuk 2004; Wilson 2008).

1.4.3 Taxonomic gaps

A marked limitation of the current state of the field is that the taxonomic coverage of quantitative genetics in the wild, and hence of this book, is heavily biased towards vertebrates, and especially birds and mammals. By way of example, Postma's extensive literature review (Chapter 2) is restricted to vertebrates because of the paucity of data on other taxa. Even within vertebrates, the great majority of heritability estimates are provided by studies of wild birds (n = 1228 estimates out of 1618; 76%) and mammals (n = 344; 21%). The main explanation for this taxonomic bias is no doubt the availability of suitable datasets. For many organisms, and especially small ones, gathering individual-level phenotypic and genetic data in natural conditions remains extremely challenging. As discussed above, much quantitative genetics in the wild has exploited long-term animal studies, and these have been heavily biased towards birds and mammals because of their ease of monitoring (Clutton-Brock & Sheldon, 2010); similar biases exist in studies of behavioural ecology or life-history evolution in wild populations. We hope that this situation changes in the coming years. Two taxon-specific chapters were included in this book as incentives to develop quantitative genetic approaches in currently under-represented taxa (arthropods, Chapter 9, Zajitschek & Bonduriansky; and plants, Chapter 8, Stinchcombe), and the rapid progresses in molecular biology (Chapter 13, Jensen *et al.*) will hopefully facilitate expansion to wider taxonomic coverage. A second limitation in coverage is in the types of traits considered: Postma's survey (Chapter 2) reveals a strong predominance of morphological (n = 1169 out of 1618, 72%) or life-history (n = 347, 21%) traits. Quantitative genetic analyses of other types of traits—for example, behavioural, immunological, physiological—are accumulating, and will be especially valuable, as will estimates of the covariances between different types of traits.

1.5 Emerging topics

Thirty-one evolutionary ecologists based in eleven different countries have contributed to this book, and many discuss future research avenues within their respective chapters. However there are several emerging topics that have not been the subject of a chapter, but which we envisage becoming exciting topics in the near future; we discuss these briefly here.

1.5.1 Non-genetic inheritance

Increasing evidence has emerged that environmental effects can induce transgenerational effects generating heritable variation (Rossiter 1996). This 'non-genetic inheritance'—defined as effects on offspring phenotype brought about by vertical transmission of factors other than DNA sequences—can be of epigenetic, developmental, parental, ecological and cultural origins, and can have a major impact on the evolution of phenotypic diversity (Bonduriansky & Day 2009). The good news is that standard quantitative genetic models can be readily parameterised to include parental effects of environmental or genetic origin (see Chapter 6, McAdam *et al.*), and an extension of these models to incorporate any other source of variation and their interactions is easily conceivable. The real challenge remains in collecting the data required to build matrices of the shared non-genetic information.

1.5.2 Heritable symbionts and host evolution

Symbionts can be considered as heritable biological traits of the host, since they are transferred horizontally from mother to offspring and can influence the host's development, survival, and reproductive abilities: for example, endosymbiotic bacteria can be important drivers of insect adaptation (Fellous *et al.* 2011), and studies have shown that maternally transmitted symbionts can play a major role in the host's ecology and evolution (Moran *et al.* 2008). Incorporating a quantitative genetic approach in such studies could contribute substantially to our understanding of evolutionary conflicts between different sources of biological information (e.g. nuclear genes and symbionts, Fellous *et al.* 2011). However although the ubiquitous importance of symbiosis may change the ways we study inheritance, taking these studies outside controlled laboratory environments and into the field will constitute a substantial challenge.

1.5.3 Human evolution

Historically, some of the earliest developments in quantitative genetics, such as Galton's work on the transmission of trait values from parents to offspring, were based on human data. However, tests of microevolutionary hypotheses in humans lagged behind those for other vertebrate species, in part because of the general opinion that our modern industrialised societies 'protect' humans from evolutionary processes. Tests of evolutionary quantitative genetic questions using human data are now emerging (e.g. Kirk *et al.* 2001; Milot *et al.* 2011), and evolutionary anthropology has become a thriving field with dedicated journals.

1.5.4 Quantitative genetics of proximate mechanisms

Although it is still largely the case that proximate and ultimate biological mechanisms are the focus of different disciplines, in some particular cases, such as for the study of phenology and seasonal timing, a unifying framework is emerging to integrate results from evolutionary ecology, physiology, chronobiology and molecular genetics (Visser *et al.* 2010). In such interdisciplinary integration, quantitative genetics is a major element for the emerging field of evolutionary physiology (Feder *et al.* 2000). Within the perspectives of assessing quantitative genetic features of physiological mechanisms in wild populations, studies of immunological processes and host–parasite interactions appear particularly promising, since genes implicated in the immune system, such as MHC genes, are well identified, which offers the possibility to integrate genomic knowledge in an 'eco-devo' framework (Sultan 2007).

1.6 Summary

We have outlined here ten big questions which we see as central to current evolutionary quantitative genetics, and five reasons for addressing them in wild populations experiencing natural environments. The application of quantitative genetics analyses to wild populations is a field that has expanded rapidly in recent years, motivated by these questions. The following chapters showcase this recent work, and illustrate how quantitative genetic analyses applied to the study of wild populations have improved our understanding of life-history evolution and evolutionary ecology.

Box 1.1 Glossary of terms

Additive genetic covariance/correlation (r_A): The additive genetic covariance between two traits is the covariance in additive genetic effects on the traits, expected to arise through linkage among pairs of genes or through pleiotropy (the effect of the same genes influencing multiple traits). The genetic correlation is a standardised measure of covariance, calculated as the ratio of the genetic covariance to the square root of the product of their respective additive genetic variances, and taking values between −1 and 1.

Additive genetic variance (V_A): The component of phenotypic variance among individuals in a trait that can be attributed to breeding values, i.e. additive effects of alleles that are independent of other alleles or loci.

Animal model: A form of mixed model in which an individual's phenotype for a trait is partitioned into a linear sum of different effects. The model includes as a random effect the breeding value of each individual, and the additive genetic variance is estimated based on the comparison of phenotypes of relatives (see Box 1.2).

Breeders' equation: The predicted change over one generation in the mean of a trait, in response to selection, defined as the product of the heritability of the trait and its selection differential: $R = h^2S$.

Breeding value: The additive effect of an individual's genotype on a trait, expressed relative to the population mean phenotype; equal to twice the deviation of the expected phenotype of its progeny from the population mean.

Dominance genetic deviations/variance (V_D): The deviation of an individual's genotypic value for a trait from its breeding value due to within-locus interactions. The variance of these deviations in a population is the *dominance variance*.

Environmental effects/variance (V_E): The magnitude of phenotypic variance among genetically identical individuals in a trait, or the component of phenotypic variance due to environmental effects. In practice, this variation might be due to different environmental conditions or to stochastic noise; it is sometimes referred to as 'residual variance' (V_R).

Interaction genetic deviations/variance—also called epistatic variance (V_I): Non-additive deviations resulting from interactions between alleles at different loci and causing deviation from the phenotype expected from additive and dominance effects; their variance is the *interaction variance*.

G-matrix: A variance–covariance matrix composed of the additive genetic variances of multiple traits on its diagonal and additive genetic covariances among traits in the off-diagonal elements.

Genotype-by-environment ($G \times E$) interactions (including genotype-by-age or $G \times Age$): Differential performance of genotypes as a function of the environment (or age) in which they are expressed. These changes can result in different levels of genetic variance for a given trait across different environmental conditions (or different ages).

Heritability (narrow-sense, h^2): The extent to which a phenotypic trait is determined by additive genetic effects. Defined as the ratio of additive genetic variance (V_A) to the total phenotypic variance (V_P), the heritability of a trait lies between 0 and 1, and indicates the degree of resemblance between relatives in a population.

Inbreeding/inbreeding depression: Inbreeding is the occurrence of mating among relatives, measured by the inbreeding coefficient. Inbreeding depression is a decline in the mean value of a trait observed in inbred progeny, usually defined relative to outbred progeny.

Indirect genetic effects: Effects on the phenotype of a focal individual caused by the genotype of one or more other individuals.

Maternal effects/variance (V_M): Environmental and/or genetic effects of a mother's phenotype on her offspring's phenotype, distinct from those due to the genes it has inherited from her, and the variance in these effects. If genetically based, maternal effects are a type of indirect genetic effect.

Permanent environmental effects/variance (V_{PE}): Environmental effects on an individual's phenotype which are constant across repeated measurements of this individual, and their variance.

Phenotypic variance (V_P): The total amount of variance (sum of all components) for a given trait in a particular population.

Phenotypic plasticity: Occurs when the same genotype produces different phenotypes in different environments. The function describing the relationship between the phenotype and an environmental gradient within the same genotype is called a reaction norm.

continued

Box 1.1 *Continued*

Random regression: A form of mixed-effect model in which individual phenotypes are modelled as a continuous function of a covariate. The intercepts and slopes of individuals' functions are fitted as random effects.

Selection: The process by which variation between individuals in a trait causes variation in their fitness (for example due to differences in fecundity or survival), generating a change in the distribution of the trait in the population within a generation. Selection is usually estimated from the relationship between the trait and an estimate of fitness: for example, the selection differential S is the covariance between trait and relative fitness.

Box 1.2 Outline of an 'animal model'

The animal model is a type of mixed model, a linear regression containing a mixture of both 'fixed' and 'random' effects (see for example McCulloch & Searle 2001). Fixed effects have predictable, repeatable effects on the mean of a trait (for example, sex or age). Random effects are used to describe factors with multiple levels sampled from a population, for which the analysis provides an estimate of the *variance* for which the effects are responsible.

In the animal model, an additive genetic effect is fitted for each individual (or animal, hence the name). In its simplest form, the phenotype y of individual i is given by:

$$y_i = \mu + a_i + e_i \qquad \text{(B1.2.1)}$$

where μ is the population mean (and no other fixed effects are fitted), a_i is the additive genetic merit of individual i and e_i is a random residual error. As with all mixed models, each random effect is assumed to have come from a specific distribution with zero mean and unknown variance: here, the random effects a_i are defined as having variance V_A, the additive genetic variance, the residual errors will have variance V_R and the total phenotypic variance in y will be $V_A + V_R$.

A more general mixed model would be given in matrix form by:

$$\mathbf{y} = \boldsymbol{X\beta} + \boldsymbol{Z}\mathbf{u} + \mathbf{e} \qquad \text{(B1.2.2)}$$

where $\mathbf{y}$ is a vector of phenotypic measures on all individuals, $\boldsymbol{\beta}$ is a vector of fixed effects, $\boldsymbol{X}$ and $\boldsymbol{Z}$ are design matrices (of 0s and 1s) relating the appropriate fixed and random effects to each individual, $\mathbf{u}$ is a vector of random effects, for example additive genetic effects, and $\mathbf{e}$ is a vector of residual errors. For the simple model in equation (B1.2.1), the matrix form is therefore:

$$\mathbf{y} = \mu + \mathbf{a} + \mathbf{e} \qquad \text{(B1.2.3)}$$

where $\boldsymbol{\beta} = \mu$, $\mathbf{X}$ is a vector of 1s, $\mathbf{Z}$ is the identity matrix, and $\mathbf{a}$ is the vector of additive genetic effects. The variance–covariance matrix for the vector $\mathbf{u}$ can then be derived from the expectations of the covariance between relatives in additive genetic effects. For any pair of individuals i, j, the additive genetic covariance between them is $2\Theta_{ij}V_A$, where Θ_{ij} is the coefficient of coancestry, the probability that an allele drawn at random from individual i will be identical by descent to an allele drawn at random from individual j (equal to, for example, 0.25 for parents and offspring, so the additive genetic covariance between parents and offspring is $\frac{1}{2}V_A$). The matrix of covariances between all pairs of individuals in the population is therefore given by $\boldsymbol{A}V_A$, where $\boldsymbol{A}$ is the additive genetic relationship matrix with individual elements $A_{ij} = 2\Theta_{ij}$. Most models assume that the errors $\mathbf{e}$ are independent, in which case the corresponding covariance matrix for the vector $\mathbf{e}$ is just $\boldsymbol{R} = \mathbf{I}V_R$ (where $\mathbf{I}$ is the identity matrix).

Estimates of the variance components (here, $\hat{V}_A$ and $\hat{V}_R$) can then be derived using either frequentist or Bayesian approaches (see e.g. Sorensen & Gianola 2002 for details). Depending on the approach used, the statistical support for non-zero values of $\hat{V}_A$ can be assessed via a likelihood ratio test, as twice the difference in log-likelihood between models with and without it included will approximate to a χ^2 distribution (on one degree of freedom), or, in a Bayesian MCMC framework, from the posterior distribution of the estimate $\hat{V}_A$. For the model in equation (3), the heritability of a trait can be estimated as $h^2 = \hat{V}_A/(\hat{V}_A + \hat{V}_R)$. These estimates of components of variance are for a 'base' population from which all other individuals in the population are descended. Estimates of variance components are unbiased by any effects of finite population size, assortative mating, selection or inbreeding in subsequent generations (Thompson 1973; Sorenson & Kennedy 1984).

continued

Box 1.2 *Continued*

Animal models are easily extended to include, first, other fixed effects: for example, it may be necessary to correct for an individual's age, sex, date of sampling and so on. Second, additional random effects can also be incorporated to account for further causes of similarity: for example, maternal or common environment effects will generate correlations within groups of individuals (Kruuk & Hadfield 2007). The statistical significance of including additional random effects can then be assessed as for the additive genetic variance. Multivariate analyses of more than one trait can be used to obtain estimates of genetic and other covariances—the relatedness matrix also defines a covariance structure for the respective additive genetic effects of different traits (Lynch & Walsh 1998)—and thereby to generate estimates of ***G***-matrices for multiple traits.

References

Arnold, S.J., Burger, R., Hohenlohe, P.A., Ajie, B.C. & Jones, A.G. (2008) Understanding the evolution and stability of the G-matrix. *Evolution*, **62**, 2451–2461.

Blows, M.W. (2007) A tale of two matrices: multivariate approaches in evolutionary biology. *Journal of Evolutionary Biology*, **20**, 1–8.

Boag, P.T. & Grant, P.R. (1978) Heritability of external morphology in Darwin's finches. *Nature*, **274**, 793–794.

Bonduriansky, R. & Day, T. (2009) Nongenetic inheritance and its evolutionary implications. *Annual Review of Ecology, Evolution, and Systematics*, **40**, 103–125.

Bouwhuis, S., Charmantier, A., Verhulst, S. & Sheldon, B.C. (2010) Individual variation in rates of senescence: natal origin effects and disposable soma in a wild bird population. *Journal of Animal Ecology*, **79**, 1251–1261.

Charmantier, A. & Garant, D. (2005) Environmental quality and evolutionary potential: lessons from wild populations. *Proceedings of the Royal Society B-Biological Sciences*, **272**, 1415–1425.

Charmantier, A., McCleery, R.H., Cole, L.R., Perrins, C., Kruuk, L.E.B. & Sheldon, B.C. (2008) Adaptive phenotypic plasticity in response to climate change in a wild bird population. *Science*, **320**, 800–803.

Charmantier, A. & Réale, D. (2005) How do misassigned paternities affect the estimation of heritability in the wild? *Molecular Ecology*, **14**, 2839–2850.

Clutton-Brock, T. & Sheldon, B.C. (2010) Individuals and populations: the role of long-term, individual-based studies of animals in ecology and evolutionary biology. *Trends in Ecology & Evolution*, **25**, 562–573.

Crnokrak, P. & Roff, D.A. (1995) Dominance variance: associations with selection and fitness. *Heredity*, **75**, 530–540.

Endler, J.A. (1986) *Natural selection in the wild*. Princeton University Press, Princeton.

Falconer, D.S. & Mackay, T.F.C. (1996) *Introduction to quantitative genetics*. Longman, New York.

Feder, M.E., Bennett, A.F. & Huey, R.B. (2000) Evolutionary physiology. *Annual Review of Ecology and Systematics*, **31**, 315–341.

Fellous, S., Duron, O. & Rousset, F. (2011) Adaptation due to symbionts and conflicts between heritable agents of biological information. *Nature Reviews Genetics*, **12**, 663–663.

Fisher, R.A. (1918) The correlation between relatives on the supposition of Mendelian inheritance. *Transactions of the Royal Society of Edinburgh*, **52**, 399–433.

Garant, D., Hadfield, J.D., Kruuk, L.E.B. & Sheldon, B.C. (2008) Stability of genetic variance and covariance for reproductive characters in the face of climate change in a wild bird population. *Molecular Ecology*, **17**, 179–188.

Hadfield, J.D. (2008) Estimating evolutionary parameters when viability selection is operating *Proceedings of the Royal Society B-Biological Sciences*, **275**, 723–734.

Hadfield, J.D., Wilson, A.J., Garant, D., Sheldon, B.C. & Kruuk, L.E.B. (2010) The misuse of BLUP in ecology and evolution. *American Naturalist*, **175**, 116–125.

Hansen, T.F., Pélabon, C. & Houle, D. (2011) Heritability is not evolvability. *Evolutionary Biology*, **38**, 258–277.

Henderson, C.R. (1975) Best linear unbiased estimation and prediction under a selection model. *Biometrics*, **31**, 423–447.

Hendry, A.P. & Kinnison, M.T. (2001) An introduction to microevolution: rate, pattern, process. *Genetica*, **112**, 1–8.

Hendry, A.P., Letcher, B.H. & Gries, G. (2003) Estimating natural selection acting on stream-dwelling Atlantic salmon: implications for the restoration of extirpated populations. *Conservation Biology*, **17**, 795–805.

Hill, W.G. (2012) Quantitative genetics in the genomics era. *Current Genomics*, **13**, 196–206.

Hill, W.G. & Kirkpatrick, M. (2010) What animal breeding has taught us about evolution. *Annual Review of Ecology, Evolution, and Systematics*, **41**, 1–19.

Hoffmann, A. & Merilä, J. (1999) Heritable variation and evolution under favourable and unfavourable conditions. *Trends in Ecology and Evolution*, **14**, 96–101.

Hoffmann, A.A. (2000) Laboratory and field heritabilities: some lessons from *Drosophila*. *Adaptive genetic variation in the wild* (eds T.A. Mousseau, B. Sinervo & J.A. Endler), pp. 200–218.

Houle, D. (1992) Comparing evolvability and variability of quantitative traits. *Genetics*, **130**, 195–204.

Husby, A., Nussey, D.H., Wilson, A.J., Sheldon, B.C., Visser, M.E. & Kruuk, L.E.B. (2010) Contrasting patterns of phenotypic plasticity in reproductive traits in two great tit populations. *Evolution*, **64**, 2221–2237.

Keller, L.F. & Waller, D.M. (2002) Inbreeding effects in wild populations. *Trends in Ecology and Evolution*, **17**, 230–241.

Kingsolver, J.G., Hoekstra, H.E., Hoekstra, J.M., Berrigan, D., Vignieri, S.N., Hill, C.E., Hoang, A., Gibert, P. & Beerli, P. (2001) The strength of phenotypic selection in natural populations. *The American Naturalist*, **157**, 245–261.

Kirk, K.M., Blomberg, S.P., Duffy, D.L., Heath, A.C., Owens, I.P.F. & Martin, N.G. (2001) Natural selection and quantitative genetics of life-history traits in western women: A twin study. *Evolution*, **55**, 423–435.

Konigsberg, L.W. & Cheverud, J.M. (1992) Uncertain paternity in primate quantitative genetic studies. *American Journal of Primatology*, **27**, 133–143.

Kruuk, L.E.B. (2004) Estimating genetic parameters in wild populations using the 'animal model'. *Philosophical Transactions of the Royal Society of London Series B, Biological Sciences*, **359**, 873–890.

Kruuk, L.E.B., Clutton-Brock, T.H., Slate, J., Pemberton, J.M., Brotherstone, S. & Guinness, F.E. (2000) Heritability of fitness in a wild mammal population. *Proceedings of the National Academy of Sciences of the United States of America*, **97**, 698–703.

Kruuk, L.E.B. & Hadfield, J.D. (2007) How to separate genetic and environmental causes of similarity between relatives. *Journal of Evolutionary Biology*, **20**, 1890–1903.

Kruuk, L.E.B., Slate, J. & Wilson, A.J. (2008) New answers for old questions: the evolutionary quantitative genetics of wild animal populations. *Annual Review of Ecology, Evolution and Systematics*, **39**, 525–548.

Lack, D. (1964) A long term study of the great tit (*Parus major*). *Journal of Animal Ecology*, **33**, 159–173.

Lande, R. & Arnold, S.J. (1983) The measurement of selection on correlated characters. *Evolution*, **37**, 1210–1226.

Lush, J.L. (1937) *Animal breeding plans*. Iowa State University Press.

Lynch, M. & Walsh, B. (1998) *Genetics and analysis of quantitative traits*. Sinauer, Sunderland, Massachusetts.

McCleery, R.H., Pettifor, R.A., Armbruster, P., Meyer, K., Sheldon, B.C. & Perrins, C.M. (2004) Components of variance underlying fitness in a natural population of the great tit *Parus major*. *The American Naturalist*, **164**, E642–E742.

McCulloch, C.E. & Searle, S.R. (2001) *Generalized, linear, and mixed models*. Wiley-Interscience, New York.

Merilä, J. & Sheldon, B.C. (2001) Avian quantitative genetics. *Current Ornithology*, **16**, 179–255.

Merilä, J., Sheldon, B.C. & Kruuk, L.E.B. (2001) Explaining stasis: microevolutionary studies in natural populations. *Genetica*, **112–113**, 199–222.

Milner, J.M., Brotherstone, S., Pemberton, J.M. & Albon, S.D. (2000) Variance components and heritabilities of morphometric traits in a wild ungulate population. *Journal of Evolutionary Biology*, **13**, 804–813.

Milot, E., Mayer, F.M., Nussey, D.H., Boisvert, M., Pelletier, F. & Reale, D. (2011) Evidence for evolution in response to natural selection in a contemporary human population. *Proceedings of the National Academy of Sciences of the United States of America*, **108**, 17040–17045.

Moran, N.A., McCutcheon, J.P. & Nakabachi, A. (2008) Genomics and evolution of heritable bacterial symbionts. *Annual Review of Genetics*, pp. 165–190.

Morrissey, M.B. & Hadfield, J.D. (2011) Directional selection in temporally replicated studies is remarkably consistent. *Evolution*, **66**, 435–442.

Morrissey, M.B., Kruuk, L.E.B. & Wilson, A.J. (2010) The danger of applying the breeder's equation in observational studies of natural populations. *Journal of Evolutionary Biology*, **23**, 2277–2288.

Morrissey, M.B. & Sakrejda, K. (2013) Unification of regression-based methods for the analysis of natural selection. *Evolution*, **67**, 2094–2100.

Morrissey, M.B. & Wilson, A.J. (2010) pedantics: an r package for pedigree-based genetic simulation and pedigree manipulation, characterization and viewing. *Molecular Ecology Resources*, **10**, 711–719.

Mousseau, T.A. & Fox, C.W. (1998) The adaptive significance of maternal effects. *Trends in Ecology and Evolution*, **13**, 403–407.

Mousseau, T.A. & Roff, D.A. (1987) Natural selection and the heritability of fitness components. *Heredity*, **59**, 181–197.

Owens, I.P.F. (2006) Where is behavioural ecology going? *Trends in Ecology & Evolution*, **21**, 356–361.

Pelletier, F., Garant, D. & Hendry, A.P. (2009) Eco-evolutionary dynamics. *Philosophical Transactions of the Royal Society of London Series B, Biological Sciences*, **364**, 1483–1489.

Perrins, C.M. & Jones, P.J. (1974) The inheritance of clutch size in the great tit (*Parus major*). *Condor*, **76**, 225–228.

Postma, E. (2006) Implications of the difference between true and predicted breeding values for the study of natural selection and micro-evolution. *Journal of Evolutionary Biology*, **19**, 309–320.

Räsänen, K. & Kruuk, L.E.B. (2007) Maternal effects and evolution at ecological time-scales. *Functional Ecology*, **21**, 408–421.

Rausher, M.D. (1992) The measurement of selection on quantitative traits: biases due to environmental covariances between traits and fitness. *Evolution*, **46**, 616–626.

Réale, D., Festa-Bianchet, M. & Jorgenson, J.T. (1999) Heritability of body mass varies with age and season in wild bighorn sheep. *Heredity*, **83**, 526–532.

Roff, D.A. (1997) *Evolutionary quantitative genetics*. Chapman & Hall, New York.

Roff, D.A. (2007) A centennial celebration for quantitative genetics. *Evolution*, **61**, 1017–1032.

Rossiter, M. (1996) Incidence and consequences of inherited environmental effects. *Annual Reviews of Ecology and Systematics*, **27**, 451–476.

Santure, A.W., de Cauwer, I., Robinson, M.R., Poissant, J., Sheldon, B.C. & Slate, J. (2013) Genomic dissection of variation in clutch size and egg mass in a wild great tit (*Parus major*) population. *Molecular Ecology*, **22**, 3949–3962.

Siepielski, A.M., DiBattista, J.D. & Carlson, S.M. (2009) It's about time: the temporal dynamics of phenotypic selection in the wild. *Ecology Letters*, **12**, 1261–1276.

Simons, A.M. & Roff, D.A. (1994) The effect of environmental variability on the heritabilities of traits of a field cricket. *Evolution*, **48**, 1637–1649.

Sorensen, D.A. & Gianola, D. (2002) *Likelihood, Bayesian and MCMC methods in quantitative genetics*. Springer-Verlag.

Sorenson, D.A. & Kennedy, B.W. (1984) Estimation of genetic variances from unselected and selected populations. *Journal of Animal Science*, **58**, 1097–1106.

Sultan, S.E. (2007) Development in context: the timely emergence of eco-devo. *Trends in Ecology & Evolution*, **22**, 575–582.

Szulkin, M. & Sheldon, B.C. (2007) The environmental dependence of inbreeding depression in a wild bird population. *Plos One*, **2**, e1027.

Thompson, R. (1973) The estimation of variance and covariance components with an application when records are subject to culling. *Biometrics*, **29**, 527–550.

van der Jeugd, H.P. & McCleery, R. (2002) Effects of spatial autocorrelation, natal philopatry and phenotypic plasticity on the heritability of laying date. *Journal of Evolutionary Biology*, **15**, 380–387.

Visscher, P.M., Hill, W.G. & Wray, N.R. (2008) Heritability in the genomics era—concepts and misconceptions. *Nature Reviews Genetics*, **9**, 255–266.

Visser, M.E., Caro, S.P., van Oers, K., Schaper, S.V. & Helm, B. (2010) Phenology, seasonal timing and circannual rhythms: towards a unified framework. *Philosophical Transactions of the Royal Society of London Series B, Biological Sciences*, **365**, 3113–3127.

Wade, M.J. & Kalisz, S. (1990) The causes of natural selection. *Evolution*, **44**, 1947–1955.

Walsh, B. & Blows, M.W. (2009) Abundant genetic variation plus strong selection = multivariate genetic constraints: a geometric view of adaptation. *Annual Review of Ecology, Evolution, and Systematics*, **40**, 41–59.

Wilson, A.J. (2008) Why h^2 does not always equal Va/Vp? *Journal of Evolutionary Biology*, **21**, 647–650.

Wilson, A.J., Reale, D., Clements, M.N., Morrissey, M.M., Postma, E., Walling, C.A., Kruuk, L.E.B. & Nussey, D.H. (2010) An ecologist's guide to the animal model. *Journal of Animal Ecology*, **79**, 13–26.

Wright, S. (1921) Systems of mating. *Genetics*, **6**, 111–178.

CHAPTER 2

Four decades of estimating heritabilities in wild vertebrate populations: improved methods, more data, better estimates?

Erik Postma

2.1 Introduction

The relative importance of nature (often interpreted as genes) and nurture (the environment), i.e. a trait's heritability (Boxes 2.1, 2.2 and 2.3; Falconer & Mackay 1996; Lynch & Walsh 1998; Visscher *et al.* 2008), has been, and in some fields still is, the subject of controversy, especially when applied to human characteristics (Box 2.3; e.g. Kempthorne 1978; Gould 1996; Rose 2006). Among modern-day evolutionary biologists, however, the consensus is that both genes and the environment are responsible for the ubiquitous amounts of morphological, behavioural, life-history and physiological variation that exists, among species and populations, as well as among individuals belonging to the same population. Indeed, it has been argued that, provided sample sizes are sufficiently large, nearly all studies reveal the existence of additive genetic variation underlying the quantitative traits under investigation, and that at least in univariate analyses, the only question deserving serious attention regards the absolute and relative magnitude of the various genetic and non-genetic components of variance (Lynch 1999). So, after over four decades of attempting to disentangle the role of genetic and the various sources of environmental variation, what have we learned about their absolute and relative roles in wild vertebrate populations?

The pioneering studies (e.g. Perrins & Jones 1974; Boag & Grant 1978; van Noordwijk *et al.* 1981a; 1981b) in the seventies and early eighties that first applied quantitative genetic methods to long-term individual-based data from free-living vertebrate populations were based on small sample sizes, relied on observational data to infer relatedness, and may only to some degree have been able to separate environmental and genetic sources of resemblance (Box 2.2). Nevertheless, by the end of the 20th century, the steadily increasing size of several individual-based long-term datasets, the use of cross-fostering, and in some cases the assignment of parentage using molecular markers, suggested that in wild populations almost all quantitative traits considered were heritable to some degree (e.g. Lynch & Walsh 1998; Merilä & Sheldon 2001). However, limited statistical power generally prevented studies from going beyond this question. This changed with the advent and application of quantitative genetic mixed model approaches to data from studies of wild populations (Box 2.2; Kruuk 2004). Their more efficient use of all available data allows for maximum exploitation of the information in complex but frequently incomplete multigenerational pedigrees.

Quantitative Genetics in the Wild. Edited by Anne Charmantier, Dany Garant, and Loeske E. B. Kruuk

Box 2.1 What heritability is, and what it is not

The heritability is the most frequently estimated quantitative genetic parameter and within the field of quantitative genetics it has a specific and well-defined meaning (Visscher *et al.* 2008). Assuming an absence of non-additive genetic variance (Chapter 7, Wolak & Keller; Hill *et al.* 2008), it provides a measure of the relative importance of additive genetic versus environmental variation (V_A and V_E, respectively) in shaping phenotypic variation (V_P). In the simplest case, V_P is equal to the sum of V_A and V_E, and the heritability is equal to V_A / V_P. The latter is typically abbreviated as h^2 (Falconer & Mackay 1996; Lynch & Walsh 1998).

Why h^2? To illustrate the principle of path analysis, Wright (1921; Lynch & Walsh 1998, pp. 827–829) described the phenotypic resemblance between parents and offspring using a set of path coefficients (standardised partial regression coefficients). He used *e* for the path from environmental to phenotypic value, *g* for the path from genotypic to gametic value, and *h* for the path from genotypic to phenotypic value. From this he derived that, in the absence of assortative mating, the correlation coefficient between a parent and his or her offspring equals h^2 / 2. Furthermore, assuming additivity and the absence of genotype–environment interactions and correlations, the equation of complete determination of an individual's phenotype is given by $h^2 + e^2 = 1$. Since then, the proportion of the phenotypic variance attributable to additive genetic variance (i.e. the narrow-sense heritability) has been indicated with h^2.

Although the concept of heritability is apparently simple, many misconceptions exist, including among evolutionary biologists. Here I will briefly address some of them (also see Visscher *et al.* 2008).

- Except for the special case of $h^2 = 0$, the heritability provides little information on the absolute amount of additive genetic variation. Because $h^2 = V_A / V_P = V_A / (V_A + V_E)$, differences in heritability for the same trait in different populations, or for different traits within the same population, can be due to differences in either V_A or V_E. Similarly, heritabilities can be the same, despite differences in V_A. Indeed, the correlation between h^2 and a mean-standardised measure of V_A (the coefficient of additive genetic variance, CV_A) is effectively zero (this chapter; Hansen *et al.* 2011).
- Although the breeders' equation states that the response to selection equals the product of selection differential and heritability (i.e. $R = h^2S$), in the absence of any information on the strength of selection, the heritability provides a poor measure of the 'evolvability' of a trait (Hansen *et al.* 2011). For example, highly heritable traits may well show very little phenotypic variation for selection to act upon if V_P is extremely small.
- The heritability of a trait provides no information on the detailed genetic architecture of a trait. So, a high heritability does not imply that a trait is influenced by many genes.
- The fact that a trait is not heritable does not mean that it is not genetically determined. Indeed, the number of fingers on our hands is genetically determined, but the great majority of variation that exist is non-genetic in origin (e.g. due to accidents). Hence the heritability of number of fingers in humans is close to zero.
- Although the heritability says something about the role of genes and environment in shaping variation within populations, it cannot be extrapolated to variation between populations (Brommer 2011). Differences in phenotypic means in a highly heritable trait between two populations can be the result of phenotypic plasticity only. The other way around, a trait may have a heritability of zero within each population, but the difference between two populations can be entirely genetic in origin.
- Just as heritabilities say nothing about the nature of genetic differences between populations, they cannot be applied to individual phenotypes. So if I am ten centimetres taller than average, a heritability of height of 0.8 does not mean that of these ten centimetres, eight are genetic and two are environmental.

It is beyond doubt that these new methods have opened up possibilities that were not feasible before, and thereby have provided new insights into the evolutionary genetics of wild populations (see Kruuk *et al.* 2008). For example, rather than separately estimating univariate heritabilities and pairwise genetic correlations, they made it possible to simultaneously estimate genetic variances and covariances for two and more traits (the so-called **G**-matrix) (e.g. Garant *et al.* 2008; Björklund *et al.* 2012). Also, they enabled modelling variance components as a function of some covariate (known as random regression mixed models) (Nussey *et al.* 2007), and allowed for the separation of individual

Box 2.2 Heritability estimation

There are many excellent books (Falconer & Mackay 1996; Lynch & Walsh 1998) and papers (e.g. Visscher *et al.* 2008) that discuss in detail the classic methods of heritability estimation, such as parent–offspring regression, full-sib/half-sib breeding design, as well as the animal model. Furthermore, a number of more recent papers provide an overview of the possibilities and pitfalls associated with the application of the animal model to data for natural populations in particular (Kruuk 2004; Postma & Charmantier 2007; Kruuk *et al.* 2008; Wilson *et al.* 2010). I refer the interested reader to these resources, and the references therein, for more details. Here I will just provide a very general and brief primer.

Unlike population genetics, which usually deals with allele frequencies for a single locus with two alleles, quantitative genetics deals with variances. Importantly, quantitative genetics does not require any information on which and what kind of genes are involved. Instead it makes the assumption that there is a large (effectively infinite) number of genes involved, all with a small effect (i.e. the infinitesimal model) (Hill 2010). Assuming related individuals share nothing but genes, the degree of phenotypic resemblance (measured as the covariance) between relatives will provide an estimate of the absolute (V_A) and the relative (h^2) amount of additive genetic variance underlying the phenotypic variance for a trait (V_P).

In principle, this can be done for all types of relatedness. In practice, however, some provide more precise and/or accurate estimates than others. For example, the resemblance between parents and their offspring may at least partly be driven by parental and common environment, rather than additive genetic effects. The same is true for full-sibs, who share the same parents and the same environment, but on top of that they resemble each other because of non-additive genetic effects (dominance and epistasis). These sources of resemblance disappear when comparing (paternal) half-sibs instead, at least in species in which fathers provide nothing but genes to their offspring.

Indeed, paternal half-sib breeding designs are commonly employed in controlled laboratory studies, as in many species it is relatively easy to mate the same male to multiple females, and to obtain large enough sample sizes (i.e. paternal half-sib families) that will result in sufficiently precise estimates. However, such approaches are typically impossible in a natural setting, where who mates with who is beyond the control of the researcher. Hence, for a long time parent–offspring regression was the most commonly used method in natural populations. To account for parental effects, offspring were sometimes cross-fostered, but this can usually only be done for a few years at most and is only possible in some species. Alternatively, one could compare offspring to their grandparents instead, but because of their more distant relatedness and because grandparents are often unknown, statistical power is lower.

Both parent–offspring regression and full-sib/half-sib analysis limit themselves to quantifying the phenotypic resemblance among one type of relatives (either among (grand-)parents and offspring, among paternal half-sibs, or among full-sibs). However, in both cases one often has phenotypic data on both parents, offspring and (half-)sibs. For example, in a parent–offspring regression, offspring phenotypes are typically averaged to obtain a single value per nest or litter. Similarly, in a half-sib breeding experiment, any phenotypic measurements that may be available for the parents are ignored. In natural populations, for which pedigrees are typically very complex, this inefficient use of the data is particularly obvious.

The animal model, a specific type of mixed model, uses the phenotypic resemblance among individuals of all types of relatedness. Additionally, it allows one to model various other sources of resemblance, like permanent environment, maternal and common environment effects. At the same time, it is possible to account for, for example, systematic differences among age classes, sexes, or years. By being able to make use of all data available, heritability estimates from an animal model are potentially more precise, and because non-genetic sources of resemblance can be accounted for, they may also be more accurate. However, whether in practice this is also the case, fully depends on the amount and structure of the data that is available (Quinn *et al.* 2006; Kruuk & Hadfield 2007).

phenotypes into their genetic and environmental components (the breeding value and environmental deviation, respectively) (but see Postma 2006; Hadfield *et al.* 2010). Finally, their increased precision allowed for the parameterisation of theoretical models of, for example, the evolution of mate choice, enabling a direct and quantitative comparison of alternative models (Charmantier & Sheldon 2006).

Although methodological advancements (Kruuk 2004; Nussey, Wilson *et al.* 2007; Kruuk *et al.* 2008) have played an important role in making

Box 2.3 Criticism of the heritability concept

Although it is the most well-known and estimated quantitative genetic parameter, the concept of heritability, and in particular its interpretation continues to be subject of debate. For example, some evolutionary geneticists emphasise the fact that heritability is a variance-standardised measure of genetic variation, and thereby provides no information on a trait's evolvability, arguing instead for mean-standardised measures such as the coefficient of additive genetic variation (see Box 2.1; Houle 1992; Hansen *et al.* 2011).

Furthermore, traits are rarely genetically independent, and the heritability does not capture the positive and negative genetic correlations that may exist with other traits (Blows 2007; Walsh & Blows 2009). Indeed, what we call a trait may sometimes be a poor approximation of what selection is acting upon. In line with this, multivariate quantitative genetic analyses that simultaneously estimate genetic variances and covariances for a large number of traits may often be more appropriate, providing insight into the genetic architecture of a trait, and a much better understanding of its evolutionary potential (Blows 2007).

Also, by definition the narrow-sense heritability measures the proportion of the phenotypic variance attributable to additive genetic variance only. This emphasis on additive genetic effects can to some extent be justified by the fact that it is only these that are passed on to the next generation, and thereby determine the response to selection. Furthermore, although non-additive genetic (dominance and epistatic) effects may sometimes be substantial (Huang *et al.* 2012), they will often behave in an additive manner and result in only small amounts of non-additive genetic variance (Hill *et al.* 2008). For an in-depth discussion of the estimation of non-additive genetic variance, as well as its biological implications, see Chapter 7, Wolak and Keller.

Finally, some have criticised the heritability concept itself, especially when applied to human characteristics. For example, Kempthorne (1978) argued forcefully that because of the correlational and observational nature of most estimation methods, heritability estimates can tell us nothing about the genetic causation of traits. Furthermore, they provide an oversimplification of reality, hampering our understanding of the link between gene, genome, and phenotype. Similarly, Rose (2006) states that '[h]eritability estimates are attempts to impose a simplistic and reified dichotomy (nature/nurture) on non-dichotomous processes.' Therefore, [. . .] heritability is a useless quantity'.

It should be noted however, that most of the problems outlined above do not so much criticise the concept of heritability per se, but rather its interpretation. Although there are many things heritability may not be able to tell us, it still provides a convenient measure of the relative importance of genetic and environmental variation. Furthermore, if phenotypic means and variances are also provided, other and maybe more appropriate measures of a trait's evolvability can easily be calculated. Finally, by additionally estimating the genetic correlations among traits, multivariate approaches do still estimate a heritability (or additive genetic variance) for each trait, and thereby provide an extension rather than a fundamentally different approach.

the quantitative genetics of natural populations the diverse and dynamic field it is now, the latter may also be attributable to a natural maturation of the field, with researchers building onto previous work to ask increasingly sophisticated questions. Furthermore, additional years are being added to datasets, which results in more phenotypic and pedigree data. Quantitative genetic methods particularly benefit from this, as they deal with the estimation of (co)variances rather than means, and are therefore notoriously data-hungry.

This raises the question: 'what are the consequences of methodological advancements for the estimates of parameters obtained in wild quantitative genetics over the last decade?' For example, when we limit ourselves to the best-known quantitative genetic parameter, the narrow-sense heritability (h^2; Box 2.1), do we find that heritabilities have changed over time? A decline could be expected if heritability estimates are becoming less confounded with other sources of similarity between relatives (and therefore less upwardly biased), because common environment and maternal effects can better be accounted for (Kruuk & Hadfield 2007; but see de Villemereuil, Gimenez & Doligez 2013; Chapter 14, Morrissey *et al.*). Alternatively, increased statistical power may result in the increased publication of small but significant estimates. Also, because errors in pedigrees (for example due to extra-pair paternity) have been shown to result in a downward bias of estimates of genetic variance (Charmantier &

Réale 2005), more accurate pedigree information could result in an increase in estimates over time. On the other hand, correcting for extra-pair paternity may allow for a better separation of genetic and common environment effects, which might increase heritability estimates. Furthermore, even if the heritability estimates themselves have not changed significantly over time, has at least their precision increased (i.e. have the standard errors around the estimates decreased)? And if precision has increased over time, can this be attributed to better methods that make better use of the available data, or are there simply more data available?

In this chapter, I aim to address these questions in a quantitative manner. I explore how new methodological developments, and the application of quantitative genetic mixed model approaches in particular, have changed and shaped quantitative genetic studies of natural populations. In particular, I assess whether estimates of the absolute and relative amounts of genetic and environmental variation, as well as their precision, are affected by the method employed, whether they have changed over time, and if they have, whether these changes can be explained by an increase in sample size or the application of new methods. In doing this, I provide an overview of what we do and do not know about the relative role of genes and the environment in shaping phenotypic variation in nature and make some recommendations for future research.

2.2 Methods

2.2.1 The dataset

A search of the literature for studies that estimated heritabilities from individual-based data on free-living populations was performed using the Web of Science database (http://apps.webofknowledge.com). Specifically, I used the following search terms in the 'Topic' field: ('wild population*' OR 'natural population*') AND ('heritabil*' OR 'genetic* estimate*') to identify relevant papers. For all publications returned by these searches I subsequently checked whether they indeed fulfilled the criteria (i.e. based on longitudinal individual-based data on wild vertebrates). Furthermore, I included all studies cited in Merilä and Sheldon (2001), and went through all publications that cited Kruuk (2004), Wilson *et al.* (2010) and Hadfield (2010). Note that this may have resulted in an increased coverage of the period after 2004. The analyses presented here are based upon estimates published up until 2011. Although this search was not exhaustive, it should have provided a reasonably representative sample of quantitative genetics studies of natural populations.

In all the studies included, parentage was assigned either on the basis of behavioural observation or on the basis of molecular markers, using software like CERVUS (Marshall *et al.* 1998), COLONY (Jones & Wang 2010) or the R package MasterBayes (Hadfield *et al.* 2006). Estimates using molecular estimates of relatedness (Ritland 1996; Ritland 2000; Thomas *et al.* 2002; Frentiu *et al.* 2008) were not included. Although these methods in theory allow for the estimation of quantitative genetic parameters in a much wider range of populations and species (Moore & Kukuk 2002), as of yet they have rarely been applied, and when they have, estimates typically were inaccurate and/or imprecise (Postma *et al.* 2002; Thomas *et al.* 2002; Pemberton 2008). However, with the number of genetic markers rapidly increasing, also for genetic non-model species, this may well change in the near future. Indeed, if the number of markers is sufficiently large, these will provide a direct measure of the proportion of genes shared between two (related or unrelated) individuals, whereas the pedigree only provides an expectation (Visscher *et al.* 2006).

I recorded all estimates of the additive genetic variance (V_A), narrow-sense heritability (h^2) and the coefficient of additive genetic variance (CV_A) (Box 2.1). If V_A was not provided, I calculated it as the product of the phenotypic variance and the heritability. If h^2 was not provided, it was calculated from the ratio of the additive genetic and phenotypic variance. If the phenotypic variance was not provided, whenever possible it was calculated from the phenotypic standard deviation or from 95% confidence intervals of the trait mean. The latter may have introduced an error, as phenotypic variances (or phenotypic standard deviations or confidence intervals) provided may often include variance that is typically excluded from the heritability estimation (Wilson 2008), which uses a denominator of phenotypic variance after correcting for relevant fixed effects (also see Discussion).

To obtain an estimate of the precision with which heritabilities have been estimated, I also recorded their standard errors. If only 95% confidence or credible intervals (in the case of estimates based on Bayesian Markov Chain Monte Carlo (MCMC) analyses) were available, an approximate standard error was calculated from these. To infer whether an estimate was significant at the 5% level, I calculated the z-ratio by dividing the heritability by its standard error. Estimates where $z \geq 1.96$ were considered significantly different from 0. Only for those estimates for which no standard error was available did I use the *p*-value provided in the paper. Significance based on the *z*-ratio was generally, but not always, in agreement with the significance reported in the paper because of, for example, some studies correcting for multiple testing or testing one-sided rather than two-sided.

For all studies for which an estimate of both the trait mean and of V_A was available, I calculated CV_A as the square root of V_A, divided by the trait mean, times 100. Although a large number of estimates of CV_A in the literature have been found to be erroneous (Garcia-Gonzalez *et al.* 2012), comparing estimates provided by the authors of the 26 studies that reported CV_A to those I calculated revealed very few discrepancies. Often these were the result of not expressing CV_A as a percentage, resulting in a 100-fold smaller estimate. Minor deviations were most likely the result of using a trait mean for the calculation of CV_A that was slightly different from the mean provided, or because I inferred V_A from the product of V_P and h^2 (see above). Whenever possible, I therefore used the estimate of CV_A as provided in the paper (expressed as a percentage). In those cases where CV_A was not provided, I used the estimate I calculated from the mean and V_A. Note that CV_A is meaningless in the case of traits without a natural zero point (e.g. laying date), proportions (survival probability), or traits with a mean of zero (e.g. principal components for size) (Houle 1992; Hansen *et al.* 2011). Although some studies presented CV_A for such traits, these estimates were excluded here.

To quantify the amount of information on which an estimate was based, the first and last years of phenotypic data used in the analysis were recorded. In the few cases where the last year was not provided, it was assumed to be the year of publication minus one. Whenever possible, the sample size used in the study was also recorded. However, it should be noted that what the sample size exactly refers to varies among species and studies, and especially among methods. For example, it may refer to the number of observations, the number of individuals, the number of parent–offspring pairs, the number of families, etc. At least to some degree these differences will be accounted for by the additional inclusion of, for example, the method that was used (see below). However, any test of an effect of sample size will be relatively conservative.

Heritabilities have been estimated for a very wide range of traits. To be able to test for systematic differences among trait types, traits were classified as morphological (e.g. bill size, tarsus length, body weight), life-history (e.g. date of first egg, age at first reproduction, lifespan, litter size, annual and lifetime reproductive success), behavioural (e.g. dispersal behaviour, helping behaviour, personality) and physiological (e.g. immune response, parasite load, yolk testosterone content). Although such classifications are made commonly, they are to some degree arbitrary, and various traits could be included in multiple categories.

Finally, the method used to estimate heritability was recorded, distinguishing between various forms of parent–offspring and grandparent–offspring regression (e.g. father–son, mid-parent–offspring), full- and half-sib analysis, and animal model analysis (using restricted maximum likelihood (REML) or MCMC techniques) (see Box 2.2 for a brief overview of methods).

2.2.2 Statistics

To explain variation in h^2 and CV_A, as well as in the precision of h^2, measured as its standard error, a series of general linear mixed models was fitted. Fixed effects included were: *trait type* (morphological, life-history, behavioural, physiological), *method* (parent–offspring regression, grandparent–offspring regression, full-sib analysis of variance, half-sib analysis of variance, animal model), *pedigree* at least partly based on genetic data (yes or no), as well as the following covariates: *publication year*, *study length* in years, and *sample size*. When analysing variation in the precision of the estimates, h^2 was included as an additional covariate to test whether small heritability estimates with large standard errors are less likely to be published. Finally, to account for the non-independence of estimates from the same study, species and population, these factors were included as random effects in models of h^2 and CV_A. Note that these analyses treat species as independent units and thus ignore the phylogenetic relationships among them. When analysing variation in the precision of h^2, only study ID was included as a random effect as there are no reasons to expect systematic differences in estimate precision among species or study areas not captured by study ID.

Statistical analyses were performed in JMP 9.0.0 (SAS Institute Inc., Cary, NC, 1989–2007). Statistics for the non-significant terms ($p > 0.05$) presented are based on the full model. These terms were subsequently removed from the model, starting with the least significant variable. Statistics for the significant terms are based on the final model. Removed non-significant terms were subsequently entered one-by-one in the minimum adequate model, but this never resulted in a significantly improved fit. Because of the structured nature of the data, random effects were retained in the model, irrespective of whether or not they explained a significant proportion of the variance.

In addition to the analyses involving the whole dataset outlined above, heritabilities for avian clutch size, laying date and tarsus length, based on either parent–offspring regression or an animal model, were analysed separately. These are among the traits for which most heritabilities are available, and they therefore provide a useful illustration of how our understanding of the relative roles of genes and the environment in shaping individual traits has changed over the past decades.

2.3 Results

2.3.1 Descriptive statistics

The complete database, including all references, is available as supplementary online material at www.oup.co.uk/companion/charmantier. In total, 1618 heritability estimates from 171 studies, 76 populations and 53 species were obtained. This makes an average (*SD*) of 9.5 (13.7) estimates per study, either for different traits, using different methods, or for males and females separately. Overall, 59 of these studies (containing a total of 488 estimates) used genetic data to assign parentage for at least part of the pedigree, whereas all others were based on relatedness inferred from behavioural observations only. The average (*SD*) number of years on which an estimate was based was 11.9 (12.2).

The earliest study included in the database was published in 1974 (Perrins & Jones 1974), reporting the heritability of great tit (*Parus major*) clutch size in Wytham Woods, using parent–offspring regression. Since then the number of studies and the number of estimates has steadily increased (Figure 2.1). Although there appears to have been a rapid increase in the number of studies around 2000, as pointed out above, this increase may to some degree be attributable to the more exhaustive search for this period.

The great majority of heritabilities is based on data from wild bird populations (n = 1228), followed by mammals (n = 344), and then fishes

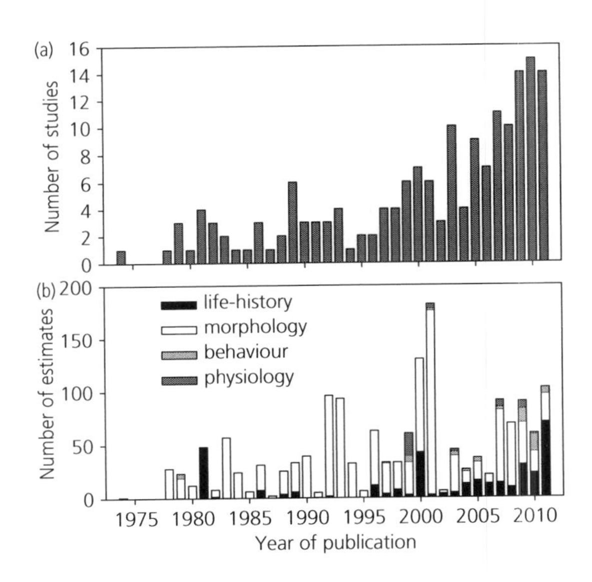

Figure 2.1 Number of (a) studies and (b) heritability estimates published per year. In (b), the number of estimates per trait type (life-history, morphology, behaviour, physiology) is indicated.

and reptiles ($n = 36$ and $n = 10$, respectively) (Figure 2.2a). There were no estimates for amphibians. Estimates covered a total of 53 species, with the first four positions occupied by birds (great tit (*Parus major*): $n = 210$, 13% of all estimates; collared flycatcher (*Ficedula albicollis*): $n = 175$; medium ground finch (*Geospiza fortis*): $n = 164$; barnacle goose (*Branta leucopsis*): $n = 101$); in total, these four species accounted for 40% of all estimates), followed by two sheep species (Soay sheep (*Ovis aries*): $n = 83$; bighorn sheep (*Ovis canadensis*): $n = 82$) (Figure 2.2a). It should be noted that for most of these species, estimates all come from a few populations at most. For example, 84 of the great tit heritability estimates come from the Hoge Veluwe population (The Netherlands), and 85 come from the Wytham Woods population (United Kingdom). Similarly, 173 of the 175 collared flycatcher estimates come from the Gotland (Sweden) population. Furthermore, the 101 barnacle goose heritabilities all come from a single population (also from Gotland, Sweden) and just three publications.

Estimates were obtained using a range of methods, including various versions of parent–offspring and grandparent–offspring regression ($n = 970$ and $n = 24$, respectively), full- and half-sib analysis ($n = 73$ and $n = 11$, respectively), and animal model methodology ($n = 540$) using either (restricted) maximum likelihood ($n = 509$) or MCMC ($n = 31$) statistical methods (Figure 2.2b).

Although heritabilities have been estimated for a very wide range of traits, the great majority was obtained for morphological traits ($n = 1169$), followed by life-history traits ($n = 347$), behavioural traits ($n = 60$) and finally physiological traits ($n = 42$). Whereas until 2000, the relative number of estimates for life-history traits was relative low (16% of all estimates), during the past decade (2001–2011) this proportion has increased to 28% (Figure 2.1b). Although heritability estimates of a correlate of annual fitness (i.e. life-history traits like clutch size, survival, annual production of fledglings) are relatively common, heritability estimates of (an aspect of) lifetime reproductive success are much rarer ($n = 28$, 1.7%), and range from -0.02 (Gustafsson 1986) to 0.9 (Kelly 2001).

2.3.2 Temporal trends

Overall and without accounting for any other sources of variation (e.g. trait type, estimation method), I find highly significant declines in both heritability estimates ($b \pm SE = -8.35 \times 10^{-3} \pm 1.79 \times 10^{-3}$, $F_{1,127.6} = 21.9$, $p < 0.001$; Figure 2.3a) and their standard errors ($b \pm SE = -5.62 \times 10^{-3} \pm 1.10 \times 10^{-3}$, $F_{1,146.6} = 26.0$, $p < 0.001$; Figure 2.3b), whereas z-ratios have increased over time ($b \pm SE = 41.4 \times 10^{-3} \pm 19.7 \times 10^{-3}$, $F_{1,127.4} = 4.43$, $p = 0.037$; Figure 2.3c). Nevertheless, the (arcsine transformed) proportion of significant estimates did not change significantly over time (weighted regression: $b \pm SE = -2.24 \times 10^{-3} \pm 5.11 \times 10^{-3}$, $F_{1,33} = 0.19$, $p = 0.66$; Figure 2.3d). Although estimates of coefficients of additive genetic variation cover a relatively short period (first estimate from 1989, and 94% estimates from 1999 or later), they do not show a systematic

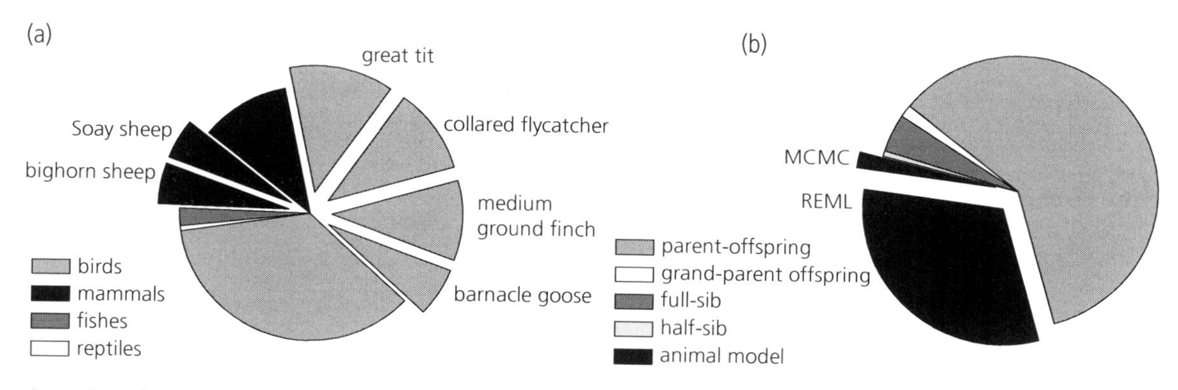

Figure 2.2 Relative number of heritability estimates per (a) taxonomic group and (b) estimation method.

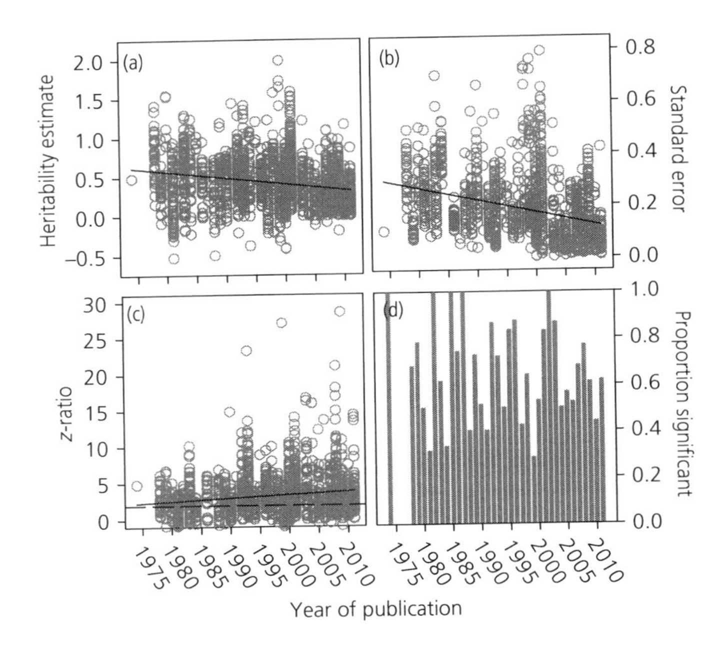

Figure 2.3 Temporal changes in (a) heritability estimates, (b) their standard error and (c) the accompanying z-ratio, as well as (d) the proportion of estimates significant at the 5% level. In (b), one extremely large standard error (2.88; Potti 1999) is not depicted. In (c), the dashed line indicates $z = 1.96$.

change over time ($b \pm \text{SE} = 0.54 \pm 0.31$, $F_{1,27.8} = 3.02$, $p = 0.093$). Below I explore these temporal trends in more detail.

2.3.3 Variation in heritability estimates

For statistical details, see Table 2.1. In short, significant differences in the size of heritability estimates exist among methods, with heritabilities from animal model analyses being lower than those from parent–offspring regressions ($F_{1,305.8} = 13.4$, $p < 0.001$; Figure 2.4a). Furthermore, heritabilities differ significantly among trait types, with morphological (mean ± SE: 0.56 ± 0.035), behavioural (0.52 ± 0.058) and physiological (0.49 ± 0.072) traits having higher heritabilities than life-history traits (mean ± SE: 0.33 ± 0.038, $F_{1,495.4} = 30.2$, $p < 0.001$; Figure 2.4b). Also, although it did not reach statistical significance in the final model, heritabilities based on pedigrees that are at least partly based on genetic data were 0.040 ± 0.024 lower than estimates based on social pedigrees only ($F_{1,140.3} = 3.71$, $p = 0.056$).

There is no systematic difference among birds, mammals, fish and reptiles. Also, there is no effect of the number of years on which the estimate is based, or of the sample size on which the estimate is based. Finally, when accounting for the effects of method and trait type, the decline in heritability estimates over time is greatly reduced ($b = -3.19 \times 10^{-3}$ vs -8.35×10^{-3}) and no longer reaches statistical significance (Table 2.1a).

2.3.4 Variation in heritability estimate precision

As expected, heritabilities based on more years and larger samples sizes had smaller standard errors (Table 2.2b). Furthermore, estimates based on animal models had the smallest standard errors, followed by estimates based on parent–offspring regression (Figure 2.4c). Finally, standard errors declined with increasing heritabilities. Again, after accounting for the effects of estimation method and sample size, heritability estimate precision did not change over time, nor did it differ among taxonomic classes or trait types (Table 2.2a, Figure 2.4d).

2.3.5 Variation in estimates of coefficients of additive genetic variation

There was a highly significant effect of trait type on CV_A estimates, with morphological traits (mean ± SE: 4.48 ± 1.46) having smaller CV_As than

Table 2.1 Variation in heritability (h^2) estimates. a) The full model and b) the minimum adequate model. For fixed effects with two levels and covariates, parameter estimates (*b*) and standard errors (SE) are provided. Study length is measured in years. Sample size is measured in various ways (see Section 2.2 for more details)

(a)

Fixed effect	***b* (SE)**	**d.f.**	***F***	***p***
Year of publication ($\times 10^{-3}$)	−3.19 (2.41)	1, 138.2	1.75	0.19
Method		4, 1242	6.71	<0.001
Taxonomic class		3, 63.9	0.70	0.57
Study length ($\times 10^{-4}$)	7.25 (16.49)	1, 189.8	0.19	0.66
Sample size ($\times 10^{-6}$)	−7.81 (7.03)	1, 434.9	1.24	0.27
Genetic pedigree [no] ($\times 10^{-3}$)	24.03 (26.06)	1, 142.2	0.85	0.36
Trait type		3, 433	26.9	<0.001
Random effect	**Variance (SE)**	**95% lower**	**95% upper**	**% explained**
Study ID ($\times 10^{-3}$)	18.4 (4.7)	9.2	27.6	19.2
Species ($\times 10^{-3}$)	9.7 (5.4)	−0.9	20.3	10.1
Population ($\times 10^{-3}$)	9.1 (6.0)	−2.8	20.9	9.5

(b)

Fixed effect	***b* (SE)**	**d.f.**	***F***	***p***
Method ($\times 10^{-3}$)		4, 980.4	12.5	<0.001
Trait type ($\times 10^{-3}$)		3, 408.1	29.2	<0.001
Random effect	**Variance (SE)**	**95% lower**	**95% upper**	**% explained**
Study ID ($\times 10^{-3}$)	13.6 (3.5)	6.8	20.4	15.5
Species ($\times 10^{-3}$)	6.9 (3.6)	−0.0	13.9	7.9
Population ($\times 10^{-3}$)	6.8 (4.5)	−2.0	15.6	7.8

life-history (mean ± SE: 15.54 ± 1.83), behavioural (mean ± SE: 13.35.11 ± 5.32) and physiological (mean ± SE: 10.69 ± 3.88) traits ($F_{1,234.5} = 13.8$, $p < 0.001$; Figures 2.4e and 2.4f; Table 2.3).

On a different note, the correlation between CV_A and h^2 (excluding cases where $h^2 = CV_A = 0$) was weak but negative and statistically significant when tested non-parametrically (Pearson's $r = -0.067$, $p = 0.21$; Spearman's $\rho = -0.18$, $p < 0.001$). This negative correlation is largely driven by morphological traits having high heritabilities and low coefficients of additive genetic variation, and residual h^2 and CV_A based on the models in Tables 2.1b and 2.3b are in fact weakly positively correlated (Pearson's $r = 0.13$, $p = 0.012$; Spearman's $\rho = 0.059$, $p = 0.27$).

2.3.6 Heritability of avian tarsus length, clutch size and laying date

Similar to the results based on the complete dataset, animal model estimates of heritability of avian tarsus length (n = 22), clutch size (n = 15) and timing of reproduction (n = 58) are slightly lower than those based on parent–offspring regression (tarsus length: n = 169; clutch size: n = 22; laying date: n = 15; method: b = 0.055 ± 0.026; $F_{1,73.3} = 4.37$, $p = 0.040$; method × trait: $F_{2,158} = 0.84$,

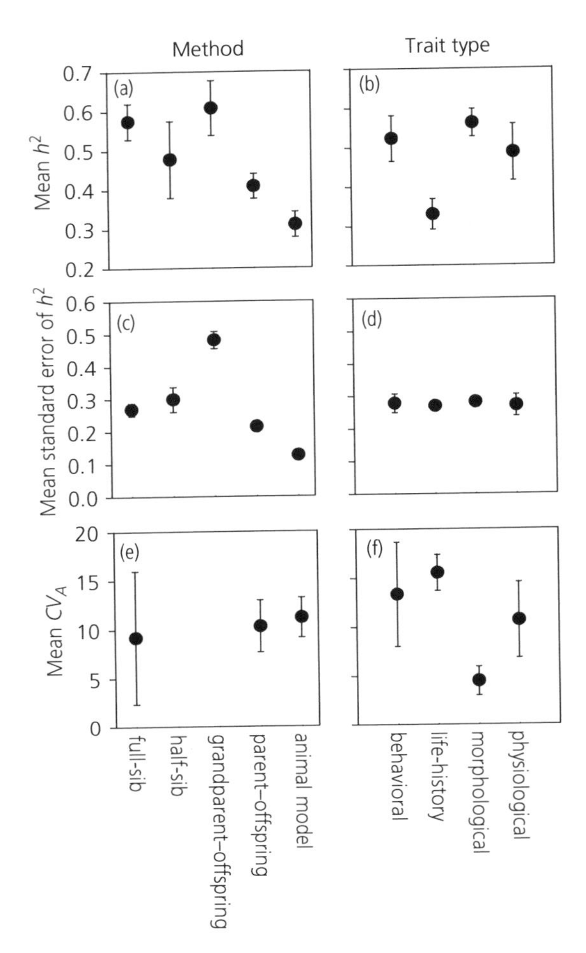

Figure 2.4 Variation in heritability estimates (h^2) (a, b), their standard error (c, d), and estimates of the additive genetic coefficient of variation (CV_A) (e, f), for different methods (a, c, e) and trait types (b, d, f). Error bars indicate the standard errors of the parameter estimates from the linear models described in the main text.

$p = 0.43$; Figure 2.5). Furthermore, the mean heritability differs significantly among the three traits ($F_{2,151.8} = 19.80$, $p < 0.001$), with heritabilities for tarsus length (mean ± SE: 0.54 ± 0.033) being larger than those for clutch size (mean ± SE: 0.25 ± 0.049) and timing of reproduction (mean ± SE: 0.21 ± 0.046). Furthermore, when accounting for estimation method and trait, there is no evidence for any temporal trends (publication year: b ± SE: $2.72 \times 10^{-3} \pm 3.67 \times 10^{-3}$, $F_{1,54.9} = 0.55$, $p = 0.46$; publication year × trait: $F_{2,90.8} = 0.37$, $p = 0.70$; Figure 2.5).

2.4 Discussion

Over the past four decades, heritability estimates have declined. This decline is to a large part accounted for by the rapid increase in the number of heritability estimates based on animal model analyses during the last decade, which tend to be smaller than those based on other methods. Relatively low heritabilities provided by animal models have also been shown in other studies that directly compared parent–offspring and animal model estimates (e.g. Kruuk 2004; Quinn *et al.* 2006; Åkesson *et al.* 2008). For example, Åkesson *et al.* (2008) found that in a Swedish population of great reed warblers (*Acrocephalus arundinaceus*), animal model estimates of heritability were on average 22% lower than those based on parent–offspring regression, at least when repeated observations were not averaged across individuals. This is very similar to the 24% difference reported here across studies and traits.

One possible explanation for the lower heritabilities provided by the animal model is that by using information across all levels of relatedness, this method provides estimates that are less biased by environmental sources of covariance among, for example, parents and offspring (van der Jeugd & McCleery 2002). Furthermore, the animal model readily allows for an explicit modelling of such sources of environmental covariance among full- and half-sibs (Kruuk 2004; Kruuk & Hadfield 2007).

In some cases, the lower heritabilities may also be the result of a systematic difference between the two methods in how heritability estimates are calculated. For instance, distinct ways of accounting for known environmental sources of phenotypic variation may explain part of the difference in mean heritability estimates (Wilson 2008). To account for environmental differences in the mean trait value among years, parent–offspring regressions are typically performed on phenotypic data that are corrected for systematic differences among years. Similarly, nest or litter effects are accounted for by using litter or nest mean value of traits, and when repeated measurements are available, these are usually averaged. In an animal model, on the other hand, these sources of variation are typically accounted for by including year (estimating variance among years), nest or litter

Table 2.2 Variation in the precision of the heritability estimates, as represented by the size of the standard error. a) The full model and b) the minimum adequate model. For further details, see Table 2.1

(a)

Fixed effect	*b* (SE)	d.f.	*F*	*p*
Year of publication ($\times 10^{-4}$)	0.64 (14.4)	1, 170	0.00	0.96
Method		4, 1237	38.02	<0.001
Taxonomic class		3, 123.4	0.22	0.88
Study length ($\times 10^{-3}$)	−2.41 (0.83)	1, 294	8.45	0.004
Sample size ($\times 10^{-6}$)	−8.52 (3.53)	1, 269.5	5.82	0.017
Genetic pedigree [no]	0.021 (0.013)	1, 157.1	2.42	0.12
Trait type		3, 682.5	0.50	0.68
Heritability	−0.026 (0.010)	1, 1333	6.48	0.011
Random effect	**Variance (SE)**	**95% lower**	**95% upper**	**% explained**
Study ID ($\times 10^{-3}$)	11.9 (1.8)	8.2	15.5	58.5

(b)

Fixed effect	*b* (SE)	d.f.	*F*	*p*
Method		4, 1084	41.89	<0.001
Study length ($\times 10^{-3}$)	−2.37 (0.72)	1, 244.9	10.72	<0.001
Sample size ($\times 10^{-6}$)	−7.59 (3.39)	1, 277.9	5.01	0.026
Heritability ($\times 10^{-3}$)	−23.0 (10.1)	1, 1348	5.16	0.023
Random effect	**Variance (SE)**	**95% lower**	**95% upper**	**% explained**
Study ID ($\times 10^{-3}$)	11.1 (1.7)	7.8	14.4	56.8

ID (estimating common environment variance) and individual ID (estimating permanent environment variance) as additional random effects. By doing this, these variance components remain part of the phenotypic variance by which the estimate of the additive genetic variance is divided to obtain the heritability.

The above can not only explain differences between estimates based on animal model and parent–offspring regression, but it might also be a source of variation among studies that use the same method but that divide V_A by different estimates of the phenotypic variance (with or without variance accounted for various fixed or random effects). Furthermore, if there are changes over time with regards to ideas on which fixed and random effects should be included, it could provide an additional source of temporal change. On the whole, this emphasizes that heritabilities should be interpreted in the context of the statistical model used to estimate them. Which model is the right one will depend on the purpose of the estimate. To facilitate interpretation and comparisons among studies, it is crucial that authors explicitly state how h^2 was calculated, and to provide estimates of all variance components, as well as an estimate of the phenotypic variance before accounting for any sources of variation. For an in-depth discussion of this issue, see Wilson (2008).

Note that although model structure can explain differences in heritability, it would be less likely to result in different absolute amount of genetic variation, measured as the coefficient of additive genetic variation. Indeed, I find no difference in

Table 2.3 Variation in absolute level of additive genetic variance, measured as the coefficient of additive genetic variance (CV_A). a) The full model, and b) the minimum adequate model. For further details, see Table 2.1

(a)

Fixed effect	*b* (SE)	d.f.	*F*	*p*
Year of publication	0.41 (0.32)	1, 29.8	1.69	0.20
Method		2, 296	0.05	0.95
Taxonomic class		2, 19.6	0.22	0.81
Study length	0.10 (0.13)	1, 28.0	0.70	0.41
Sample size ($\times 10^{-5}$)	6.17 (32.0)	1, 105.3	0.04	0.85
Genetic pedigree [no]	0.45 (1.85)	1, 23.0	0.06	0.81
Trait type		3, 243.8	14.16	<0.001
Random effect	**Variance (SE)**	**95% lower**	**95% upper**	**% explained**
Study ID	38.8 (17.3)	4.6	72.4	24.4
Species	5.6 (16.0)	−25.8	36.9	3.5
Population	16.6 (12.5)	−7.9	41.1	10.5

(b)

Fixed effect	*b* (SE)	d.f.	*F*	*p*
Trait type		3, 238.4	16.6	<0.001
Random effect	**Variance (SE)**	**95% lower**	**95% upper**	**% explained**
Study ID	34.9 (15.2)	5.1	64.7	24.6
Species	−2.7 (9.5)	−21.2	15.9	−1.9
Population	15.6 (10.6)	−5.1	36.3	11.0

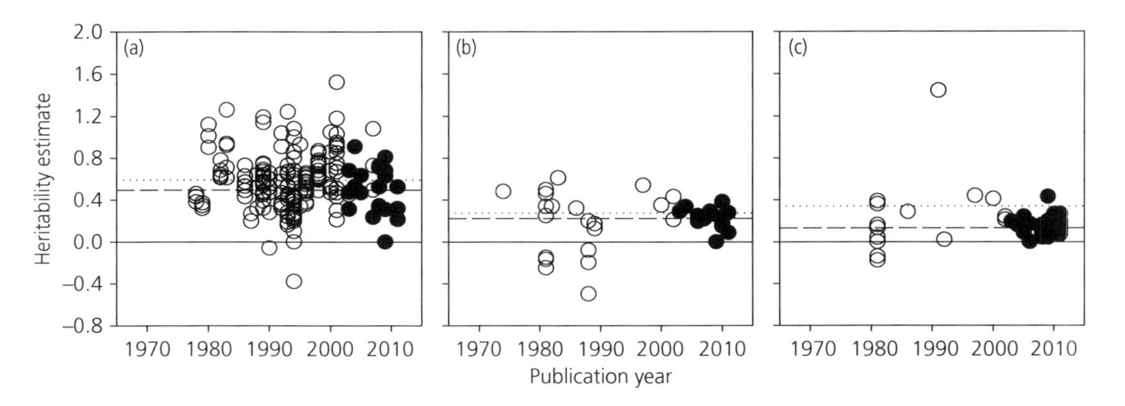

Figure 2.5 Heritability estimates for avian tarsus length (a), clutch size (b) and laying date (c). Open and closed circles indicate estimates based on parent–offspring regression and animal model, respectively. Dotted and dashed lines indicate the least-square mean heritability across all years based on parent–offspring regression and animal model, respectively.

CV_A among parent–offspring regression and animal model estimates.

Analyses revealed that life-history traits had significantly smaller heritabilities than morphological traits. As the relative number of heritability estimates for life-history traits has increased over time (Figure 2.1b), this further contributed to the observed reduction in the mean heritability over time. Several studies have found that, across species, life-history traits have lower heritabilities than morphological traits (Mousseau & Roff 1987; Houle 1992; Tschirren & Postma 2010). Similarly, within populations and species, traits more closely related to fitness tend to have lower heritabilities (Gustafsson 1986; Merilä & Sheldon 2000; McCleery *et al.* 2004; Teplitsky *et al.* 2009). Although this pattern has been interpreted as support for natural selection having eroded genetic variation in fitness-traits, heritabilities often provide a poor measure of the absolute amount of additive genetic variation, as suggested by the weak correlation between heritability and coefficients of additive genetic variance found here and elsewhere (Hansen *et al.* 2011). Indeed, life-history traits tend to have higher and morphological traits lower CV_As, both within (e.g. Merilä & Sheldon 2000; Coltman *et al.* 2005; Tschirren & Postma 2010) (but see McCleery *et al.* 2004; Teplitsky *et al.* 2009) and across species and populations (Houle 1992; this chapter).

While the effects of method and trait type documented here are in line with previous findings, the finding that heritability estimates partly based on relatedness inferred from molecular markers tend to be smaller than those based solely on behavioural observations, even after correcting for other methodological differences and temporal trends, is more surprising. Indeed, one would expect misassigned paternities or maternities, due to for example extra-pair copulations or egg dumping, to introduce errors in the pedigree and thus reduce rather than increase heritabilities (Charmantier & Réale 2005). Alternatively, being able to distinguish between within- and extra-pair offspring in a nest would allow for a better separation of additive genetic and common environment effects, possibly resulting in less biased and hence lower heritabilities.

The pattern observed could also be an artefact due to the increased use of molecular markers to assign parentage over time, and due to the fact that they are used much more commonly in mammalian than in avian studies. If there are systematic differences in the types of traits studied in birds and mammals, or between avian and mammalian heritabilities, this could introduce a correlation between the effects of trait type and pedigree correction, and thereby a spurious effect of pedigree correction. Indeed, the effect of pedigree correction is no longer significant in the full model including trait type (Table 2.1a).

Finally, it should also be noted that the quality of pedigrees based on genetic data has increased over time, with some of the earlier genetic pedigrees having error rates as high as 20% (Kruuk 2004). As a consequence, the difference between estimates based on genetic vs social pedigrees may have changed over time. However, I do not find evidence for this here (interaction with publication year: $F_{1,128.7} = 0.12$, $p = 0.73$). On the whole, this suggests that the effect of pedigree errors on estimates of heritability may be more complex than typically assumed.

The temporal decline in heritability estimates is accompanied by a decline in their standard errors. This decline can again be accounted for by an increased use of the animal model, which by making more efficient use of the available data results in more precise estimates (also see Kruuk 2004; Åkesson *et al.* 2008). This trend can further be explained by the fact that datasets are increasing in size and detail with time. Interestingly, there also is a negative association between heritability and its standard error. In other words, smaller heritabilities tend to be estimated less precisely. This finding speaks against a large publication bias, which would result in small heritabilities with relatively large standard errors being less likely to get published, resulting in a positive association. Note that the absence of an effect of sample size or study length on heritability provides further support for the absence of a strong publication bias (Table 2.1a). Unfortunately, lacking a measure of sample size that is comparable across populations and methods, application of classical meta-analytical tools to detect publication biases (e.g. funnel plots) was not possible here.

Although heritability estimates are getting increasingly precise, as the estimates are at the same time getting smaller, this has not resulted in more significant estimates appearing in the literature (Figure 2.3). Instead, whereas estimates for traits like great tit tarsus length, for which now tremendous amounts of data are available, are getting increasingly precise, at the same time estimates for traits for which much less data are available are appearing in the literature, like personality (e.g. Quinn *et al.* 2009; Réale *et al.* 2009) and physiological traits (e.g. Bonneaud *et al.* 2009; Nilsson *et al.* 2009). Similarly, increasingly powerful methods will make it possible to obtain estimates for a wider range of species and populations. Given the current overrepresentation of a few species and populations (Figure 2.2), this is a very welcome development.

At this point it should be noted that studies such as these suffer from a number of important limitations. First of all, as pointed out above, the various variables used to explain variation in h^2, its standard error, and CV_A, are to some degree correlated, making it difficult to disentangle their effects. Studies comparing methods and trait types within a single population (e.g. McCleery *et al.* 2004; Åkesson *et al.* 2008) are far less susceptible to this problem. On the other hand, it is less clear to what degree findings for a single population can be extrapolated to other populations and species. Second, we have to assume that the estimates reported in the literature are estimated correctly. Finally, although the non-independence of estimates made on the same species was accounted for here, different species were treated as independent as the phylogenetic relationships among them were ignored. There is however some evidence for systematic differences in the heritability estimates among species (see Table 2.1), and it would be interesting to test whether this variation shows evidence of, for example, a phylogenetic signal. Indeed, comparative analyses of heritability are currently lacking.

2.4.1 Conclusion

Based on the above, we can conclude that the temporal decline we observe in heritability can be attributed to an increased adoption of the animal model, and more studies estimating heritabilities for traits more closely related to fitness. Although heritability estimates for a given trait are getting increasingly precise, on average they show very little change. In other words, our understanding of the relative roles of genes and the environment in shaping phenotypic variation in natural populations has become more complete (as heritability estimates for more and more traits, species and populations become available), and these estimates are becoming more and more precise (smaller standard errors), yet it has not changed in a fundamental manner.

The evidence suggests that the great majority of quantitative traits are heritable to some degree, and hence that most traits have the potential to respond to selection. However, a quantitative prediction of the response to selection requires not only a precise and unbiased estimate of the heritability of a trait, but also its (genetic) correlations with other traits, and fitness in particular. Therefore, the adoption of new methods such as the animal model, which at least in theory can provide such estimates, has been a major step forward in our understanding of the evolutionary genetics and dynamic of natural populations. On the other hand, however, it does not provide a substitute for large amounts of high-quality data.

We have now largely moved beyond solely estimating heritabilities. Instead, they are a means to an end, providing a small but important piece of the bigger puzzle of understanding the ubiquitous amounts of variation we see around us. Although methodological advances are a major driver of progress, the field of wild quantitative genetics will need to continue to put questions ahead of methods. Only this way the decades to come will be as productive and exciting as those that have passed.

Acknowledgments

I would like to thank Anja Bürkli and Franziska Lörcher, who created the database that formed the basis of this chapter, the University of Zurich and the Swiss National Science Foundation for funding my research, and Peter Grant, John Stinchcombe and the editors for comments that greatly improved previous versions of this chapter.

References

Åkesson, M., Bensch, S., Hasselquist, D., Tarka, M. & Hansson, B. (2008) Estimating heritabilities and genetic correlations: comparing the 'animal model' with parent-offspring regression using data from a natural population. *PLoS ONE*, **3**, e1739.

Björklund, M., Husby, A. & Gustafsson, L. (2012) Rapid and unpredictable changes of the **G**-matrix in a natural bird population over 25 years. *Journal of Evolutionary Biology*, **26**, 1–13.

Blows, M.W. (2007) A tale of two matrices: multivariate approaches in evolutionary biology. *Journal of Evolutionary Biology*, **20**, 1–8.

Boag, P.T. & Grant, P.R. (1978) Heritability of external morphology in Darwin finches. *Nature*, **274**, 793–794.

Bonneaud, C., Sinsheimer, J.S., Richard, M., Chastel, O. & Sorci, G. (2009) MHC polymorphisms fail to explain the heritability of phytohaemagglutinin-induced skin swelling in a wild passerine. *Biology Letters*, **5**, 784–787.

Brommer, J.E. (2011) Whither P_{ST}? The approximation of Q_{ST} by P_{ST} in evolutionary and conservation biology. *Journal of Evolutionary Biology*, **24**, 1160–1168.

Charmantier, A. & Réale, D. (2005) How do misassigned paternities affect the estimation of heritability in the wild? *Molecular Ecology*, **14**, 2839–2850.

Charmantier, A. & Sheldon, B.C. (2006) Testing genetic models of mate choice evolution in the wild. *Trends in Ecology & Evolution*, **21**, 417–419.

Coltman, D.W., O'Donoghue, P., Hogg, J.T. & Festa-Bianchet, M. (2005) Selection and genetic (co)variance in bighorn sheep. *Evolution*, **59**, 1372–1382.

de Villemereuil, P., Gimenez, O. & Doligez, B. (2013) Comparing parent—offspring regression with frequentist and Bayesian animal models to estimate heritability in wild populations: a simulation study for Gaussian and binary traits. *Methods in Ecology and Evolution*, **4**, 260–275.

Falconer, D.S. & Mackay, T.F.C. (1996) *Introduction to quantitative genetics*, 4 edn. Longman, Harlow.

Frentiu, F.D., Clegg, S.M., Chittock, J., Burke, T., Blows, M.W. & Owens, I.P.F. (2008) Pedigree-free animal models: the relatedness matrix reloaded. *Proceedings of the Royal Society B-Biological Sciences*, **275**, 639–647.

Garant, D., Hadfield, J.D., Kruuk, L.E.B. & Sheldon, B.C. (2008) Stability of genetic variance and covariance for reproductive characters in the face of climate change in a wild bird population. *Molecular Ecology*, **17**, 179–188.

Garant, D., Sheldon, B.C. & Gustafsson, L. (2004) Climatic and temporal effects on the expression of secondary sexual characters: genetic and environmental components. *Evolution*, **58**, 634–644.

Garcia-Gonzalez, F., Simmons, L.W., Tomkins, J.L., Kotiaho, J.S. & Evans, J.P. (2012) Comparing evolvabilities: Common errors surrounding the calculation and use of coefficients of additive genetic variation. *Evolution*, **66**, 2341–2349.

Gould, S.J. (1996) *The mismeasure of man*, 2nd edn. W.W. Norton & Company, Inc., New York.

Gustafsson, L. (1986) Lifetime reproductive success and heritability: Empirical support for Fisher's fundamental theorem. *American Naturalist*, **128**, 761–764.

Hadfield, J.D. (2010) MCMC methods for multi-response generalised linear mixed models: The MCMCglmm R package. *Journal of Statistical Software*, **33**, 1–22.

Hadfield, J.D., Richardson, D.S. & Burke, T. (2006) Towards unbiased parentage assignment: combining genetic, behavioural and spatial data in a Bayesian framework. *Molecular Ecology*, **15**, 3715–3730.

Hadfield, J.D., Wilson, A.J., Garant, D., Sheldon, B.C. & Kruuk, L.E.B. (2010) The misuse of BLUP in ecology and evolution. *American Naturalist*, **175**, 116–125.

Hansen, T.F., Pelabon, C. & Houle, D. (2011) Heritability is not evolvability. *Evolutionary Biology*, **38**, 258–277.

Hill, W.G. (2010) Understanding and using quantitative genetic variation. *Philosophical Transactions of the Royal Society of London Series B, Biological Sciences*, **365**, 73–85.

Hill, W.G., Goddard, M.E. & Visscher, P.M. (2008) Data and theory point to mainly additive genetic variance for complex traits. *PLoS Genetics*, **4**, e1000008.

Houle, D. (1992) Comparing evolvability and variability of quantitative traits. *Genetics*, **130**, 195–204.

Huang, W., Richards, S., Carbone, M.A., Zhu, D., Anholt, R.R.H., Ayroles, J.F., Duncan, L., Jordan, K.W., Lawrence, F., Magwire, M.M., Warner, C.B., Blankenburg, K., Han, Y., Javaid, M., Jayaseelan, J., Jhangiani, S.N., Muzny, D., Ongeri, F., Perales, L., Wu, Y.-Q., Zhang, Y., Zou, X., Stone, E.A., Gibbs, R.A. & Mackay, T.F.C. (2012) Epistasis dominates the genetic architecture of *Drosophila* quantitative traits. *Proceedings of the National Academy of Sciences of the United States of America*, **109**, 15553–15559.

Jones, O.R. & Wang, J.L. (2010) COLONY: a program for parentage and sibship inference from multilocus genotype data. *Molecular Ecology Resources*, **10**, 551–555.

Kelly, M.J. (2001) Lineage loss in Serengeti cheetahs: consequences of high reproductive variance and heritability of fitness on effective population size. *Conservation Biology*, **15**, 137–147.

Kempthorne, O. (1978) Logical, epistemological and statistical aspects of nature-nurture data interpretation. *Biometrics*, **34**, 1–23.

Kruuk, L.E.B. (2004) Estimating genetic parameters in natural populations using the 'animal model'. *Philosophical*

Transactions of the Royal Society of London Series B-Biological Sciences, **359**, 873–890.

Kruuk, L.E.B. & Hadfield, J.D. (2007) How to (reliably) separate genetic and environmental causes of similarity between relatives. *Journal of Evolutionary Biology*, **20**, 1890–1903.

Kruuk, L.E.B., Slate, J. & Wilson, A.J. (2008) New answers for old questions: the evolutionary quantitative genetics of wild animal populations. *Annual Review of Ecology, Evolution, and Systematics*, **39**, 525–548.

Lynch, M. (1999) Estimating genetic correlations in natural populations. *Genetical Research*, **74**, 255–264.

Lynch, M. & Walsh, B. (1998) *Genetics and analysis of quantitative traits*. Sinauer, Sunderland, MA.

Marshall, T.C., Slate, J., Kruuk, L.E.B. & Pemberton, J.M. (1998) Statistical confidence for likelihood-based paternity inference in natural populations. *Molecular Ecology*, **7**, 639–655.

McCleery, R.H., Pettifor, R.A., Armbruster, P., Meyer, K., Sheldon, B.C. & Perrins, C.M. (2004) Components of variance underlying fitness in a natural population of the great tit *Parus major*. *American Naturalist*, **164**, E1–E11.

Merilä, J. & Sheldon, B.C. (2000) Lifetime reproductive success and heritability in nature. *American Naturalist*, **155**, 301–310.

Merilä, J. & Sheldon, B.C. (2001) Avian quantitative genetics. *Current Ornithology*, **16**, 179–255.

Moore, A.J. & Kukuk, P.F. (2002) Quantitative genetic analysis of natural populations. *Nature Reviews Genetics*, **3**, 971–978.

Mousseau, T.A. & Roff, D.A. (1987) Natural selection and the heritability of fitness components. *Heredity*, **59**, 181–197.

Nilsson, J.A., Akesson, M. & Nilsson, J.F. (2009) Heritability of resting metabolic rate in a wild population of blue tits. *Journal of Evolutionary Biology*, **22**, 1867–1874.

Nussey, D.H., Wilson, A.J. & Brommer, J.E. (2007) The evolutionary ecology of individual phenotypic plasticity in wild populations. *Journal of Evolutionary Biology*, **20**, 831–844.

Pemberton, J.M. (2008) Wild pedigrees: the way forward. *Proceedings of the Royal Society B-Biological Sciences*, **275**, 613–621.

Perrins, C.M. & Jones, P.J. (1974) The inheritance of clutch size in the great tit (*Parus major*). *Condor*, **76**, 225–228.

Postma, E. (2006) Implications of the difference between true and predicted breeding values for the study of natural selection and micro-evolution. *Journal of Evolutionary Biology*, **19**, 309–320.

Postma, E. & Charmantier, A. (2007) What 'animal models' can and cannot tell ornithologists about the genetics of wild populations. *Journal of Ornithology*, **148**, 633–642.

Postma, E., Kruuk, L.E.B., Merilä, J., Van Noordwijk, A.J. & Sheldon, B.C. (2002) Quantitative genetic analysis of natural populations: old wine in a new but defective bottle? *Nature Reviews Genetics, online only*.

Potti, J. (1999) Maternal effects and the pervasive impact of nestling history on egg size in a passerine bird. *Evolution*, **53**, 279–285.

Quinn, J.L., Charmantier, A., Garant, D. & Sheldon, B.C. (2006) Data depth, data completeness, and their influence on quantitative genetic estimation in two contrasting bird populations. *Journal of Evolutionary Biology*, **19**, 994–1002.

Quinn, J.L., Patrick, S.C., Bouwhuis, S., Wilkin, T.A. & Sheldon, B.C. (2009) Heterogeneous selection on a heritable temperament trait in a variable environment. *Journal of Animal Ecology*, **78**, 1203–1215.

Réale, D., Martin, J., Coltman, D.W., Poissant, J. & Festa-Bianchet, M. (2009) Male personality, life-history strategies and reproductive success in a promiscuous mammal. *Journal of Evolutionary Biology*, **22**, 1599–1607.

Ritland, K. (1996) Marker-based method for inferences about quantitative inheritance in natural populations. *Evolution*, **50**, 1062–1073.

Ritland, K. (2000) Marker-inferred relatedness as a tool for detesting heritability in nature. *Molecular Ecology*, **9**, 1195–1204.

Rose, S.P.R. (2006) Commentary: Heritability estimates—long past their sell-by date. *International Journal of Epidemiology*, **35**, 525–527.

Teplitsky, C., Mills, J.A., Yarrall, J.W. & Merilä, J. (2009) Heritability of fitness components in a wild bird population. *Evolution*, **63**, 716–726.

Thomas, S.C., Coltman, D.W. & Pemberton, J.M. (2002) The use of marker-based relationship information to estimate the heritability of body weight in a natural population: a cautionary tale. *Journal of Evolutionary Biology*, **15**, 92–99.

Tschirren, B. & Postma, E. (2010) Quantitative genetics research in Zebra Finches: where we are and where to go. *Emu*, **110**, 268–278.

van der Jeugd, H.P. & McCleery, R.H. (2002) Effects of spatial autocorrelation, natal philopatry and phenotypic plasticity on the heritability of laying date. *Journal of Evolutionary Biology*, **15**, 380–387.

van Noordwijk, A.J., Van Balen, J.H. & Scharloo, W. (1981a) Genetic and environmental variation in clutch size of the great tit (*Parus major*). *Netherlands Journal of Zoology*, **31**, 342–372.

van Noordwijk, A.J., Van Balen, J.H. & Scharloo, W. (1981b) Genetic variation in the timing of reproduction in the great tit. *Oecologia*, **49**, 158–166.

Visscher, P.M., Hill, W.G. & Wray, N.R. (2008) Heritability in the genomics era—concepts and misconceptions. *Nature Reviews Genetics*, **9**, 255–266.

Visscher, P.M., Medland, S.E., Ferreira, M.A.R., Morley, K.I., Zhu, G., Cornes, B.K., Montgomery, G.W. & Martin, N.G. (2006) Assumption-free estimation of heritability from genome-wide identity-by-descent sharing between full siblings. *PLoS Genetics*, **2**, e41.

Walsh, B. & Blows, M.W. (2009) Abundant genetic variation plus strong selection = multivariate genetic constraints: a geometric view of adaptation. *Annual Review of Ecology Evolution, and Systematics*, **40**, 41–59.

Wilson, A.J. (2008) Why h^2 does not always equal V_A/V_P? *Journal of Evolutionary Biology*, **21**, 647–650.

Wilson, A.J., Réale, D., Clements, M.N., Morrissey, M.M., Postma, E., Walling, C.A., Kruuk, L.E.B. & Nussey, D.H. (2010) An ecologist's guide to the animal model. *Journal of Animal Ecology*, **79**, 13–26.

Wright, S. (1921) Correlation and causation. *Journal of Agricultural Research*, **20**, 557–585.

www.oup.co.uk/companion/charmantier

CHAPTER 3

Quantitative genetic approaches to understanding sexual selection and mating system evolution in the wild

Jane M. Reid

3.1 Introduction

Sexual selection stems from variation in mating success, and more specifically from variation in access to opposite-sex gametes (Andersson 1994; Kokko *et al.* 2006). It is a potent force that can drive the evolution of complex mating systems and associated behaviours, physiologies and morphologies including seemingly extravagant ornamental traits, and also influence population persistence and speciation (Lande 1981; Andersson 1994; Birkhead & Pizzari 2002; Kokko & Brooks 2003; Shuster & Wade 2003; Arnqvist & Rowe 2005; Ritchie 2007; Eberhard 2009; Pizzari & Wedell 2013). Understanding the causes and consequences of variation in mating success, and of resulting sexual selection and trait and mating system evolution, are therefore longstanding and ongoing ambitions in evolutionary biology (Andersson 1994; Kokko et al. 2003, 2006; Shuster & Wade 2003; Jones & Ratterman 2009). In this chapter, I review the opportunities for quantitative genetic analyses of wild population data to answer key questions regarding sexual selection and associated trait and mating system evolution, and illustrate these opportunities using two case studies: the coevolution of female mating preferences and male ornamentation in collared flycatchers (*Ficedula albicollis*), and extra-pair reproduction in song sparrows (*Melospiza melodia*).

3.2 Foundations of sexual selection

Sexual selection, and resulting trait and mating system evolution, require some form of systematic variation in mating success with respect to phenotypic traits and any underlying genetic variation. The necessary variation in mating success could arise through multiple non-exclusive mechanisms. Direct competition for access to mates or associated resources such as breeding sites can create substantial variation in mating success among competing same-sex individuals. Sexual selection on traits that underlie success in this intra-sex competition will result (Andersson 1994). Variation in mating success can also stem from some form of systematic mate choice among opposite-sex individuals. Immediate fitness benefits of such mate choice are sometimes evident, for example when potential mates vary in their provision of resources such as nuptial gifts or breeding sites that increase the choosing individual's immediate fecundity, reproductive success or survival to future reproductive events. Mate choice is then under direct natural selection in choosing individuals so long as resulting increases in reproductive success or survival exceed any cost of choice measured in these same currencies (Kokko *et al.* 2003). Choice simultaneously imposes sexual selection on chosen and rejected members of the opposite sex, and therefore on traits that cause

Quantitative Genetics in the Wild. Edited by Anne Charmantier, Dany Garant, and Loeske E. B. Kruuk

variation in the acquisition of the resource that creates the benefit of mate choice.

These explanations for variation in mating success and resulting sexual selection stemming from intra-sex competition and/or resource-based mate choice can, to some level, be adequately developed verbally (Andersson 1994; Kokko *et al.* 2003). They can also be tested through phenotypic experiments, for example by manipulating relevant traits or resources possessed by individuals of the competing sex and observing associated variation in their mating success, and/or in the reproductive success or survival of the choosing sex. Indeed, sexual selection stemming from intra-sex competition or resource-based mate choice underpins our understanding of the evolution of numerous traits, from weaponry to nuptial provisioning and visual and acoustic ornaments (Andersson 1994).

However, historically more contentious forms of mate choice and resulting sexual selection occur when direct natural selection on mate choice appears to be absent or negative. This occurs when there is no apparent benefit of mate choice in terms of increasing the choosing individual's own reproductive success or survival, or choice actually reduces these fitness components. The persistence of mate choice, and resulting sexual selection, is then hypothesised to stem from some 'indirect' genetic benefit that increases the fitness of resulting offspring (Andersson 1994; Kokko *et al.* 2003, 2006). Interest in the degree to which such genetic effects, and the resulting 'indirect selection', could drive mate choice and cause sexual selection gained even further impetus with the realisation that variation in effective mating success can also be generated by variation in post-copulatory fertilisation success, due to sperm competition or 'cryptic' female choice (Birkhead & Pizzari 2002; Simmons 2005; Parker 2006; Eberhard 2009). Since sperm seemingly contribute only genes to a reproductive event, direct natural selection on post-copulatory mate choice appears minimal, at least at first glance. The evolution of post-copulatory mate choice and fertilisation bias, and the causal and consequent sexual selection, is therefore hypothesised to stem from genetic variation in fitness of resulting offspring (Pizzari & Birkhead 2002; Simmons 2005; Evans & Simmons 2008).

Perhaps because of their elusive nature and associated intellectual challenge, the search for indirect genetic mechanisms that could drive pre- and post-copulatory mate choice, and hence cause sexual selection and resulting trait and mating system evolution, remains a major industry in evolutionary biology. Major objectives have been to develop theoretical frameworks that explain and predict the evolution of sexually selected traits through various forms of indirect selection, and to test resulting hypotheses regarding both pre-copulatory and post-copulatory mating biases (Jennions & Petrie 2000; Birkhead & Pizzari 2002; Kokko *et al.* 2003, 2006; Mead & Arnold 2004; Neff & Pitcher 2005; Simmons 2005; Pizzari & Wedell 2013). Some general principles, such as the potential for 'runaway' evolution of ornamental traits and associated mating preferences, are now reasonably well established. However there remain substantial uncertainties regarding the forms and magnitudes of different components of indirect selection that could and do cause or result from mate choice, their relative contributions to the evolution of ornamental traits, reproductive physiologies, behaviours and mating systems, and the mechanisms that maintain the necessary genetic variation (Mead & Arnold 2004; Simmons 2005; Kokko *et al.* 2006; Slatyer *et al.* 2012).

Initial focus on additive effects of independently acting alleles has been joined by increasing interest in situations where fitness depends on allele combinations and hence genetic dominance and/or epistasis (Tregenza & Wedell 2000; Neff & Pitcher 2005). Furthermore, it is increasingly recognised that genetic effects on traits and fitness components, and selection, may be sex-specific, creating evolutionary sexual conflict over mating systems and associated traits (Chapman *et al.* 2003; Arnqvist & Rowe 2005; Parker 2006; Bonduriansky & Chenoweth 2009). Numerous solutions to the conundrum of what maintains genetic variation under persistent directional sexual selection have been proposed, including large mutational targets formed by 'condition dependent' polygenic traits, non-additive genetic effects, indirect genetic effects stemming from interactions among individuals, sexual conflict, and frequency- and environment-dependence (Rowe & Houle 1996; Reinhold 2002; Neff & Pitcher 2005; Pischedda &

Chippindale 2006; Miller & Moore 2007; Kruuk *et al.* 2008).

Such mechanisms may act in diverse combinations, creating complex opportunities for and constraints on evolution. Hypotheses explaining mate choice and the resulting sexual selection that invoke multiple components of sex-specific genetic variation and covariation, potentially acting across multiple traits, are consequently complex, and cannot be adequately formulated through simple verbal logic (Kokko *et al.* 2006; Walsh & Blows 2009). Furthermore they cannot be adequately tested through purely phenotypic analyses because they explicitly involve within-sex and cross-sex genetic covariances, including in traits with sex-limited phenotypic expression (Mead & Arnold 2004; Pischedda & Chippindale 2006; Kruuk *et al.* 2008). Such genetic covariances cannot be readily inferred from phenotypic covariances because they may commonly be confounded by environmental covariances or, in the case of cross-sex covariances, may not be observable at the phenotypic level. Full understanding of the interacting evolutionary causes and consequences of mate choice and sexual selection will therefore require integrated theory that incorporates multiple components of direct and indirect natural and sexual selection on key traits, and that can be explicitly tested in populations experiencing natural genetic and environmental variation.

Here I summarise the value of quantitative genetic approaches in formalising hypotheses regarding the evolutionary causes and consequences of mate choice and sexual selection, and of consequent trait and mating system evolution, focussing primarily on hypotheses involving 'indirect' genetic mechanisms. I then consider the opportunity to explicitly test such quantitative genetic theory in wild populations. I do not aim to review the extensive literature on sexual selection or mating system evolution (for which see Andersson 1994; Jennions & Petrie 2000; Birkhead & Pizzari 2002; Shuster & Wade 2003; Mead & Arnold 2004; Arnqvist & Rowe 2005; Neff & Pitcher 2005; Simmons 2005; Kokko *et al.* 2006; Parker 2006; Ritchie 2007; Eberhard 2009; Jones & Ratterman 2009; Pizzari & Wedell 2013), or even to comprehensively review quantitative genetic work in these fields. Rather, I introduce the conceptual and empirical merits of a quantitative genetic approach, and use two case studies to illustrate the opportunities for wild population quantitative genetics to bring new insights, the associated challenges, and the theoretical and empirical developments that are still required.

3.3 The merits of a quantitative genetic approach to understanding sexual selection

Multiple conceptual paradigms have been harnessed to probe the potential causes of mate choice and sexual selection, and the consequences for trait and mating system evolution. Population genetic approaches explicitly consider allele frequencies at a few loci that are assumed to underlie key traits (Kirkpatrick 1982; Reinhold 2002). Game theoretical approaches identify evolutionarily stable strategies among defined competing phenotypes (Chapman *et al.* 2003; Parker 2006). Broadly defined 'behavioural ecological' approaches consider costs and benefits of trait expression in terms of phenotypic life-history components and associated behaviours (Andersson 1994; Westneat & Stewart 2003). These approaches all have advantages and can provide valuable insights.

However, in their basic forms, they also have limited ability to capture key genetic processes that could cause and arise from sexual selection. For example, they do not readily incorporate or measure multi-locus genetic covariances that are likely to arise given any form of non-random mating and reproduction with respect to polygenic quantitative traits that show additive genetic variance. Since sexual selection stems from non-random mating, and resulting genetic covariances between mating preferences and focal traits are explicitly postulated to shape evolutionary consequences, this constraint limits the ability of population genetic, game theoretical and behavioural ecological approaches to provide pertinent predictions or tests (Lande 1981; Barton & Turelli 1991; Mead & Arnold 2004; Moore & Pizzari 2005). Furthermore, these approaches do not readily allow components of indirect selection on mate choice or trait expression stemming from additive or non-additive genetic variation in offspring fitness to be compared with components stemming directly from variation in adult fecundity or survival. Nor do they explicitly evaluate the relative impacts of genetic versus environmental variation on fitness or resulting trait or mating preference evolution. In contrast, the quantitative genetic paradigm has attributes that are conducive to developing evolutionary models of mate choice and sexual selection that incorporate all these components, and that also facilitate empirical testing. I summarise these attributes below.

First, an explicit assumption underlying quantitative genetics is that genetic variation in 'quantitative' traits, which exhibit measurable continuous phenotypic variation on some scale, stems from alleles at numerous loci of small ('infinitesimal') effect (Lynch & Walsh 1998; Hill 2010). The assumption that traits are polygenic with infinitesimal allelic effects is statistically convenient; genetic effects can then be assumed to approximate multivariate normality despite Mendelian segregation at individual loci. However this assumption is also likely to be appropriate for key traits and fitness components underlying mate choice, sexual selection and mating system evolution. High-density marker studies on model species indicate that additive genetic variance in even relatively simple quantitative traits, with human height being the classic example, does stem from numerous loci of predominantly small effect (Hill 2010; Rockman 2012). Similarly high-powered genomic analyses have not yet been applied to fitness, secondary sexual traits or mating preferences in wild populations (see Chapter 13, Jensen *et al.*). However, recent analyses failed to identify any major loci underlying clutch size or egg mass in great tits (*Parus major*) and showed chromosomal effects that were proportional to chromosome size, indicating that these traits are probably highly polygenic and influenced by widely distributed loci (Santure *et al.* 2013). Specific loci with apparently large effects on certain traits have been identified, such as major histocompatibility complex loci (Neff & Pitcher 2005), and *RXFP2* which underlies horn phenotype in Soay sheep (*Ovis aries*; Johnston *et al.* 2011), but these may be unusual (see Chapter 13, Jensen *et al.*). Furthermore, the theory of 'condition-dependent' trait expression and consequent 'genic capture' explicitly postulates that secondary sexual traits, and consequently fitness, are influenced by numerous loci, thereby explaining why genetic variation is maintained under directional sexual selection (Rowe & Houle 1996). Indeed additive genetic variance in fitness, by definition, comprises additive genetic effects stemming from all loci that are under selection at any time. The quantitative genetic paradigm, perhaps more than single- or few-locus paradigms, may therefore be appropriate to developing and testing sexual selection theory (Lande 1981).

Second, well-established general principles of evolutionary quantitative genetics state that cross-generational evolutionary change in any trait can be understood as a function of the various components of selection acting on the focal trait and genetically correlated traits across population members (Lande & Arnold 1983; Arnold & Wade 1984). Evolutionary responses to direct and indirect selection can consequently be predicted in terms of phenotypic selection gradients and/or additive genetic variances and covariances in and among traits and fitness components, at least given some assumptions (Lande & Arnold 1983; Arnold & Wade 1984; Lynch & Walsh 1998; Rausher 1992; Morrissey *et al.* 2010). General quantitative genetic frameworks also allow explicit consideration of non-additive genetic variation, various components of direct and intergenerational environmental variation, genotype-by-environment interactions, indirect genetic effects stemming from interactions among individuals, and cross-sex genetic covariances and evolutionary trade-offs among life-history components within and among individuals (Moore *et al.* 1997; Lynch & Walsh 1998; Bijma *et al.* 2007), all of which are hypothesised to shape the outcome of sexual selection. These general frameworks are potentially just as applicable to the evolution of mating preferences and associated sexually selected traits as to any other trait. They therefore have potential to encompass many seemingly disparate or conflicting hypotheses and contentions regarding mate choice and sexual selection, including classical 'runaway' and 'good genes' explanations for the evolution of exaggerated traits, 'sensory exploitation' stemming from direct selection on alleles with pleiotropic effects on mate choice, evolutionary sexual conflict stemming from sex-specific selection on genetically correlated traits, mate choice for 'compatible genes', and environment-dependence and phenotypic plasticity (Neff & Pitcher 2005; Kokko *et al.* 2006; Kruuk *et al.* 2008).

Third, and critically, quantitative genetics is an empirical science; genetic and environmental variances and covariances can be estimated from observed phenotypes of relatives, including across sexes or life-stages that do not express particular traits but through whom indirect selection might still act (Lynch & Walsh 1998; Hill 2010; Hill & Kirkpatrick 2010). This empirical basis means that quantitative genetic models of mate choice and sexual selection should be amenable to explicit testing, including in wild populations. Indeed, quantitative genetics has a long and successful history of predicting the phenotypic outcome of mating decisions in the context of animal and plant breeding; breeders predict the genetic and phenotypic values of offspring that would result from specific crosses and orchestrate matings accordingly (Hill & Kirkpatrick 2010). Such breeding programmes often focus primarily on additive genetic effects and predicted breeding values, but can incorporate non-additive genetic effects including dominance and inbreeding, and indirect genetic effects stemming from interactions among individuals (Lynch & Walsh 1998; Bijma *et al.* 2007; Hill 2010). Breeding programmes differ from natural mating systems in that specific matings are imposed, creating highly constrained and often strong regimes of 'sexual selection'. Variation in 'fitness' is hard to interpret because individuals that are not selected to breed may be culled, while offspring may be harvested rather than allowed to mature. However there is no fundamental reason why evolutionary biologists should not use a similar quantitative genetic paradigm to that long exploited by industrial animal and plant breeders to predict and understand the consequences of mating decisions and sexual selection in nature. Such an explicit quantitative genetic approach, involving estimation of key genetic covariances among mating preferences and ornamental and reproductive traits and fitness components, remains rather under-utilised in wild populations.

3.4 Quantitative genetic theory of sexual selection

The above attributes mean that a quantitative genetic approach to understanding sexual selection has many virtues. Indeed, quantitative genetic models that explicitly or implicitly assume polygenicity have proved pivotal to formalising hypotheses regarding sexual selection and associated mate choice and trait evolution, and have provided key proofs of principle (Barton & Turelli 1991; Mead & Arnold 2004; Kokko *et al.* 2006).

Lande (1981) used a polygenic quantitative genetic approach to consider 'runaway' evolution stemming from assortative mating between individuals with heritable ornamental display traits and corresponding mating preferences. Lande's model, and numerous subsequent developments, show that exaggerated traits and corresponding preferences can in principle evolve through runaway processes resulting from genetic covariances between trait and preference, including when there is direct natural selection against trait and/or preference, and when the focal trait is also genetically correlated with other fitness components (Barton & Turelli 1991; Mead & Arnold 2004; Kokko *et al.* 2006). Moreover these models imply that genetic covariances, and hence some degree of evolutionary runaway, could arise due to linkage disequilibria given additive genetic variation in mating propensity with respect to any trait that itself shows additive genetic variation, creating indirect selection without need for pleiotropy. Resulting evolutionary dynamics are predicted to depend on the (relative) magnitudes of additive genetic variances and covariances in and among traits and associated preferences (Lande 1981; Mead & Arnold 2004; Kokko *et al.* 2006).

Such models, that were originally formulated to consider pre-copulatory sexual selection, have been re-interpreted to consider post-copulatory sexual selection. For example, it has been proposed that female multiple mating (polyandry), one widespread mating system that remains generally unexplained, might have evolved to induce fertilisation by males with high additive genetic value for success in sperm competition (analogous to an arbitrary display trait that increases mating success) and/or high additive genetic value for other fitness components that are genetically correlated with sperm competitive ability (Birkhead & Pizzari 2002; Simmons 2005; Evans & Simmons 2008).

Quantitative genetic models have also been used to formalise the hypothesis that exaggerated display traits and associated female preferences reflect sexual conflict over mating. Gavrilets *et al.* (2001) showed that female preferences could stem from direct selection on females to reduce costs of mating, while Moore and Pizzari (2005) provided a general quantitative genetic treatment of sexual conflict over reproductive decisions in a framework of indirect genetic effects among interacting individuals. Indeed, sexual conflict can, in general, be defined in terms of negative cross-sex genetic and phenotypic covariances among traits and fitness components (Kruuk *et al.* 2008; Bonduriansky & Chenoweth 2009).

These various models can appear disparate and contradictory, and have been viewed as alternatives to be distinguished (Mead & Arnold 2004; Kokko *et al.* 2003, 2006). However many underlying concepts have been, or could potentially be, unified within single quantitative genetic frameworks that predict evolutionary responses to various components of direct and indirect selection, in terms of genetic and phenotypic variances and covariances in and among relevant traits and fitness components within and among interacting males and females (Kokko *et al.* 2006; Kruuk *et al.* 2008). Such frameworks could potentially identify key quantitative genetic parameters that need to be estimated in order to understand what forces drive and constrain ongoing sexual selection and resulting trait and mating system evolution in nature.

3.5 Approaches to testing quantitative genetic theory of sexual selection

As evidenced above, quantitative genetics provides key theoretical frameworks on which our understanding of the evolutionary causes and consequences of mate choice and sexual selection, and resulting trait and mating system evolution, relies. It is also a well-established empirical science. There should consequently be ample opportunities to explicitly test evolutionary theory (Mead & Arnold 2004; Evans & Simmons 2008). Empirical studies are needed to estimate genetic variances and covariances among mating preferences and key morphological, physiological and behavioural traits and fitness components. Such studies could distinguish between hypotheses explaining the evolution and persistence of such traits and resulting mating systems and, perhaps more usefully, quantify the relative magnitudes of different components of direct and indirect selection that act on the various traits of interest (Kirkpatrick & Barton 1997; Kokko *et al.* 2003).

One powerful approach to quantitative genetic analysis, and hence to estimating genetic (co)variances that are postulated to underlie evolution, is to enforce some form of organised breeding among known parents and measure phenotypes of resulting offspring (Lynch & Walsh 1998). This approach generally requires a captive population where mating can be experimentally controlled. Such conditions can also allow environmental variation to be minimised or otherwise controlled, and facilitate measurement of key traits such as reproductive behaviours and physiologies that are difficult to quantify in free-living animals (Postma *et al.* 2006; Forstmeier *et al.* 2011). Artificial insemination can be used to eliminate selection stemming directly from mating, sperm competition and any associated maternal effects, and to create planned distributions of close relatives such as full- or half-sibs that maximise statistical power (Neff & Pitcher 2005; Simmons 2005). Well-designed breeding experiments can therefore maximise power and minimise numerous potential biases, and consequently make valuable contributions to testing sexual selection theory (Evans *et al.* 2007; Simmons & Kotiaho 2007; Bilde *et al.* 2008; Forstmeier *et al.* 2011; García-Gonzáles & Simmons 2011).

However, such breeding experiments also have drawbacks. Constraining environmental variation may remove sources of variation in mating success that shape sexual selection in nature. Furthermore, understanding the causes and evolutionary consequences of sexual selection requires multi-generational variation in fecundity and survival as well as mating success to be measured, allowing estimation of genetic (co)variances among ornamental and reproductive traits and fitness components (Kokko *et al.* 2006). These fitness components can be difficult to measure or meaningless in breeding experiments where sexual selection is imposed rather than arising naturally, and where natural selection may be negated by the breeding design or weakened in the absence of predators, parasites or resource limitation (Neff & Pitcher 2005).

Moreover, an obvious but rarely emphasised problem is that breeding experiments by definition impose some form of artificial structured mating. They could consequently alter the genetic covariances that would arise due to linkage disequilibria given natural forms of non-random mating, and that are predicted to drive or constrain the ongoing evolution of mate choice and associated traits (Lande 1981; Walsh & Blows 2009). Structured breeding experiments designed to estimate key genetic variances and covariances underlying sexual selection may therefore alter the very (co)variances that they aim to measure (Kokko *et al.* 2006). Unsurprisingly, therefore, quantitative genetic breeding experiments that deliberately impose tightly constrained artificial mating regimes may not always be an ideal means to answer questions regarding the causes and consequences of natural variation in mating.

Finally, appropriate biological inference may be challenging when populations available for experimental breeding are descended from unnatural genetic backgrounds, or have experienced numerous generations of adaptation to unnatural conditions or inbreeding (Ala-Honkola *et al.* 2011). The obvious solution to all these difficulties is to complement the powerful quantitative genetic analyses that utilise breeding experiments by estimating key parameters and testing sexual selection theory in wild populations experiencing natural environmental and genetic variation in mating success, survival and fecundity (Charmantier & Sheldon 2006; Rodriguez-Muñoz *et al.* 2010).

Field estimates of quantitative genetic parameters relevant to sexual selection theory were historically largely restricted to heritabilities of secondary sexual traits and life-history components estimated from parent-offspring regression or basic cross-fostering designs (Norris 1993; Andersson 1994; Griffith *et al.* 1999; Chapter 2, Postma). Modern mixed-model analyses, such as animal models, offer potentially more powerful and less biased estimation, with better discrimination between environmental and genetic variances and relative robustness to selection and inbreeding that occur within natural mating systems (Kruuk 2004; Kruuk & Hadfield 2007; Hill 2010; Chapter 2, Postma). They also facilitate estimation of further key parameters, such as additive genetic covariances between sex-limited traits and fitness components (Kruuk *et al.* 2008). Such analyses therefore potentially allow key genetic variances and covariances underlying sexual selection and mating system evolution to be estimated in natural reproductive systems with unbalanced variation in mating and reproductive success, as is imperative when the objective is to understand the evolutionary causes and consequences of exactly that variation.

3.6 Linking quantitative genetic theory and test

Sophisticated quantitative genetic theory pertaining to sexual selection, and sophisticated quantitative genetic analysis of wild population data, are both now well established fields. However their joint application to understanding sexual selection and associated mate choice and mating system evolution is still surprisingly restricted (Charmantier & Sheldon 2006).

Many quantitative genetic models of sexual selection were derived to formalise verbal hypotheses regarding sexual selection and prove basic principles, such as 'runaway' or 'good genes' processes, not to explicitly guide empirical tests (Lande 1981; Gavrilets *et al.* 2001; Kirkpatrick & Hall 2004; Moore & Pizzari 2005; Kokko *et al.* 2006). This may partly be because key models date from eras when genetic covariances could not be readily estimated in free-breeding populations. The process of developing evolutionary quantitative genetic theory was therefore inevitably divorced from field biology (Mead & Arnold 2004). Many models are consequently not formulated in terms of genetic variances and covariances among observable phenotypes and fitness components which, due to recent advances in statistical analysis, could now be estimated in wild populations. Instead, they involve quantities such as residual variation in fitness after accounting for mating success, naturally selected optima in display traits and mating behaviour, and shape parameters underlying response functions that are conceptually useful but empirically intractable (Lande 1981; Gavrilets *et al.* 2001; Kokko *et al.* 2006; Rowe & Day 2006).

Some basic quantitative genetic parameters that must define the evolution of mate choice and associated sexually selected traits and mating systems, and which therefore need to be estimated, can be identified from existing quantitative genetic or verbal models of sexual selection or general trait evolution. These include additive genetic variances and covariances among mating preferences, ornamental traits and fitness components (Lande 1981; Arnold & Wade 1984; Mead & Arnold 2004). However, such verbal reasoning and general theory might fail to identify specific genetic (co)variances that underlie complex mechanisms of sexual selection and resulting evolution stemming from indirect selection and sexual conflict involving multiple traits (Kirkpatrick & Barton 1997; Kokko *et al.* 2006; Rowe & Day 2006; Walsh & Blows 2009).

The historical inability to estimate key parameters, and associated paucity of explicitly testable quantitative genetic theory, means that empiricists interested in testing hypotheses regarding sexual selection in wild populations still do not readily think in explicit quantitative genetic terms or estimate key parameters, even though estimation is now technically feasible (Mead & Arnold 2004; Kokko *et al.* 2006). Field tests have historically been the preserve of behavioural ecologists, whose language of 'indirect genetic benefits' (typically interpreted to mean increased offspring fitness) does not entirely equate to the quantitative geneticist's 'indirect selection' (meaning selection stemming from genetic covariances with traits that are under direct selection). For example, numerous mating behaviours and physiologies are widely postulated to reflect evolutionary sexual conflict (Gavrilets *et al.* 2001; Westneat & Stewart 2003; Arnqvist & Rowe 2005; Rowe & Day 2006; Eberhard 2009). Such conflict can be defined in terms of negative genetic covariances among relevant traits and fitness components across males and females (Rowe & Day 2006; Kruuk *et al.* 2008; Bonduriansky & Chenoweth 2009). However an explicit quantitative genetic approach to quantifying sexual conflict is surprisingly rarely suggested or implemented in the context of wild population studies of mate choice or consequent sexual selection. Even some relatively recent reviews suggest combinations of behavioural observations, experimental evolution, trait manipulation and comparative analyses and crosses as empirical approaches, but not quantitative genetics (Chapman *et al.* 2003; Pizzari & Snook 2003; Westneat & Stewart 2003; Eberhard 2009, but see comments in Birkhead & Pizzari 2002; Rowe & Day 2006; Kruuk *et al.* 2008; Poissant *et al.* 2010).

If hypotheses explaining the evolution of mate choice and associated traits and mating systems could be explicitly formulated in terms of genetic and environmental variances and covariances among measurable phenotypic traits and fitness components, then key hypotheses explaining sexual selection could be tested through quantitative genetic analysis of wild population data. However, substantial developments in quantitative genetic theory, analysis and data are still required to achieve this goal. Here I use two case studies to illustrate the opportunities afforded and the challenges faced by an evolutionary quantitative genetic approach to understanding sexual selection in the wild. Both case studies are valuable in developing, implementing and demonstrating new approaches to answering long-standing questions in evolutionary biology (Charmantier & Sheldon 2006), but have also proved contentious in terms of theory, data or analysis.

3.7 Case study 1: Quantitative genetic insights into the evolution of ornamental traits and associated mating preferences

One central aim in evolutionary biology has been to explain the evolution of extravagant 'ornamental' secondary sexual traits (Andersson 1994). Such traits are expected to reduce their bearer's survival and/or fecundity and hence to experience negative natural selection. Some other force is consequently required to explain their evolution and persistence. Increased mating success, and resulting positive sexual selection, is widely suggested to provide such a force. The remaining objectives, then, are to understand why such traits confer increased mating success and specifically to understand the evolution of directional mating preferences for specific traits, and to quantify the (relative) forces of resulting direct and indirect selection on preference and ornamentation (Kirkpatrick & Barton 1997; Kokko *et al.* 2003, 2006).

3.7.1 A quantitative genetic model

Kirkpatrick and Barton (1997) derived a general quantitative genetic model that considers the evolution of a female mating preference for a display trait that indicates a male's additive genetic value for fitness. Their model, which focusses on indirect selection, makes relatively few assumptions regarding the exact form of female preference or underlying genetic architecture, defining female preference as 'any measurable component of a female's behaviour that influences which male she mates'. It is explicitly formulated in terms of parameters that could potentially be estimated, or at least approximated, in wild populations. Specifically, the change in mean female preference over one generation that is attributable to indirect selection (Δ_I) can be approximated (in phenotypic standard deviations) as $\Delta_I \approx {}^1/_2 \rho_{PT}\, r_{TW}\, h_T\, h^2{}_P \sqrt{G_w}$, where ρ_{PT} is the phenotypic correlation between female preference and male display trait across breeding pairs, r_{TW} is the genetic correlation between male trait and fitness, h_T is the square root of the heritability of the male trait, $h^2{}_P$ is the heritability of the female preference and G_w, strictly defined, is the variance in the lifetime fitness of a male's genotype if it were expressed in numerous males and females. This expression can be rewritten as $\Delta_I \approx r_{PT}\, r_{TW}\, h_P \sqrt{G_w}$ where r_{PT} is the genetic correlation between female preference and male trait (since $r_{PT} \approx {}^1/_2\, \rho_{PT}\, h_T\, h_P$, Kirkpatrick & Barton 1997). The per-generation evolutionary change in the female preference due to indirect selection can therefore be approximated as a function of the genetic correlations between preference, trait and fitness, the heritability of the preference and the additive genetic variance in fitness (Kirkpatrick & Barton 1997).

Kirkpatrick and Barton's (1997) model is valuable because it identifies key estimable parameters that underlie the evolution of female preferences through indirect selection, and defines a means of explicitly predicting resulting evolutionary change. However, few key parameters had been estimated in wild populations. Qvarnström *et al.* (2006a) subsequently made a valiant attempt to estimate these parameters using 24 years of data from collared flycatchers (*Ficedula alibcollis*) breeding on the Swedish island of Gotland.

3.7.2 Forehead patch size in collared flycatchers

The collared flycatcher is a socially monogamous migratory passerine (Figure 3.1). Breeding pairs

(a)

(b)

Figure 3.1 Analyses of long-term life-history and pedigree data from (a) collared flycatchers (*Ficedula albicollis*) and (b) song sparrows (*Melospiza melodia*) illustrate the opportunities for explicit quantitative genetic approaches of wild population data to inform our understanding of mate choice and mating system evolution, but also illustrate major remaining challenges. Photos: Johan Träff and Sylvain Losdat.

form after spring migration; males defend nest sites, and both sexes contribute to parental care. Males have striking white forehead patches that vary in size and influence female mate choice (Figure 3.1, Qvarnström *et al.* 2000). Chicks hatched in the Gotland study area were ringed before fledging, and the parents that reared these chicks were identified. Male forehead patch sizes were measured, and a metric of annual contribution to fitness that incorporates an individual's number of recruited offspring and its own survival to the subsequent year was calculated (Qvarnström *et al.* 2006a). The substantial pedigree data (albeit incomplete and based on social rather than genetic parentage) allowed animal models to be used to estimate additive genetic variances, covariances and correlations in and among annual fitness, male forehead patch size and the forehead patch size of a female's observed social mate. The latter trait was interpreted as a measure of a female's mating preference. The measure of annual fitness was considered a sex-specific trait because it has an estimated inter-sex genetic correlation (r_{mf}) of less than one in the study population (Qvarnström *et al.* 2006a; Brommer *et al.* 2007). This necessitated an extension to Kirkpatrick and Barton's (1997) basic model, by defining $G_W = \frac{1}{2}(G_{wm} + r_{mf}\, G_{wf})$ where G_{wm} and G_{wf} were interpreted as the coefficients of additive genetic variation for male and female fitness respectively (Qvarnström *et al.* 2006a; Brommer *et al.* 2007).

Qvarnström *et al.* (2006a) estimated substantial additive genetic variance and heritability in male forehead patch size ($h^2_T = 0.38 \pm 0.03$ *SE*, $n = 4220$ males), and small but non-zero additive genetic variance and heritability in male and female annual fitness ($h^2 = 0.03 \pm 0.01$ *SE*, $n = 3{,}019$ males and $h^2 = 0.04 \pm 0.01$ *SE*, $n = 2934$ females respectively). Furthermore, there was a positive genetic correlation between male forehead patch size and male annual fitness of $r_{TW} = 0.15 \pm 0.09$ *SE*, although this did not differ significantly from zero. There was also small additive genetic variance and heritability in the forehead patch size of a female's social mate (interpreted as a female's realised mating preference and hence as a female trait, $h^2_P = 0.03 \pm 0.01$ SE, $n = 4345$ females). The genetic correlation between this measure of female preference and male forehead patch size was small and did not differ from zero ($r_{PT} = -0.02 \pm 0.17$ *SE*). The rate of evolutionary change in the female mating preference attributable to indirect selection, Δ_I, was consequently estimated to be small. These estimates imply that some form of direct selection must underlie the evolution and persistence of the female preference; most probably increased reproductive success stemming from resources or paternal care provided by males with large forehead patches (Qvarnström *et al.* 2000, 2006a).

3.7.3 Strengths and limitations

Through their analyses, Qvarnström *et al.* (2006a) provide the first estimates of key parameters underlying the coevolution of display trait and preference through this particular component of indirect selection in a wild population. They thereby demonstrate the potential for explicit quantitative genetic analyses to test major evolutionary hypotheses (Charmantier & Sheldon 2006). However they also illustrate some practical, analytical and conceptual difficulties that are likely to afflict such studies (Postma *et al.* 2006).

First, Qvarnström *et al.* (2006a) used the forehead patch size of a female's observed social mate to measure female mating preference. This interpretation was challenged because the phenotype of a female's realised mate could deviate substantially from her intrinsic mating preference, particularly in socially monogamous species such as collared flycatchers where pairing opportunities are constrained (Postma *et al.* 2006). The use of a female's observed mate's trait value to measure mating preference therefore requires careful interpretation in the context of Kirkpatrick and Barton's (1997) original model (Qvarnström *et al.* 2006b), but may be the only practical approach when absolute rather than realised female preferences are difficult to measure.

Second, Qvarnström *et al.* (2006a) measured social mate choice and did not account for extra-pair reproduction even though *c*.15% of collared flycatcher offspring are sired by extra-pair males (Brommer *et al.* 2007). Such paternity error has several consequences. It introduces error and potentially bias into estimates of male annual fitness in relation to forehead patch size (Sheldon & Ellegren

1999). It causes pedigree error and hence error in the relatedness matrix that underlies animal model analysis, and moreover this error is not independent of the error in estimated male fitness. Finally, incorrect identification of a female's mates could also introduce error and potentially bias into estimated phenotypic and genetic associations between inferred female preference (i.e. the forehead patch size of a female's various mates) and male forehead patch size. Indeed, extra-pair reproduction is itself widely postulated to reflect indirect selection stemming from increased genetic value of a female's extra-pair male versus social mate (see Section 3.8). Unmeasured extra-pair reproduction (or other sources of systematic paternity error that commonly afflict wild population studies) could therefore bias key parameter estimates. The magnitude of such bias is not yet clear because no studies have directly compared analogous analyses run before and after correcting paternity error, or described the pattern of occurrence of paternity error in relation to trait values and relatedness in sufficient detail to support realistic simulations.

Third, Qvarnström *et al.*'s (2006a) measure of annual fitness included components of offspring survival as well as adult reproductive success and survival. This potentially impedes interpretation of estimated genetic covariances in the context of estimating evolutionary responses to indirect selection, since underlying theory pertains to changes in allele frequencies from zygote to zygote rather than adult to adult. Furthermore, estimates of G_{wm}, G_{wf} and r_{mf} changed substantially when lifetime reproductive success was subsequently considered (Brommer *et al.* 2007), illustrating that conclusions can vary markedly with the choice of fitness measure.

Finally, while Kirkpatrick and Barton's (1997) model makes few explicit assumptions regarding the form of the female mating preference, the derivation and proposed parameterisation (which relies on quasi linkage equilibrium approximations) does require assumptions that could be violated in wild populations. These include polygenicity and weak selection; primarily additive genetic variation in fitness; that genes underlying trait and preference are autosomal; and that G_{wm} and G_{wf} can be adequately approximated by estimating additive genetic variances in fitness. Some of these assumptions can be relaxed or may be reasonable, at least in relation to other sources of error and bias that are likely to afflict correlative analyses of field data (Kirkpatrick & Barton 1997; Kirkpatrick & Hall 2004). However they should be borne in mind by empiricists aiming to parameterise Kirkpatrick and Barton's (1997) model and draw quantitative conclusions.

3.8 Case study 2: Quantitative genetic insights into the evolutionary ecology of extra-pair reproduction

Socially monogamous systems such as collared flycatchers, where both paired adults provide resources and parental care, facilitate collection of observational pedigree data and are consequently common subjects for quantitative genetic analyses of wild populations. However, historically, they were not the systems where mate choice most obviously required explanation in terms of indirect selection or genetic variation in offspring fitness.

This situation was revolutionised when molecular genetic analyses revealed widespread extra-pair reproduction in socially monogamous systems; offspring are commonly sired by males other than a female's paired social mate, even though the paired male often still provides resources and paternal care to all offspring (Jennions & Petrie 2000; Griffith *et al.* 2002; Uller & Olsson 2008). Since extra-pair sires seemingly provide only gametes, there is no obvious source of direct selection for extra-pair reproduction by females. In contrast, numerous sources of direct selection against extra-pair reproduction are postulated, including reduced paternal care by the female's cuckolded social mate, and increased disease or predation risk (Birkhead & Pizzari 2002; Griffith *et al.* 2002; Westneat & Stewart 2003; Simmons 2005). The evolution and persistence of female extra-pair reproduction consequently requires explanation. One particularly influential hypothesis, which has stimulated huge empirical effort, is that female extra-pair reproduction, and other forms of polyandry that apparently lack direct resource provision, increase offspring genetic value and evolve through indirect selection (Jennions & Petrie 2000; Tregenza & Wedell 2000; Simmons 2005; Slatyer *et al.* 2012).

3.8.1 'Behavioural ecological' approaches

Attempts to understand female extra-pair reproduction in wild populations have traditionally followed the 'behavioural ecological' paradigm, where phenotypic costs and benefits to females are estimated in terms of immediate fitness components or parental behaviours (Griffith *et al.* 2002; Westneat & Stewart 2003). However, costs and benefits estimated from observational phenotypic data are prone to bias due to environmental covariation between extra-pair reproduction and fitness components (Kempenaers & Sheldon 1997; Eliassen & Kokko 2008). The experimental approach that facilitated estimation of analogous costs and benefits of secondary sexual ornamentation, where ornamentation is manipulated and effects on other traits or fitness components are recorded, is not easily applicable to extra-pair reproduction. This is because female extra-pair reproduction cannot be readily manipulated without simultaneously impacting other components of female or male life-history or behaviour (Kempenaers & Sheldon 1997; Griffith 2007). Explicit estimates of direct selection on extra-pair reproduction are consequently lacking (Arnqvist & Kirkpatrick 2005).

Extra-pair reproduction does, however, create an ad hoc opportunity to test the key hypothesis that female extra-pair reproduction is under indirect selection because a female's extra-pair offspring are of higher genetic value than the within-pair offspring she would otherwise have produced. This is widely achieved by comparing phenotypic traits or fitness components between maternal half-sibs sired by within-pair and extra-pair males and reared within the same clutch, brood or litter (Griffith *et al.* 2002; Slatyer *et al.* 2012). Assuming that such maternal half-sibs experience similar environments and maternal genetic contributions on average (i.e. that these environments and contributions do not vary systematically with paternity within a clutch or litter) maternal half-sib comparisons provide an ingenious means of assessing the difference in genetic value of offspring sired by different males when genetic value cannot be explicitly measured. This approach underpins all empirical estimates of 'indirect genetic benefits' of female extra-pair reproduction, and is probably the most widely implemented approach to estimating such benefits for any mating behaviour in the wild (Griffith *et al.* 2002; Arnqvist & Kirkpatrick 2005; Sardell *et al.* 2011; Schmoll *et al.* 2009; Slatyer *et al.* 2012).

However, the approach of inferring a difference in genetic value from basic phenotypic comparisons between maternal half-sibs has major limitations (Reid & Sardell 2012). The key assumption that environmental effects on offspring phenotype are independent of paternity may be violated, for example because paternity varies systematically with hatch or birth order within a clutch or litter (Magrath *et al.* 2009). Additive and non-additive genetic effects cannot be readily distinguished, and the relative magnitudes of genetic and environmental effects on fitness cannot be quantified. Furthermore, such maternal half-sib analyses are inevitably restricted to mixed-paternity broods and hence to functionally polyandrous females, and therefore ignore selection acting on genes underlying extra-pair reproduction through females whose offspring are all sired by either within-pair or extra-pair males. In most socially monogamous populations where extra-pair reproduction occurs, most females are in fact functionally monogamous (Griffith *et al.* 2002). Estimates of selection on extra-pair reproduction derived solely from females that produce mixed-paternity broods could consequently be severely biased if, as widely postulated and observed, extra-pair reproduction is non-random with respect to numerous attributes of females and/or their paired social mates.

Some of these limitations can, in principle, be resolved by refining the basic half-sib comparison approach (e.g. by controlling for hatch order, Magrath *et al.* 2009). However, such refinements distract from the facts that phenotypic maternal half-sib comparisons are simply a convenient ad hoc means of approximating the difference in genetic value between a female's within-pair and extra-pair offspring when genetic value cannot be measured properly, and moreover that even unbiased maternal half-sib comparisons might still provide severely biased estimates of total selection on extra-pair reproduction. Better ambitions might therefore be to explicitly estimate the difference in genetic value between maternal half-sibs, and the total indirect selection on female extra-pair reproduction due to genetic covariances with paternal genetic value, and predict resulting evolution. These objectives are potentially achievable through quantitative genetic analysis.

3.8.2 An initial quantitative genetic approach

Arnqvist and Kirkpatrick (2005) reformulated Kirkpatrick and Barton's (1997) general quantitative genetic model of mating behaviour evolution (see Section 3.7) to predict the evolutionary response to indirect selection on female extra-pair reproduction stemming from increased additive genetic value of extra-pair versus within-pair offspring. They purposely formulated their model in terms of parameters that could be estimated through phenotypic or basic quantitative genetic analysis of wild population data. They approximated the evolutionary response to indirect selection on female extra-pair reproduction (in phenotypic standard deviations per generation) as $\Delta_I = h^2_F\ \sigma_F\ d_{EW}$, where h^2_F and σ_F are the heritability and population-wide phenotypic standard deviation of female extra-pair reproduction respectively, and d_{EW} is the mean difference in phenotypic fitness between maternal half-sib extra-pair and within-pair offspring (which was assumed to approximate the difference in additive genetic value). Arnqvist and Kirkpatrick (2005) also state that the evolutionary response to direct selection can be estimated on an analogous scale as $\Delta_D = {}^1/_2\, h^2_F\ \sigma_F\ \beta_F$, where β_F is the direct selection gradient on female extra-pair reproduction.

These models are valuable in initiating an approach to quantitatively comparing predicted evolutionary responses to components of direct and indirect selection. However, Arnqvist and Kirkpatrick's (2005) parameterisation of these models was criticised because, as Arnqvist and Kirkpatrick themselves discuss, available empirical estimates of key quantities, specifically d_{EW} and β_F, were incomplete, simplistic and probably biased (see above, and Griffith 2007; Eliassen & Kokko 2008; Reid & Sardell 2012). Furthermore, h^2_F and σ_F had not been estimated in any wild population (although generic maxima can be defined). In addition, the model presentation invited debate about what female trait should be considered (Griffith 2007; Arnqvist & Kirkpatrick 2007). Arnqvist and Kirkpatrick's (2005) paper is written in terms of 'extra-pair copulations' (i.e. extra-pair matings), but it is actually extra-pair reproduction (i.e. extra-pair fertilisation) that is important in the context of indirect selection stemming from offspring additive genetic value (noted in the appendix to their paper). This is because extra-pair copulations that do not result in extra-pair offspring cannot create genetic covariances between alleles underlying extra-pair reproduction and fitness through linkage disequilibrium, or cause consequent indirect selection. Furthermore, Arnqvist and Kirkpatrick (2005) estimated β_F in terms of the reduction in paternal care associated with extra-pair 'copulations' because direct selection had never been more comprehensively measured. However, the links between postulated reductions in paternal care and extra-pair copulation vs extra-pair reproduction are not clear, and might depend on the genetic correlations among copulation frequency and fertilisation success, and on a male's ability to distinguish within-pair offspring from extra-pair offspring. In contrast there could be components of selection that do act directly on extra-pair copulation per se, for example through resulting disease or predation risk. This creates the confusing situation that the female traits and values of h^2_F and σ_F that are relevant to Arnqvist and Kirkpatrick's (2005) model for Δ_I are not necessarily the same as those that are relevant to Δ_D, and depend on what sources of direct and indirect selection are considered.

Such debates regarding ambiguous definitions and parameterisation are important (Griffith 2007; Arnqvist & Kirkpatrick 2007). However they distract attention from the conceptual merits of taking an explicit quantitative genetic approach to understanding extra-pair reproduction, and from the corresponding need to estimate key genetic and phenotypic variances and covariances underlying selection and resulting evolution. Even Arnqvist and Kirkpatrick's (2005) expression for Δ_I is (deliberately) formulated in terms of ad hoc estimation of

d_{EW} as the difference in phenotypic fitness between maternal half-sibs; it does not in fact define an explicit quantitative genetic approach to estimating indirect selection on extra-pair reproduction or predicting any consequent evolutionary response. Moving the field on, recent quantitative genetic analyses of data from a wild population of song sparrows (*Melospiza melodia*) illustrate the opportunities for new insights, and remaining impediments in terms of theory and data.

3.8.3 Extra-pair reproduction in song sparrows

As for any trait, ongoing evolution of female extra-pair reproduction requires non-zero additive genetic variation and heritability (Arnqvist & Kirkpatrick 2005). Reid *et al.* (2011) used comprehensive paternity and pedigree data from socially monogamous but genetically polygynandrous song sparrows inhabiting Mandarte Island, Canada, to estimate these quantities.

Mandarte holds a small, resident song sparrow population that has been studied intensively since 1975 (Figure 3.1). All breeding attempts have been systematically monitored, all adults and offspring have been individually colour-ringed, and the social parents of all offspring have been identified (Smith *et al.* 2006). Since 1993, virtually all chicks and adults have been genotyped, and the true genetic parents of all chicks have been identified with high statistical confidence (Sardell *et al.* 2010). A highly resolved genetic pedigree has been compiled, allowing the numbers of within-pair and extra-pair chicks produced and reared by each female and male in each year, and the relatedness between all individuals, to be estimated. Reid *et al.* (2011) used a binomial animal model to estimate a heritability of 0.18 (95% credible interval $= 0.05-0.31$, $n = 204$ females) in a female's liability to produce an extra-pair offspring as opposed to a within-pair offspring. This implies that, depending on further constraints, there is potential for continuing evolution of female extra-pair reproduction in the song sparrow population.

The same system was used to estimate the phenotypic difference in fitness between same-brood maternal half-sib extra-pair and within-pair offspring (Sardell *et al.* 2011). On average across mixed-paternity broods, a mother's extra-pair daughters were less likely to recruit than her within-pair daughters, while extra-pair sons tended to be more likely to recruit than within-pair sons (Sardell *et al.* 2011). These results could be interpreted to imply that there may be differential indirect selection on female extra-pair reproduction through extra-pair sons versus daughters.

However, such analyses cannot distinguish additive from non-additive genetic effects, or distinguish either from any confounding environmental effects on the relative phenotypes of within-pair and extra-pair maternal half-sibs (Magrath *et al.* 2009; Reid & Sardell 2012). A more explicit and unbiased approach would therefore be to directly quantify the mean difference in additive genetic value (i.e. breeding value) for fitness between a female's extra-pair and within-pair offspring. However this difference is not straightforward to estimate, especially from wild population data; individual breeding values are typically predicted with substantial uncertainty, and post hoc estimates of differences will be biased towards zero unless the difference of interest is specified in the predictive model (Hadfield *et al.* 2010).

This difficulty can be resolved by rewriting the required difference in breeding value as a genetic covariance, which can then be estimated directly. Reid and Sardell (2012) show that the mean difference in breeding value between a female's extra-pair offspring and the (hypothetical) within-pair offspring they replaced ($E[\Delta BV]$) is proportional to the genetic covariance (cov_A) between a male's 'success in extra-pair reproduction', defined as the number of extra-pair offspring sired (N_E) minus the number of offspring lost through cuckoldry (N_C), and offspring fitness (W, Reid & Sardell 2012). Specifically, $E[\Delta BV] = {}^1/_2(cov_A(N_E - N_C, W))/E[N_E]$, where $E[N_E]$ is the number of extra-pair offspring sired averaged over males (Reid & Sardell 2012).

Reid and Sardell (2012) applied a bivariate animal model to the song sparrow dataset to estimate the appropriate covariance using offspring survival to recruitment as an initial fitness proxy, and estimated $E[\Delta BV]$ as -0.27 (95% credible interval $= -0.53 - -0.02$, $n = 293$ males and 2196 offspring). This result shows that, opposite to prediction, a female's extra-pair offspring had lower additive genetic value for recruitment than the within-pair offspring they replaced on average. This pattern was consistent across sons and daughters, implying that the sex-specific effects observed in the purely phenotypic analysis reflected sex-specific environmental and/or non-additive genetic effects (Sardell *et al.* 2011).

3.8.4 Strengths and limitations

Together, the song sparrow analyses show that biased or erroneous evolutionary inferences can potentially be drawn from purely phenotypic analyses. They also demonstrate the need for new quantitative genetic theory that is explicitly designed to permit empirical test of key hypotheses. A male's 'success in extra-pair reproduction', defined as the number of extra-pair offspring sired (N_E) minus the number of offspring lost through cuckoldry (N_C), is not a biologically intuitive trait; it does not measure male fitness because it does not include the number of within-pair offspring that a male sired. Rather, it stems purely from an algebraic transformation that rewrites the difference in additive genetic value between a female's extra-pair and within-pair offspring as a genetic covariance, and thereby facilitates unbiased estimation using an animal model analysis.

However, while the genetic covariance between male 'success in extra-pair reproduction' and offspring fitness should give an unbiased estimate of the mean difference in additive genetic value between a female's extra-pair and within-pair offspring, and therefore constitutes a substantial advance on purely phenotypic analyses, it still does not adequately measure indirect selection on female extra-pair reproduction (Reid & Sardell 2012). The focus on estimating the difference in fitness (or breeding value for fitness) between maternal half-sib extra-pair and within-pair offspring originates from this being the only historically feasible means of approximating the 'indirect genetic benefit' of female extra-pair reproduction in wild populations. In fact, this difference is not what we really need to measure. This is because neither basic phenotypic comparisons, nor Reid and Sardell's (2012) comparison of breeding values, incorporate females that were phenotypically monogamous and produced only within-pair offspring. One important future objective should therefore be to estimate the genetic covariance between female extra-pair reproduction, including values pertaining to monogamous females, and paternal genetic contribution to all resulting offspring. This has not been achieved and is not necessarily straightforward. Because a female's genetic contribution to offspring fitness cannot be assumed to be independent of her genetic value for extra-pair reproduction, maternal and paternal genetic contributions to offspring fitness need to be distinguished. New theory is required to derive the key genetic covariances, which can then be estimated.

Nevertheless, the fascinating result that extra-pair song sparrow offspring were of lower additive genetic value than the (hypothetical) within-pair offspring they replaced compounds rather than resolves the overarching question of why we see the evolution and persistence of female extra-pair reproduction. Numerous other parts of the puzzle now need to be addressed, and could profitably be tackled through an explicit quantitative genetic approach. For example, one pre-eminent hypothesis is that female extra-pair reproduction represents the outcome of sexual conflict over mating rate given that direct selection for male multiple mating is widely expected (Westneat & Stewart 2003; Arnqvist & Kirkpatrick 2005). This hypothesis could be addressed by testing for negative genetic covariances among male and female extra-pair reproduction and fitness, but this has not yet been attempted in a wild population.

3.9 Quantitative genetics and sexual selection: opportunities and challenges

Key questions regarding the forms and relative impacts of direct and indirect selection on mate choice, reproductive traits and associated mating systems have long foiled evolutionary biologists. This is because key evolutionary processes, such as changes in frequencies of alleles underlying mating preferences due to genetic covariances with alleles underlying fitness, cannot (yet) be measured directly. Kirkpatrick and Barton (1997) and Arnqvist and Kirkpatrick's (2005) models, and Qvarnström *et al.* (2006a), Reid *et al.* (2011) and Reid and Sardell's (2012) initial parameterisations, are exciting because they illustrate potentially tractable approaches to answering thorny questions regarding the form and magnitude of indirect selection in wild populations experiencing natural genetic and environmental variation in mating success and fitness. Similar quantitative genetic approaches could be applied to other questions that are currently at the forefront of evolutionary biology, such as the degree to which evolutionary sexual conflict drives or constrains the evolution of reproductive traits and mating systems (Kruuk *et al.* 2008; Bonduriansky & Chenoweth 2009; Poissant *et al.* 2010). Therefore, while quantitative genetic analyses of contemporary variation may not fully inform our understanding of historical evolution of traits under sexual selection, they could provide fundamental insights into contemporary selection regimes and resulting micro-evolution or stasis. The value of estimating key quantitative genetic parameters in appropriate populations has consequently been repeatedly emphasised (Birkhead & Pizzari 2002; Kokko *et al.* 2003; Mead & Arnold 2004; Moore & Pizzari 2005; Kruuk *et al.* 2008). However, as evidenced by the two case studies, significant challenges remain in terms of theory, data, analysis and interpretation.

3.9.1 New theory

Quantitative genetic models that explicitly identify key parameters underlying the ongoing evolution of specific mating systems and associated traits need to be developed. These models need to be formulated in terms of estimable parameters that are defined and explained sufficiently unambiguously for empiricists to identify and measure the correct traits and draw confident conclusions, thereby avoiding the kinds of controversies and tangents illustrated by the collared flycatcher and song sparrow case studies. This will not necessarily require entirely new theoretical frameworks. Rather, it requires existing general frameworks (Arnold & Wade 1984; Barton & Turelli 1991; Rausher 1992; Kirkpatrick & Hall 2004; Kruuk *et al.* 2008; Kirkpatrick 2009) to be tailored to specific questions of interest to empiricists who are equipped to parameterise resulting models. This is not a trivial task; in order to ensure that appropriate parameters are identified, estimated and interpreted, empiricists will need to invest in understanding the derivation of relevant quantitative genetic theory and theoreticians will need to invest in understanding what exact quantities empiricists can and cannot measure, or estimate from resulting data. Indeed, the basic 'breeder's equation' approach, where separate estimates of phenotypic selection and genetic variation are combined to predict evolutionary responses, may prove inappropriate if the underlying assumption of sole causal effects of measured traits on fitness is invalid (Morrissey *et al.* 2010). Direct estimation of additive genetic covariances between mating behaviours and fitness components could then allow unbiased prediction of responses to direct selection when all contributing traits and environmental covariates cannot be explicitly measured or controlled (Rausher 1992). This approach was recently attempted for female extra-pair reproduction in song sparrows (see Reid 2012 for more details).

3.9.2 New data

Appropriate phenotypic, pedigree and fitness data need to be collected. Phenotyping may require temporary captivity and experiments rather than solely field observation, particularly for traits such as absolute mating preferences, mating rates and sperm performance that are not readily observable in the wild (Postma *et al.* 2006; Forstmeier *et al.* 2011). Paternity error, whether stemming from unmeasured extra-pair reproduction or from inadequate sampling or molecular genetic analysis, should no longer be ignored. Such error may have relatively little impact on estimated heritabilities of non-fitness traits (Charmantier & Réale 2005), but is surely not ignorable when the hypotheses under test involve non-random mating and resulting covariances among male trait values, fitness and/or relatedness. This applies particularly to systems with no paternal care and hence where indirect genetic effects could substantially affect sexual selection, but where full paternity assignment may be particularly challenging (Walling *et al.* 2010). Recent analyses illustrate the value of treating extra-pair reproduction as a phenotype of interest rather than ignorable error (Forstmeier *et al.* 2011; Reid *et al.* 2011; Reid & Sardell 2012). Further problems with inference may arise when, as is common, parentage cannot be assigned until after some post-conception mortality has occurred (Simmons 2005; Hadfield 2008; Reid *et al.* 2011). Furthermore, even well-specified animal models are not impervious to confounding environmental effects. Some form of cross-fostering or other manipulation may consequently be required to obtain unbiased estimates of genetic variances and covariances (Kruuk & Hadfield 2007). Quantitative genetic analyses that are sufficiently powerful and unbiased to answer key questions in sexual selection may therefore require substantial strategic data collection, and hence new field effort, rather than solely relying on existing data collected for other purposes.

3.9.3 New analyses and interpretation

Existing quantitative genetic analyses, as applied to wild population data, need to be refined or extended to consider sex-linked genetic variation, non-additive and indirect genetic effects and inbreeding. These extensions may be critical in the context of sexual selection, since sex-linkage is predicted to arise given sexual conflict (Chapman *et al.* 2003; Kirkpatrick & Hall 2004; Pischedda & Chippindale 2006) and mate choice is widely postulated to depend on interactions among individuals, non-additive genetic effects and inbreeding (Tregenza & Wedell 2000; Moore & Pizzari 2005; Neff & Pitcher 2005). These extensions will impose further challenging data requirements, and wild population datasets may often provide low power to estimate quantitative genetic parameters of interest (Hill 2010). This may be inevitable since power and potential bias depend on the distribution of relatives over environments, which in turn depends on the form of mate choice and the mating system. For example, polygamy may increase the number of half-sibs versus full-sibs compared to monogamy and spread these half-sibs across multiple rearing environments, thereby affecting power to estimate additive versus dominance genetic variances, or to distinguish either from environmental variances. The perennial questions of how to measure fitness and interpret it in the context of evolutionary quantitative genetics also remain open (Kirkpatrick 2009).

3.9.4 New opportunities

Understanding the causes and consequences of sexual selection has long motivated and challenged evolutionary biologists. Recent developments in quantitative genetic analyses offer exciting opportunities to test evolutionary theory in wild populations. If sufficient effort is invested in developing theory, data and analysis then quantitative genetics could potentially provide new and integrated understanding of components of selection acting on mate choice and associated traits in males and females, and resulting mating system evolution. These insights would complement those achievable through other approaches, such as experimental evolution, comparative analyses and detailed observations of behaviour and physiology.

These objectives are daunting, particularly for field biologists with no background in quantitative genetics. However useful progress does not necessarily require all key parameters to be identified or estimated simultaneously. Even robust estimates of single parameters such as additive genetic variances in sex-specific reproductive traits, behaviours and fitness components (e.g. Hadfield *et al.* 2006; Reid *et al.* 2011; McFarlane *et al.* 2011) are valuable starting points and should be increasingly achievable as our long-term datasets and analytical tools mature.

Acknowledgments

I thank Jarrod Hadfield and Rebecca Sardell for invaluable discussions, Jon Brommer, Anne Charmantier, Dany Garant, Mark Kirkpatrick and Loeske Kruuk for helpful comments, and the Royal Society for funding.

References

Ala-Honkola, O., Manier, M.K., Lüpold, S. & Pitnick, S. (2011) No evidence for postcopulatory inbreeding avoidance in *Drosophila melanogaster*. *Evolution*, **65**, 2699–2705.

Andersson, M. (1994) *Sexual selection*. Princeton University Press, Princeton.

Arnold, S.J. & Wade, M.J. (1984) On the measurement of natural and sexual selection: theory. *Evolution*, **38**, 709–719.

Arnqvist, G. & Kirkpatrick, M. (2005) The evolution of infidelity in socially monogamous passerines: the strength of direct and indirect selection on extrapair copulation behaviour in females. *American Naturalist*, **165**, S26–S37.

Arnqvist, G. & Kirkpatrick, M. (2007) The evolution of infidelity in socially monogamous passerines revisited: a reply to Griffith. *American Naturalist*, **169**, 282–283.

Arnqvist, G. & Rowe, L. (2005) *Sexual conflict*. Princeton University Press, Princeton.

Barton, N. & Turelli, M. (1991) Natural and sexual selection on many loci. *Genetics*, **127**, 229–255.

Bijma, P., Muir, W.M. & van Arendonk, J.A.M. (2007) Multilevel selection 1: Quantitative genetics of inheritance and response to selection. *Genetics*, **175**, 277–288.

Bilde, T., Friberg, U., Maklakov, A.A., Fry, J.D. & Arnqvist, G. (2008) The genetic architecture of fitness in a seed beetle: assessing the potential for indirect genetic benefits of female choice. *BMC Evolutionary Biology*, **8**, 295.

Birkhead, T.R. & Pizzari, T. (2002) Postcopulatory sexual selection. *Nature Reviews Genetics*, **3**, 262–273.

Bonduriansky, R. & Chenoweth, S.F. (2009) Intralocus sexual conflict. *Trends in Ecology and Evolution*, **24**, 280–288.

Brommer, J.E., Kirkpatrick, M., Qvarnström, A. & Gustafsson, L. (2007) The intersexual genetic correlation for lifetime fitness in the wild and its implications for sexual selection. *PLoS One*, **8**, e744.

Chapman, T., Arnqvist, G., Bangham, J. & Rowe, L. (2003) Sexual conflict. *Trends in Ecology and Evolution*, **18**, 41–47.

Charmantier, A. & Sheldon, B.C. (2006) Testing genetic models of mate choice evolution in the wild. *Trends in Ecology and Evolution*, **21**, 417–419.

Charmantier, A. & Réale, D. (2005) How do misassigned paternities affect the estimation of heritabilities in the wild? *Molecular Ecology*, **14**, 2839–2850.

Eberhard, W.G. (2009) Postcopulatory sexual selection: Darwin's omission and its consequences. *Proceedings of the National Academy of Sciences of the United States of America*, **106**, 10025–10032.

Eliassen, S. & Kokko, H. (2008) Current analyses do not resolve whether extra-pair paternity is male or female driven. *Behavioural Ecology & Sociobiology*, **62**, 1795–1804.

Evans, J.P., García-González, F. & Marshall, D.J. (2007) Sources of genetic and phenotypic variance in fertilization rates and larval traits in a sea urchin. *Evolution*, **61**, 2832–2838.

Evans, J.P. & Simmons, L.W. (2008) The genetic basis of traits regulating sperm competition and polyandry: can selection favour the evolution of good- and sexy-sperm? *Genetica*, **134**, 5–19.

Forstmeier, W., Martin, K., Bolund, E., Schielzeth, H. & Kempenaers, B. (2011) Female extrapair mating behaviour can evolve via indirect selection on males. *Proceedings of the National Academy of Sciences of the United States of America*, **108**, 10608–10613.

García-González, F. & Simmons, L.W. (2011) Good genes and sexual selection in dung beetles (*Onthophagus Taurus*): genetic variance in egg-to-adult and adult viability. *PLoS One*, **6**, e16233.

Gavrilets, S., Arnqvist, G. & Friberg, U. (2001) The evolution of female mate choice by sexual conflict. *Proceedings of the Royal Society B-Biological Sciences*, **268**, 531–539.

Griffith, S.C. (2007) The evolution of infidelity in socially monogamous passerines: neglected components of direct and indirect selection. *American Naturalist*, **169**, 274–281.

Griffith, S.C., Owens, I.P.F. & Thuman, K.A. (2002) Extra pair paternity in birds: a review of interspecific variation and adaptive function. *Molecular Ecology*, **11**, 2195–2212.

Griffith, S.C., Owens, I.P.F. & Burke, T. (1999) Environmental determination of a sexually selected trait. *Nature*, **400**, 358–360.

Hadfield, J.D., Burgess, M.D., Lord, A., Phillimore, A.B., Clegg, S.M. & Owens, I.P.F. (2006) Direct versus indirect sexual selection: genetic basis of colour, size and recruitment in a wild bird. *Proceedings of the Royal Society B-Biological Sciences*, **273**, 1347–1353.

Hadfield, J.D. (2008) Estimating evolutionary parameters when viability selection is operating. *Proceedings of the Royal Society B-Biological Sciences*, **275**, 723–734.

Hadfield, J.D., Wilson, A.J., Garant, D., Sheldon, B.C. & Kruuk, L.E.B. (2010) The misuse of BLUP in ecology and evolution. *American Naturalist*, **175**, 116–125.

Hill, W.G. & Kirkpatrick, M. (2010) What animal breeding has taught us about evolution. *Annual Review of Ecology, Evolution, and Systematics*, **41**, 1–19.

Hill, W.G. (2010) Understanding and using quantitative genetic variation. *Philosophical Transactions of the Royal Society of London Series B, Biological Sciences*, **365**, 73–85.

Jennions, M.D. & Petrie, M. (2000) Why do females mate multiply? A review of the genetic benefits. *Biological Reviews*, **75**, 21–64.

Johnston, S.E., McEwan, J.C., Pickering, N.K., Kijas, J.W., Beraldi, D., Pilkington, J.G., Pemberton, J.M. & Slate, J. (2011) Genome-wide association mapping identifies the genetic basis of discrete and quantitative variation in sexual weaponry in a wild sheep population. *Molecular Ecology*, **20**, 2555–2566.

Jones, A.G. & Ratterman, N.L. (2009) Mate choice and sexual selection: What have we learned since Darwin? *Proceedings of the National Academy of Sciences of the United States of America*, **106**, 10001–10008.

Kempenaers, B. & Sheldon, B.C. (1997) Studying paternity and paternal care: pitfalls and problems. *Animal Behaviour*, **53**, 423–427.

Kirkpatrick, M. (1982) Sexual selection and the evolution of female choice. *Evolution*, **36**, 1–12.

Kirkpatrick, M. (2009) Patterns of quantitative genetic inheritance in multiple dimensions. *Genetica*, **136**, 271–284.

Kirkpatrick, M. & Barton, N.H. (1997) The strength of indirect selection on female mating preferences. *Proceedings of the National Academy of Sciences of the United States of America*, **94**, 1282–1286.

Kirkpatrick, M. & Hall, D.W. (2004) Sexual selection and sex linkage. *Evolution*, **58**, 683–691.

Kokko, H. & Brooks, R. (2003) Sexy to die for? Sexual selection and the risk of extinction. *Annales Zoologici Fennici*, **40**, 207–219.

Kokko, H., Brooks, R., Jennions, M.D. & Morley, J. (2003) The evolution of mate choice and mating biases. *Proceedings of the Royal Society B-Biological Sciences*, **270**, 653–664.

Kokko, H., Jennions, M.D. & Brooks, R. (2006) Unifying and testing models of sexual selection. *Annual Review of Ecology, Evolution, and Systematics*, **37**, 43–66.

Kruuk, L.E.B. (2004) Estimating genetic parameters in natural populations using the 'animal model'. *Philosophical Transactions of the Royal Society of London Series B, Biological Sciences*, **359**, 873–890.

Kruuk, L.E.B. & Hadfield, J.D. (2007) How to separate genetic and environmental causes of similarity between relatives. *Journal of Evolutionary Biology*, **20**, 1890–1903.

Kruuk, L.E.B., Slate, J. & Wilson, A.J. (2008) New answers for old questions: the evolutionary quantitative genetics of wild animal populations. *Annual Review of Ecology, Evolution, and Systematics*, **39**, 525–548.

Lande, R. (1981) Models of speciation by sexual selection on polygenic traits. *Proceedings of the National Academy of Sciences of the United States of America*, **78**, 3721–3725.

Lande, R. & Arnold, S.J. (1983) The measurement of selection on correlated characters. *Evolution*, **37**, 1210–1226.

Lynch, M. & Walsh, B. (1998) *Genetics and analysis of quantitative traits*. Sinauer, Sunderland.

Magrath, M.J.L., Vedder, O., van der Velde, M. & Komdeur, J. (2009) Maternal effects contribute to the superior performance of extra-pair offspring. *Current Biology*, **19**, 792–797.

McFarlane, S.E., Lane, J.E., Taylor, R.W., Gorrell, J.C., Coltman, D.W., Humphries, M.M., Boutin, S. & McAdam, A.G. (2011) The heritability of multiple male mating in a promiscuous mammal. *Biology Letters*, **7**, 268–371.

Mead, L.S. & Arnold, S.J. (2004) Quantitative genetic models of sexual selection. *Trends in Ecology and Evolution*, **19**, 264–271.

Miller, C.W. & Moore, A.J. (2007) A potential resolution to the lek paradox through indirect genetic effects. *Proceedings of the Royal Society B-Biological Sciences*, **274**, 1279–1286.

Moore, A.J., Brodie, E.D. & Wolf, J.B. (1997) Interacting phenotypes and the evolutionary process: I. Direct and indirect genetic effects of social interactions. *Evolution*, **51**, 1352–1362.

Moore, A.J. & Pizzari, T. (2005) Quantitative genetic models of sexual conflict based on interacting phenotypes. *American Naturalist*, **165**, S88–S97.

Morrissey, M.B., Kruuk, L.E.B. & Wilson, A.J. (2010) The danger of applying the breeder's equation in observational studies of natural populations. *Journal of Evolutionary Biology*, **23**, 2277–2288.

Neff, B.D. & Pitcher, T.E. (2005) Genetic quality and sexual selection: an integrated framework for good genes and compatible genes. *Molecular Ecology*, **14**, 19–38.

Norris, K. (1993) Heritable variation in a plumage indicator of viability in male great tits *Parus major*. *Nature*, **362**, 537–539.

Parker, G.A. (2006) Sexual conflict over mating and fertilization: an overview. *Philosophical Transactions of the Royal Society of London Series B, Biological Sciences*, **361**, 235–259.

Pischedda, A. & Chippindale, A.K. (2006) Intralocus sexual conflict diminishes the benefits of sexual selection. *PLoS Biology*, **4**, 2099–2103.

Pizzari, T. & Snook, R.R. (2003) Sexual conflict and sexual selection: chasing away paradigm shifts. *Evolution*, **57**, 1223–1236.

Pizzari, T. & Wedell, N. (2013) The polyandry revolution. *Philosophical Transactions of the Royal Society of London Series B, Biological Sciences*, **368**, 20120041.

Poissant, J., Wilson, A.J. & Coltman, D.W. (2009) Sex-specific genetic variance and the evolution of sexual dimorphism: a systematic review of cross-sex genetic correlations. *Evolution*, **64**, 97–107.

Postma, E., Griffith, S.C. & Brooks, R. (2006) Evolution of mate choice in the wild. *Nature*, **444**, E16.

Qvarnström, A., Brommer, J.E. & Gustafsson, L. (2006a) Testing the genetics underlying the co-evolution of mate choice and ornament in the wild. *Nature*, **441**, 84–86.

Qvarnström, A., Brommer, J.E. & Gustafsson, L. (2006b) Evolution of mate choice in the wild: reply. *Nature*, **444**, E16–E17.

Qvarnström, A., Pärt, T. & Sheldon B.C. (2000) Adaptive plasticity in mate preference linked to differences in reproductive effort. *Nature*, **405**, 344–347.

Rausher, M.D. (1992) The measurement of selection on quantitative traits: biases due to environmental covariances between traits and fitness. *Evolution*, **46**, 616–626.

Reid, J.M. (2012) Predicting evolutionary responses to selection on polyandry in the wild: additive genetic covariances with female extra-pair reproduction. *Proceedings of the Royal Society of London Series B, Biological Sciences*, **279**, 4652–4660.

Reid, J.M., Arcese, P., Sardell, R.J. & Keller, L.F. (2011) Heritability of female extra-pair paternity rate in song sparrows (*Melospiza melodia*). *Proceedings of the Royal Society B-Biological Sciences*, **278**, 1114–1120.

Reid, J.M. & Sardell, R.J. (2012) Indirect selection on female extra-pair reproduction? Comparing the additive genetic value of maternal half-sib extra-pair and within-pair offspring. *Proceedings of the Royal Society B-Biological Sciences*, **279**, 1700–1708.

Reinhold, K. (2002) Modelling the evolution of female choice strategies under inbreeding conditions. *Genetica*, **116**, 189–195.

Ritchie, M.G. (2007) Sexual selection and speciation. *Annual Review of Ecology, Evolution, and Systematics*, **38**, 79–102.

Rockman, M.V. (2012) The QTN program and the alleles that matter for evolution: all that's gold does not glitter. *Evolution*, **66**, 1–17.

Rodriguez-Muñoz, R., Bretman, A., Slate, J., Walling, C.A. & Tregenza, T. (2010) Natural and sexual selection in a wild insect population. *Science*, **328**, 1269–1272.

Rowe, L. & Day, T. (2006) Detecting sexual conflict and sexually antagonistic coevolution. *Philosophical Transactions of the Royal Society of London Series B, Biological Sciences*, **361**, 277–285.

Rowe, L. & Houle, D. (1996) The lek paradox and the capture of genetic variance by condition dependent traits. *Proceedings of the Royal Society of London Series B, Biological Sciences*, **263**, 1415–1421.

Santure, A.W., De Cauwer, I., Robinson, M.R., Poissant, J., Sheldon, B.C. & Slate, J. (2013) Genomic dissection of variation in clutch size and egg mass in a wild great tit (*Parus major*) population. *Molecular Ecology*, **22**, 3949–3962.

Sardell, R.J., Keller, L.F., Arcese, P., Bucher, T. & Reid, J.M. (2010) Comprehensive paternity assignment: genotype, spatial location and social status in song sparrows, *Melospiza melodia*. *Molecular Ecology*, **19**, 4352–4364.

Sardell, R.J., Keller, L.F., Arcese, P. & Reid, J.M. (2011) Sex-specific differential survival of within-pair and extra-pair offspring in song sparrows. *Proceedings of the Royal Society B-Biological Sciences*, **278**, 3251–3259.

Schmoll, T., Schurr, F.M., Winkel, W., Epplen, J.T. & Lubjuhn, T. (2009) Lifespan, lifetime reproductive performance and paternity loss of within-pair and extra-pair offspring in the coal tit *Periparus ater*. *Proceedings of the Royal Society B-Biological Sciences*, **276**, 337–345.

Sheldon, B.C. & Ellegren, H. (1999) Sexual selection resulting from extra-pair paternity in collared flycatchers. *Animal Behaviour*, **57**, 285–298.

Shuster, S. & Wade, M.J. (2003) *Mating systems and strategies*. Princeton University Press, Princeton.

Simmons, L.W. (2005) The evolution of polyandry: sperm competition, sperm selection and offspring viability. *Annual Review of Ecology, Evolution, and Systematics*, **35**, 125–146.

Simmons, L.W. & Kotiaho, J.S. (2007) Quantitative genetic correlation between trait and preference supports a sexually selected sperm process. *Proceedings of the National Academy of Sciences of the United States of America*, **104**, 16604–16608.

Slatyer, R.A., Mautz, B.S., Backwell, P.R.Y. & Jennions, M.D. (2012) Estimating genetic benefits of polyandry from experimental studies: a meta-analysis. *Biological Reviews*, **87**, 1–33.

Smith, J.N.M., Keller, L.F., Marr, A.B. & Arcese, P. (2006) *Conservation and biology of small populations: the song sparrows of Mandarte Island*. Oxford University Press, New York.

Tregenza, T. & Wedell, N. (2000) Genetic compatibility, mate choice and patterns of parentage: invited review. *Molecular Ecology*, **9**, 1013–1027.

Uller, T. & Olsson, M. (2008) Multiple paternity in reptiles: patterns and processes. *Molecular Ecology*, **17**, 2566–2580.

Walling, C.A., Pemberton, J.M., Hadfield, J.D. & Kruuk, L.E.B. (2010) Comparing parentage inference software: a reanalysis of a red deer pedigree. *Molecular Ecology*, **19**, 1914–1928.

Walsh, B. & Blows, M.W. (2009) Abundant genetic variation + strong selection = multivariate genetic constraints: a geometric view of adaptation. *Annual Review of Ecology, Evolution, and Systematics*, **40**, 41–59.

Westneat, D.F. & Stewart, I.R.K. (2003) Extra-pair paternity in birds: causes, correlates and conflict. *Annual Review of Ecology, Evolution, and Systematics*, **34**, 365–396.

CHAPTER 4

Individual behaviour: behavioural ecology meets quantitative genetics

Niels J. Dingemanse and Ned A. Dochtermann

4.1 Behavioural ecology meets quantitative genetics

In the field of behavioural ecology, classic and current research paradigms consider behaviour from an optimality perspective. This approach assumes that animals have evolved to make adaptive 'decisions' given the constraints and trade-offs with which they are faced. For example, as predicted by theory, individual great tits (*Parus major*) narrow or widen their diet as a function of the abundance of the most profitable prey (Krebs *et al.* 1977), and change patch residence times during foraging as a function of travel time between alternative foraging patches (Cowie 1977).

Within this paradigm, behavioural ecology as a field has for a long time focussed on the average level of a behavioural response across all sampled individuals while regarding other sources of variation as noise around an adaptive mean (Dall, Houston & McNamara 2004). Likewise, research on behavioural plasticity has focused on population averages in plasticity. Interestingly, the repeated observations of the same individuals that are required to quantify an individual's plasticity often simultaneously reveal substantial between-individual differences in behaviour (meta-analysis: Bell *et al.* 2009) (Figure 4.1). For example, Ural owl (*Strix uralensis*) mothers change their level of nest defence aggression across years as a function of vole density but some mothers are simultaneously consistently more aggressive compared to others in all years (Kontiainen *et al.* 2009). Consequently, populations harbour variation *among* individuals (in average behaviour) and *within* individuals (in how the behaviour expressed at each instance deviates from the individual's mean) (Figure 4.2). In the field of behavioural ecology such observations have recently stimulated the development of theoretical models that propose adaptive explanations for the existence of between-individual variation in behaviour per se (Dall *et al.* 2004; Dingemanse & Wolf 2010; Wolf & Weissing 2010). Specifically: what are the conditions leading to natural or sexual selection favouring individual repeatability in behaviour rather than individual plasticity alone? Empirical testing of this novel adaptive theory requires researchers to estimate and compare the respective magnitude of variation between vs within individuals (Dingemanse & Dochtermann 2013).

Understanding how variation is distributed between vs within individuals is achieved by 'partitioning' phenotypic variance to estimate the contribution of specific 'variance components' (see Figure 4.2 for a graphical illustration of this idea). For example, sexual selection has been suggested to favour males that provision offspring at a stable rate (Schuett *et al.* 2010). The observed behavioural variation within individuals in provisioning

Quantitative Genetics in the Wild. Edited by Anne Charmantier, Dany Garant, and Loeske E. B. Kruuk

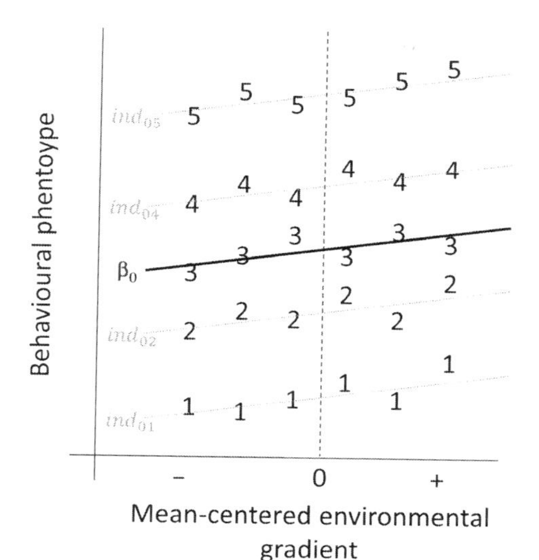

Figure 4.1 Reaction norm plot showing variation between and within each of five individuals assayed seven times over an environmental gradient. The response variable (y-axis) might represent aggressiveness, and the predictor variable, population density (x-axis). The black solid line gives the population–average reaction norm; its intercept (β_0) lies at the position where the mean-centred environmental gradient has the value zero, and its slope (β_1) is positive. The grey dotted lines indicate each individual's reaction norm (numbered 1–5); individual reaction norms differ from the population–average reaction in their intercept ($+ind_{0j}$), where ind_{0j} is either negative (individuals 1 and 2), zero (individual 3) or positive (individuals 4 and 5). There is thus variance among their intercepts (between-individual variance: V_{ind_0}; Figure 4.2) resulting in nonzero values of individual repeatability. Each observation also deviates slightly from the individual's estimate reaction norm ($+e_{0ij}$) leading to within-individual residual variance (V_{e_0}).

rate should thus be relatively low for males compared to females, and repeatability (the *proportion* of variation explained by differences between individuals; Falconer & Mackay 1996) consequently be higher for males compared to females—a prediction empirically supported in some species of birds (Nakagawa *et al.* 2007). While relatively new in behavioural ecology, such splitting of phenotypic variation into different components constitutes the backbone of quantitative genetics analyses (Lynch & Walsh 1998) leading to an increasing appreciation, and exchange of ideas, between the two fields. This exchange has also led to a growing interest in the importance of quantitative genetics in the study of

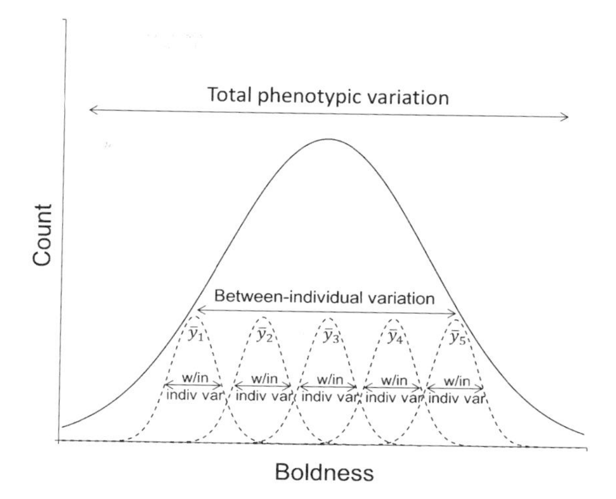

Figure 4.2 Conceptualisation of how total phenotypic variation in the behaviour 'boldness' corresponds to between-individual variation and within-individual variation. The solid curve corresponds to the overall distribution of boldness responses irrespective of individual identity. The dashed curves correspond to distributions of boldness responses by specific individuals (j: 1 – 5). The variation amongst individual means ($\bar{y}_j$) relative to the population mean is between-individual variation (V_{ind_0}) or personality variation when referring specifically to behaviours. Within-individual variation (arrows within dashed curves) can be expressed as the distribution from individual–mean values ($y_{ij} - \bar{y}_j$). This representation of individual variation follows that of Bolnick *et al.* (2003).

behavioural evolution (e.g. Smiseth *et al.* 2008) on the part of behavioural ecologists (called 'evolutionary behavioural ecology'; Westneat & Fox 2010), and places research on individuality firmly in an evolutionary framework (Dochtermann & Roff 2010; Dochtermann 2011; Dochtermann & Dingemanse 2013).

In this chapter, we detail how both fields might benefit from a further exchange of ideas. We do this by mathematically detailing patterns that interest behavioural ecologists studying the adaptive nature of between-individual (co)variation in behaviour under the labels of 'animal personalities' and 'behavioural syndromes' (defined below). We explicitly formulate behavioural patterns of current interest in a statistical language that is familiar to quantitative geneticists in an attempt to avoid any misunderstanding due to the usage of field-specific jargon

Box 4.1 Adaptive explanations for individuality in behaviour

We provide here a brief sketch of adaptive explanations for animal personality and behavioural syndromes developed by behavioural ecologists (for reviews see also Dall *et al.* 2004; Dingemanse & Wolf 2010; Wolf & Weissing 2010; Dingemanse & Wolf 2013). Three classes of explanation (cf. Wolf & Weissing 2010) have been proposed: i) models that investigate how differences in 'state' can lead to individual repeatability and between-individual correlations, ii) models that investigate the role of feedbacks between state and behaviour in stabilising initial individual differences in behaviour, and iii) models based on alternative explanations (Dingemanse & Wolf 2010). We discuss the former two types of models in more detail. In these models, state is defined as a feature of the organism that affects the balance between costs and benefits of behavioural actions (Houston & McNamara 1999), representing either a genetic or environmentally induced quality of the individual (e.g. amount of body reserves). A key example of a state-dependent model is from the study by Wolf et al. (2007), in which the authors proposed a general explanation for the presence of both i) repeatable variation in single behaviours across time and situations (personality), and ii) between-individual correlations (behavioural syndromes) in the context of 'risky' behaviours (defined as behaviours that facilitate the acquisition of resources at the expense of increased risk of mortality, predation, or parasitism). This model was designed to explain the commonly reported aggressiveness–boldness syndrome (Garamszegi *et al.* 2012), and shows how individual differences in (the state-variable) 'residual reproductive value' can lead to selection favouring risky personalities. This insight is based on the asset protection principle, which predicts risk should only be taken when there is little to lose. The model does not provide any prediction on whether such state-dependent personality/syndrome structure is due to genetic or environmental factors (Dingemanse & Wolf 2010), though both mechanisms have now been empirically supported. First, in wild ungulates there exists genetic variation in life-history strategies that is genetically correlated with risky behaviours, suggesting that additive genetic variation (V_A) in risky behaviour might also be underpinned by adaptive state-dependence (Réale *et al.* 2009). Second, wild passerine birds show substantial between-individual variance in exploratory behaviour (Dingemanse *et al.* 2012b), which is largely attributable to permanent environment variance (V_{PE}) (Quinn *et al.* 2009; Nicolaus *et al.* 2012). An experimental study revealed that manipulation effects on future fitness expectations (annual adult survival) were associated with long-term changes in this risky behaviour in the predicted direction, implying that repeatable variation in behaviour was possibly of environmental origin (Nicolaus *et al.* 2012). Many state-variables are, however, inherently unstable (Wolf & Weissing 2010), and any state-dependent between-individual variation in behaviour might consequently erode over time. State-dependent personality and syndrome structure is nevertheless predicted to persist in situations where state and behaviour feedback positively, stabilising initial differences in state, hence behaviour, over time (e.g. Luttbeg & Sih 2010).

(cf. Dingemanse & Dochtermann 2013). We also sketch adaptive theory that behavioural ecologists have developed for a range of between-individual variance components and highlight exemplary associated pieces of empirical support (Box 4.1). We further discuss how the current focus on individuality has contributed to a deeper understanding of the quantitative genetics background of behaviour. These sections jointly illustrate how concepts developed in quantitative genetics have benefited modern behavioural ecology paradigms. We end by briefly discussing our views on how both quantitative genetics and behavioural ecology might benefit from a continued exchange of ideas.

4.2 Understanding individual variation in behaviour

4.2.1 Why behavioural ecologists study individuality

Growing interest in understanding variation in behavioural ecology has gone hand in hand with the study of 'adaptive animal personalities' (Réale *et al.* 2010). This new field of behavioural ecology has emerged for two key reasons. First, most behavioural traits appear to show substantial levels of individual repeatability (meta-analysis: Bell *et al.* 2009), and functionally distinct behaviours are often phenotypically correlated (meta-analysis:

Garamszegi *et al.* 2012). Aggressiveness and activity are, for example, both repeatable and correlated in wild North American red squirrels (*Tamiasciurus hudsonicus*): some individuals are consistently more aggressive, and relatively aggressive individuals are also relatively explorative and active (Boon *et al.* 2007). Second, the existence of repeatable variation in phenotypic attributes whose expression is potentially highly plastic (e.g. behaviour, or physiology) is not readily predicted by standard evolutionary theory (Dall *et al.* 2004). Whilst classic evolutionary theory predicts that negative frequency-dependent selection can maintain phenotypic variation within single populations (Maynard Smith 1982), it does not necessarily predict that individuals should play fixed—as opposed to mixed (i.e. conditional or plastic)—strategies (Dall *et al.* 2004). For example, in foraging games, what are the conditions that generate certain individuals to consistently play 'scrounger' and others to consistently play 'producer' (i.e. pure strategies) rather than all individuals playing scrounger vs producer with a certain frequency (i.e. mixed strategies; Dubois *et al.* 2012)? Growing awareness for the need of adaptive theory explaining such (repeatable) between-individual variation has led to a surge of empirical and theoretical research (detailed in Box 4.1), particularly in the context of behaviour (reviewed by Dall *et al.* 2004; Dingemanse & Wolf 2010; Dingemanse & Wolf 2013).

4.2.2 Animal personality

Behavioural ecologists studying individual variation focus on four main types of variation verbally termed i) *animal personality*, ii) *individual differences in plasticity*, iii) *personality-related plasticity* and iv) *behavioural syndromes*. Quantitative geneticists are familiar with the statistical counterparts of these terms as variance components (see Wolf & Weissing 2010; Dingemanse & Dochtermann 2013) and as attributes of the well-known phenotypic equation (Box 4.2; cf. Nussey *et al.* 2007). The most simple of these sources of variation, animal personality, is incorporated in the following mathematical equation:

$$y_{ij} = (\beta_0 + ind_{0j}) + \beta_1 x_{ij} + e_{0ij} \qquad (4.1)$$

This equation describes the phenotypic response (y_{ij}), for example level of aggressiveness, of individual j produced at instance i as a function of an environmental gradient (covariate x_{ij} , e.g. population density experienced by individual j at instance i). The equation features four distinct elements that jointly predict each observation of aggressiveness (y_{ij}): i) the population mean reaction norm intercept (β_0; the grand mean value of average individual responses where the covariate (x_{ij}) equals zero), ii) the population mean reaction slope (β_1; for simplicity estimated as a linear slope), iii) the individual's average deviation ($+ind_{0j}$) from the population mean intercept (β_0), and iv) the instance's deviation from the individual's reaction norm ($+e_{0ij}$). Here, the environmental gradient is assumed to be centred on its mean value, which results in the intercept being placed at the position where the environment gradient has its average value (Figure 4.1). Within the framework of Eqn. 4.1, between-individual differences, i.e. differences in ind_{0j}, can be thought of as differences between individual averages (e.g. $\bar{y}_1 - \bar{y}_2$; Figure 4.2). We can then conceptualise between-individual variance (V_{ind_0}) as the overall variance in these individual means, relative to the population mean. Within an individual, repeated behavioural responses will not always be exactly the same, giving rise to within-individual differences (Figure 4.2). Across all individuals, these within-individual differences correspond to so-called residual or within-individual variance (V_{e_0}). These two variance components, V_{ind_0} and V_{e_0}, are routinely estimated jointly using mixed-effect modelling approaches (e.g. Pinheiro & Bates 2000). Elsewhere we detail their application in the context of personality research (Dingemanse & Dochtermann 2013).

Animal personality, which is often verbally defined as 'consistent individual differences in behaviour across time or contexts' (Réale *et al.* 2007), exists when individuals differ in their reaction norm intercept (i.e. $V_{ind_0} > 0$), where personality or 'behavioural type' (Bell 2007) refers to each individual's specific behavioural reaction norm intercept (ind_{0j}) (Dingemanse *et al.* 2010b). Thus, V_{ind_0} represents personality variation.

Defining personality variation as V_{ind_0} is increasingly common in behavioural ecology. For example, this general approach has been used to understand

Box 4.2 Understanding complex patterns of individual variation in behaviour

Multiple factors can contribute to individual variation in behaviours and other flexible traits. In Eqn. 4.1, we discussed how 'personality' or 'behavioural type' (ind_{0j}) can be incorporated into statistical linear models. However, individuals can vary in more complicated ways than that represented by Eqn. 4.1. Nussey *et al.* (2007) discuss these complications using what is known amongst quantitative geneticists as 'the phenotypic equation'. Here we describe how the phenotypic equation applies to behavioural variation and define the equation's parameters in terms familiar to behavioural ecologists. Specifically, parameters of the phenotypic equation can be translated as i) *animal personality*, ii) *individual plasticity*, and iii) *personality-related plasticity*.

The phenotypic equation describes the relationship of a phenotypic response to a number of factors:

$$y_{ij} = (\beta_0 + ind_{0j}) + (\beta_1 + ind_{1j})x_{ij} + e_{0ij} \qquad \text{(B4.2.1)}$$

where y_{ij} is the behavioural response of interest, for example level of aggressiveness, of individual j produced at instance i as a function of an environmental gradient (covariate x_{ij}, e.g. population density experienced by individual j at instance i). This phenotypic response (y_{ij}) can be usefully decomposed into five distinct elements: i) the population mean reaction norm intercept (β_0; the grand mean value of average individual responses); ii) the population mean reaction slope (β_1; the coefficient relating x_{ij} to y_{ij}), and, importantly, this covariate should be centred as in Figure 4.1 if population level parameters are to be interpreted as means; iii) the individual's deviation in reaction norm intercept (ind_{0j}) from the population-mean intercept (β_0); iv) the individual's deviation in reaction norm slope (ind_{1j}) from the population mean slope (β_1); and v) the instance's deviation from the individual's reaction norm (e_{0ij}) (cf. Dingemanse *et al.* 2010b; Westneat *et al.* 2011; Dingemanse & Dochtermann 2013). These parameters are typically estimated through the use of a type of mixed-effect model called 'random regression' (Henderson 1982; Kirkpatrick & Heckman 1989). Random regression models individual-specific deviations from the population mean value with respect to intercepts (ind_{0j}) and slopes (ind_{1j}). The variance amongst individuals in intercepts (V_{ind_0}), in slopes (V_{ind_1}), and the covariance among individuals between intercepts and slopes (Cov_{ind_0,ind_1}) are the key variance components relating to behavioural ecological parameters.

From this framework we can now begin to formally define relevant behavioural ecological concepts. As we have previously discussed, *animal personality* refers to each individual's specific behavioural reaction norm intercept (ind_{0j}) and V_{ind_0} represents personality variation (Dingemanse *et al.* 2010b). *Individual plasticity* exists when individuals differ in behavioural reaction norm slope (i.e. $V_{ind_1} > 0$), and *personality-related plasticity* (Mathot *et al.* 2012) when the intercept and slope of the same behavioural reaction norm are correlated (i.e. $r_{ind_0,ind_1} = Cov_{ind_0,ind_1} / \sqrt{V_{ind_0} V_{ind_1}} \neq 0$) or when the intercept of one behavioural reaction norm is correlated with the slope of another (Mathot *et al.* 2012). Behavioural ecologists increasingly focus on such differences in behavioural plasticity between personality types (Wolf, van Doorn & Weissing 2008; Mathot *et al.* 2011), a topic that goes beyond the scope of this chapter but is extensively reviewed elsewhere (Mathot *et al.* 2012; Dingemanse & Wolf 2013).

individual variation in aggression of sex-changing reef fish (Sprenger *et al.* 2012) and kangaroo rats (Dochtermann *et al.* 2012), sex and species differences in how individual kangaroo rats store food (Jenkins 2011), population differences in V_{ind_0} in exploratory behaviour in great tits (Dingemanse *et al.* 2012b), and how personality and plasticity interact in sparrows (Westneat *et al.* 2011).

4.2.3 Quantitative genetics of personality

While individual variation in behaviour can be interesting in its own right (Bell *et al.* 2009; Wolf & Weissing 2012), defining and estimating personality from the framework outlined above allows for direct consideration of personalities in terms of quantitative genetic parameters. Considering personalities (and other aspects of individual variation) in quantitative genetic terms is necessary to properly understand the evolutionary implications of behavioural variation. The definition of personality variation provided by Eqn. 4.1 creates an immediate, albeit underappreciated, bridge between personality research and quantitative genetics.

This bridge exists because personality variation, V_{ind_0}, can be further broken down explicitly in terms of quantitative genetic parameters:

$$V_{ind_0} = V_{\mathrm{A}} + V_{\mathrm{D}} + V_{\mathrm{PE}} \tag{4.2}$$

where $V_{\mathrm{A}}$is additive genetic variance, V_{D} is variance in the effects of genetic dominance, and V_{PE} is variability due to permanent environmental effects (cf. Falconer & Mackay 1996). V_{A} corresponds to underlying variation in how many genes of small effect additively contribute to phenotypic variation. For example, appreciable V_{A} explains 12% of observed variation in aggression of red squirrels (Taylor *et al.* 2012). V_{PE} represents variance between individuals over the time span within which the repeated measures were taken, i.e. they do not necessarily imply environmental effects with long-lasting effects on behaviour (Wilson *et al.* 2010). We further note that other variance components might be added to this equation, such as interactions between genetic and environmental effects, as has been observed to occur with predation risk and sociability in three-spined sticklebacks *Gasterosteus aculeatus* (Dingemanse *et al.* 2009). Unfortunately, the relative importance of these quantitative genetics parameters has not been extensively investigated in animal personality research (but see van Oers *et al.* 2004; Dingemanse *et al.* 2012a).

In defining personality variation according to Eqn. 4.2, the ratio of personality variation to total phenotypic variation (V_{P}) is then synonymous with repeatability:

$$repeatability = \frac{V_{ind_0}}{V_{\mathrm{P}}} = \frac{V_{\mathrm{A}} + V_{\mathrm{D}} + V_{\mathrm{PE}}}{V_{\mathrm{P}}} \tag{4.3}$$

This relationship is familiar to both behavioural ecologists and quantitative geneticists and relates readily to narrow-sense heritability (h^2) which is defined as:

$$h^2 = \frac{V_{\mathrm{A}}}{V_{\mathrm{P}}}. \tag{4.4}$$

That repeatability and narrow-sense heritability differ due to the inclusion of V_{D} and V_{PE} in the numerator of the former has been long recognised by evolutionary ecologists and has led many researchers to suggest that repeatability can be used as a proxy for heritability (Boake 1989; Hayes & Jenkins 1997). So what then can we say about behavioural repeatabilities and heritabilities? Specifically, what do behavioural repeatabilities and heritabilities reveal about selection and evolution of behaviour and individuality?

Mousseau and Roff (1987), Stirling, Réale, and Roff (2002), Bell *et al.* (2009), and van Oers and Sinn (2013) have previously described patterns in behavioural heritabilities and repeatabilities (see also Chapter 2, Potsma). Mousseau and Roff (1987) concluded that behaviours generally had similar heritabilities to physiological traits but that they were more heritable than life-history traits and less heritable than morphological traits. Likewise, Stirling *et al.* (2002) found that physiological and behavioural traits had comparable heritabilities but were less heritable than morphological traits. Across the two reviews, heritability of behaviour was estimated to be approximately 0.3. More recently Potsma (Chapter 2) reported a mean heritability for behaviours, across 60 studies, of ~0.5 while van Oers and Sinn (2013) found an average heritability across 209 estimates of 0.26. Across trait types, field studies typically produce higher heritability estimates, a result also found for behaviours: wild populations had average behavioural heritabilities of 0.36 vs 0.24 for domestic populations (van Oers & Sinn 2013). Likewise, Bell *et al.* (2009) in a review of 759 estimates found that field studies typically yielded higher estimates of repeatability than did lab studies. In the case of heritabilities, this pattern is likely to be the result of greater conflation of V_{PE} with V_{A} in the field than the lab. For example, an experimental study in zebra finches (*Taeniopygia guttata)* revealed that offspring resembled their foster parents' exploratory behaviour (Schuett et al. 2013). This implies that genetic vs environmental parental effects should best be separated by means of cross-fostering experiments to avoid biased estimates of V_{A} (see also Kruuk & Hadfield 2007). For repeatabilities, a potential cause of differences between lab and wild is less readily apparent.

One key question addressed in Mousseau and Roff's (1987) comparison of heritabilities was whether different types of traits might have experienced historical selection of different relative strength, and whether some types of traits are more closely associated with fitness. Fisher (1930) inferred that traits under greater selection will exhibit less additive genetic variation than traits under weaker selection. More specifically, directional and stabilising selection will deplete available additive genetic variation. Fisher's inference has led to a general expectation that the traits most closely related to fitness will exhibit the lowest heritabilities (Mousseau & Roff 1987). Generally, it has been concluded that behaviours have likely been under selection pressures of similar strength as observed for physiological traits (Mousseau & Roff 1987; Stirling *et al.* 2002) and, possibly as life-history traits (Stirling *et al.* 2002). However, as behavioural traits might be expected to be measured with greater error, this result should be viewed as tentative. Unfortunately, neither Mousseau and Roff (1987) nor Stirling *et al.* (2002) asked whether differences in historical selective pressures might be inferred for different behaviours. A more recent review by van Oers and Sinn (2013) provides the opportunity to ask whether particular behaviours have been under greater selective pressure.

For wild populations, aggression had the lowest average heritability across studies (0.28) while exploratory behaviour had the highest (0.58) with activity and boldness having relatively moderate heritabilities (0.39 and 0.31 respectively). Adopting the rationale of Mousseau and Roff (1987), aggression would thus be inferred to have been under the greatest selective pressure. Unfortunately, van Oers and Sinn (2013) restricted their analysis to particular behaviours and excluded behaviours such as those related to mate choice, narrowing the scope of inferences that might be drawn. However, if we assume that repeatability can be used as a proxy for heritability (e.g. Boake 1989; Hayes & Jenkins 1997), Bell *et al.*'s (2009) review of behavioural repeatabilities might produce additional insights.

If, following the above discussion, behaviours with lower repeatabilities might be inferred to have been under greater historical selection, then Bell *et al.*'s (2009) review suggests that migration, mate preferences, and activity have been under the greatest selective pressure while mating activity, habitat selection, and aggression have been under the least. This conclusion contradicts that based on the heritabilities reported by van Oers and Sinn (2013) and warrants further investigation.

4.2.4 Quantitative genetics of behavioural syndromes

Besides animal personality, behavioural ecologists studying individual variation often also focus on *behavioural syndromes* (Sih *et al.* 2004). Behavioural syndromes have been identified in a wider variety of taxa (meta-analysis: Garamszegi *et al.* 2013) and refer to when an individual's average behaviour in one context is correlated with its average behaviour in another context. For example, in three-spined sticklebacks, aggression and exploratory behaviour in a novel environment are positively correlated as part of a behavioural syndrome for individuals from ponds with predators (Dingemanse *et al.* 2007). Similarly, aggression and boldness are positively correlated as part of a behavioural syndrome in kangaroo rats (Dochtermann *et al.* 2012). Statistically, behavioural syndromes refer to non-zero correlations between two behavioural traits at the between-individual level (Dingemanse *et al.* 2012c), i.e. correlations between the parts of behavioural traits that represent personality. Behavioural syndromes can be estimated using bi- or multivariate implementations of Eqn. 4.1 (Dingemanse & Dochtermann 2013). In its simplest form, this requires a bivariate equivalent of the phenotypic equation:

$$\begin{aligned} y_{ij} &= (\beta_{0y} + ind_{0yj}) + e_{0yij} \\ z_{ij} &= (\beta_{0z} + ind_{0zj}) + e_{0zij} \end{aligned} \tag{4.5}$$

where y and z represent phenotypic traits that vary both within and between individuals; for example, aggression and boldness (e.g. Dochtermann et al. 2012). As detailed above (Eqn. 4.1), this type of equation captures between-individual (V_{ind_0}) and within-individual (V_{e_0}) variation in each of two behavioural traits. Provided that these behaviours were measured repeatedly for the same set of individuals and at the same point in time (Dingemanse & Dochtermann 2013), one may additionally ask whether there is covariance between the behaviours both between individuals ($COV_{ind_{0y},ind_{0z}}$) and within individuals ($COV_{e_{0y},e_{0z}}$). In other words, does an individual's mean value for one trait ($\bar{y}_j$) covary with its mean value for another ($\bar{z}_j$)? And, does an individual's deviation from its mean value for one trait ($y_{ij} - \bar{y}_j$) covary with its deviation from its mean value for another ($z_{ij} - \bar{z}_j$)?

For example, in house sparrows (*Passer domesticus*), individuals that were relatively explorative over many days were on average also relatively inactive (i.e. $COV_{ind_{0y},ind_{0z}} < 0$), but on specific days where those individuals were relatively explorative compared to their own mean they were also relatively active (i.e. $COV_{e_{0y},e_{0z}} > 0$) (Mutzel *et al.* 2011). Recently, there has been considerable attention in the separation of within- and between-individual correlations and its associated statistical tools (Dingemanse *et al.* 2012c; Brommer 2013; Dingemanse & Dochtermann 2013). Here we simply detail how between- ($r_{ind_{0y},ind_{0z}}$) and within-individual ($r_{e_{0y},e_{0z}}$) correlations are defined:

$$r_{ind_{0y},ind_{0z}} = \frac{COV_{ind_{0y},ind_{0z}}}{\sqrt{V_{ind_{0y}} V_{ind_{0z}}}} \tag{4.6a}$$

$$r_{e_{0y},e_{0z}} = \frac{COV_{e_{0y},e_{0z}}}{\sqrt{V_{e_{0y}} V_{e_{0z}}}} \tag{4.6b}$$

where $r_{ind_{0y},ind_{0z}} \neq 0$ indicates behavioural syndrome structure (Dingemanse *et al.* 2012c). In other words, behavioural syndromes do not simply reflect phenotypic correlations between behavioural traits, but phenotypic correlations between the repeatable parts of behavioural traits and require a sampling design where each behaviour is assayed repeatedly for the same set of individuals (detailed in Dingemanse & Dochtermann 2013).

By defining a behavioural syndrome as $r_{ind_{0y},ind_{0z}}$ (Eqn. 4.6a), behavioural syndromes represent between-individual correlations between two repeatable behaviours which, as we have discussed, can be readily parsed in terms familiar to quantitative geneticists. As was the case for personality variation, we can breakdown a behavioural syndrome into its constituent quantitative genetic components:

$$\begin{aligned} r_{ind_{0y},ind_{0z}} &= r_{A_{0y},A_{0z}} \sqrt{\frac{V_{A_y}}{V_{P_y}} \frac{V_{A_z}}{V_{P_z}}} \\ &+ r_{D_{0y},D_{0z}} \sqrt{\frac{V_{D_y}}{V_{P_y}} \frac{V_{D_z}}{V_{P_z}}} \\ &+ r_{PE_{0y},PE_{0z}} \sqrt{\frac{V_{PE_y}}{V_{P_y}} \frac{V_{PE_z}}{V_{P_z}}} \end{aligned} \tag{4.7}$$

where $r_{A_{0y},A_{0Z}}$ is the additive genetic correlation, $r_{D_{0y},D_{0z}}$ is the correlation between two behaviours due to pleiotropic effects of dominance, and $r_{PE_{0y},PE_{0z}}$ is the correlation between two behaviours due to the permanent environment. V_A, V_D, V_{PE}, and V_P refer, respectively, to the additive genetic variance, dominance variance, permanent environmental variance, and total phenotypic variance of behaviours y and z. Framed in this manner, behavioural syndromes represent the correlation between two behavioural traits due to the joint effects of additive genetic effects of genes, genetic dominance, and the effect of shared permanent environmental effects on two (or more) traits. Unfortunately behavioural syndromes have until recently rarely been estimated as $r_{ind_{0y},ind_{0z}}$ (Dingemanse & Dochtermann 2013).

While there are few instances in which $r_{ind_{0y},ind_{0z}}$ has been estimated for behaviours, considerable research has been directed toward understanding whether and how well phenotypic correlations approximate genetic correlations (Cheverud 1988; Roff 1996; Kruuk *et al.* 2008; Dochtermann 2011). Indeed, behavioural correlations at the level of the phenotype have recently been compared to behavioural correlations at the additive genetic level (Dochtermann 2011). This review found that the sign of genetic correlations was well estimated by phenotypic correlations. Moreover, the magnitude of genetic correlations was generally well estimated by phenotypic correlations ($\rho = 0.87$, Dochtermann 2011).

4.2.5 Behavioural syndromes and evolutionary constraints

One area of considerable speculation within the behavioural ecological literature involves whether behavioural syndromes might act as evolutionary constraints (Bell 2005). Generally, single traits possess sufficient additive genetic variation and thus are not likely to pose substantial constraints from a univariate perspective (Blows & Hoffmann 2005; Chapter 10, Kruuk *et al.*). The correlations within syndromes, however, create the possibility that genetic covariances amongst behaviours might create evolutionary constraints (Walsh & Blows 2009). In this book, Kruuk *et al.* (Chapter 10) and Teplitsky and colleagues (Chapter 12) discuss constraints, a topic which has produced a rich literature with competing methods of estimation and, indeed, competing philosophies as to what constraint means.

An informal definition of constraint as it pertains to behavioural syndromes would be the degree to which behavioural correlations reduce evolutionary responses relative to if all correlations were zero (Dochtermann & Dingemanse 2013). More specifically, this can be considered in all possible directions of selection. Others have suggested competing definitions focusing on the response in the direction selection has previously been observed to act upon in a population (e.g. Schluter 1996; Agrawal & Stinchcombe 2009). The definition we have suggested might be more general in allowing for inferences to be made not only about current selection but also about how populations might be able to respond under changing selective regimes (Dochtermann & Dingemanse 2013). Hansen & Houle (2008) provide a standardised metric for the estimation of constraint defined in this manner which they have termed 'autonomy' ($\bar{a}$). Autonomy represents the ability of a population to respond in multivariate space relative to the response in the absence of correlations (Hansen & Houle 2008), ranges between 0 and 1, and can be interpreted as:

- an autonomy approaching 1 means that the covariances do not constrain evolutionary responses in k-dimensional phenotypic space;
- an autonomy approaching 0 means that the covariances wholly constrain evolutionary responses in k-dimensional phenotypic space in at least one direction;
- an intermediate autonomy suggests intermediate constraints on evolutionary responses in k-dimensional phenotypic space relative to if covariances were zero.

Recently, we estimated the degree of evolutionary constraint imposed by behavioural syndrome structure by calculating $\bar{a}$ for 35 published **G**-matrices (a tabulation of additive genetic variances and additive genetic covariances), which included studies documenting genetic correlations between behaviours regardless of their significance (i.e. presence of syndrome structure). This analysis, which represents the first attempt to evaluate whether

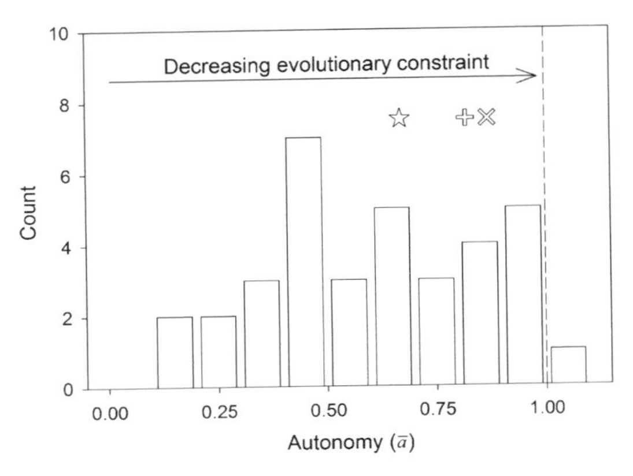

Figure 4.3 Distribution of estimates of autonomy ($\bar{a}$) from 35 genetic covariance matrices (see Dochtermann & Dingemanse 2013 for additional details). The star corresponds to the restricted maximum likelihood estimated autonomy for behaviours. The symbols + and × correspond to autonomy estimates for life-history traits and morphology traits, respectively. While the autonomy estimate was based on published additive genetic variance–covariance matrices, the autonomy estimates for life-history and morphology were based on average genetic correlations for these types of traits as reported by Roff (1996) and thus require further investigation.

syndromes might generally constrain evolutionary responses, revealed that—on average—additive genetic behavioural correlations have the potential to constrain evolutionary response by 33% (Figure 4.3; Dochtermann & Dingemanse 2013). While autonomy has not been estimated for other types of traits, average genetic correlations between morphological traits and between life-history traits suggest that behavioural syndromes might generate relatively strong evolutionary constraints (Dochtermann & Dingemanse 2013).

4.3 What quantitative geneticists can learn from behavioural ecologists

We end our discussion by briefly detailing our views on how both quantitative genetics and behavioural ecology might benefit from a continued exchange of ideas. Behavioural ecologists seek to understand behavioural variation from an adaptive perspective. They have often assumed i) that populations are at their adaptive peaks, ii) that individual variation represents noise around an adaptive mean, and iii) that functionally distinct behaviours may readily be studied in isolation. Quantitative genetics, in contrast, poses i) that populations continuously evolve in response to selection, and ii) that both heritable individual variation as well as trait associations are key to predicting phenotypic evolution. This view dates back to some of the earliest formulation of quantitative genetics such as the well-known breeder's equation, extends to the modern era of quantitative genetics and the development of multivariate versions of the breeder's equation (Lande 1979; Lande & Arnold 1983), and is at the heart of current efforts at understanding evolutionary constraints (Walsh & Blows 2009; Kruuk, this volume; Dochtermann & Dingemanse 2013). The study of animal personality, which is, as discussed in this chapter, currently revealing growing evidence for the presence of both substantial heritable variation in behaviour and genetic correlations between functionally distinct behaviours, will inherently lead behavioural ecologists to reconsider their assumptions and thereby place their research more firmly into an evolutionary framework (Westneat & Fox 2010). Specifically, there is growing awareness that (heritable) individual variation matters for ecological processes (Wolf & Weissing 2012) and that multivariate approaches to the study of behavioural adaptation are generally advisable (cf. Dochtermann & Dingemanse 2013).

At the same time, behavioural ecologists have various insights to offer to quantitative geneticists. Specifically, behavioural ecology as a field has a long tradition of generating testable hypotheses concerning behavioural adaptation. Their research has typically made a version of 'the phenotypic gambit' (Grafen 1984), that assumes that phenotypic distributions reliably describe additive genetic distributions. Explanations and predictions made at the phenotypic level can thus generate hypotheses and predictions that extend to the additive genetic level. This is particularly the case because adaptive models for between-individual (co)variation in behaviour are generally constructed to explain repeatable behaviour regardless of its specific proximate (e.g. genetic vs environmental) underpinning (Box 4.1). Consider two examples: first, quantitative geneticists often estimate genetic correlations between suites of phenotypic traits and ask whether the observed phenotypic traits may be collapsed into fewer underlying principle components (e.g. Phillips & Arnold 1999). Behavioural ecologists instead are increasingly applying statistical approaches that a priori consider hypotheses regarding correlation structure between suites of traits (Dochtermann & Jenkins 2007; Dingemanse *et al.* 2010a). For example, novel adaptive theory predicts that life-history trade-offs select for between-individual correlations between specific behaviours ('risky behaviours'), namely those that increase resource acquisition at the expense of increased risk of mortality (Wolf *et al.* 2007). Now imagine that one assayed four behavioural traits (v, w, y, z), three of which (v, w, z) represented risky behaviours while one (y) represented another type of behaviour. Instead of simply estimating their covariance matrix $\boldsymbol{\Omega}_{ind}$ at the between-individual level:

$$\boldsymbol{\Omega}_{ind} = \begin{bmatrix} V_{ind_{0vj}} & & & \\ Cov_{ind_{0vj},ind_{0wj}} & V_{ind_{0wj}} & & \\ Cov_{ind_{0vj},ind_{0yj}} & Cov_{ind_{0wj},ind_{0yj}} & V_{ind_{0yj}} & \\ Cov_{ind_{0vj},ind_{0zj}} & Cov_{ind_{0wj},ind_{0zj}} & Cov_{ind_{0yj},ind_{0zj}} & V_{ind_{0zj}} \end{bmatrix} \tag{4.8}$$

or asking whether this matrix could be collapsed into fewer principle components, one might instead design a specific statistical test to ask how well the data fitted the model's prediction that the non-risky behaviour was not part of a behavioural syndrome. This could, for example, be achieved by asking how this 'unconstrained' model (Eqn. 4.8) would fit compared to one where covariances with the non-risky behaviour were constrained to zero:

$$\boldsymbol{\Omega}_{ind} = \begin{bmatrix} V_{ind_{0vj}} & & & \\ Cov_{ind_{0vj},ind_{0wj}} & V_{ind_{0wj}} & & \\ 0 & 0 & V_{ind_{0yj}} & \\ Cov_{ind_{0vj},ind_{0zj}} & Cov_{ind_{0wj},ind_{0zj}} & 0 & V_{ind_{0zj}} \end{bmatrix} \tag{4.9}$$

Such hypothesis testing exercises can efficiently be implemented as structural equation models (see also Roff & Fairbairn 2011) and reveal considerable insight into the structure of correlation matrices (see Sprenger *et al.* 2012 for a worked example). In other words, $\boldsymbol{G}$-matrices can be used as the raw material for testing predictions of adaptive theory.

Second, behavioural ecologists have proposed that certain social environments harbour multiple social niches whereas others do not. For example in cooperative (group-living) fish species, where group members specialise in specific tasks (Bergmüller & Taborsky 2010), adaptive hypotheses come with general, testable, predictions about variance components that may readily be translated to the additive genetic level. First, the amount of between- and within-individual variance is expected to vary between environments (e.g. populations, species, or years). Second, the amount of between-individual variance should negatively covary with the amount of within-individual variance. Third, social environments that elicit diversifying selection on socially expressed behaviours (e.g. aggression, nest cleaning, anti-predator vigilance) should harbour increased between- and decreased within-individual variance in behaviour. In other words, such adaptive theory comes with specific predictions about variance components (Figure 4.2) that may readily be applied to quantitative genetics.

In summary, the behavioural ecologists' current interest in adaptive explanations for individuality explicitly concerns variance components, a specific focus of quantitative genetics. Their adaptive approach may thus be considered a resource that may be used to better understand patterns phenotypic variation observed in nature, whether genetic or environmental in origin.

Acknowledgments

We thank the editors for encouraging us to write this chapter. NAD was supported in part by the NDSU EPSCoR program. NJD was supported by the Max Planck Society. We also thank Jon Brommer, Francesca Santostefano, and the editors for helpful commentary on previous versions of this chapter.

References

Agrawal, A.F. & Stinchcombe, J.R. (2009) How much do genetic covariances alter the rate of adaptation? *Proceedings of the Royal Society B-Biological Sciences*, **276**, 1183–1191.

Bell, A.M. (2005) Behavioral differences between individuals and two populations of stickleback (*Gasterosteus aculeatus*). *Journal of Evolutionary Biology*, **18**, 464–473.

Bell, A.M. (2007) Future directions in behavioural syndromes research. *Proceedings of the Royal Society of London Series B, Biological Sciences*, **274**, 755–761.

Bell, A.M., Hankison, S.J. & Laskowski, K.L. (2009) The repeatability of behaviour: a meta-analysis. *Animal Behaviour*, **77**, 771–783.

Bergmüller, R. & Taborsky, M. (2010) Animal personality and social niche specialisation. *Trends in Ecology and Evolution*, **25**, 504–511.

Blows, M.W. & Hoffmann, A.A. (2005) A reassessment of genetic limits to evolutionary change. *Ecology*, **86**, 1371–1384.

Boake, C.R.B. (1989) Repeatability: its role in evolutionary studies of mating behavior. *Evolutionary Ecology*, **3**, 173–182.

Bolnick, D.I., Svanback, R., Fordyce, J.A., Yang, L.H., Davis, J.M., Hulsey, C.D. & Forister, M.L. (2003) The ecology of individuals: incidence and implications of individual specialization. *American Naturalist*, **161**, 1–28.

Boon, A.K., Réale, D. & Boutin, S. (2007) The interaction between personality, offspring fitness and food abundance in North American red squirrels. *Ecology Letters*, **10**, 1094–1104.

Brommer, J.E. (2013) On between-individual and residual (co)variances in the study of animal personality: are you willing to take the phenotypic gambit? *Behavioural Ecology and Sociobiology*, **67**, 1027–1032.

Cheverud, J.M. (1988) A comparison of genetic and phenotypic correlations. *Evolution*, **42**, 958–968.

Cowie, R.J. (1977) Optimal foraging in great tits (*Parus major*). *Nature*, **268**, 137–139.

Dall, S.R.X., Houston, A.I. & McNamara, J.M. (2004) The behavioural ecology of personality: consistent individual differences from an adaptive perspective. *Ecology Letters*, **7**, 734–739.

Dingemanse, N.J., Barber, I., Wright, J. & Brommer, J.E. (2012a) Quantitative genetics of behavioural reaction norms: genetic correlations between personality and behavioural plasticity vary across stickleback populations. *Journal of Evolutionary Biology*, **25**, 485–496.

Dingemanse, N.J., Bouwman, K.M., van de Pol, M., van Overveld, T., Patrick, S.C., Matthysen, E. & Quinn, J.L. (2012b) Variation in personality and behavioural plasticity across four populations of the great tit *Parus major*. *Journal of Animal Ecology*, **81**, 116–126.

Dingemanse, N.J. & Dochtermann, N.A. (2013) Quantifying individual variation in behaviour: mixed-effect modelling approaches. *Journal of Animal Ecology*, **82**, 39–54.

Dingemanse, N.J., Dochtermann, N.A. & Nakagawa, S. (2012c) Defining behavioural syndromes and the role of 'syndrome deviation' to study its evolution. *Behavioral Ecology and Sociobiology*, **66**, 1543–1548.

Dingemanse, N.J., Dochtermann, N.A. & Wright, J. (2010a) A method for exploring the structure of behavioural syndromes to allow formal comparison within and between datasets. *Animal Behaviour*, **79**, 439–450.

Dingemanse, N.J., Kazem, A.J.N., Réale, D. & Wright, J. (2010b) Behavioural reaction norms: where animal personality meets individual plasticity. *Trends in Ecology and Evolution*, **25**, 81–89.

Dingemanse, N.J., van der Plas, F., Wright, J., Réale, D., Schrama, M., Roff, D.A., van der Zee, E. & Barber, I. (2009) Individual experience and evolutionary history of predation affect expression of heritable variation in fish personality and morphology. *Proceedings of the Royal Society of London Series B, Biological Sciences*, **276**, 1285–1293.

Dingemanse, N.J. & Wolf, M. (2010) A review of recent models for adaptive personality differences. *Philosophical Transactions of the Royal Society of London Series B, Biological Sciences*, **365**, 3947–3958.

Dingemanse, N.J. & Wolf, M. (2013) Between-individual differences in behavioural plasticity: causes and consequences. *Animal Behaviour*, **85**, 1031–1039.

Dingemanse, N.J., Wright, J., Kazem, A.J.N., Thomas, D.K., Hickling, R. & Dawnay, N. (2007) Behavioural syndromes differ predictably between 12 populations of stickleback. *Journal of Animal Ecology*, **76**, 1128–1138.

Dochtermann, N.A. (2011) Testing Cheverud's conjecture for behavioral correlations and behavioral syndromes. *Evolution*, **65**, 1814–1820.

Dochtermann, N.A. & Dingemanse, N.J. (2013) Behavioral syndromes as evolutionary constraints. *Behavioral Ecology*, **24**, 806–811.

Dochtermann, N.A. & Jenkins, S.H. (2007) Behavioural syndromes in Merriam's kangaroo rats (*Dipodomys merriami*): a test of competing hypotheses. *Proceedings of the Royal Society of London Series B-Biological Sciences*, **274**, 2343–2349.

Dochtermann, N.A., Jenkins, S.H., Swartz, M.J. & Hargett, A.C. (2012) The roles of competition and environmental heterogeneity in the mainteance of behavioral variation and covariation. *Ecology*, **93**, 1330–1339.

Dochtermann, N.A. & Roff, D.A. (2010) Applying a quantitative genetics framework to behavioural syndrome research. *Philosophical Transactions of the Royal Society of London Series B, Biological Sciences*, **365**, 4013–4020.

Dubois, F., Giraldeau, L.A. & Réale, D. (2012) Frequency-dependent payoffs and sequential decision-making favour consistent tactic use. *Proceedings of the Royal Society B-Biological Sciences*, **279**, 1977–1985.

Falconer, D.S. & Mackay, T.F.C. (1996) *Introduction to quantitative genetics*. Longman, New York.

Fisher, R.A. (1930) *The genetical theory of natural selection*. Oxford University Press, Oxford.

Garamszegi, L.Z., Marko, G. & Herczeg, G. (2012) A meta-analysis of correlated behaviours with implications for behavioural syndromes: mean effect size, publication bias, phylogenetic effects and the role of mediator variables. *Evolutionary Ecology*, **26**, 1213–1235.

Grafen, A. (1984) Natural selection, kin selection and group selection. In: *Behavioural ecology: An evolutionary approach* (ed. J. R. Krebs & N. B. Davies), pp. 62–84. Blackwell Scientific, Oxford.

Hansen, T.F. & Houle, D. (2008) Measuring and comparing evolvability and constraint in multivariate characters. *Journal of Evolutionary Biology*, **21**, 1201–1219.

Hayes, J.P. & Jenkins, S.H. (1997) Individual variation in mammals. *Journal of Mammalogy*, **78**, 274–293.

Henderson, C.R. (1982) Analysis of covariance in the mixed model: higher-level, nonhomogeneous, and random regressions. *Biometrics*, **38**, 623–640.

Houston, A.I. & McNamara, J.M. (1999) *Models of adaptive behaviour*. Cambridge University Press, Cambridge.

Jenkins, S.H. (2011) Sex differences in repeatability of food-hoarding behaviour of kangaroo rats. *Animal Behaviour*, **81**, 1155–1162.

Kirkpatrick, M. & Heckman, N. (1989) A quantitative genetic model for growth, shape, reaction norms, and other infinite-dimensional characters. *Journal of Mathematical Biology*, **27**, 429–450.

Kontiainen, P., Pietiäinen, H., Huttunen, K., Karell, P., Kolunen, H. & Brommer, J.E. (2009) Aggressive Ural owl mothers recruit more offspring. *Behavioral Ecology*, **20**, 789–796.

Krebs, J.R., Erichsen, J.T., Webber, M.I. & Charnov, E.L. (1977) Optimal prey selection in the great tit (*Parus major*). *Animal Behaviour*, **25**, 30–38.

Kruuk, L.E.B. & Hadfield, J.D. (2007) How to separate genetic and environmental causes of similarity between relatives. *Journal of Evolutionary Biology*, **20**, 1890–1903.

Kruuk, L.E., Slate, J. & Wilson, A.J. (2008) New answers for old questions: the evolutionary quantitative genetics of wild animal populations. *Annual Review of Ecology Evolution and Systematics*, **39**, 525–548.

Lande, R. (1979) Quantitative genetics analysis of multivariate evolution, applied to brain:body size allometry. *Evolution*, **33**, 402–416.

Lande, R. & Arnold, S.J. (1983) The measurement of selection on correlated characters. *Evolution*, **37**, 1210–1226.

Luttbeg, B. & Sih, A. (2010) Risk, resources, and state-dependent adaptive behavioural syndromes. *Philosophical Transactions of the Royal Society of London Series B, Biological Sciences*, **365**, 3977–3990.

Lynch, M. & Walsh, B. (1998) *Genetics and analysis of quantitative traits*. Sinauer, Sunderland, MA.

Mathot, K., van den Hout, P., Piersma, T., Kempenaers, B., Réale, D. & Dingemanse, N.J. (2011) Disentangling the roles of frequency- versus state-dependence in generating individual differences in behavioural plasticity. *Ecology Letters*, **14**, 1254–1262.

Mathot, KJ., Wright, J., Kempenaers, B. & Dingemanse, N.J. (2012) Adaptive strategies for managing uncertainty may explain personality-related differences in behavioural plasticity. *Oikos*, **121**, 1009–1020.

Maynard Smith, J. (1982) *Evolution and the theory of games*. Cambridge University Press, Cambridge.

Mousseau, T.A. & Roff, D.A. (1987) Natural selection and the heritability of fitness components. *Heredity*, **59**, 181–197.

Mutzel, A., Kempenaers, B., Laucht, S., Dingemanse, N.J. & Dale, J. (2011) Circulating testosterone levels do not affect exploration in house sparrows: observational and experimental tests. *Animal Behaviour*, **81**, 731–739.

Nakagawa, S., Gillespie, D., Hatchwell, B. & Burke, T. (2007) Predictable males and unpredictable females: sex difference in repeatability of parental care in a wild bird population. *Journal of Evolutionary Biology*, **20**, 1674–1681.

Nicolaus, M., Tinbergen, J.M., Bouwman, K.M., Michler, S.P.M., Ubels, R., Both, C., Kempenaers, B. & Dingemanse, N.J. (2012) Experimental evidence for adaptive personalities in a wild passerine bird. *Proceedings of the Royal Society of London Series B, Biological Sciences*, **279**, 4885–4892.

Nussey, D.H., Wilson, A.J. & Brommer, J.E. (2007) The evolutionary ecology of individual phenotypic plasticity in wild populations. *Journal of Evolutionary Biology*, **20**, 831–844.

Phillips, P.C. & Arnold, S.J. (1999) Hierarchical comparison of genetic variance-covariance matrices. I. Using the Flury hierarchy. *Evolution*, **53**, 1506–1515.

Pinheiro, J.C. & Bates, D.M. (2000) *Mixed effect models in S and S-PLUS*. Springer, New York.

Quinn, J.L., Patrick, S.C., Bouwhuis, S., Wilkin, T.A. & Sheldon, B.C. (2009) Heterogeneous selection on a heritable temperament trait in a variable environment. *Journal of Animal Ecology*, **78**, 1203–1215.

Réale, D., Dingemanse, N.J., Kazem, A.J.N. & Wright, J. (2010) Evolutionary and ecological approaches to the study of personality. *Philosophical Transactions of the Royal Society of London Series B, Biological Sciences*, **365**, 3937–3946.

Réale, D., Martin, J., Coltman, D.W., Poissant, J. & Festa-Bianchet, M. (2009) Male personality, life-history strategies and reproductive success in a promiscuous mammal. *Journal of Evolutionary Biology*, **22**, 1599–1607.

Réale, D., Reader, S.M., Sol, D., McDougall, P. & Dingemanse, N.J. (2007) Integrating temperament in ecology and evolutionary biology. *Biological Reviews*, **82**, 291–318.

Roff, D. & Fairbairn, D. (2011) Path analysis of the genetic integration of traits in the sand cricket: a novel use of BLUPs. *Journal of Evolutionary Biology*, **24**, 1857–1869.

Roff, D.A. (1996) The evolution of genetic correlations: an analysis of patterns. *Evolution*, **50**, 1392–1403.

Schluter, D. (1996) Adaptive radiation along genetic lines of least resistance. *Evolution*, **50**, 1766–1774.

Schuett, W., Dall, S.R.X., Wilson, A.J. & Royle, N.J. (2013) Environmental transmission of a personality trait: foster parent behaviour predicts offspring exploration behaviour in zebra finches. *Biology Letters*, **9**, 20130120.

Schuett, W., Tregenza, T. & Dall, S.R.X. (2010) Sexual selection and animal personality. *Biological Reviews*, **85**, 217–246.

Sih, A., Bell, A. & Johnson, J.C. (2004) Behavioral syndromes: an ecological and evolutionary overview. *Trends in Ecology and Evolution*, **19**, 372–378.

Smiseth, P.T., Wright, J. & Kölliker, M. (2008) Parent-offspring conflict and co-adaptation: behavioural ecology meets quantitative genetics. *Proceedings of the Royal Society of London Series B, Biological Sciences*, **275**, 1823–1830.

Sprenger, D.S., Dingemanse, N.J., Dochtermann, N.A., Theobald, J. & Walker, S.P.W. (2012) Aggressive females become aggressive males in a sex-changing reef fish. *Ecology Letters*, **15**, 986–992.

Stirling, D.G., Réale, D. & Roff, D.A. (2002) Selection, structure and the heritability of behaviour. *Journal of Evolutionary Biology*, **15**, 277–289.

Taylor, R.W., Boon, A.K., Dantzer, B., Reale, D., Humphries, M.M., Boutin, S., Gorrell, J.C., Coltman, D.W. & Mcadam, A.G. (2012) Low heritabilities, but genetic and maternal correlations between red squirrel behaviours. *Journal of Evolutionary Biology*, **25**, 614–624.

van Oers, K., Drent, P.J., de Jong, G. & van Noordwijk, A.J. (2004) Additive and nonadditive genetic variation in avian personality traits. *Heredity*, **93**, 496–503.

van Oers, K. & Sinn, D.L. (2013) Quantitative and molecular genetics of animal personality. In: *Animal personalities: behavior, physiology, and evolution* (ed. C. Carere & D. Maestripieri), pp. 149–200. University of Chicago Press, Chicago and London.

Walsh, B. & Blows, M.W. (2009) Abundant genetic variation plus strong selection = multivariate genetic constraints: a geometric view of adaptation. *Annual Review of Ecology Evolution, and Systematics*, **40**, 41–59.

Westneat, D.F. & Fox, C.W. (2010) *Evolutionary behavioural ecology*. Oxford University Press, New York.

Westneat, D.F., Hatch, M.I., Wetzel, D.P. & Ensminger, A.L. (2011) Individual variation in parental care reaction norms: integration of personality and plasticity. *American Naturalist*, **178**, 652–667.

Wilson, A.J., Réale, D., Clements, M.N., Morrissey, M.M., Postma, E., Walling, C.A., Kruuk, L.E.B. & Nussey, D.H. (2010) An ecologist's guide to the animal model. *Journal of Animal Ecology*, **79**, 13–26.

Wolf, M., van Doorn, G.S., Leimar, O. & Weissing, F.J. (2007) Life-history trade-offs favour the evolution of animal personalities. *Nature*, **447**, 581–585.

Wolf, M., van Doorn, G.S. & Weissing, F.J. (2008) Evolutionary emergence of responsive and unresponsive personalities. *Proceedings of the National Academy of Sciences of the United States of America*, **105**, 15825–15830.

Wolf, M. & Weissing, F.J. (2010) An explanatory framework for adaptive personality differences. *Philosophical Transactions of the Royal Society of London Series B, Biological Sciences*, **365**, 3959–3968.

Wolf, M. & Weissing, F.J. (2012) Animal personalities: consequences for ecology and evolution. *Trends in Ecology and Evolution*, **27**, 452–461.

CHAPTER 5

The quantitative genetics of senescence in wild animals

Anne Charmantier, Jon E. Brommer and Daniel H. Nussey

5.1 Introduction

Over the course of their lifespan, animals often display profound and complex changes in their physiology and in fitness-related phenotypes. An appreciation of age-related variation in phenotypic traits associated with demographic rates and fitness is fundamental to our understanding of evolutionary and ecological dynamics (Charlesworth 1980; Caswell 2001). A common pattern in animal taxa is an increase in performance- and fitness-related traits through development and early adulthood, followed by a plateau in long-lived species during so-called prime adulthood, after which comes a decline that is commonly attributed to senescence (Caughley 1966; Gaillard *et al.* 1998). Although much attention has been directed at understanding processes and patterns of development and improvement through early life, a growing amount of attention in evolutionary genetics and ecology is being paid to the deterioration observed in later life (reviews in Monaghan *et al.* 2008; Nussey *et al.* 2008a; Wilson *et al.* 2008). From an evolutionary standpoint, senescence is defined as the late-life decline of contributions to individual fitness through a decrease in reproduction or survival (Rose 1991; Stearns 1992). Initially, the evolutionary processes through which senescence could emerge *de novo* presented a substantial theoretical challenge to evolutionary biologists. Surely, natural selection would oppose genetic variants associated with age-related declines in fitness during adulthood? The answer to this puzzle emerged through the now seminal works of verbal and then mathematical evolutionary theory by Medawar, Williams, Hamilton and Charlesworth (Rose 1991). This work showed how age-independent forces of mortality from environmental causes would lead to a reduction in the strength of selection with age and thereby allow late-acting deleterious mutations to persist in natural populations (Medawar 1952; Hamilton 1966).

With the evolution and maintenance of senescence *de novo* explained, the potentially much grander challenge for evolutionary biologists is now to use theory and empirical data to explain the astonishing variation in lifespan and patterns of ageing and senescence among species, populations and individuals in natural environments (Williams *et al.* 2006; Baudisch 2008; Nussey *et al.* 2013). There is now good evidence that artificial selection, genetic manipulation and alternation of diet in short-lived laboratory models can have profound influences on both lifespan and patterns of senescence (Kirkwood & Austad 2000; Fontana *et al.* 2010). Although laboratory studies have provided crucial support for evolutionary theories of senescence and insights into the mechanisms of ageing, environmental conditions will ultimately influence both the expression and the manner in which selection acts on the ageing process. Indeed few would dispute that standard laboratory conditions are quite unlike those under which the ageing process actually evolved (Kawasaki *et al.* 2008; Carey 2011). Therefore, understanding the way in which genes

Quantitative Genetics in the Wild. Edited by Anne Charmantier, Dany Garant, and Loeske E. B. Kruuk

influence age-dependent phenotypic variation and how this, in turn, influences fitness in natural settings is a central challenge in evolutionary biology (Promislow *et al.* 2006; Wilson *et al.* 2008).

There is now abundant evidence for age-related declines in reproductive performance traits and survival in wild birds, mammals, and fishes, with evidence also emerging at a slower rate in reptiles and insects (reviewed in Nussey *et al.* 2013). Longitudinal studies of wild vertebrates are also increasingly documenting among-individual variation in senescence rates, although the evolutionary and ecological causes and consequences of this variation remain poorly understood (Reznick *et al.* 2004; Bouwhuis *et al.* 2012; Nussey *et al.* 2013). Despite many of these studies also having pedigree or genomic information available, it is notable that there are still very few published attempts to quantify genetic contributions to variation in lifespan and ageing patterns in wild animals. This chapter builds on a previous review from around five years ago, which synthesised work to date examining quantitative genetic studies of senescence (equally called ageing throughout this chapter) in the wild (Wilson *et al.* 2008). We begin by briefly outlining the history of evolutionary theories of ageing, highlighting the need to focus on robustly and reliably quantifying genotype-by-age interactions and, ultimately, their association with age-specific components of fitness. We then describe available frameworks for undertaking such analyses using pedigree-based quantitative genetic methods, focussing on the most widely used approach, a reaction-norm based random-regression framework. We discuss the rapidly emerging literature demonstrating individual differences in senescence rates in wild vertebrates, and go on to review the much smaller literature which has sought to model age-dependent patterns of genetic (co)variation in such systems. This has inevitably involved the application of rather complicated statistical models, and here we aim to provide a critical and personal view of this work, which is necessarily of a rather technical nature. We conclude with recommendations for how to overcome some of the inconsistencies and issues in the existing literature, as well as identifying several important avenues for future work in this area.

5.2 The evolutionary theory of ageing

The first theoretical arguments explaining the origin and maintenance of senescence trace back to the work of Alfred Russel Wallace (1823–1913) and August Weismann (1834–1914), pre-dating Charles Darwin's published work on the theory of natural selection (Darwin 1859). Their arguments were largely based on the reasoning that senescence reduces longevity which in turn accelerates the renewal of generations, thus facilitating adaptation to changing environments at the population or species level (Wallace 1865; Weismann 1889). The group selectionist nature of these theories did not survive the scrutiny of subsequent evolutionary theoreticians, and the ideas were superseded by the now classical evolutionary theory of senescence, first laid out by Peter Medawar. His simple but profoundly influential idea was that even in the absence of senescence, the pressure of age-independent extrinsic mortality means that genetic mutations expressed in early life will have a higher impact on an individual's fitness and hence will be under stronger selection compared to mutations expressed later in life (Medawar 1952). This idea of declining selection with age gave lead to the proposition of two main population genetic mechanisms explaining the *de novo* evolution of senescence: mutation accumulation (MA) and antagonistic pleiotropy (AP). Under MA, the mutation-selection balance allows for the accumulation of deleterious mutations acting in late life, leading to a decrease of physiological functions in old age (Medawar 1952). AP is based on optimality models of evolution (Maynard Smith 1978) and posits that, because selection is much stronger in early life than late life, late-acting deleterious mutations can be actively selected for if they display pleiotropic beneficial effects in early life (Williams 1957). The oft-quoted disposable soma theory (DS; Kirkwood 1977) can be viewed as a physiological variant of AP based on the balance in the individual allocation of energy between maintenance and reproduction. From a quantitative genetics viewpoint, DS and AP are hard to distinguish and are often grouped under the umbrella term of 'life-history' theories of ageing (Partridge & Barton 1993).

The theoretical and empirical application of quantitative genetics in this field has focussed considerable energy on the possibility of discriminating the contributions of AP and MA to the ageing process. Theoretical work of Brian Charlesworth and colleagues formalised the idea that comparing genetic (co)variances across age classes could allow testing for the relative contributions of MA and AP mechanisms (Charlesworth 1990; Charlesworth & Hughes 1996; Charlesworth 2001). Early models predicted an increase with age in additive genetic variance V_A, dominance variance and inbreeding depression under MA (Charlesworth & Hughes 1996). However, subsequent work has shown these patterns can also arise under AP if certain simplistic assumptions of those original models are relaxed (Moorad & Promislow 2009). A recent study provides predictions that may be diagnostic of MA rather than AP under conditions of interbreeding among populations (Escobar *et al.* 2008). However, generally speaking the only pattern of genetic (co)variation that has been routinely and robustly used in empirical quantitative genetic studies to support one mechanism over the other has been identification of negative genetic correlations between early and later life, which are expected under AP but not pure MA (Rose & Charlesworth 1981; Tatar *et al.* 1996). It is important to note that alleles with beneficial early life effects but detrimental late-life effects could readily become fixed by natural selection, leaving no trace of genetic variation or covariation for the empirical research to detect. Thus, the absence of negative genetic correlations between early and late life cannot be used to infer that AP has played no role in the evolution of ageing in the evolutionary past.

There is ample empirical evidence from model laboratory study organisms supporting antagonistic genetic correlations consistent with AP or DS. These include artificial selection experiments on *Drosophila melanogaster* providing strong evidence for negative additive genetic correlations between early fecundity and longevity (Rose & Charlesworth 1981; Tatar *et al.* 1996; Kirkwood & Austad 2000), and genetic manipulations that extend lifespan in several short-lived model organisms, with negative effects on growth or reproduction (Kirkwood & Austad 2000; Nussey *et al.* 2013). There is also emerging support from wild vertebrate systems that individuals showing increased investment in early-life reproduction show more rapid subsequent rates of senescence, consistent with life-history theories of ageing (e.g. Charmantier *et al.* 2006b; Nussey *et al.* 2006; Reed *et al.* 2008; Bouwhuis *et al.* 2010). None of this provides proof that MA is not also important, since AP and MA are not mutually exclusive mechanisms. In fact, several authors have argued that the debate over whether AP or MA is more important in the evolution of ageing represents something of a blind alley for evolutionary geneticists (Snoke & Promislow 2003; Wilson *et al.* 2008). However, whilst quantitative genetics might not allow us to easily distinguish the roles of AP and MA in the evolution of senescence, it does provide a conceptual and statistical framework through which we can capture age-dependent patterns of genetic (co)variation and link these to age-dependent changes in natural selection.

Early studies of quantitative genetics in the wild tended to consider genetic variances and covariances as constant with respect to age, whilst both evolutionary theory and available evidence from the laboratory would predict a genetic contribution to variation in the senescence process and age-dependent patterns of genetic (co)variation. The application of models capable of testing for and estimating such age-dependent variation to data collected in natural settings is of fundamental importance in addressing whether and how senescence is capable of evolving over future generations. For instance, is there heritable variation in ageing rates? How strongly is selection on fitness-associated traits in later life constrained by antagonistic selection and pleiotropy acting in earlier life? We currently have only a very limited understanding of the genetic architecture underlying variation in senescence in natural populations, which hampers both our ability to predict their evolutionary dynamics and to make inferences about how natural selection has shaped genetic variation underpinning the ageing process across populations and species (Promislow *et al.* 2006). To achieve this ambitious goal, we must first properly quantify the age-specific **G**-matrix, and it is the analytical challenge this poses that we turn to next.

5.3 Character state and function-valued trait models

In this section, we outline how quantitative genetic approaches can assess the pattern of change in genetic (co)variances with age, and why this is a useful goal to pursue on data collected in natural populations. Perhaps the most robust and intuitive approach is to estimate the additive genetic variance for each age class, and the genetic covariances between the ages. However, this so-called *character state approach* clearly requires the estimation of a large number of parameters. The covariance matrix for n age classes will contain $n(n+1)/2$ elements, leading to frequent parameter model convergence problems when applied to the small samples that are often characteristic of field studies, relative to laboratory and agricultural data sets. Alternative approaches view the focal performance trait as a function-valued (also called infinite-dimensional) trait, and focus on estimation of the function or functions that describe how that trait changes with age at the individual or genetic level (Kirkpatrick & Heckman 1989). This *function-valued trait approach* has the advantage of requiring the estimation of fewer parameters, the number of which depends on the complexity of the function (e.g. if linear, only two parameters are needed), rather than on the number of age classes. However, it comes with the cost of assuming that variances and covariances vary in accordance with the chosen function(s). If the assumed functions provide a poor fit to the real biological pattern then the models will be at best non-informative and at worst incorrect and misleading. It is important to note that both character state and function-valued trait models can and are increasingly applied in an analogous manner to study phenotypic plasticity and genotype-by-environment interactions ($G \times E$) in natural populations, and we refer the reader to Gienapp and Brommer's chapter (Chapter 15) of this book for further insight into such work.

The most widely applied function-valued trait approach in wild quantitative genetics is *random regression* (Henderson 1982), which is based on the assumption that breeding values follow an nth-order orthogonal polynomial function of age. In Box 5.1, we provide further technical details of this approach and its application. When applied to longitudinal data, random regression can provide a direct test of whether or not individuals vary in their ageing rates (as described by a linear or polynomial reaction norm), and then subsequently whether that among-individual variation has an additive genetic basis. The parameters estimated can be used to predict how the genetic covariance structure varies within and among ages (Kirkpatrick *et al.* 1990). It is important to note that a variety of different function-valued approaches exist beyond linear and polynomial functions. For instance, autocorrelation functions have been suggested as an appropriate way to describe the additive genetic covariance between ages, and require estimation of only one or a few parameters (Pletcher & Geyer 1999). However, to our knowledge, they have not been used in studies of natural populations, and standard autoregressive functions cannot estimate the negative correlations among ages which are explicitly predicted by life-history theories of ageing. Each function-valued trait approach has its specific assumptions and limitations, but they can all provide a statistical description of the age-specific **G**-matrix. Which approach is most appropriate will depend on the question and data in hand (Stinchcombe *et al.* 2012).

Figure 5.1 provides an example of a simple case of a function-valued trait approach where the ageing function is linear, and illustrates three very different and evolutionarily significant ways in which patterns of age-dependent genetic variation might differ. For simplicity, these reaction norms reflect phenotypes that, on average, are showing a steady decline with age. Note that it is common to estimate reaction norms in terms of deviations from the mean trait expression for each age, by including age as a fixed effect in random-regression models (Eq. B5.1.1, Box 5.1). In applying a random regression animal model (RRAM; Eq. B5.1.1b in Box 5.1), we are able to test whether genotypes differ in not just their average trait expressed across the lifespan (elevation), but also their ageing rate (slope), and then extrapolate from the estimated pattern of genetic variation in reaction norms to predict how genetic variance changes with age and the magnitude and direction of genetic correlations among ages (left to right in Figure 5.1). Under scenario 1

Box 5.1 The anatomy of a Random Regression Animal Model (RRAM)

We here provide detail of the most widely applied function-valued trait approach, the RRAM, which hinges on the infinite-dimensional model (Kirkpatrick, Lofsvold & Bulmer 1990; Gomulkiewicz & Kirkpatrick 1992). In a random regression model, a trait (z) in individual (i) is expressed as a function of its age (age), at the phenotypic level as follows:

$$z_{i,age} = \mu + AGE_F + f_{ind}(x, age) + \varepsilon_{i,age} \qquad \text{(B5.1.1a)}$$

Here, μ denotes the fixed effect for the overall population mean, and AGE_F is a factorial fixed effect which captures variation in the trait mean across all age groups. The $f_{ind}(x, age)$ is a random regression function describing an orthogonal polynomial of order x that varies at the level of the individual. For example, a first-order polynomial for the individual effect would be $f_{ind} = ind_0 + ind_1 \times age$ and thus comprise an elevation and a linear slope term. The random regression parameter is a random effect with a mean of zero and normally distributed (co)variances, and a first-order polynomial function will result in the estimation of three (co)variance terms: the variation in the individual elevation, the variation in slope and the covariance between them. Importantly, the inclusion of the fixed factor for age means that the individual variance estimates are conditioned on age-specific means. In other words, this tests for variation in individual elevations and slopes, independent of age-related variation at the population level.

The individual level random regression functions can be partitioned further to estimate the contributions of genetic and non-genetic (so-called permanent environment effects) as follows:

$$z_{i,age} = \mu + AGE_F + f_a(x, age) + f_{pe}(x, age) + \varepsilon_{i,age} \qquad \text{(B5.1.1b)}$$

where functions $f_a(x, age)$ and $f_{pe}(x, age)$ describe random regression functions at the level of additive genetic and permanent environment (i.e. permanent between-individual differences estimated from repeated measures), respectively. The f_a function will provide estimates of the additive genetic variance components of the ageing functions. For instance, a second-order polynomial of effect f_a would represent the function $a_{0,i} + a_{1,i} \times age + a_{2,i} \times age^2$ at the additive genetic level. This would produce estimates of three additive genetic variances (V) and three genetic covariances (COV), as illustrated in the age-specific additive genetic matrix ($\boldsymbol{K}_a$) below:

$$\boldsymbol{K}_a = \begin{bmatrix} V_{a0} & COV_{a0,1} & COV_{a0,2} \\ COV_{a0,1} & V_{a1} & COV_{a1,2} \\ COV_{a0,2} & COV_{a1,2} & V_{a2} \end{bmatrix}$$

A similar matrix $\boldsymbol{K}_{pe}$ would simultaneously be estimated for the pe_y effects. In laboratory systems, where inbred or clonal lines are likely to be the group or unit of experimentation, quantitative genetic studies can directly test genetic variation at these grouping levels. In longitudinal studies in the wild (as discussed in the context of phenotypic plasticity in Nussey *et al.* 2007), the objective has to be to first establish whether individuals vary in their age-specific trends (i.e. the extent of the individual-by-age interaction or $I \times A$), by fitting $f_{ind}(x, age)$ in equation B5.1.1a. For instance, if a model with a non-zero-order polynomial function outperforms a model with a zero-order polynomial (under which changes in the trait with age cannot differ among individuals), then this constitutes evidence that individuals differ in their ageing rates. Having found evidence for $I \times A$, the next step would be to implement equation 1b and partition $I \times A$ into constituent permanent environment ($PE \times A$) and genetic ($G \times A$) components. Dissection of $I \times A$ into $G \times A$ and $PE \times A$ is typically achieved based on the resemblance of population-wide relatives using a pedigree-based 'animal model' (Lynch & Walsh 1998), and these models are hence referred to RRAMs.

Once parameters have been estimated, age-specific (co)variances can be derived from the final additive genetic and permanent environment random regression covariance matrices $\boldsymbol{K}_a$ and $\boldsymbol{K}_{pe}$, respectively (Meyer 1998), as well as their confidence intervals (Fischer *et al.* 2004). For example, the age-specific $\boldsymbol{G}$-matrix is given by $\boldsymbol{G} = \boldsymbol{\Phi K_a \Phi'}$, where $\boldsymbol{\Phi}$ is a column vector denoting the ages where one wants to evaluate the random regression estimates and $\boldsymbol{\Phi'}$ is the transposed matrix $\boldsymbol{\Phi}$. One important additional consideration here is the structure of the residuals (ε) with respect to age. The assumption that residuals will be constant with age is unlikely to be met, and in Eq. B5.1.1b, residuals are viewed as age specific. Indeed, models including age-dependent variation in residuals are better supported in all field studies where this has been examined. Some previous studies have fitted unstructured age-dependent residuals in RRAM models—directly estimating all possible variances and covariances at this level within and among ages or age classes (Wilson *et al.* 2005b; Wilson *et al.* 2007b; Robinson *et al.* 2008). Simpler, diagonal residual error structures (residual variances can vary with age, but residual covariances among ages assumed to be zero) have also been used, often when models including unconstrained residual error structures would not converge (e.g. Wilson *et al.* 2007a; Nussey *et al.* 2008b; Kim *et al.* 2011). It should be quite clear that failing to account suitably for age-related heterogeneity in the residual variance structure of the model is likely to result in biased or misleading estimates at the individual or genetic level.

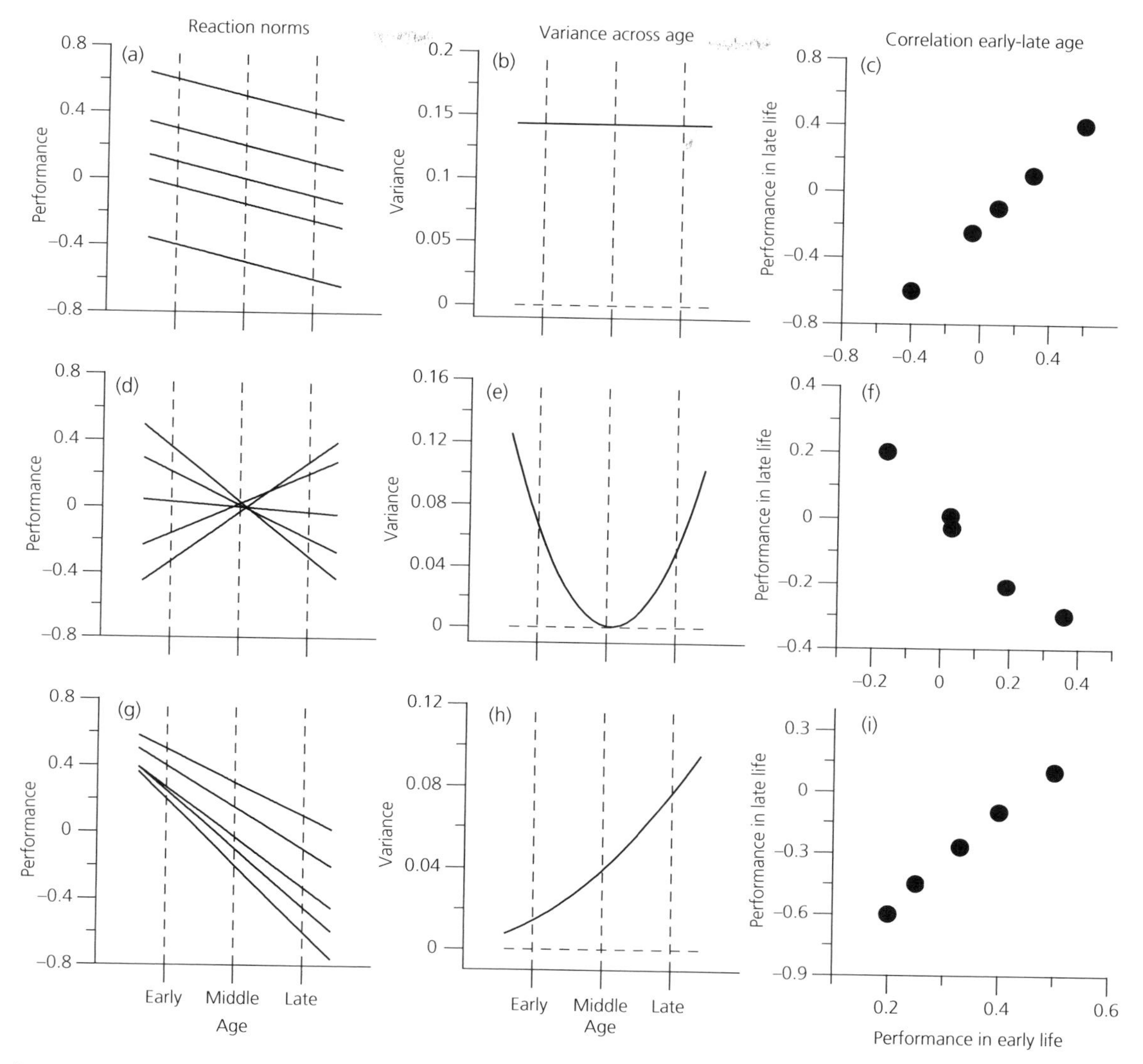

Figure 5.1 A reaction norm perspective on the quantitative genetics of ageing, and its connection to a character state approach. A performance trait varies over age classes. Lines in the left-hand side panels represent age-specific reaction norms for five different genotypes in old age, thereby denoting variation between genotypes ($G \times A$); dots in the right-hand panel represent the five different genotypes. Alternatively, the value of the performance traits can be evaluated at the character states 'early', 'middle', and 'late' life (dotted vertical lines). Middle panels show the additive genetic variance across the ages associated with the patterns of reaction norms. The right-hand side panels show the genetic correlation in the performance trait between the character states 'early' and 'late' life. Scenario 1: (a) All reaction norms decline at an equal rate, there is no $G \times A$, leading to (b) constant additive genetic variance across ages, and (c) tight positive genetic correlation in performance at early and late life. Scenario 2: (d) Reaction norms cross at middle age, there is $G \times A$, with (e) a nonlinear change in additive genetic variance with age. There is (f) a negative genetic covariance between performance in early and late life (change in ranking of the genotypes). Scenario 3: (g), Reaction norms 'fan' out, but their ranking is maintained across the ages, leading to (h) increase in additive genetic variance with age, and (i) positive genetic correlation between early and late life.

(Figure 5.1a–c), there is no genetic variation in the rate of ageing: any selection on senescence therefore can have no evolutionary consequences. In the other two scenarios, there is genetic variation in the rate of ageing ($G \times A$), either because the reaction norms cross (scenario 2: Figure 5.1d–f) or because they 'fan out' (scenario 3: Figure 5.1g–i). When reaction norms cross, selection on ageing may be constrained, because there are negative correlations between a genotype's performance in early and in later life (Figure 5.1f), so natural selection favouring slower senescence might be offset by associated fitness costs in early life. Under scenario 3, genetic variation in later life (Figure 5.1h) and positive pleiotropy exist across ages (Figure 5.1i), which would be predicted to facilitate an evolutionary response to selection favouring slower senescence (assuming other environmental components of variation remained constant with age).

5.4 Quantitative genetic studies of ageing in wild animals: a brief review

Case studies where individuals differ in their rate of ageing in wild vertebrate populations have emerged in the last decade (see recent reviews in Nussey *et al.* 2008a; Nussey *et al.* 2013). A number of studies have now used random regression to formally test and quantify $I \times A$ in wild vertebrates (Table 5.1). All studies we are aware of that have attempted this have found evidence for significant $I \times A$, although this may reflect publication bias (Brommer *et al.* 2007; Wilson *et al.* 2007a; Nussey *et al.* 2008b; Brommer *et al.* 2010). Many other studies of wild vertebrates provide indirect evidence of $I \times A$: rather than explicitly quantifying individual variation in ageing rates as random effects, these studies document interactions between ageing terms and environmental or early life-history in the fixed effects structure of their models. Several studies of wild vertebrates have demonstrated that individual ageing rates depend on the environmental conditions experienced earlier in life (e.g. Mysterud *et al.* 2001; Nussey *et al.* 2007; Hayward *et al.* 2009; Bouwhuis *et al.* 2010). This constitutes evidence for $I \times A$, but also explicitly identifies a non-genetic driver of this among-individual variation in ageing. Additionally, there is mounting evidence from wild vertebrates linking increased investment in reproduction in early adulthood with reduced lifespan, later-life reproductive performance or more rapid reproductive senescence (e.g. Descamps *et al.* 2006; Nussey *et al.* 2006; Reed *et al.* 2008; Bouwhuis *et al.* 2010; Massot *et al.* 2011). Investigations in other taxa are much scarcer but evidence for similar trade-offs is slowly building in invertebrates as well (see e.g. Bonduriansky & Brassil 2005 and Chapter 9, Zajitschek & Bonduriansky). This line of research provides important support for life-history theories of ageing (AP and DS) as well as evidence of $I \times A$. However, only two studies to our knowledge have tested for a genetic basis to an observed trade-offs (although see Charmantier *et al.* 2006b; Nussey *et al.* 2008b).

Previously, Charmantier *et al.* (2006a) had summarised studies which had tested for changes in heritability of a trait with age in wild vertebrates. The vast majority of the studies undertaken at that point considered only changes in quantitative genetic parameters across the early part of life, and mostly used parent–offspring regression or full-sib analysis of variance (Charmantier *et al.* 2006a). Subsequently, Wilson *et al.* (2008) reviewed studies of wild animals that tested for patterns of change in quantitative genetic parameters with age, which were at the time considered potentially diagnostic indicators of the action of either MA or AP. As discussed above, two 'indicators' of MA (increasing additive genetic variance V_A and inbreeding depression with age) have subsequently been demonstrated to also potentially occur under AP (Moorad & Promislow 2009). Here, we aim to summarise studies of wild vertebrates which have attempted to test for $G \times A$ using pedigree-based animal models and character state and/or random-regression approaches (Table 5.1). We consider only studies that have considered a substantial proportion of the lifespan of the organism in question (i.e. at least from birth to prime adulthood).

Table 5.1 outlines the details of the 11 published studies on eight different species we could find that met these criteria. Note that one study analyses data from two different species (Wilson *et al.* 2007a), but several of the studies deal with different traits in the same population (e.g. Soay sheep on St Kilda:

Table 5.1 Studies investigating the quantitative genetics of ageing in wild animals. Trait expression was considered across all or most age classes occurring in the population and the between-individual variance ($I \times A$) and genetic variance ($G \times A$) in the rate of ageing was considered as variance after correcting for fixed-effect differences between individuals. All studies used pedigree-based animal models and 'Method' refers to character state (CS) or RRAM statistical approaches. "V_R structure" refers to whether the models estimated all residual (co)variances among age classes ('Hetero'), assumed a diagonal residual variance structure with respect to age ('Diag') or assumed a constant residual across ages ('Homo'). We report whether a statistical test was reported which supported the presence of $G \times A$ at $p < 0.05$, and what the predicted direction of change in V_A was with age, regardless of statistical significance ('+': increase; '–': decrease, 'NR': not reported). Whether relevant variance components for $I \times A$, $PE \times A$ and $G \times A$ terms in Eq. B5.1.1a and B5.1.1b in the main text were reported in the paper are listed, with reason why not if one was given by author ('N/A': not applicable either because only CS was used or because all residual (co)variances among age classes were estimated and the $PE \times A$ term was never fitted)

Species	Trait	Main method	V_R structure	Significant $G \times A$ (P<0.05)?	Direction of change in V_A	$I \times A$ split into $G \times A$ & $PE \times A$?	Reference
Bighorn sheep (*Ovis canadensis*)	Body mass	CS	Diag	Yes	+	N/A	(Réale & Festa-Bianchet 2000)
Bighorn sheep (*Ovis canadensis*)	Body mass	CS + RRAM	Hetero	Yes	+	N/A	(Wilson *et al.* 2005b)
Mute swan (*Cygnus olor*)	Egg-laying date	CS	Diag	Yes	+	N/A	(Charmantier *et al.* 2006a)
Collared flycatcher (*Ficedula albicollis*)	Annual fitness	RRAM	Homo	No	+	No#	(Brommer *et al.* 2007)
Soay sheep (*Ovis aries*)	Body mass	RRAM	Hetero	Yes	+	N/A	(Wilson *et al.* 2007b)
Soay sheep (*Ovis aries*)	Annual fitness	CS + RRAM	Diag	No	+	No$	(Wilson *et al.* 2007a)
Soay sheep (*Ovis aries*)	Horn growth	RRAM	Hetero	Yes	–	N/A	(Robinson *et al.* 2008)
Red deer (*Cervus elaphus*)	Annual fitness	CS + RRAM	Diag	Yes	+	No$	(Wilson *et al.* 2007a)
Red deer (*Cervus elaphus*)	Offspring birth mass	RRAM	Diag	Yes	+	No$	(Nussey *et al.* 2008b)
Pre-industrial human (*Homo sapiens*)	Fecundity	CS + RRAM	Diag	No	+	No*	(Pettay *et al.* 2008)
Common gull (*Larus canus*)	Annual fitness	RRAM	Diag	No	NR	Yes	(Brommer *et al.* 2010)
Blue-footed booby (*Sula nebouxii*)	Offspring's age at recruitment	RRAM	Diag	Yes	–	No*	(Kim *et al.* 2011)

$: $PE \times A$ dropped from final model; *: no reason given for not including $PE \times A$; #: model including $PE \times A$ did not converge.

Wilson *et al.* 2007a; Wilson *et al.* 2007b; Robinson *et al.* 2008). Two early studies in our list used a character state approach only, but presented evidence that both V_A and heritability of body mass in bighorn sheep (*Ovis canadensis*) and egg-laying date in mute swans (*Cygnus olor*) were higher in the oldest age classes than at other ages (Réale & Festa-Bianchet 2000; Charmantier *et al.* 2006a). The results imply some form of $G \times A$ is present, but note that these studies fitted separate sequential univariate animal models for different age groups, rather than the multivariate character state approach that has been subsequently advocated and applied in the study of genotype-by-environment interactions. The first study, to our knowledge, to apply RRAM to test for $G \times A$ in a wild animal population investigated body mass in bighorn sheep but restricted itself to the first five years of life, which is before senescence is evident in this system (Wilson *et al.* 2005b). This and several subsequent studies adopting a similar RRAM approach have put forward evidence for significant $G \times A$ in a variety of wild vertebrates (Wilson *et al.* 2005b; Wilson *et al.* 2007a; Wilson *et al.* 2007b; Nussey *et al.* 2008b; Robinson *et al.* 2008; Kim *et al.* 2011). In most cases, RRAM predicted an increase in V_A with age (see Table 5.1). To date, only two published studies applying a similar methodology failed to provide evidence for significant $G \times A$ (Brommer *et al.* 2007; Brommer *et al.* 2010), whilst another study presented evidence for an increase in V_A with age in pre-industrial humans but was unable to formally assess the significance of the $G \times A$ effect (Pettay *et al.* 2008). Although this list of studies remains too short to perform a proper meta-analysis, their results seem to point to common $G \times A$, with increasing V_A expressed in old ages.

5.5 Quantitative genetic studies of ageing in wild animals: a critique

As explained in detail in Box 5.1, we would argue that studies of the quantitative genetic basis of ageing using RRAM should begin by testing for and quantifying the patterns of differential rates of ageing among individuals ($I \times A$). If $I \times A$ is significant, it would then subsequently be decomposed into variation associated with additive genetic ($G \times A$) and permanent environmental ($PE \times A$) causes. Five aspects of the studies presented in Table 5.1 hamper our interpretation of similarities and differences in both the presence of $G \times A$ and the patterns of change in genetic covariances with age in wild animals. We discuss these in turn below, but we first begin this critique with a heartfelt admission of *mea culpa*, as the authors of this chapter either lead or were heavily involved in many of the studies listed here. We hope that by highlighting issues evident in these studies and from our own experiences in applying quantitative genetic models to questions related to ageing in the wild, we can identify some rules of best practice and help promote greater consistency in analytical methodology and reporting of results in this important area.

First, despite the use of the same approach (RRAM) in the majority of studies, there remains *little consensus in model fitting procedure*. Astonishingly, only a single study in Table 5.1 actually presents statistical tests explicitly comparing a model similar to that in Eq. B5.1.1a testing $I \times A$, with one similar to Eq. B5.1.1b testing $PE \times A$ and $G \times A$, as well as full details of all estimated variance components and their associated standard errors (Brommer *et al.* 2010). In particular, there is a lack of consensus on how to deal with the major statistical problem of dramatic sample sizes and power reduction associated with older age. This reduction can not only lead to models that do not converge, but if not accounted for appropriately, can also lead to biased estimates of age-specific performance and age-specific genetic variance components (Wilson *et al.* 2008). Some studies, which have focussed on age-related variation through to prime adulthood and thus avoided this power issue, have actually fitted full age-dependent residual variance covariance structures and argued this obviates the need to fit a $PE \times A$ term and allows $G \times A$ to be directly estimated without the prior step (e.g. Wilson *et al.* 2005a; Robinson *et al.* 2008). Other studies, which have been interested in considering the full range of adult ages including those associated with senescence, have simply lacked the sample sizes for later ages to adopt this approach. Instead, they have adopted the approach advocated in Box 5.1 and fitted an age-dependent diagonal residual variance structure which allows

V_R to vary with age but assumes no covariance among ages (Table 5.1). However, in one study of collared flycatchers (*Ficedula albicollis*), models including either $PE \times A$ as well as $G \times A$, or a diagonal V_R structure, would not converge (Brommer *et al.* 2007). In studies of Soay sheep (*Ovis aries*) and red deer (*Cervus elaphus*), the $PE \times$ A term was dropped from the final model as it was found to be non-significant (Wilson *et al.* 2007a; Nussey *et al.* 2008b), which could inflate estimates of $G \times A$ in the same way as not fitting a permanent environment effect will inflate estimates of V_A in a simple animal model (Kruuk & Hadfield 2007). A more recent study of offspring recruitment age as a function of parental age in blue-footed boobies (*Sula nebouxii*) did not fit a $PE \times A$ at the parental level but fitted a RRAM age function to a discretised three-level age term, rather than treating age as continuous (Kim *et al.* 2011). We would advocate retaining $PE \times A$ terms of the same order of any fitted $G \times A$ terms in RRAM models, unless a full, unconstrained residual variance matrix can be fitted with respect to age. This will both provide greater consistency of approach among studies and mean that the proportion of $I \times A$ attributable to PE and A components can be assessed by the reader (Brommer *et al.* 2010).

Second, there is an evident *lack of studies of the heritability of lifespan and age-dependent variation in survival probability* in wild populations. Although two of the studies in Table 5.1 have looked at senescence in annual fitness, survival comprises only part of this annual measure. This lack is certainly down to the major statistical challenges involved. First, one must deal with the dichotomous nature of survival. Although classical survival analysis techniques, such as Cox proportional hazard models, have been adapted to allow pedigree-based estimation of additive genetic variation, these models estimate V_A in lifespan rather than in age-related declines in survival (Ducrocq & Solkner 1998). One possible approach may be to treat survival as a threshold trait (see e.g. Moorad & Promislow 2011), but this has not been attempted in a wild population to our knowledge. A second key issue lies in accounting for imperfect detectability or capture of individuals within quantitative genetics models. Although again pedigree-based estimation of quantitative genetic parameters have recently been implemented within capture-mark-recapture models (Papaïx *et al.* 2010), no one has attempted to capture genetic variation in age-dependent survival probabilities in such a framework. Although these issues can only be dealt in more complicated models, it is essential to understand the quantitative genetics of actuarial senescence in concert with reproductive senescence, especially since it is likely that both are correlated via life-history trade-offs.

Third, the *standard of reporting model statistics is poor*. The only study in Table 5.1 reporting estimates of variance and standard errors for $I \times A$, followed by those for $G \times A$ and $PE \times A$, found that about 20% of the $I \times A$ variance was due to $G \times A$ (Brommer *et al.* 2010). Studies in Soay sheep and red deer reported a lack of significant $PE \times A$ (Wilson *et al.* 2007a; Nussey *et al.* 2008b), which implies that $G \times A$ was a proportionally large part of the overall $I \times A$ in these study systems, but quantitative estimates required to determine this were not included in the publication. We strongly recommend that researchers not fitting a fully age-structured residual variance (as in Wilson *et al.* 2005b), which is actually unlikely to be feasible when considering all adult ages, report all variance components – as well as associated confidence or credible intervals – estimated in models equivalent to those in Eq. B5.1.1a and Eq. B5.1.1b. The aim should be to allow the reader to judge for themselves how the models are decomposing $I \times A$ variance into $G \times A$ and $PE \times A$ components (cf. Brommer *et al.* 2010).

Fourth, as we suggest in Section 5.3, we detect a risk of *over-reliance on RRAM approaches* in the present literature. Although this is a versatile approach, other function-valued trait approaches exist (Pletcher & Geyer 1999) and are being developed (e.g. Ragland & Carter 2004; Izem & Kingsolver 2005), although they remain largely untested within the context of wild populations. RRAM studies in the wild have typically fitted quite low order polynomial functions (linear or quadratic), partly because models with higher-order polynomials did not converge, thereby hindering a full evaluation of their applicability to the data. Caution regarding higher-order polynomial functions is sensible, where concerns about edge effects (e.g. strong changes in V_A at late ages where data are scarce) are particularly warranted. However, linear and quadratic polynomial RRAM functions make quite specific and restrictive assumptions about the form of change in covariances among and with ages, which is not the case for character state approaches. For instance, a linear function assumes a quadratic change in V_A with age, which may not be biologically reasonable. We would strongly encourage exploration of the other function-valued trait approaches (Stinchcombe *et al.* 2012), which might provide useful points of comparison as they make rather different assumptions.

We note that several studies in Table 5.1 have sought to validate results suggesting $G \times A$ from RRAM by running them side-by-side with character state models. Studies considering ages up to prime adulthood have been able to fit full multivariate age-dependent character state models, which provide good support for the patterns of change in V_A with age predicted by RRAM (Wilson *et al.* 2005a; Wilson *et al.* 2007b). In studies considering full adult age ranges, this has typically involved running multivariate models across broad age classes as sample sizes in later ages are too small to support fully age-dependent character state models. In general, where this has been done it provides quantitative support for $G \times A$ and qualitative support to the patterns predicted by RRAM (Wilson *et al.* 2007a). However, the matchup is not always clear cut: in a study of a pre-industrial human population, the standard errors on V_A estimates for two broad age classes were very large, suggesting no significant V_A in either age group, although RRAM suggested an increase in V_A with age (Pettay *et al.* 2008). In a study of blue-footed boobies, character state models also provide little evidence for any significant V_A in any age group, although RRAM suggested a significant decline in V_A with age (Kim *et al.* 2011). In general, backing up results from RRAM with a character state model seems to us a good practice, but authors need to be clearer about where the two sets of results agree and disagree and exactly what any discrepancy means for the interpretation of their results.

Fifth and finally, we are quite concerned about the *lack of recent studies and momentum* in applying quantitative genetic approaches to study ageing in wild populations. Since the initial studies using RRAM emerged in the mid-2000s, the number of new studies has been accumulating at a low rate (Table 5.1). Although these kinds of analyses are technically quite challenging and demand both high-quality longitudinal phenotypic data and high-quality pedigrees from natural systems, there is no doubt there are other wild systems offering opportunities to look for evidence of $I \times A$ and $G \times A$. The worry is that there may be an emerging reporting and publication bias towards reporting only significant $I \times A$ and/or $G \times A$ which may taint the available evidence to date. Alternatively, the demanding requirements on the data and the statistics to analyse them may be hindering further studies. We would urge researchers not to be deterred from undertaking and publishing rigorous studies that fail to detect $I \times A$ or $G \times A$. Ultimately, our ability to address crucial evolutionary questions about the evolution of ageing at the among-species level hinges on the availability of high-quality, consistently and clearly reported results of just this kind. Studies of senescence in different taxa in natural settings are likely to be crucial in addressing questions including: Are there conserved evolutionary and physiological mechanisms underpinning variation in senescence and lifespan? Is ageing rate generally heritable outside the laboratory? And how does life-history and ecology influence the evolution of ageing?

5.6 Looking forward

In this section, we highlight four areas which we feel merit particular attention in future studies of the quantitative genetics of ageing in natural populations.

1) *Statistical power:* The sample sizes and data structures required to detect and reliably estimate $G \times A$ using function-valued approaches is an important and largely unexplored issue in the evolutionary literature (Stinchcombe *et al.* 2012). Power is likely to be mainly determined by the pedigree structure and by the extent of repeated measures of the performance trait over the ages of related individuals, especially where measures made on related animals at old ages will limit the power of the analysis. There is a real possibility that the data structure of wild populations may rarely provide

sufficient statistical power to partition $G \times A$ from non-genetic individual level variation, yet more studies are needed to assess this risk. Our personal experiences suggests that often, whilst models are able to partition $I \times A$ into $G \times A$ and $PE \times A$ (see equations in Box 5.1), doing so with data sets from field studies rarely results in an improvement of model fit, and errors around the $G \times A$ and $PE \times A$ components are often very large indeed. The issue is whether this is because ageing rates are not heritable in many wild systems or because of a lack of statistical power (see discussion on a similar issue in Chapter 15, Gienapp & Brommer). Developing simulation-based power analyses with the aim of finding the minimal effect size of $G \times A$ which can be detected in typical datasets of wild populations is a priority in future research. Such studies have already been conducted at the individual level based on random-regression approaches and suggest very large sample sizes and detailed longitudinal data may be required to detect significant $I \times A$, let alone $G \times A$ (Martin *et al.* 2011; van de Pol 2012). However, the extension of such analyses into quantitative genetic models is likely to be important for understanding the feasibility of addressing the quantitative genetics of ageing in the wild.

2) Multivariate approaches and natural selection on ageing: Most studies to date in the wild have considered the quantitative genetics of ageing on one or two traits. It has become increasingly clear that different traits associated with fitness do not show the same patterns of age-related decline in wild vertebrates (Nussey *et al.* 2009; Evans *et al.* 2011), although whether these divergent patterns of senescence have a genetic basis remains unknown. Furthermore, although some studies have looked at the genetics of ageing in annual fitness measures (Table 5.1), the questions of how ageing rates in different phenotypic traits map to reproductive fitness and/or how age-dependent selection on phenotype might change with age have not been explored in natural populations. Multivariate quantitative genetic models provide a framework to investigate the genetic correlations in ageing patterns among traits as well as exploring age-specific trade-offs among traits and how such trade-offs might change with age. Likewise, a multivariate approach could be used to jointly consider the selective process and the quantitative genetic parameters (Wilson *et al.* 2008). Provided genotypes vary in their rate of senescence (i.e. $G \times A$ is present), including an estimate of lifetime fitness in a multivariate random-regression model provides estimates of selection on reaction-norm elevation and slope (Brommer *et al.* 2012). Such analyses require that selection and inheritance are measured on the same quantities in order to properly account for the selection bias inherent in studies of ageing (Wilson *et al.* 2008). That said, the power issues alluded to above will obviously be exacerbated under a multivariate framework; so it remains to be determined whether analyses of this kind in wild populations are feasible.

3) Sexual selection and senescence: A major avenue of research which has been largely overlooked in natural populations is the interplay between sexual selection and the evolution of senescence. Although theory and empirical evidence in the laboratory point to sex-specific trade-offs and ageing patterns (Promislow 2003; Partridge *et al.* 2005), the role of sexual conflicts in shaping lifespan and ageing is largely unknown (but see e.g. Carranza *et al.* 2004; Penn & Smith 2007; Tuljapurkar *et al.* 2007). As advocated by Bonduriansky and colleagues, further investigation of this interplay would greatly benefit our understanding both of senescence, and of sexual strategies (Bonduriansky *et al.* 2008). Insects such as the seed beetle (*Callosobruchus maculatus*) (Maklakov *et al.* 2009) or the antler fly (*Protopiophila litigata*; Bonduriansky & Brassil 2005) have recently proven excellent model species to test for trade-offs between reproduction and somatic maintenance. Although conducting quantitative genetic analyses on such small and fast-reproducing species in the wild has often been considered problematic, there clearly are some systems which offer good opportunities for field-research on ageing (see Chapter 9, Zajitschek & Bonduriansky). In this context, studies outside the artificial environment of laboratories will provide particularly precious data since it is well known that ageing patterns, natural and sexual selection pressures, and estimates of quantitative genetic parameters are all sensitive to the environment (Service & Rose 1985; Charmantier & Garant 2005; Partridge & Gems 2007).

4) *Genomics:* Relatively inexpensive high-throughput genomic techniques are increasingly being applied to both non-model organisms and field study systems (see Chapter 13, Jensen *et al.*). Although we are not aware of any study specifically looking to map genes associated with variation in lifespan or senescence in a wild vertebrate, this is surely a promising avenue for future research. In model systems of ageing like yeast, nematodes, flies and rodents there is great excitement surrounding the discovery of apparently conserved genes and pathways that appear to regulate lifespan in the laboratory across these diverse taxa (Fontana *et al.* 2010; Partridge 2010). Indeed, genomic research on senescence based on quantitative trait loci (QTL) approaches or genome-wide association (GWA) studies have already been conducted for some time in laboratory model species. These studies have highlighted that many diseases associated with senescence are regulated by multiple interacting QTLs, but they have also identified genetic variants associated with increased longevity in a wide range of taxa (Kenyon 2010). A crucial open question is whether the genes identified in laboratory systems actually show any allelic variation in natural populations and, if they do, how such variation influences life-history traits, lifespan and the onset and rate of senescence. Extending our genomic understanding of ageing from laboratory to field could provide crucial insights into how natural selection shapes and maintains genetic variation underpinning lifespan and ageing processes (Nussey *et al.* 2013).

We have argued here that the challenges facing quantitative geneticists interested in ageing in natural populations are to develop and apply robust methods to determine the degree to which among-individual variation in ageing patterns has an underlying genetic basis, and then to quantify the age-dependent genetic architecture (i.e. age-specific **G**-matrix) of focal traits. This will provide crucial insight into past evolutionary pressures shaping the ageing process, and also the evolutionary potential of traits in the future, as well as the role of life-history trade-offs in the maintenance of genetic variation in natural systems. Admittedly, this requires high-quality, long-term, individual-based studies that are highly demanding and costly to gather. Furthermore, high-quality pedigrees and advanced statistical approaches are needed to obtain insight into the quantitative genetics underlying patterns of ageing. That said, we have shown that potentially powerful conceptual and statistical frameworks are available and increasingly widely used in evolutionary genetics and animal breeding that would allow such analyses to be pursued in the wild should sufficient longitudinal phenotypic and pedigree data be available. We believe that future studies, when taking into account points raised in Section 5.5., will increase our understanding of the quantitative genetics of ageing in the wild. In particular, we hope that higher standards of analysis and reporting should facilitate meta-analysis and thus generalisation across populations and species.

Acknowledgments

We are grateful to Alastair Wilson, Felix Zajitschek, Russell Bonduriansky, Daniel Promislow, Dany Garant and Loeske Kruuk for helpful comments and discussion. AC was funded by an Overseas Fellowship from the Service pour la Science et la Technologie de l'Ambassade de France au Royaume-Uni and the ANR (grant 12-ADAP-0006-02), and she thanks the LARG at the University of Cambridge for hosting her. DHN was supported by a BBSRC David Phillips fellowship. Authors contributed equally to this work and order of authorship was decided at random.

References

Baudisch, A. (2008) *Inevitable aging?: Contributions to evolutionary-demographic theory*. Springer-Verlag, Berlin.

Bonduriansky, R. & Brassil, C.E. (2005) Reproductive ageing and sexual selection on male body size in a wild population of antler flies (*Protopiophila litigata*). *Journal of Evolutionary Biology*, **18**, 1332–1340.

Bonduriansky, R., Maklakov, A., Zajitschek, F. & Brooks, R. (2008) Sexual selection, sexual conflict and the evolution of ageing and life span. *Functional Ecology*, **22**, 443–453.

Bouwhuis, S., Charmantier, A., Verhulst, S. & Sheldon, B.C. (2010) Individual variation in rates of senescence: natal origin effects and disposable soma in a wild bird population. *Journal of Animal Ecology*, **79**, 1251–1261.

Bouwhuis, S., Choquet, R., Sheldon, B.C. & Verhulst, S. (2012) The forms and fitness cost of senescence: age-specific recapture, survival, reproduction, and reproductive value in a wild bird population. *American Naturalist*, **179**, E15–E27.

Brommer, J.E., Kontiainen, P. & Pietiainen, H. (2012) Selection on plasticity of seasonal life-history traits using random regression mixed model analysis. *Ecology and Evolution*, **3**, 695–704.

Brommer, J.E., Rattiste, K. & Wilson, A. (2010) The rate of ageing in a long-lived bird is not heritable. *Heredity*, **104**, 363–370.

Brommer, J.E., Wilson, A.J. & Gustafsson, L. (2007) Exploring the genetics of aging in a wild passerine bird. *American Naturalist*, **170**, 643–650.

Carey, J.R. (2011) Biodemography of the Mediterranean fruit fly: aging, longevity and adaptation in the wild. *Experimental Gerontology*, **46**, 404–411.

Carranza, J., Alarcos, S., Sanchez-Prieto, C.B., Valencia, J. & Mateos, C. (2004) Disposable-soma senescence mediated by sexual selection in an ungulate. *Nature*, **432**, 215–218.

Caswell, H. (2001) *Matrix population models: construction, analysis and interpretation*. Sinauer, Sunderland, MA.

Caughley, G. (1966) Mortality patterns in mammals. *Ecology*, **47**, 906–918.

Charlesworth, B. (1980) *Evolution in age-structured populations*. Cambridge University Press, Cambridge.

Charlesworth, B. (1990) Optimization models, quantitative genetics, and mutation. *Evolution*, **44**, 520–538.

Charlesworth, B. (2001) Patterns of age-specific means and genetic variances of mortality rates predicted by the mutation-accumulation theory of ageing. *Journal of Theoretical Biology*, **210**, 47–65.

Charlesworth, B. & Hughes, K.A. (1996) Age-specific inbreeding depression and components of genetic variance in relation to the evolution of senescence. *Proceedings of the National Academy of Sciences of the United States of America*, **93**, 6140–6145.

Charmantier, A. & Garant, D. (2005) Environmental quality and evolutionary potential: lessons from wild populations. *Proceedings of the Royal Society B-Biological Sciences*, **272**, 1415–1425.

Charmantier, A., Perrins, C., McCleery, R.H. & Sheldon, B.C. (2006a) Age-dependent genetic variance in a life-history trait in the mute swan. *Proceedings of the Royal Society B-Biological Sciences*, **273**, 225–232.

Charmantier, A., Perrins, C., McCleery, R.H. & Sheldon, B.C. (2006b) Quantitative genetics of age at reproduction in wild swans: support for antagonistic pleiotropy models of senescence. *Proceedings of the National Academy of Sciences of the United States of America*, **103**, 6587–6592.

Darwin, C. (1859) *The origin of species*. Murray, London.

Descamps, S., Boutin, S., Berteaux, D. & Gaillard, J.M. (2006) Best squirrels trade a long life for an early reproduction. *Proceedings Of The Royal Society B-Biological Sciences*, **273**, 2369–2374.

Ducrocq, V. & Solkner, J. (1998) *"Survival Kit-V3.0" a package for large analyses of survival data*. University of New England, Armidale, Australia.

Escobar, J.S., Jarne, P., Charmantier, A. & David, P. (2008) Outbreeding alleviates senescence in hermaphroditic snails as expected from the mutation-accumulation theory. *Current Biology*, **18**, 906–910.

Evans, S.R., Gustafsson, L. & Sheldon, B.C. (2011) Divergent patterns of age-dependence in ornamental and reproductive traits in the collared flycatcher. *Evolution*, **65**, 1623–1636.

Fischer, T.M., Gilmour, A.R. & van der Werf, J.H.J. (2004) Computing approximate standard errors for genetic parameters derived from random regression models fitted by average information REML. *Genetics Selection Evolution*, **36**, 363–369.

Fontana, L., Partridge, L. & Longo, V.D. (2010) Extending healthy life span—from yeast to humans. *Science*, **328**, 321–326.

Gaillard, J.-M., Festa-Bianchet, M. & Yoccoz, N.G. (1998) Population dynamics of large herbivores: variable recruitment with constant adult survival. *Trends in Ecology and Evolution*, **13**, 58–63.

Gomulkiewicz, R. & Kirkpatrick, M. (1992) Quantitative genetics and the evolution of reaction norms. *Evolution*, **46**, 390–411.

Hamilton, W.D. (1966) The moulding of senescence by natural selection. *Journal of Theoretical Biology*, **12**, 12–45.

Hayward, A.D., Wilson, A.J., Pilkington, J.G., Pemberton, J.M. & Kruuk, L.E.B. (2009) Ageing in a variable habitat: environmental stress affects senescence in parasite resistance in St Kilda Soay sheep. *Proceedings of the Royal Society B-Biological Sciences*, **276**, 3477–3485.

Henderson, C.R. (1982) Analysis of covariance in the mixed model:higher-level, nonhomogeneous, and random regressions. *Biometrics*, **38**, 623–640.

Izem, R. & Kingsolver, J.G. (2005) Variation in continuous reaction norms: quantifying directions of biological interest. *American Naturalist*, **166**, 277–289.

Kawasaki, N., Brassil, C.E., Brooks, R.C. & Bonduriansky, R. (2008) Environmental effects on the expression of life span and aging: an extreme contrast between wild and captive cohorts of *Telostylinus angusticollis* (Diptera: Neriidae). *American Naturalist*, **172**, 346–357.

Kenyon, C.J. (2010) The genetics of ageing. *Nature*, **467**, 622–622.

Kim, S.Y., Drummond, H., Torres, R. & Velando, A. (2011) Evolvability of an avian life history trait declines with father's age. *Journal of Evolutionary Biology*, **24**, 295–302.

Kirkpatrick, M. & Heckman, N. (1989) A quantitative genetic model for growth, shape, reaction norms, and other infinite-dimensional characters. *Journal of Mathematical Biology*, **27**, 429–450.

Kirkpatrick, M., Lofsvold, D. & Bulmer, M. (1990) Analysis of the inheritance, selection and evolution of growth trajectories. *Genetics*, **124**, 979–993.

Kirkwood, T.B.L. (1977) Evolution of aging. *Nature*, **270**, 301–304.

Kirkwood, T.B.L. & Austad, S.N. (2000) Why do we age? *Nature*, **408**, 233–238.

Kruuk, L.E.B. & Hadfield, J.D. (2007) How to separate genetic and environmental causes of similarity between relatives. *Journal of Evolutionary Biology*, **20**, 1890–1903.

Lynch, M. & Walsh, B. (1998) *Genetics and analysis of quantitative traits*. Sinauer Associates, Inc., Sunderland, Mass.

Maklakov, A.A., Bonduriansky, R. & Brooks, R.C. (2009) Sex differences, sexual selection, and ageing: an experimental evolution approach. *Evolution*, **63**, 2491–2503.

Martin, J.G.A., Nussey, D.H., Wilson, A.J. & Réale, D. (2011) Measuring individual differences in reaction norms in field and experimental studies: a power analysis of random regression models. *Methods in Ecology and Evolution*, **2**, 362–374.

Massot, M., Clobert, J., Montes-Poloni, L., Haussy, C., Cubo, J. & Meylan, S. (2011) An integrative study of ageing in a wild population of common lizards. *Functional Ecology*, **25**, 848–858.

Maynard Smith, J. (1978) Optimization theory in evolution. *Annual Review of Ecology and Systematics*, **9**, 31–56.

Medawar, P.B. (1952) *An unsolved problem of biology*. H.K. Lewis & Co., London.

Meyer, K. (1998) Estimating covariance functions for longitudinal data using a random regression model. *Genetics Selection Evolution*, **30**, 221–240.

Monaghan, P., Charmantier, A., Nussey, D.H. & Ricklefs, R.E. (2008) The evolutionary ecology of senescence. *Functional Ecology*, **22**, 371–378.

Moorad, J.A. & Promislow, D.E.L. (2009) What can genetic variation tell us about the evolution of senescence? *Proceedings of the Royal Society B-Biological Sciences*, **276**, 2271–2278.

Moorad, J.A. & Promislow, D.E.L. (2011) Evolutionary demography and quantitative genetics: age-specific survival as a threshold trait. *Proceedings of the Royal Society B-Biological Sciences*, **278**, 144–151.

Mysterud, A., Yoccoz, N.G., Stenseth, N.C. & Langvatn, R. (2001) Effects of age, sex and density on body weight of Norwegian red deer: evidence of density-dependent senescence. *Proceedings of the Royal Society B-Biological Sciences*, **268**, 911–919.

Nussey, D.H., Coulson, T., Festa-Bianchet, M. & Gaillard, J.M. (2008a) Measuring senescence in wild animal populations: towards a longitudinal approach. *Functional Ecology*, **22**, 393–406.

Nussey, D.H., Froy, H., Lemaitre, J.-F., Gaillard, J.M. & Austad, S.N. (2013) Senescence in natural populations of animals: widespread evidence and its implications for bio-gerontology. *Ageing Research Reviews*, **112**, 214–225.

Nussey, D.H., Kruuk, L.E.B., Donald, A., Fowlie, M. & Clutton-Brock, T.H. (2006) The rate of senescence in maternal performance increases with early-life fecundity in red deer. *Ecology Letters*, **9**, 1342–1350.

Nussey, D.H., Kruuk, L.E.B., Morris, A., Clements, M.N., Pemberton, J.M. & Clutton-Brock, T.H. (2009) Inter- and intra-sexual variation in ageing patterns across reproductive traits in a wild red deer population. *American Naturalist*, **174**, 342–357.

Nussey, D.H., Kruuk, L.E.B., Morris, A. & Clutton-Brock, T.H. (2007) Environmental conditions in early life influence ageing rates in a wild population of red deer. *Current Biology*, **17**, R1000–R1001.

Nussey, D.H., Wilson, A.J. & Brommer, J.E. (2007) The evolutionary ecology of individual phenotypic plasticity in wild populations. *Journal of Evolutionary Biology*, **20**, 831–844.

Nussey, D.H., Wilson, A.J., Morris, A., Pemberton, J., Clutton-Brock, T. & Kruuk, L.E.B. (2008b) Testing for genetic trade-offs between early- and late-life reproduction in a wild red deer population. *Proceedings of the Royal Society B-Biological Sciences*, **275**, 745–750.

Papaïx, J., Cubaynes, S., Buoro, M., Charmantier, A., Perret, P. & Gimenez, O. (2010) Combining capture-recapture data and pedigree information to assess heritability of demographic parameters in the wild. *Journal of Evolutionary Biology*, **23**, 2176–2184.

Partridge, L. (2010) The new biology of ageing. *Philosophical Transactions of the Royal Society of London Series B-Biological Sciences*, **365**, 147–154.

Partridge, L. & Barton, N.H. (1993) Optimality, mutation and the evolution of aging. *Nature*, **362**, 305–311.

Partridge, L. & Gems, D. (2007) Benchmarks for ageing studies. *Nature*, **450**, 165–167.

Partridge, L., Gems, D. & Withers, D.J. (2005) Sex and death: what is the connection? *Cell*, **120**, 461–472.

Penn, D.J. & Smith, K.R. (2007) Differential fitness costs of reproduction between the sexes. *Proceedings of the National Academy of Sciences of the United States of America*, **104**, 553–558.

Pettay, J.E., Charmantier, A., Wilson, A.J. & Lummaa, V. (2008) Age-specific genetic and maternal effects in fecundity of preindustrial Finnish women. *Evolution*, **62**, 2297–2304.

Pletcher, S.D. & Geyer, C.J. (1999) The genetic analysis of age-dependent traits: Modeling the character process. *Genetics*, **153**, 825–835.

Promislow, D. (2003) Mate choice, sexual conflict, and evolution of senescence. *Behavior Genetics*, **33**, 191–201.

Promislow, D.E.L., Fedorka, K.M. & Burger, J.M.S. (2006) Evolutionary biology of aging: future directions. *Handbook of the Biology of Aging* (eds E.J. Masoro & S.N. Austad), pp. 217–242. Academic Press, Burlington, MA.

Ragland, G.J. & Carter, P.A. (2004) Genetic covariance structure of growth in the salamander *Ambystoma macrodactylum*. *Heredity*, **92**, 569–578.

Réale, D. & Festa-Bianchet, M. (2000) Quantitative genetics of life-history traits in a long-lived wild mammal. *Heredity*, **85**, 593–603.

Reed, T., Kruuk, L.E.B., Wanless, S., Frederiksen, M., Cunningham, E.J.M. & Harris, M.P. (2008) Reproductive senescence in a long-lived seabird: rates of decline in late life performance are associated with varying costs of early reproduction. *American Naturalist*, **171**, E89–E101.

Reznick, D.N., Bryant, M.J., Roff, D., Ghalambor, C.K. & Ghalambor, D.E. (2004) Effect of extrinsic mortality on the evolution of senescence in guppies. *Nature*, **431**, 1095–1099.

Robinson, M.R., Pilkington, J.G., Clutton-Brock, T.H., Pemberton, J.M. & Kruuk, L.E.B. (2008) Environmental heterogeneity generates fluctuating selection on a secondary sexual trait. *Current Biology*, **18**, 751–757.

Rose, M.R. (1991) *Evolutionary biology of aging*. Oxford University Press, New York.

Rose, M.R. & Charlesworth, B. (1981) Genetics of life history in *Drosophila melanogaster* .1. Sib analysis of adult females. *Genetics*, **97**, 172–186.

Service, P.M. & Rose, M.R. (1985) Genetic covariation among life history components: the effect of novel environments. *Evolution*, **39**, 943–945.

Snoke, M.S. & Promislow, D.E.L. (2003) Quantitative genetic tests of recent senescence theory: age-specific mortality and male fertility in *Drosophila melanogaster*. *Heredity*, **91**, 546–556.

Stearns, S.C. (1992) *The evolution of life histories*. Oxford University Press, Oxford.

Stinchcombe, J.R., Kirkpatrick, M. & Function-valued Traits Working Group (2012) Genetics and evolution of function-valued traits: understanding environmentally responsive phenotypes. *Trends in Ecology & Evolution*, **27**, 637–647.

Tatar, M., Promislow, D.E.I., Khazaeli, A.A. & Curtsinger, J.W. (1996) Age-specific patterns of genetic variance in *Drosophila melanogaster*. 2. Fecundity and its genetic covariance with age-specific mortality. *Genetics*, **143**, 849–858.

Tuljapurkar, S.D., Puleston, C.O. & Gurven, M.D. (2007) Why Men Matter: Mating patterns drive evolution of human lifespan. *Plos One*, **2**.

van de Pol, M. (2012) Quantifying individual variation in reaction norms: how study design affects the accuracy, precision and power of random regression models. *Methods in Ecology and Evolution*, **3**, 268–280.

Wallace, A.R. (1865) The action of natural selection in producing old age, decay, and death. In: *Essays upon heredity and kindred biological problems* (ed. A. Weismann). Clarendon Press, Oxford.

Weismann, A. (ed.) (1889) *Essays upon heredity and kindred biological problems*. Clarendon Press, Oxford.

Williams, G.C. (1957) Pleiotropy, natural selection, and the evolution of senescence. *Evolution*, **11**, 398–411.

Williams, P.D., Day, T., Fletcher, Q. & Rowe, L. (2006) The shaping of senescence in the wild. *Trends in Ecology & Evolution*, **21**, 458–463.

Wilson, A.J., Charmantier, A. & Hadfield, J.D. (2008) Evolutionary genetics of ageing in the wild: empirical patterns and future perspectives. *Functional Ecology*, **22**, 431–442.

Wilson, A.J., Coltman, D.W., Pemberton, J.M., Overall, A.D.J., Byrne, K.A. & Kruuk, L.E.B. (2005a) Maternal genetic effects set the potential for evolution in a free-living vertebrate population. *Journal of Evolutionary Biology*, **18**, 405–414.

Wilson, A.J., Kruuk, L.E.B. & Coltman, D.W. (2005b) Ontogenetic patterns in heritable variation for body size: using random regression models in a wild ungulate population. *American Naturalist*, **166**, E177–E192.

Wilson, A.J., Nussey, D.H., Pemberton, J.M., Pilkington, J.G., Morris, A., Pelletier, F., Clutton-Brock, T.H. & Kruuk, L.E.B. (2007a) Evidence for a genetic basis of aging in two wild vertebrate populations. *Current Biology*, **17**, 2136–2142.

Wilson, A.J., Pemberton, J.M., Pilkington, J.G., Clutton-Brock, T.H., Coltman, D.W. & Kruuk, L.E.B. (2007b) Quantitative genetics of growth and cryptic evolution of body size in an island population. *Evolutionary Ecology*, **21**, 337–356.

CHAPTER 6

The effects of others' genes: maternal and other indirect genetic effects

Andrew G. McAdam, Dany Garant and Alastair J. Wilson

6.1 Introduction

Evolutionary quantitative genetics seeks to quantify sources of genetic variation in fitness-related traits to assess the potential for traits to respond to future natural selection and to infer how previous patterns of natural selection might have shaped contemporary genetic architecture (Chapter 1, Kruuk *et al.*). Using contemporary patterns of genetic variation to make prospective and retrospective inferences requires that estimates of genetic variation are both unbiased and comprehensive. More specifically, appropriate inferences depend on our ability 1) to quantify sources of genetic variation in a way that is not biased by confounding effects (e.g. of shared environments), and 2) to properly account for all sources of genetic variation that can contribute to evolutionary change.

Social interactions, in which the attributes of an individual are likely to be affected by interactions with conspecifics, are widespread in nature. Often this 'social environment' will play a major role in determining an individual's phenotype. Here we use 'social' to describe any effect on an individual's phenotype that is caused by the attributes of interacting conspecifics. These include effects arising from direct interactions (e.g. parent–offspring interaction, aggression) as well as effects that do not result from direct physical or behavioural interactions (e.g. effects of maternal oviposition location on offspring, intraspecific competition). Effects of the social environment on individual phenotype have the potential to bias our estimates of additive genetic variation. Second, they can also include a hidden genetic component, arising from 'indirect genetic effects', with major consequences for evolutionary trajectories. These two issues have led to increasing interest in trying to quantify the influence of social interactions on genetic variation and to understand its implications for evolutionary dynamics.

In this chapter, we provide a conceptual overview of indirect genetic effects in general and maternal genetic effects as a specific example of them. After outlining their broad importance, we provide more detailed coverage of methodologies that can be used to quantify maternal effects in the wild. We highlight two complementary approaches that have traditionally been used to study maternal effects in the wild as well as a new hybrid approach that might prove useful in some systems. We briefly discuss some challenges associated with the interpretation of empirical estimates of maternal effects, before finally discussing what seem to us to be the most exciting avenues of future research in this field.

We have not attempted to provide a comprehensive review of indirect genetic effects in general or and maternal effects in particular. In places, we refer the reader to other more thorough reviews of certain aspects of this broad topic. However, we hope that

Quantitative Genetics in the Wild. Edited by Anne Charmantier, Dany Garant, and Loeske E. B. Kruuk

our chapter provides a useful introduction to this exciting area of research as well as more detailed descriptions of some of the tools that can be used to study indirect and maternal genetic effects in the wild. While we have tried to make the main messages of this chapter accessible to a broad audience, including students that are new to the field, there are sections that are necessarily technical. We hope that the modular nature of the chapter will allow readers to skip over some of the technical details while still appreciating the general importance of the broader concepts.

6.1.1 Indirect genetic effects

Indirect genetic effects (IGEs) can be defined as occurring any time the phenotype of an individual is causally influenced by the genotype of one or more others (Moore *et al.* 1997). This broad definition encompasses effects that can arise from social interactions of any type in groups of any relatedness structure. It is also worth noting that social effects are not confined to animals. For example, a plant might respond to being shaded by neighbours by increasing its vertical growth. Differences in growth rates among plants might, therefore, be determined in part by the degree of shading they receive from their neighbours. Although shading is experienced by the focal plant as an environmental effect, different neighbours might vary in the amount of shading they cause because of heritable traits such as leaf area. In this way genes expressed by neighbours (e.g. for leaf area) influence phenotypic expression of a focal individual's trait (e.g. growth rate).

Interactions among conspecific individuals are fundamental to the expression of many traits studied by evolutionary ecologists. For example, social dominance, mate choice and cooperation are social traits that can only be observed when individuals interact. However, as the hypothetical example above illustrates, IGEs arising from competition could have impacts on growth, or resource-dependent life-history traits not usually considered 'social traits' (Wilson 2013). Importantly, IGEs can also affect phenotypic evolution, with the magnitude of the effect dependent on the strength of the indirect effect, selection on affected traits, and the relatedness structure among interacting individuals. Under certain circumstances, IGEs may facilitate rapid selection responses, including responses to multilevel (i.e. among-group) selection in kin and non-kin structured groups (Bijma & Wade 2008). In other situations, IGEs may act to constrain evolutionary responses to selection. For example, when individuals compete over a limited resource, IGEs are expected to reduce the amount of genetic variance in resource-dependent traits upon which selection can act (Wolf *et al.* 2008; Hadfield *et al.* 2011).

Although IGEs, or 'associative effects' as they are also known, have long been recognised by quantitative geneticists interested in selection responses in livestock (Griffing 1967; Muir 2005; Bijma *et al.* 2007), empirical studies of IGEs in wild animal populations have been quite limited to date. One exception to this has been in the particular context of maternal effects, which—if they have a genetic basis of variation—represent a specific type of IGE. Since they are the best-known and most widely studied form of IGEs, we focus this chapter principally on maternal effects. However, we note that expanding the scope of empirical research to include other types of IGEs—though certainly challenging in practice—is likely to yield great insights into how traits respond to selection and why they sometimes do not (Wilson 2013). Efforts to do this have now begun, and we return to this theme of non-maternal IGEs later in the chapter.

6.1.2 Maternal effects

In many wild organisms, some of the most important conspecific interactions are those that occur between offspring and their parents. In many taxa (most notably mammals) interactions between mothers and their offspring are particularly important, frequently having widespread consequences for offspring development and fitness (Mousseau & Fox 1998b). Maternal effects[1] have been defined as i) the influence of maternal attributes on offspring phenotypes above and beyond the direct

[1] The predominance of mother–offspring interactions across taxa, and the historical inertia of the term 'maternal effects', have led to its widespread usage to describe parental effects in general.

inheritance of offspring genes from mother to offspring (Mousseau & Fox 1998a), and ii) the causal influence of the maternal genotype or phenotype on the offspring phenotype (Wolf & Wade 2009). Though essentially equivalent, these two definitions differ slightly in their emphasis in a manner that corresponds with two complementary empirical methodologies for investigating maternal effects, which we detail below.

Offspring size provides a good example of a trait that is influenced by maternal effects in many species. In pied flycatchers, for example, offspring size is known to be heritable (Potti & Merino 1994)—offspring size depends on genes inherited from their mother and father. However, mothers also affect offspring size through egg size and lay date. These maternal traits contribute to variation in offspring size, with females that lay larger eggs earlier in the season raising larger offspring (Potti & Merino 1994). Of course parental effects are not limited to those contributed by mothers. In systems where fathers make important contributions to parental care, paternal effects on offspring phenotype may also be present. In fact, paternal energy expenditure also affects the size of nestling pied flycatchers (Moreno *et al.* 1997). This would be an example of a paternal effect.

Maternal effects have been widely quantified in diverse taxa (Roach & Wulff 1987; Mousseau & Dingle 1991; Mousseau & Fox 1998a; Räsänen & Kruuk 2007; Maestripieri & Mateo 2009). The importance of interactions between mothers and their offspring for offspring development means that maternal effects can often be quite large. As might be expected, maternal effects are often strongest on traits expressed early in life (Bernardo 1996a; b; Wilson & Réale 2006; Lindholm *et al.* 2006; Harris & Uller 2009), but they can also persist into adulthood (e.g. Fox *et al.* 2003; Kerr *et al.* 2007).

Maternal effects arise because of variation among females in traits that influence offspring phenotype. This among-female variation itself may arise from environment processes (maternal environmental effects), genetic variation (maternal genetic effects) or interactions between genes and the environment ($G \times E$). For example, in mammals, variation in offspring post-natal growth can arise because of differences in the quantity of milk produced by nursing mothers. Milk production would, therefore, represent a source of maternal effects on offspring growth, but why do females differ? It could be that milk production is affected by the food quality or quantity available to mothers. This would give rise to maternal environmental effects on offspring. However, some variation among females could be due to the underlying genetics of milk production. These genetic effects on milk production represent maternal genetic effects on offspring growth because maternal genes for milk production affect the phenotypes of offspring. Finally, differences among females in how they respond to changes in food availability could be genetically based. That is, some genotypes might increase milk production more for a given rise in food availability. This would represent a genotype-by-environment interaction for milk production, and a maternal-genotype by maternal-environment interaction on offspring growth rates.

Maternal genetic effects occur when the phenotype of one individual (offspring) depends on the genotype of another (the mother). They thus represent a particular example of an IGE. In general, the evolutionary consequences of IGEs on traits under selection will depend on the relatedness structure among interacting individuals (Bijma & Wade 2008). Clearly mothers and their offspring are related, so genes influencing maternal and offspring traits will be present within the same genome. Strong behavioural and physiological interactions between mother and offspring can, therefore, lead to the co-adaptation of maternal and offspring effects (Kölliker *et al.* 2005) with important consequences for evolutionary dynamics (see below).

It is important to remember, however, that the only maternal effects that are IGEs are those attributable to genetic variation among mothers. Maternal-environment effects will not contribute to evolutionary dynamics, but could present us with a misleading view of how much genetic variance is present for offspring phenotypes. Because maternal effects are usually shared by all offspring of a given mother (e.g. maternal milk production affects all offspring; but see Crean & Marshall 2009), they tend to cause phenotypic similarity among siblings that can mistakenly be attributed to additive genetic effects (Kruuk & Hadfield 2007). Initial interest in maternal effects was, therefore, motivated primarily by a desire to control for them as a potential source of upward bias in estimates of additive genetic variance and heritability. However, evolutionary biologists became progressively more interested in maternal effects when it became more widely recognised that maternal effects could themselves be genetically based (Mousseau *et al.* 2009).

6.1.3 Maternal effects and evolutionary dynamics

One way to conceptualise maternal effects is as effects on offspring phenotype that arise from the environment provided by its mother. While we don't normally think of 'environments' evolving, this 'maternal environment' can evolve if it is determined, at least in part, by maternally expressed genes. Thus, if there are maternal genetic effects then the maternal environment provided to offspring can evolve, which will alter our expectations of how offspring traits under selection will change through time (Kirkpatrick & Lande 1989; Wolf *et al.* 1998; 2008). Maternal genetic effects therefore, provide an additional source of genetic variation, which is expected to affect the rate of evolution (Kirkpatrick & Lande 1989).

Maternal genetic effects have been extensively studied in captive animals, in particular domestic ungulates (Wilson & Réale 2006). There is also ample evidence of a genetic basis to many maternal traits that likely have important consequences for offspring phenotypes (reviewed in Räsänen & Kruuk 2007). For example, clutch size and lay date are commonly found to be heritable maternal traits in birds (Sheldon *et al.* 2003; Garant *et al.* 2008) that also affect offspring traits such as body mass (e.g. Wilkin *et al.* 2006). Nevertheless, the genetic basis to variation in maternal traits is rarely studied with respect to its indirect effects on specific offspring traits (Räsänen & Kruuk 2007), so the evolutionary consequences of maternal genetic effects remains largely unexplored. As one exception, McAdam *et al.* (2002) used cross-fostering experiments to quantify the magnitude of maternal effects on offspring growth in red squirrels (*Tamiasciurus hudsonicus*) and inferred their genetic basis from the heritability of specific maternal traits that were found to be important to offspring growth. Similarly, Wilson *et al.* (2005) directly quantified maternal genetic effects on three traits in Soay sheep (*Ovis aries*) using an 'animal model' analysis of a multigenerational pedigree. In both examples, maternal genetic effects substantially increased the evolutionary capacity of the offspring traits that were studied.

In addition to providing an additional source of genetic variation on which selection can act, maternal effects can also introduce an evolutionary time lag where the response to selection in the offspring generation will depend on the selection (strength and direction) acting in the maternal generation (Kirkpatrick & Lande 1989). In this case, genes having an effect on the phenotypes of offspring are not only expressed by a different individual (the mother), but they are being expressed in a different generation, which results in the evolutionary time lag (see Kirkpatrick & Lande 1989 for details). Whether this lagged response further accelerates or decelerates the response to selection will depend on fluctuations in selection across generations. While these predictions have not been widely tested, McAdam and Boutin (2004) tracked changes in offspring growth rates of red squirrels across 12 years and found that maternal genetic effects increased the evolutionary response to current selection, and introduced an evolutionary time lag in which the response to selection in the current generation also depended on the strength of selection in the previous generation.

It is generally expected that genetic variance can change with environmental conditions as a result of $G \times E$ interactions (Charmantier & Garant 2005). In principle, this expectation applies to variance resulting from maternal genetic effects as well as to variance attributable to direct genetic effects (i.e. the influence of an individual's own genotype). If $G_M \times E$ is present then the importance of maternal genetic effects varies with the environment in which mothers and offspring find themselves. The presence of $G_M \times E$ also implies that there is genetic variation among mothers in plasticity for the maternal trait (or traits) that influence offspring phenotype (see Chapter 15, Gienapp & Brommer for discussion of the correspondence between $G \times E$ and genetic variation in plasticity). This is important because

$G_M \times E$ is required for the evolution of transgenerational plasticity.

6.1.4 Maternal effects as adaptive plasticity

The ability of mothers to adjust offspring phenotype so as to match the environmental conditions they will experience at independence is referred to as 'adaptive transgenerational phenotypic plasticity' (e.g. Fox *et al.* 1997; Galloway & Etterson 2007) or 'anticipatory maternal effects' (see Marshall & Uller 2007). For such adaptive plasticity to evolve, there must be genetic variation in maternal plasticity ($G_M \times E$), but $G_M \times E$ has rarely been quantified in the wild to date (see Section 6.4.2). Females must also be able to obtain information about forthcoming local environmental conditions. Additionally, the fitness costs (to the mother) of obtaining this information and plastically adjusting the phenotypic expression of those traits giving rise to maternal effects (e.g. parental care) must be relatively low (see Uller 2008 for an integrative framework for the evolution of such effects). These conditions are most likely to be met if mothers and offspring experience environmental conditions that are correlated and/or if the cues used to predict future conditions are reliable in space or time. While conceptually appealing and supported empirically in a few systems (e.g. Galloway & Etterson 2007; Dantzer *et al.* 2013), Uller *et al.* (2013) found that overall empirical support for adaptive transgenerational plasticity was weak. Adaptive benefits of transgenerational phenotypic plasticity could also be conditional on other variables such as mate quality (Harris & Uller 2009), offspring sex (Badyaev *et al.* 2006), habitat quality (Räsänen *et al.* 2005), population dynamics or a combination of these factors (see Koskela *et al.* 2004 for example). The influence of maternal effects plasticity is thus highly dynamic across time and space.

Although experimental evidence is not overwhelming (Uller *et al.* 2013), studies of breeding phenology in natural populations provide some of the best-documented observational examples of anticipatory maternal effects. Although typically heritable (Sheldon *et al.* 2003; Garant *et al.* 2008), female traits such as laying date also show high levels of phenotypic plasticity, particularly in strongly seasonal environments. For instance, in many passerine birds in seasonal environments laying too early or too late will result in a timing mismatch between the period of peak resource abundance and the period during which mothers (and potentially fathers) are provisioning the young. Such timing mismatches will adversely affect offspring traits such as growth rate or body mass, and ultimately reduce offspring fitness. In great tits (*Parus major*), for example, females that are able to lay their eggs earlier in warmer springs typically enjoy greater offspring survival than others (Nussey *et al.* 2005; but see also Charmantier *et al.* 2008; Husby *et al.* 2010). Plastic adjustments in laying date in response to warmer spring temperature, therefore, have important effects on offspring phenotype and survival given the availability of food resources at later reproductive stages.

Maternal adjustment of dispersal phenotypes is another type of maternal effect that can also affect population dynamics and range expansion. In western bluebirds (*Sialia mexicana*), mothers produce more sons earlier in the egg-laying sequence when resources are scarce (Duckworth 2009). These males are more aggressive and are more prone to disperse and acquire a new territory elsewhere, which is a better strategy when local resources are declining or when population density is high. In contrast, sons produced later in the egg-laying sequence (less aggressive) will be favoured under abundant resources scenarios, as these males are more philopatric. This adjustment of phenotypes presumably arises through manipulation of hormones, although exact mechanisms are still unclear.

6.2 Methods—how are maternal effects measured?

Given the importance of maternal effects, both as a confounding factor when we want to estimate direct genetic variation, and as a potential source of adaptive potential in their own right, it is worth carefully considering how best to identify and quantify them. Below we outline two broad empirical strategies that can be adopted. Which strategy is more appropriate will depend on the goals of the study and on what data are available to a researcher (or can be collected). In what follows, we note that some complexity is unavoidable if we want to provide sufficient detail for a reader to properly implement a given approach. It is our hope, however, that this detail does not impair the non-specialist's ability to appreciate the general approaches that can be implemented.

The two broad strategies for investigating maternal effects on offspring phenotypic traits can be termed 'variance partitioning' and 'trait-based' (see Figure 6.1a and 6.1b, respectively). They follow from the best-known quantitative genetic models of selection response in the presence of maternal (genetic) effects presented by Willham (1963; 1972) and Kirkpatrick and Lande (1989), respectively. These two approaches roughly follow the two conceptual definitions of i) maternal effects as influences of maternal attributes above and beyond direct genetic inheritance (Mousseau & Fox 1998a) vs ii) the causal influence of a maternal genotype or phenotype on the offspring phenotype (Wolf & Wade 2009), as outlined in Section 6.1.2. Recently, Hadfield (2012) compared and critiqued these models and we refer the reader to this work for a detailed treatment of the underlying theory and the models' respective assumptions.

6.2.1 Variance partitioning strategy

Under a variance partitioning strategy, maternal effects are quantified by estimating the amount of variance in an offspring trait that is explained by maternal identity (Figure 6.1A; see Chapter 1, Kruuk *et al.* for a basic description of variance component partitioning). This typically requires that phenotypic data are available from more than one offspring for each mother. Remembering the definitions of maternal effects given above, this maternal variance component (V_M) should be estimated from

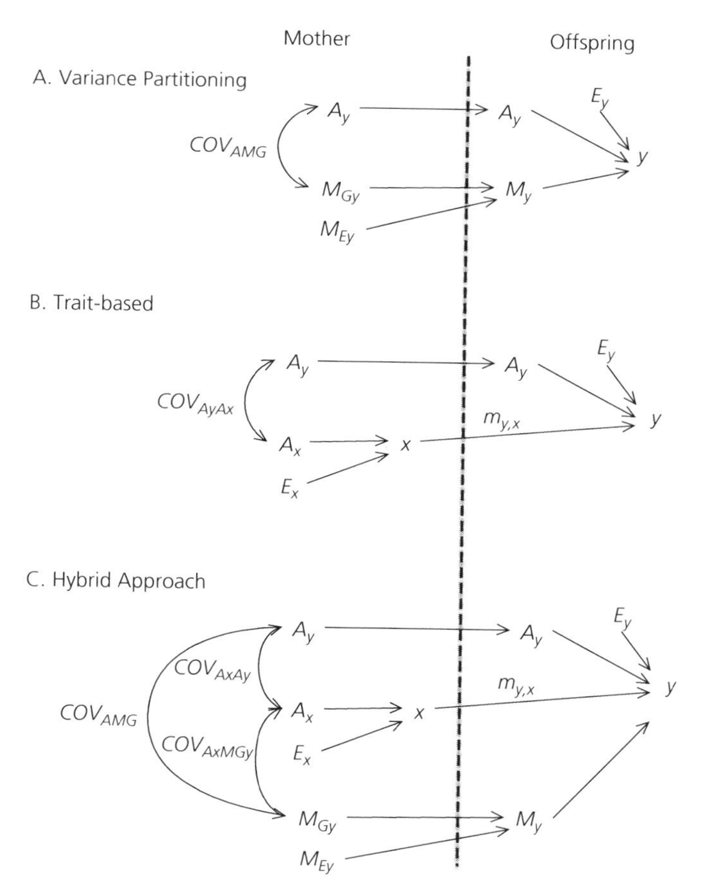

Figure 6.1 Path models for two general approaches to estimating maternal effects that have been used in the past (A and B) and a hybrid approach that we propose here (C). Single-headed arrows represent causal effects while double-headed arrows indicate correlations. The vertical dashed line separates the maternal generation from the offspring generation. In the variance partitioning approach (*A*), variation in some offspring trait, y, is partitioned into environmental (E_y), genetic (A_y) and maternal sources (M_y) without specific reference to any particular maternal trait. This maternal effect (M_y) measures 'maternal performance' for the offspring trait and can be further partitioned into genetic (M_{Gy}) and environmental sources (M_{Ey}). In the trait-based approach (B), the effects of a specific maternal trait (x) or multiple traits is modelled. Here the effect of this maternal trait on the offspring trait is represented by $m_{y,x}$. In the hybrid approach (C), the effects of specific traits can be modelled while still accounting for the remaining variation in maternal performance for the offspring trait.

a statistical model that accounts for any additive inheritance of the offspring trait. Failure to do this might result in the direct additive genetic variance (V_A) causing upward bias of V_M (for exactly the same reasons as a failure to model maternal effects can result in upwardly biased estimates of V_A and h^2). It is worth noting that variance partitioning strategies are usually based on the assumption that attributes of the mother cause her offspring to be more similar phenotypically to one another. These approaches are ill-equipped to assess maternal effects that increase phenotypic variance within a brood (reviewed in Crean & Marshall 2009).

Under a classical full-sib/half-sib breeding design, the presence of maternal effects might be inferred, at least indirectly, from the relative magnitude of sire and dam variance components estimated using a nested ANOVA (Lynch & Walsh 1998). If the latter is significantly larger than the former, it implies the presence of maternal effects (and or dominance effects) in addition to additive inheritance (for an opposite example see Chapter 14, Morrissey *et al.*). However, it is rarely possible to perform such controlled breeding designs in the wild with sufficient sample size to have reasonable power (Merilä & Sheldon 2001), so other approaches based on natural matings are often required for estimating maternal effects in such conditions.

6.2.1.1 Partitioning variance using cross-fostering

In some study systems where offspring are accessible and can be exchanged between nests, cross-fostering experiments have proved a valuable tool to allow statistical separation of maternal variance from other genetic and environmental components (see Roff 1997 Table 7.16; see also Merilä & Sheldon 2001 for examples in birds). For instance, partial reciprocal cross-fostering (see Rutledge *et al.* 1972), where only parts of broods/litters are marked and reciprocally exchanged among females (Figure 6.2), may be used to assess environmental and genotypic influences on traits.

In a reciprocal cross-fostering design, phenotypic variation in an offspring trait, y, can be partitioned using a two-way nested ANOVA represented by the linear model

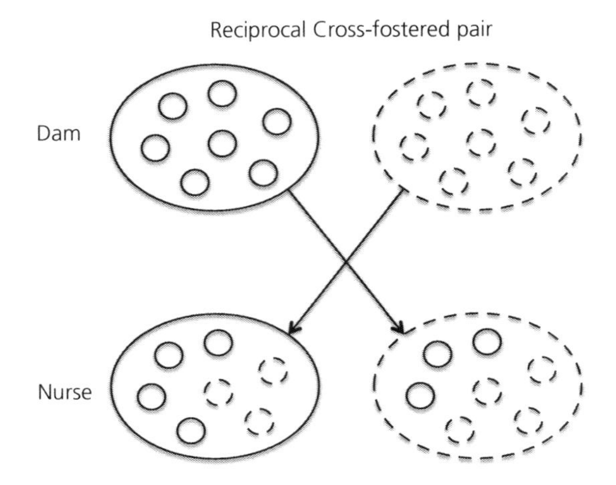

Figure 6.2 In a reciprocal cross-fostering design, approximately half of one brood is exchanged with an equal number of offspring from the paired brood. The manipulation results in some offspring being raised by their genetic dam and others by a foster nurse.

$$y_{hijk} = \mu + P_h + D_{i(h)} + N_{j(h)} + DN_{ij(h)} + e_{hijk} \quad (6.1)$$

where y_{hijk} is the phenotype of the kth individual born to the dam i, but raised by nurse j, in the hth pair of reciprocally crossed litters (Riska *et al.* 1985); μ is the population mean; and $D_{i(h)}$ and $N_{j(h)}$ are the effects of the ith dam and j^{th} nurse within the hth pair, respectively. Variance among dams reflects direct genetic variance (V_A) as well as some dominance variance (Lynch & Walsh 1998). Maternal effects (environmental, genetic and dominance) can be estimated from the among-nurse variance.

In the past, there has been some confusion about how to best estimate the genetic covariance between direct and maternal genetic effects (COV_{AMG}) from this breeding design. Rutledge *et al.* (1972) initially estimated COV_{AMG} directly from the interaction variance in this model (see also Lynch & Walsh 1998). This assumes the interaction variance represents the covariance among full-sibs raised by their own mothers minus the variance among dams and among nurses. This is true in a North Carolina II design (Lynch & Walsh 1998), but in a reciprocal cross-fostering design, offspring raised by an unrelated dam receive maternal effects that are independent from direct genetic effects. As a result, COV_{AMG} is not included in the phenotypic variance in fostered offspring, and the interaction variance has no genetic expectation (Riska *et al.* 1985). COV_{AMG} can instead be estimated from two separate ANOVAs as the difference in the among-litter variance from one model that considers only offspring that were raised by their genetic mother and the among-litter variance from a second model that considers only offspring that were raised by an unrelated dam (Riska *et al.* 1985). In this case COV_{AMG} needs to be added to the other variance components in order to calculate total phenotypic variance. Standard errors for COV_{AMG} can be estimated by jackknifing at the level of the cross-fostered pair of litters (e.g. McAdam *et al.* 2002).

Cross-fostering studies have been conducted in the wild in several birds (see Merilä & Sheldon 2001 for examples) and some mammals (McAdam *et al.* 2002; Mappes & Koskela 2004). Although cross-fostering coupled with an ANOVA analysis can detect maternal effects, it does not readily allow these to be further decomposed into maternal genetic and maternal environmental effects. McAdam et al. (2002) used reciprocal cross-fostering to estimate the magnitude of maternal effects on offspring growth rates in red squirrels and then inferred the genetic basis for these maternal effects by attempting to identify specific maternal traits responsible for 'maternal performance' for offspring growth. In this reciprocal cross-fostering design, maternal performance was measured as the difference between mothers in a cross-fostered pair in the average growth rate of all of their nursed offspring. McAdam *et al.* (2002) correlated differences in 'maternal performance' within cross-fostered pairs with differences in maternal traits (e.g. litter size) within each pair to estimate the relative importance of specific traits to overall maternal performance for offspring growth. They then used the heritability of these specific traits (estimated using an animal model approach; Réale *et al.* 2003) to provide an indirect measure of maternal genetic effects on growth acting through these specific maternal traits. Indirect approaches such as this are the only way to estimate maternal genetic effects from cross-fostering data when the relatedness among mothers is unknown.

Another limitation of cross-fostering experiments is that they cannot control for maternal effects acting prior to cross-fostering. As a result, any maternal effects acting prior to cross-fostering (e.g., differential maternal investment in eggs or prenatal hormone effects in mammals) will still be confounded with direct genetic effects since these will be contained in the effects of the birth mother. As far as we are aware, studies in the wild using embryo transfer experiments (e.g. Moore *et al.* 1970) or similar techniques to disentangle prenatal from post-natal maternal effects have yet to be performed.

6.2.1.2 Partitioning maternal variance using an animal model

Perhaps the most common method for studies of wild animal populations is to include maternal identity as an additional random effect in an animal model. This yields an estimate of the maternal variance component (V_M) that is conditioned on additive inheritance. Often this variance is expressed as a proportion of total phenotypic variance (V_P), which can be interpreted as the repeatability of 'maternal performance' for the offspring trait. For example, Wilson *et al.* (2005) estimated that maternal effects accounted for 23% of the total phenotypic variation in the birth weight of Soay sheep.

Given a pedigree structure with at least three generations' depth, the animal model can be further extended to partition V_M into genetic (V_{MG}) and environment (V_{ME}) components, and to estimate the covariance between direct additive and maternal genetic effects (COV_{AMG}) (see Wilson *et al.* 2009b for practical guidance on how to do this). In their study of the birth weight of Soay sheep, Wilson *et al.* (2005) estimated that 12% of the total phenotypic variation was due to maternal genetic effects, which was greater than direct genetic effects on this offspring trait. At least three generations of pedigree data are required because V_{MG} can only be estimated if individual mothers have (female) relatives in the data set that are also informative for maternal performance. For some pedigree structures, direct genetic effects might still be confounded by maternal effects even when each is modelled explicitly (Kruuk & Hadfield 2007), and parameter bias may also occur in the presence of pedigree errors (Morrissey *et al.* 2007). Ideally, including cross-fostering experiments within a larger pedigree analysis is probably the best way to completely disentangle these two influences on the phenotypes of offspring when maternal effects are strong (Kruuk & Hadfield 2007). Where this is not possible, simulation tools such as the R-package pedantics can usefully be applied to assess whether the structure of available data is likely to allow accurate estimation of both maternal genetic and additive sources of variance (Morrissey *et al.* 2007; Morrissey & Wilson 2010).

6.2.2 Trait-based strategy

An alternative approach is to consider a trait-based strategy following Kirkpatrick and Lande (1989), in which maternal effects are quantified not as the amount of variance explained by maternal identity, but rather as the strength of a (causal) relationship between maternal trait (x) and offspring phenotype (y; Figure 6.1B). The strength of this effect is represented by the maternal effect coefficient, which is commonly denoted m. In the simplest case, where data are available for a set of mothers each with one offspring, then linear regression of y on x could be used to estimate the maternal effect coefficient. In practice more complex statistical models (e.g., mixed effect models) will often be preferred to estimate this regression coefficient because they offer greater flexibility to deal with real-world data structures (e.g. multiple offspring per female, and repeated measures of maternal phenotype). Regardless of the analytical method, estimating m requires that both maternal and offspring phenotypes are measured, and that the maternal trait that causally influences offspring is known (or assumed).

If $m_{y,x}$ (now explicitly defined to be the effect of maternal trait x on offspring trait y) is non-zero, then we can conclude that there are maternal effects on offspring trait y mediated by maternal trait x; and if x is heritable, then these will include a genetic component. For example, experimental manipulations of clutch size in side-blotched lizards (*Uta stansburiana*) revealed a causal effect of maternal clutch size on egg size and offspring size (Sinervo & Licht 1991; Sinervo *et al.* 1992). Clutch size is heritable in side-blotched lizards (Sinervo & McAdam 2008), so there are maternal genetic effects on offspring size acting through maternal clutch size. Note, however, that $m_{y,x} = 0$ does not necessarily mean that there are no maternal effects on y, only that there are none attributable to the maternal trait x.

6.2.3 Which strategy is actually more useful?

For empiricists, a useful feature of the variance partitioning approach is that a researcher can address the question of whether maternal (genetic) effects are an important source of offspring trait variance without knowing (or observing) the maternal trait(s) involved. However, an obvious corollary of this is that variance partitioning is limited precisely because it provides no biological insight into the mechanisms that give rise to the maternal effects. Conversely, a trait-based approach can test mechanism, particularly if experimentation is possible to demonstrate causality, but has higher data requirements in that both maternal and offspring trait phenotypes are needed. The scale of maternal effect coefficients from a trait-based approach will also depend on the scale on which the maternal and offspring trait have been measured. In contrast, the variance partitioning approach generates a dimensionless measure of the importance of maternal effects to phenotypic variation (V_M / V_P) that can be more easily compared across traits and studies.

If the ultimate goal of a maternal effect study is to predict the evolutionary dynamics of phenotypes under natural selection, then there are additional considerations. Following Willham (1963; 1972) the 'total heritability' of an offspring trait h^2_T can be estimated as (V_A + 1.5 $COV_{A,MG}$ + 0.5 V_{MG}) / V_P. For the case of selection on a single (offspring) trait, this total heritability can be estimated using a variance partitioning strategy and used in place of h^2 to predict a selection response by the breeder's equation. However, this prediction is not appropriate if there is selection through maternal fitness—i.e. if maternal effect traits also influence the mother's fitness—a situation that is expected for many scenarios of interest to evolutionary ecologists (Hadfield 2012). For instance, maternal effects on offspring traits or fitness components (e.g. early survival) are expected to arise from variation in parental care. However, if maternal care is costly (Royle *et al.* 2012), then it must be negatively selected via maternal fitness. In the presence of this selection on maternal care through maternal fitness, the dynamics of the offspring trait will not be adequately predicted by the total heritability and selection acting directly on the offspring trait (Hadfield 2012).

Selection on maternal performance (through maternal fitness) can be incorporated either by using extensions of Willham's model as proposed by Cheverud (1984) or by using a trait-based model (Kirkpatrick & Lande 1989). Though providing the most complete and general description of evolutionary dynamics in the presence of maternal genetic effects, the Kirkpatrick and Lande model has not been widely used by empiricists to date (but see McGlothlin & Galloway in press; McAdam & Boutin 2004). This is in part because the model is inherently difficult to understand (but see Hadfield 2012 for an accessible treatment of the primary literature), and in part because the data requirements to parameterise it are very high. In practice, it is extremely difficult to quantify all the parameters in the Kirkpatrick and Lande model for a wild species. Nevertheless, McAdam and Boutin (2004) found empirical support for maternal effect evolution of offspring growth rates in red squirrels, despite making some assumptions about unmeasured parameters in the Kirkpatrick and Lande (1989) model.

6.2.4 A hybrid strategy for empiricists

Variance partitioning and trait-based strategies for measuring maternal effects, therefore, differ in their relative advantages and limitations. If a researcher is interested both in the magnitude of maternal effects on an offspring phenotype and in the pathways by which they arise (i.e. which maternal traits are involved), then a hybrid strategy may prove insightful. There are at least two ways in which this might be achieved.

First, one could expand a univariate animal model of some offspring trait y to include fixed effects of a maternal trait x. For example, we might initially model trait y expressed by individual i with mother j as

$$y_i = \mu + a_i + M_{Gj} + M_{Ej} + \varepsilon_i \qquad (6.2)$$

where μ is the population mean, a_i is the offspring's (direct) genetic merit, M_{Gj} is the maternal genetic effect, M_{Ej} is the maternal (permanent) environment effect, and ε_i is a residual term, assumed to capture unspecified environmental (and non-additive genetic) effects as well as measurement error (please refer to Chapter 1, Kruuk *et al.*, for an introduction to this notation and terminology). Here a_i, M_{Gj}, M_{Ej} and ε_i are assumed to be normally distributed with means of zero and variances, to be estimated as V_A, V_{MG}, V_{ME} and V_R, respectively. As described above (see Section 6.2.1.2), the total phenotypic variance can be partitioned into these components given an appropriate data structure (i.e. ideally with repeated offspring by individual mothers and a pedigree of ≥ 3 generations). It is important to note that the inclusion of M_{Ej} is needed when observations from multiple offspring per female are considered. Otherwise, estimates of V_{MG} will be inflated, just as estimates of direct genetic variance (V_A) are inflated when there are multiple observations per individual unless permanent environmental effects (V_{PE}) are explicitly modelled (Kruuk & Hadfield 2007).

However, if maternal trait x is also measured then the model could be expanded to include the fixed regression of x on y (Figure 6.1C) such that

$$y_i = \mu + m\,x_j + a_i + M_{Gj} + M_{Ej} + \varepsilon_i \qquad (6.3)$$

where m is the maternal effect regression coefficient. Under this formulation the estimates of V_{MG} and V_{ME} should be interpreted as estimates of the maternal variance components conditional on the maternal trait x. If x is the sole source of maternal effects on y then we expect V_{MG} and V_{ME} to collapse to zero in this second model. Conversely, if the regression coefficient m was non-zero but maternal variance components also remained significant, we might conclude that additional maternal effects also influence y. Multiple regression could be used to test a suite of maternal traits, assessing the importance of each from its corresponding partial regression coefficient, and from the reduction in maternal variance obtained by its inclusion in the model.

A second approach would be to build a bivariate model of x and y in which we specify the traits expressed by individual i with mother j as:

$$x_i = \mu_x + a_{xi} + pe_{xi} + \varepsilon_{xi}$$
$$y_i = \mu_y + a_{yi} + M_{Gyj} + M_{Eyj} + \varepsilon_{yi} \quad (6.4)$$

where pe_{xi} is the permanent environment effect on maternal trait x and all other effects are as defined previously (relating to trait x or y as denoted by the corresponding subscript). By formulating a bivariate model, we can estimate a genetic variance–covariance matrix (**G**) for x and y that includes the contributions of both direct additive and maternal genetic effects. Here we have direct additive effects on both traits with maternal effects on y only such that **G** can be specified and estimated as

$$\mathbf{G} = \begin{bmatrix} V_{A_x} & COV_{A_xA_y} & COV_{A_xMG_y} \\ COV_{A_xA_y} & V_{A_y} & COV_{A_yMG_y} \\ COV_{A_xMG_y} & COV_{A_yMG_y} & V_{MG_y} \end{bmatrix} \quad (6.5)$$

In this matrix, the variance terms (on the diagonal) are as defined above (note that V_{MGy} refers to the maternal genetic variance on trait y) while the three covariance terms are worth explaining. The first, $COV_{A_xA_y}$, is simply the direct additive covariance between the two traits, which when standardised would represent the genetic correlation between the two traits. The third, $COV_{A_yMG_y}$, is the (within-individual) genetic covariance between trait y and maternal performance on trait y. This is the covariance term that appears in Willham's model as described earlier (referred to as COV_{AMG} above). However, the second term, $COV_{A_xMG_y}$, is also of particular interest as it describes the relationship between genetic merit for trait x and genetic merit for maternal performance on trait y. Provided that the relationship between maternal x and offspring y is indeed causal (and quite probably even if it is not), then the sign of $COV_{A_xMG_y}$ will reflect that of the phenotypic regression coefficient m.

In principle, if genetic variation among mothers at trait x accounted for all maternal genetic effects on y, then COV_{AxMG_y} would scale to a genetic correlation of +1 or −1 (depending on the sign of m). By the same logic, if x is the source of all maternal effects, we would also predict a within-individual correlation between pe_{xi} and M_{Eyj} of unity. In practice, the uncertainty around any correlation estimate may limit the utility of this as a null model against which to test for additional unrecognised maternal effect pathways. However, at least in a qualitative sense, a slightly more nuanced picture may be obtained by estimating the proportion of V_{MGy} that is genetically independent of trait x. This can be determined as

$$1 - V_{MGy:x} / V_{MG} \quad (6.6)$$

where $V_{MGy:x}$ is the maternal genetic variance in y conditional on x (which can be obtained from **G** following the methods described in Hansen *et al.* 2003). As with the first hybrid approach described, extending to a set of n maternal traits to evaluate their relative contributions to maternal effects on y would be straightforward in principle. In practice, the limitation is likely to be the ability to parameterise a **G**-matrix of dimension $n + 2$.

6.3 Interpretation

Given sufficient data and appropriate analyses, there still remains the problem of drawing appropriate conclusions from studies of maternal effects. The complexity of concepts and terminology can sometimes lead to confusion regarding the interpretation of results. Here we highlight and try to clarify two such issues that arise in studies of maternal effects.

6.3.1 Maternal effects vs brood effects

Depending on the particular data collected and the method of analysis employed, the biological interpretation of what maternal effects are and what might cause them can differ widely. For example, in a cross-fostering experiment, maternal effects estimated under a variance partitioning approach account for similarity in phenotype among offspring raised in a common brood. Brood effects need not be heritable or even repeatable at the level of individual females. In fact some of the similarity among offspring within a brood might be due to common environmental factors that are not causally dependent on the mother at all (Kruuk & Hadfield 2007). Animal model analyses that estimate maternal effects by fitting a random effect for maternal identity include only those maternal influences that are repeatable across her lifetime. This latter approach is perhaps justifiable from an evolutionary perspective, but will clearly miss some important sources of variation. For instance, female body condition may sometimes be an important driver of maternal effects, but could vary a lot among reproductive episodes and hence among broods within any given female (thus having a low repeatability) (Kruuk & Hadfield 2007). Inconsistent brood effects within a female's lifetime might be particularly important when breeding events occur across large fluctuations in environmental conditions. Both a brood effect and a maternal identity effect could be fitted within the same model but require observations of multiple offspring per brood and multiple broods within a female's lifetime. These demands on the data structure will place limitations on the wild systems for which both of these sources of variation can be estimated.

6.3.2 What are maternal correlations among traits?

Maternal effects may impact multiple aspects of offspring phenotype. Consequently, they can be an important source of correlation structure among offspring traits. For example, hormonal maternal effects in side-blotched lizards adaptively integrate specific throat colours, dorsal back patterning and antipredatory escape behaviours (Lancaster *et al.* 2007; Lancaster *et al.* 2010). Such maternal effects on trait combinations can also be revealed through multivariate models. For example, in addition to estimating genetic correlations between traits, a bivariate animal model expanded to include a random effect of maternal identity estimates a maternal variance for each trait and the covariance (and hence correlation) between traits that is explained by maternal identity. Using this approach, Taylor *et al.* (2012) found evidence of both a genetic correlation and a maternal effect correlation between red squirrel activity and aggression. This means that some maternal attribute simultaneously affected both the activity and aggression of offspring. These findings also suggest that correlations between offspring traits can result from either pleiotropic direct genetic effects or maternal effects, which could result from physiological trade-offs or the integrating effects of some maternal trait, such as maternal hormones.

6.4 Future Directions

While the evolutionary significance of maternal genetic effects and other types of IGEs is well established in theory, there is much empirical work that needs to be done to understand just how important these effects of genes in the environment are for evolution in nature. In this final section, we outline what we believe are the principal and most exciting avenues of future research that will provide this understanding.

6.4.1 Maternal genetic effects in the wild

Central to the appreciation of maternal effects as important contributors to evolutionary dynamics was the recognition that, although maternal effects are experienced by offspring as an environmental effect, they are caused by both environmental and genetic sources of variation, like any other trait. Yet, unlike most other traits, our exploration of the quantitative genetics of maternal effects on offspring phenotype in the wild remains in its infancy. Quantifying parameters that are commonplace for other traits is still rarely accomplished for maternal effects, so we have few empirical examples, which makes general conclusions challenging. For example, what is the relative importance of environmental and genetic contributions to individual variation in maternal effects (i.e. what is their heritability)? From a variance partitioning perspective, the magnitudes of maternal effects have frequently been quantified, but this has been done largely to control for this potential bias in estimates of direct genetic variance. More rarely have maternal genetic effects been directly quantified in the wild (e.g. Wilson *et al.* 2005). In contrast, from the trait-based perspective, the quantitative genetics of important maternal traits have frequently been investigated in the wild (Räsänen & Kruuk 2007), but this is usually done because of an interest in these maternal traits themselves rather than because of their effects on offspring traits of interest. As a result, maternal effect coefficients, which provide a crucial link between maternal and offspring traits in trait-based models of maternal effect evolution, are often measured separately from quantitative genetic investigations of maternal traits or are often not measured at all. As a result, we still know remarkably little about the genetic basis to maternal effects in the wild, despite the widespread importance of phenotypic maternal effects and the important evolutionary consequences of maternal effects if they are genetically based.

6.4.2 Maternal effect plasticity

It is now commonplace to quantify individual reaction norms and their genetic basis in wild populations (Nussey *et al.* 2007), yet we know relatively little about how maternal effects respond to environmental gradients ($M \times E$). This maternal effect plasticity lies at the heart of the concept of adaptive transgenerational plasticity. However, in order for such maternal effect plasticity to evolve it must be genetically based ($M_G \times E$). There are remarkably few studies that have quantified the genetic basis to maternal effect plasticity (but see Wilson *et al.* 2006). There have also been many studies of $G \times E$ in maternal traits (see Husby *et al.* 2010 for example), but as above these are often done in isolation from the offspring traits that are affected by the maternal traits being studied. Quantifying $M_G \times E$ in a range of systems, particularly in response to environmental gradients that are known to correlate with natural selection (MacColl 2011), will be particularly important for advancing our understanding of the evolution of transgenerational plasticity. Similarly maternal effects might exhibit important changes with age ($M \times A$). Given the potential importance of maternal effects for offspring fitness, changes in maternal effects with age could be an important component of senescence in the wild (Nussey *et al.* 2008).

6.4.3 Integrating insights from captivity

Our understanding of maternal effects in natural populations could benefit from integrating knowledge obtained from previous research performed in experimental conditions in the laboratory (see Fox *et al.* 1997; Pakkasmaa *et al.* 2003 for examples). In particular, much has been learned about the stability/variability of maternal effects over time and across different environments by using controlled laboratory conditions under which the replication of populations is easier and the power to separate maternal and environmental effects is greater. For instance, Plaistow and Benton (2009) used soil mites (*Sancassani berlesei*) as a model species to assess population level consequences of maternal effects under different environmental contexts. They generated 20 'populations' in the lab derived from young or old mothers and studied them at high or low density. They showed a context-dependent influence of maternal effects: the provisioning of eggs by young and old females caused differences in offspring life-history (greater variation in egg size) at low density (high resources) that affected transient population dynamics. These effects were not observed at high density under which competition was high and egg-size variation constrained, which reduced the maternal effect on population dynamics. Their results imply that in fluctuating populations, there might be delayed density dependence that does not always translate into persistent population cycles.

A possible compromise between laboratory studies performed in controlled but often ecologically unrealistic environments and natural environments, in which populations are difficult to replicate and follow accurately, is the use of semi-natural environments. This would allow some of the conditions to be controlled while others are left naturally variable. For example, Venturelli *et al.* (2010) used naturalised pond experiments to study three experimental populations of walleye (*Sander vitreus*), each with 25 unique families. They showed that juvenile survival in walleye was affected by maternal effects through egg size (greater in older, larger females), which translated into large changes in population maximum reproductive rate. Interestingly, maximum reproductive rate was twice as high when older females were abundant as compared to when they were rare. These results further improve our fundamental understanding of maternal effects, but could also help improve management strategies of exploited populations.

6.4.4 Genomic approaches

With the advent of genomic methods and the increased availability of numerous molecular markers for several biological systems, genomic mapping approaches are now being applied to the detection of maternal effects. For example, Wolf *et al.* (2011) used genotype information from parents and offspring in combination with cross-fostering in an experimental population of mice to detect quantitative trait loci (QTL) linked to maternal effects. They managed to separate prenatal and post-natal maternal genetic effects when identifying QTL affecting offspring body weight and weight gain. More recently, Wolf and Cheverud (2012) developed a conceptual framework they called 'statistical cross-fostering', again based on marker information from parents and offspring. Interestingly, this approach allows the separation of direct and maternal effects but without experimental manipulations. The fundamental problem to be solved by each approach is that direct and maternal genetic effects on offspring are confounded. Whereas the physical moving of offspring between nests in a cross-fostering experiment breaks apart these associations, Wolf and Cheverud (2012) proposed using marker data to restrict the statistical analysis to only those individuals for which these effects are not confounded at a particular locus. For example, heterozygous females produce offspring with different genotypes. Direct genetic effects on offspring phenotype can, therefore, be estimated free from confounding maternal genetic effects by comparing the phenotypes of offspring with different genotypes at a given locus, but which were all raised by mothers with the same heterozygous genotype at that locus. Similarly, offspring that are heterozygous are genetically identical at that locus, but can be produced by mothers with different maternal genotypes. Maternal genetic effects of that locus can, therefore, be quantified by measuring phenotypic differences among offspring that are all heterozygous, but which were born to mothers of differing genotypes (see Wolf & Cheverud 2012 for more details). Up until now, these approaches have been largely applied to model species in controlled environments but application and development of functional genomics tools and databases mean that methods for wild populations are now emerging (Ekblom & Galindo 2011; Pavey *et al.* 2012).

6.4.5 Maternal genetic effects on offspring fitness

Given that traits must genetically covary with fitness to evolve (Price 1970), there has been a call for more estimates of genetic variance in fitness as a prerequisite to adaptive evolution (Morrissey *et al.* 2010). While there are some empirical estimates of the heritability of fitness in the wild, no study has yet to quantify maternal genetic effects on fitness. Maternal effects on fitness have previously been quantified (Kruuk *et al.* 2000; Foerster *et al.* 2007; Schroeder *et al.* 2012), but their genetic basis has not yet been estimated. Maternal genetic effects are a potentially significant reservoir of genetic variance in fitness because they are assumed to be sheltered from the erosive effects of selection while carried by males (Wade 1998). This is likely a reasonable assumption although we are not aware of an empirical test of this prediction. It is possible that maternal genetic effects for fitness might experience some selection in males (despite males not expressing maternal phenotypes) if they have pleiotropic effects on other male attributes (e.g., Sinervo & McAdam 2008). When quantifying maternal genetic effects on fitness, appropriate definitions of fitness become critical (Wolf & Wade 2001). For example, if there are maternal genetic effects on early offspring survival, then these would mistakenly appear as direct genetic effects on fitness if early offspring survival were assigned to the fitness of the mother (e.g. number of offspring surviving to one year of age).

6.4.6 Non-maternal IGEs

As noted earlier, maternal genetic effects are just one type—albeit the best studied type—of IGEs. IGEs can occur whenever individuals within a population interact in ways that influence their phenotypic expression (Moore *et al.* 1997). As with maternal genetic effects, the evolutionary importance of IGEs stems from the fact that they represent an additional, but often unrecognised, source of genetic variation that will contribute to selection responses. In recent years, there has been rapid development of both quantitative genetic theory and practical modelling approaches that means studying non-maternal IGEs is becoming more feasible for empiricists (Kölliker *et al.* 2005; Bijma & Wade 2008; Bailey 2012).

Just as for maternal genetic effects, models may be formulated under trait-based (Moore *et al.* 1997) or variance partitioning (Bijma *et al.* 2007) approaches, although 'translation' between these views is possible (McGlothlin & Brodie 2009) and hybrid strategies (such as those outlined above) may also be useful to empiricists (Wilson 2013). In trait-based models, the effect of a (heritable) trait expressed by a social partner on some aspect of focal phenotype is captured by a coefficient usually denoted as Ψ. This can be thought of as a generalised version of the maternal effect coefficient m, which is described above. Although normally treated as a static property of a population, there is also evidence that Ψ itself can be genetically variable (e.g., among strains in captive bred guppies, Bleakley & Brodie 2009) and can evolve in response to selection acting on it (Chenoweth *et al.* 2010). Under a variance partitioning approach, the trait(s) expressed by social partners that influence focal phenotype are not known, but the effect of IGEs on focal traits are nonetheless detectable from a suitable data structure. Following earlier work (Griffing 1967; 1976; Muir 2005), Bijma *et al.* (2007) presented a model in which a trait (y) expressed by a focal individual (i) in a group of n interacting conspecifics (j) depends on the phenotypic mean (μ), the direct additive effect of its own genes (a_D) plus the sum of (indirect) phenotypic effects from its social partners (p_{IND}) and a residual term (ε). The indirect effects of social partner phenotypes on focal expression of y can themselves be decomposed into genetic (a_{IND}) and environmental (e_{IND}) terms such that

$$\begin{aligned} y_i &= \mu + a_{Di} + \sum_{i \neq j}^{n} p_{INDj} + \varepsilon_i \\ &= \mu + a_{Di} + \sum_{i \neq j}^{n} a_{INDj} + \sum_{i \neq j}^{n} e_{INDj} + \varepsilon_i \end{aligned} \tag{6.7}$$

This equation makes clear that an individual's phenotype is determined by the sum of all indirect effects from interacting conspecifics. Instead of measuring the indirect effects of conspecifics on a focal individual's phenotype, it is possible to instead measure the overall indirect effects of a focal individual on all conspecifics. The variance among individuals in their influence on the population phenotypic mean through both direct and indirect genetic contributions is referred to as the 'total breeding value' and is given as

$$\sigma^2_{TBV} = \sigma^2_{A(D)} + 2(n-1)\sigma_{A(D,IND)} + (n-1)^2\sigma^2_{A(IND)} \tag{6.8}$$

where the direct ($\sigma^2_{A(D)}$) and indirect ($\sigma^2_{A(IND)}$) genetic variances, as well as the direct–indirect covariance term ($\sigma_{A(D,IND)}$) can be empirically estimated using simple extensions of the animal model. This requires that the pedigree structure for the population of interest spans groups (i.e. individuals in each group must have relatives in other groups). It is important to highlight that σ^2_{TBV} is not a straight replacement for the (direct) additive variance in predictive models of selection response (e.g. the breeders' equation). This is because the extent to which the indirect genetic variance and the direct–indirect genetic covariance terms contribute to selection response depends on both the relatedness within groups and the extent of multilevel (i.e. among-group) selection (see Bijma *et al.* 2007). A full discussion of multilevel selection models and the implications of IGEs for selection responses is beyond the scope of the present chapter, and we refer the reader elsewhere for accessible treatments of the literature (Bijma & Wade 2008; Wilson 2013).

Although empirical work has been limited to date, recent studies of captive animal populations (e.g., Wilson *et al.* 2009a; Chenoweth *et al.* 2010) and wild vertebrate systems (e.g. Brommer & Rattiste 2008; Teplitsky *et al.* 2010; Wilson *et al.* 2011) suggest that a wider recognition of IGEs should provide major insights. For example, Brommer and Rattiste (2008) showed that while lay date in common gulls (*Larus canus*) is a heritable female trait, it is also subject to an indirect genetic effect from the male partner in a pair, and there is a negative correlation between direct (female) and indirect (male) genetic effects. This negative cross-sex correlation is expected to limit the response to strong selection for earlier laying in this species (Brommer & Rattiste

2008). A role for IGEs as a source of (absolute) evolutionary constraint was also demonstrated in a study of dominance in red deer (*Cervus elaphus*; Wilson *et al.* 2011). In general, it is expected that when individuals compete over limited resources, any IGEs arising from those interactions will act so as to reduce the potential of resource-dependent traits (e.g. growth, fecundity) to respond to directional selection on them (Wilson 2013). Nevertheless, there are many other scenarios and types of interaction in which IGEs are expected to arise. In at least some cases, including the evolution of cooperative behaviours, theory predicts that IGEs could facilitate faster selection responses than anticipated from considering direct genetic effects only. The technical and practical challenges of applying IGE models more widely in natural populations need to be addressed (Wilson 2013), but there is no doubt that detecting and quantifying non-maternal IGEs will enhance our understanding of social evolution in the broadest sense.

6.5 Conclusions

Despite substantial growth in interest in maternal effects since the landmark contributions of Kirkpatrick and Lande (1989), Bernardo (1996b) and Mousseau and Fox (1998a), there remain many important, but as yet rarely explored, research directions with respect to their quantitative genetics in the wild. Given their ubiquity and possible implications for evolutionary dynamics, maternal effects could have a potentially large role in feedbacks between ecology and evolution (Pelletier *et al.* 2009). Understanding the importance of maternal effects to evolutionary dynamics, however, requires an understanding of their genetic basis. We still have much to learn about patterns of maternal genetic variation and covariation—within and among traits. In particular, we note that there is a great need for empirical studies that explore the possibility of interactions between maternal genetic effects and environmental gradients. Such interactions would provide the basis for the adaptive evolution of transgenerational phenotypic plasticity. We also need to know more about the extent to which maternal genetic effects are an important source of variation in, and covariation between, maternal and offspring fitness. Finally, while there is still much to learn about maternal effects, the last few years has seen rapid development of more general IGE models. Hopefully this will encourage empiricists to broaden their focus and use quantitative genetic approaches to explore the evolutionary implications of social interactions across a wide range of contexts and scenarios.

Acknowledgments

We thank the editors for the invitation to write this chapter. Gustavo Betini, Phillip Gienapp, Loeske Kruuk and an anonymous reviewer provided helpful comments on an earlier version of this chapter. McAdam and Garant were supported by the Natural Sciences and Engineering Research Council of Canada (NSERC) and Wilson by a BBSRC David Phillips fellowship during the writing of this chapter.

References

Badyaev, A.V., Acevedo Seaman, D., Navara, K.J., Hill, G.E. & Mendonca, M.T. (2006) Evolution of sex-biased maternal effects in birds: III. Adjustment of ovulation order can enable sex-specific allocation of hormones, carotenoids, and vitamins. *Journal of Evolutionary Biology*, **19**, 1044–1057.

Bailey, N.W. (2012) Evolutionary models of extended phenotypes. *Trends in Ecology & Evolution*, **27**, 561–569.

Bernardo, J. (1996a) The particular maternal effect of propagule size, especially egg size: patterns, models, quality of evidence and interpretations. *American Zoologist*, **36**, 216–236.

Bernardo, J. (1996b) Maternal effects in animal ecology. *Integrative and Comparative Biology*, **36**, 83–105.

Bijma, P. & Wade, M.J. (2008) The joint effects of kin, multilevel selection and indirect genetic effects on response to genetic selection. *Journal of Evolutionary Biology*, **21**, 1175–1188.

Bijma, P., Muir, W.M. & Van Arendonk, J.A.M. (2007) Multilevel selection 1: Quantitative genetics of inheritance and response to selection. *Genetics*, **175**, 277–288.

Bleakley, B.H. & Brodie, E.D. (2009) Indirect genetic effects influence antipredator behavior in guppies: estimates of the coefficient of interaction psi and the inheritance of reciprocity. *Evolution*, **63**, 1796–1806.

Brommer, J.E. & Rattiste, K. (2008) 'Hidden' reproductive conflict between mates in a wild bird population. *Evolution*, **62**, 2326–2333.

Charmantier, A. & Garant, D. (2005) Environmental quality and evolutionary potential: Lessons from wild populations. *Proceedings of the Royal Society B-Biological Sciences*, **272**, 1415–1425.

Charmantier, A., McCleery, R.H., Cole, L.R., Perrins, C., Kruuk, L.E.B. & Sheldon, B.C. (2008) Adaptive phenotypic plasticity in response to climate change in a wild bird population. *Science*, **320**, 800–803.

Chenoweth, S.F., Rundle, H.D. & Blows, M.W. (2010) Experimental evidence for the evolution of indirect genetic effects: changes in the interaction effect coefficient, psi, due to sexual selection. *Evolution*, **64**, 1849–1856.

Cheverud, J.M. (1984) Evolution by kin selection: a quantitative genetic model illustrated by maternal performance in mice. *Evolution*, **38**, 766–777.

Crean, A.J. & Marshall, D.J. (2009) Coping with environmental uncertainty: Dynamic bet hedging as a maternal effect. *Philosophical Transactions of the Royal Society B-Biological Sciences*, **364**, 1087–1096.

Dantzer, B., Newman, A.E.M., Boonstra, R., Palme, R., Boutin, S., Humphries, M.M. & McAdam, A.G. (2013) Density triggers maternal hormones that increase adaptive offspring growth in a wild mammal. *Science*, **340**, 1215–1217.

Duckworth, R.A. (2009) Maternal effects and range expansion: A key factor in a dynamic process? *Philosophical Transactions of the Royal Society B-Biological Sciences*, **364**, 1075–1086.

Ekblom, R. & Galindo, J. (2011) Applications of next generation sequencing in molecular ecology of non-model organisms. *Heredity*, **107**, 1–15.

Foerster, K., Coulson, T., Sheldon, B.C., Pemberton, J.M., Clutton-Brock, T.H. & Kruuk, L.E.B. (2007) Sexually antagonistic genetic variation for fitness in red deer. *Nature*, **447**, 1107–1110.

Fox, C.W., Bush, M.L. & Wallin, W.G. (2003) Maternal age affects offspring lifespan of the seed beetle, *Callosobruchus maculatus*. *Functional Ecology*, **17**, 811–820.

Fox, C.W., Thakar, M.S. & Mousseau, T.A. (1997) Egg size plasticity in a seed beetle: an adaptive maternal effect. *American Naturalist*, **149**, 149–163.

Galloway, L.F. & Etterson, J.R. (2007) Transgenerational plasticity is adaptive in the wild. *Science*, **318**, 1134–1136.

Garant, D., Hadfield, J.D., Kruuk, L.E.B. & Sheldon, B.C. (2008) Stability of genetic variance and covariance for reproductive characters in the face of climate change in a wild bird population. *Molecular Ecology*, **17**, 179–188.

Griffing, B. (1967) Selection in reference to biological groups. I. Individual and group selection applied to populations of unordered groups. *Australian Journal of Biological Sciences*, **20**, 127–139.

Griffing, B. (1976) Selection in reference to biological groups.VI. Use of extreme forms of nonrandom groups to increase selection efficiency. *Genetics*, **82**, 723–731.

Hadfield, J.D. (2012) The quantitative genetic theory of parental effects. In: *The evolution of parental care* (ed. N.J. Royle, P.T. Smiseth & M. Kölliker), p. 267. Oxford University Press, USA.

Hadfield, J.D., Wilson, A.J. & Kruuk, L.E.B. (2011) Cryptic evolution: does environmental deterioration have a genetic basis? *Genetics*, **187**, 1099–1113.

Hansen, T.F., Pélabon, C., Armbruster, W.S. & Carlson, M.L. (2003) Evolvability and genetic constraint in *Dalechampia* blossoms: components of variance and measures of evolvability. *Journal of Evolutionary Biology*, **16**, 754–766.

Harris, W.E. & Uller, T. (2009) Reproductive investment when mate quality varies: Differential allocation versus reproductive compensation. *Philosophical Transactions of the Royal Society B-Biological Sciences*, **364**, 1039–1048.

Husby, A., Nussey, D.H., Visser, M.E., Wilson, A.J., Sheldon, B.C. & Kruuk, L.E.B. (2010) Contrasting patterns of phenotypic plasticity in reproductive traits in two great tit (*Parus major*) populations. *Evolution*, **64**, 2221–2237.

Kerr, T.D., Boutin, S., LaMontagne, J.M., McAdam, A.G. & Humphries, M.M. (2007) Persistent maternal effects on juvenile survival in North American red squirrels. *Biology Letters*, **3**, 289–291.

Kirkpatrick, M. & Lande, R. (1989) The evolution of maternal characters. *Evolution*, **43**, 485–503.

Koskela, E., Huitu, O., Koivula, M., Korpimäki, E. & Mappes, T. (2004) Sex-biased maternal investment in voles: importance of environmental conditions. *Proceedings of the Royal Society B-Biological Sciences*, **271**, 1385–1391.

Kölliker, M., Brodie, E.D., III & Moore, A.J. (2005) The coadaptation of parental supply and offspring demand. *The American Naturalist*, **166**, 506–516.

Kruuk, L.E.B. & Hadfield, J.D. (2007) How to separate genetic and environmental causes of similarity between relatives. *Journal of Evolutionary Biology*, **20**, 1890–1903.

Kruuk, L.E.B., Clutton-Brock, T., Slate, J. & Pemberton, J.M. (2000) Heritability of fitness in a wild mammal population. *Proceedings of the National Academy of Sciences of the United States of America*, **97**, 698–703.

Lancaster, L.T., McAdam, A.G. & Sinervo, B. (2010) Maternal adjustment of egg size organizes alternative escape behaviors, promoting adaptive phenotypic integration. *Evolution*, **64**, 1607–1621.

Lancaster, L.T., McAdam, A.G., Wingfield, J.C. & Sinervo, B.R. (2007) Adaptive social and maternal induction of antipredator dorsal patterns in a lizard with alternative social strategies. *Ecology Letters*, **10**, 798–808.

Lindholm, A.K., Hunt, J. & Brooks, R. (2006) Where do all the maternal effects go? Variation in offspring body size through ontogeny in the live-bearing fish *Poecilia parae*. *Biology Letters*, **2**, 586–589.

Lynch, M. & Walsh, B. (1998) *Genetics and analysis of quantitative traits*. Sinauer Associates, Inc., Sunderland, MA.

MacColl, A.D.C. (2011) The ecological causes of evolution. *Trends in Ecology & Evolution*, **26**, 514–522.

Maestripieri, D. & Mateo, J.M. (eds). (2009) *Maternal effects in mammals*. University of Chicago Press, Chicago.

Mappes, T. & Koskela, E. (2004) Genetic basis of the trade-off between offspring number and quality in the bank vole. *Evolution*, **58**, 645–650.

Marshall, D.J. & Uller, T. (2007) When is a maternal effect adaptive? *Oikos*, **116**, 1957–1963.

McAdam, A.G. & Boutin, S. (2004) Maternal effects and the response to selection in red squirrels. *Proceedings of the Royal Society B-Biological Sciences*, **271**, 75–79.

McAdam, A.G., Boutin, S., Réale, D. & Berteaux, D. (2002) Maternal effects and the potential for evolution in a natural population of animals. *Evolution*, **56**, 846–851.

McGlothlin, J.W. & Brodie, E.D., III. (2009) How to measure indirect genetic effects: the congruence of trait-based and variance-partitioning approaches. *Evolution*, **63**, 1785–1795.

McGlothlin, J.W. & Galloway, L.F. *in press*. The contribution of maternal effects to selection response: an empirical test of competing models. *Evolution*. doi:10.1111/evo.12235

Merilä, J. & Sheldon, B.C. (2001) Avian quantitative genetics. *Current Ornithology*, **16**, 179–255.

Moore, A.J., Brodie, E.D., III & Wolf, J.B. (1997) Interacting phenotypes and the evolutionary process: I. Direct and indirect genetic effects of social interactions. *Evolution*, **51**, 1352–1362.

Moore, R.W., Eisen, E.J. & Ulberg, L.C. (1970) Prenatal and postnatal maternal influences on growth in mice selected for body weight. *Genetics*, **64**, 59–68.

Moreno, J., Potti, J. & Merino, S. (1997) Parental energy expenditure and offspring size in the pied flycatcher *Ficedula hypoleuca*. *Oikos*, **79**, 559–567.

Morrissey, M.B. & Wilson, A.J. (2010) pedantics: an r package for pedigree-based genetic simulation and pedigree manipulation, characterization and viewing. *Molecular Ecology Resources*, **10**, 711–719.

Morrissey, M.B., Kruuk, L.E.B. & Wilson, A.J. (2010) The danger of applying the breeder's equation in observational studies of natural populations. *Journal of Evolutionary Biology*, **23**, 2277–2288.

Morrissey, M.B., Wilson, A.J., Pemberton, J.M. & Ferguson, M.M. (2007) A framework for power and sensitivity analyses for quantitative genetic studies of natural populations, and case studies in Soay sheep (*Ovis aries*). *Journal of Evolutionary Biology*, **20**, 2309–2321.

Mousseau, T.A. & Dingle, H. (1991) Maternal effects in insect life histories. *Annual Review of Entomology*, **36**, 511–534.

Mousseau, T.A. & Fox, C.W. (1998a) *Maternal effects as adaptations*. Oxford University Press, USA.

Mousseau, T.A. & Fox, C.W. (1998b) The adaptive significance of maternal effects. *Trends in Ecology & Evolution*, **13**, 403–407.

Mousseau, T.A., Uller, T., Wapstra, E. & Badyaev, A.V. (2009) Evolution of maternal effects: past and present. *Philosophical Transactions of the Royal Society B-Biological Sciences*, **364**, 1035–1038.

Muir, W.M. (2005) Incorporation of competitive effects in forest tree or animal breeding programs. *Genetics*, **170**, 1247–1259.

Nussey, D.H., Coulson, T., Festa-Bianchet, M. & Gaillard, J.M. (2008) Measuring senescence in wild animal populations: towards a longitudinal approach. *Functional Ecology*, **22**, 393–406.

Nussey, D.H., Postma, E., Gienapp, P. & Visser, M.E. (2005) Selection on heritable phenotypic plasticity in a wild bird population. *Science*, **310**, 304–306.

Nussey, D.H., Wilson, A.J. & Brommer, J.E. (2007) The evolutionary ecology of individual phenotypic plasticity in wild populations. *Journal of Evolutionary Biology*, **20**, 831–844.

Pakkasmaa, S., Merilä, J. & O'hara, R. (2003) Genetic and maternal effect influences on viability of common frog tadpoles under different environmental conditions. *Heredity*, **91**, 117–124.

Pavey, S.A., Bernatchez, L., Aubin-Horth, N. & Landry, C.R. (2012) What is needed for next-generation ecological and evolutionary genomics? *Trends in Ecology & Evolution*, **27**, 673–676.

Pelletier, F., Garant, D. & Hendry, A.P. (2009) Eco-evolutionary dynamics. *Philosophical Transactions of the Royal Society Of London Series B-Biological Sciences*, **364**, 1483–1489.

Plaistow, S.J. & Benton, T.G. (2009) The influence of context-dependent maternal effects on population dynamics: an experimental test. *Philosophical Transactions of the Royal Society B-Biological Sciences*, **364**, 1049–1058.

Potti, J. & Merino, S. (1994) Heritability estimates and maternal effects on tarsus length in pied flycatchers, *Ficedula hypoleuca*. *Oecologia*, **100**, 331–338.

Price, G. (1970) Selection and covariance. *Nature*, **227**, 520–521.

Räsänen, K. & Kruuk, L.E.B. (2007) Maternal effects and evolution at ecological time-scales. *Functional Ecology*, **21**, 408–421.

Räsänen, K., Laurila, A. & Merilä, J. (2005) Maternal investment in egg size: environment- and population-specific effects on offspring performance. *Oecologia*, **142**, 546–553.

Réale, D., Berteaux, D., McAdam, A.G. & Boutin, S. (2003) Lifetime selection on heritable life-history traits in a natural population of red squirrels. *Evolution*, **57**, 2416–2423.

Riska, B., Rutledge, J. & Atchley, W. (1985) Covariance between direct and maternal genetic effects in mice, with a model of persistent environmental influences. *Genetical Research*, **45**, 287–297.

Roach, D. & Wulff, R. (1987) Maternal effects in plants. *Annual Review of Ecology and Systematics*, **18**, 209–235.

Roff, D.A. (1997) *Evolutionary quantitative genetics*. Chapman and Hall, New York.

Royle, N.J., Smiseth, P.T. & Kölliker, M. (2012) *The evolution of parental care*. Oxford University Press, USA.

Rutledge, J., Robinson, O., Eisen, E. & Legates, J. (1972) Dynamics of genetics and maternal effects in mice. *Journal of Animal Science*, **35**, 911–918.

Schroeder, J., Burke, T., Mannarelli, M.-E., Dawson, D.A. & Nakagawa, S. (2012) Maternal effects and heritability of annual productivity. *Journal of Evolutionary Biology*, **25**, 149–156.

Sheldon, B.C., Kruuk, L.E.B. & Merilä, J. (2003) Natural selection and inheritance of breeding time and clutch size in the collared flycatcher. *Evolution*, **57**, 406–420.

Sinervo, B. & Licht, P. (1991) Proximate constraints on the evolution of egg size, number, and total clutch mass in lizards. *Science*, **252**, 1300–1302.

Sinervo, B. & McAdam, A.G. (2008) Maturational costs of reproduction due to clutch size and ontogenetic conflict as revealed in the invisible fraction. *Proceedings of the Royal Society B-Biological Sciences*, **275**, 629–638.

Sinervo, B., Doughty, P., Huey, R. & Zamudio, K. (1992) Allometric engineering: a causal analysis of natural selection on offspring size. *Science*, **258**, 1927–1930.

Taylor, R.W., Boon, A.K., Dantzer, B., Réale, D., Humphries, M.M., Boutin, S., Gorrell, J.C., Coltman, D.W. & McAdam, A.G. (2012) Low heritabilities, but genetic and maternal correlations between red squirrel behaviours. *Journal of Evolutionary Biology*, **25**, 614–624.

Teplitsky, C., Mills, J.A., Yarrall, J.W. & Merilä, J. (2010) Indirect genetic effects in a sex-limited trait: the case of breeding time in red-billed gulls. *Journal of Evolutionary Biology*, **23**, 935–944.

Uller, T. (2008) Developmental plasticity and the evolution of parental effects. *Trends in Ecology & Evolution*, **23**, 432–438.

Uller, T., Nakagawa, S. & English, S. (2013) Weak evidence for anticipatory parental effects in plants and animals. *Journal of Evolutionary Biology*, **26**, 2161–2170.

Venturelli, P.A., Murphy, C.A., Shuter, B.J., Johnston, T.A., van Coeverden de Groot, P.J., Boag, P.T., Casselman, J.M., Montgomerie, R., Wiegand, M.D. & Leggett, W.C. (2010) Maternal influences on population dynamics: evidence from an exploited freshwater fish. *Ecology*, **91**, 2003–2012.

Wade, M.J. (1998) The evolutionary genetics of maternal effects. In: *Maternal effects as adaptations* (ed. T.A. Mousseau & C.W. Fox). Oxford University Press, USA.

Wilkin, T.A., Garant, D., Gosler, A.G. & Sheldon, B.C. (2006) Density effects on life history traits in a wild population of the great tit *Parus major*: analyses of long term data with GIS techniques. *Journal of Animal Ecology*, **75**, 604–615.

Willham, R.L. (1963) The covariance between relatives for characters composed of components contributed by related individuals. *Biometrics*, **19**, 18–27.

Willham, R.L. (1972) The role of maternal effects in animal breeding: III. Biometrical aspects of maternal effects in animals. *Journal of Animal Science*, **35**, 1288–1293.

Wilson, A.J. (2013) Competition as a source of constraint on life history evolution in natural populations. *Heredity*, **112**, 70–78.

Wilson, A.J. & Réale, D. (2006) Ontogeny of additive and maternal genetic effects: lessons from domestic mammals. *American Naturalist*, **167**, E23–E38.

Wilson, A.J., Coltman, D.W., Pemberton, J.M., Overall, A.D.J., Byrne, K.A. & Kruuk, L.E.B. (2005) Maternal genetic effects set the potential for evolution in a free-living vertebrate population. *Journal of Evolutionary Biology*, **18**, 405–414.

Wilson, A.J., Gelin, U., Perron, M.C. & Réale, D. (2009a) Indirect genetic effects and the evolution of aggression in a vertebrate system. *Proceedings of the Royal Society B-Biological Sciences*, **276**, 533–541.

Wilson, A.J., Morrissey, M.B., Adams, M.J., Walling, C.A., Guinness, F.E., Pemberton, J.M., Clutton-Brock, T.H. & Kruuk, L.E.B. (2011) Indirect genetics effects and evolutionary constraint: an analysis of social dominance in red deer, *Cervus elaphus*. *Journal of Evolutionary Biology*, **24**, 772–783.

Wilson, A.J., Pemberton, J.M., Pilkington, J.G., Coltman, D.W., Mifsud, D.V., Clutton-Brock, T.H. & Kruuk, L.E.B. (2006) Environmental coupling of selection and heritability limits evolution. *PLoS biology*, **4**, 1270–1275.

Wilson, A.J., Réale, D., Clements, M.N., Morrissey, M.B., Postma, E., Walling, C.A., Kruuk, L.E.B. & Nussey, D.H. (2009b) An ecologist's guide to the animal model. *Journal of Animal Ecology*, **79**, 13–26.

Wolf, J. & Cheverud, J.M. (2012) Detecting maternal-effect loci by statistical cross-fostering. *Genetics*, **191**, 261–277.

Wolf, J.B. & Wade, M.J. (2001) On the assignment of fitness to parents and offspring: whose fitness is it and when does it matter? *Journal of Evolutionary Biology*, **14**, 347–356.

Wolf, J.B. & Wade, M.J. (2009) What are maternal effects (and what are they not)? *Philosophical Transactions of the Royal Society B-Biological Sciences*, **364**, 1107–1115.

Wolf, J.B., Brodie, E.D., III, Cheverud, J.M., Moore, A.J. & Wade, M.J. (1998) Evolutionary consequences of indirect genetic effects. *Trends in Ecology & Evolution*, **13**, 64–69.

Wolf, J.B., Harris, W.E. & Royle, N.J. (2008) The capture of heritable variation for genetic quality through social competition. *Genetica*, **134**, 89–97.

Wolf, J.B., Leamy, L.J., Roseman, C.C. & Cheverud, J.M. (2011) Disentangling prenatal and postnatal maternal genetic effects reveals persistent prenatal effects on offspring growth in mice. *Genetics*, **189**, 1069–1082.

CHAPTER 7

Dominance genetic variance and inbreeding in natural populations

Matthew E. Wolak and Lukas F. Keller

7.1 Introduction

Phenotypic variation of traits studied in wild populations is the result of many different genes acting in concert, often in combination with environmental variation introduced during development and growth (Boag and van Noordwijk 1987). Following this logic, the basic quantitative genetic variance partitioning framework breaks down the phenotypic distribution of a trait in a population (i.e. phenotypic variance, σ_P^2) into contributions from additive genetic (σ_A^2), non-additive genetic, and environmental (σ_E^2) variances. Non-additive genetic variance can be further partitioned into sources of variance from dominance (σ_D^2) and epistatic (σ_I^2) allelic interactions. The following equation summarises the basic components of phenotypic variance:

$$\sigma_P^2 = \sigma_A^2 + \sigma_D^2 + \sigma_I^2 + \sigma_E^2 \qquad (7.1)$$

(Falconer & Mackay 1996, p.122). Long-term studies of natural populations have allowed numerous estimations of the additive genetic component in this equation for a range of characters (see Chapter 2, Postma; Kruuk *et al.* 2008). In contrast, little is known about the relative contribution of the other components to the phenotypic variance in wild populations.

The variance partitioning framework also yields insights into how a trait will respond to selection. Specifically, it allows the derivation of a prediction of the change in mean phenotype for the generation after selection. This is modelled using the breeder's equation,

$$R = h^2 S = (\sigma_A^2 / \sigma_P^2) S \qquad (7.2)$$

a fundamental equation in quantitative genetics (Roff 1997; for limitations in wild populations see Morrissey *et al.* 2010). The predicted response to selection from one generation to the next, R, is a function of the heritability, h^2 (defined as the ratio: σ_A^2/σ_P^2), of a particular trait and the strength of selection, S. Since all components on the right-hand side of Eqn. 7.1 contribute to phenotypic variance (denominator of the heritability in Eqn. 7.2), they also factor into the estimation of R in Eqn. 7.2 (Roff 2006). Beyond additive genetic variance, the importance of other genetic variance components (e.g. dominance and epistasis) for predicting the evolution of adaptive traits is generally not studied in wild populations. The lack of non-additive genetic variance estimates from studies of wild populations is a consequence of both the difficulty empiricists have encountered to estimate non-additive variances and the long-standing practice of assuming non-additive genetic effects are unimportant to the overall trajectory of trait evolution.

The assumption that non-additive dynamics are unimportant stems from Fisher's (1958) argument that non-additive genetic effects would be irrelevant in large populations because these effects depend on the genetic background in which they

Quantitative Genetics in the Wild. Edited by Anne Charmantier, Dany Garant, and Loeske E. B. Kruuk

are expressed. If every possible background is present in a population, the non-additive genetic effects average out to zero. This happens as population size approaches infinity, an assumption of Fisher's work. The genetic background also determines allele frequencies which, in turn, affect the magnitude of non-additive genetic variance components. Hill *et al.* (2008) show, for example, that extreme allele frequencies drive the dominance variance quite low (dominance variance is a function of both the dominance genetic effect of alleles at a locus and their frequencies in the population; see Section 7.2). Their theoretical, biallelic models show that only when allele frequencies are near 0.5 should we expect to see dominance variance contribute most, or an equivalent contribution with additive variance, to the overall genetic variance. Selection for alleles with dominance effects, however, tends to drive the frequency in the population to an extreme value (i.e. close to one or close to zero).

Contrary to Fisher, Wright considered non-additive effects an important factor in the genetic basis of evolutionary change, because they have the potential to cause rapid phenotypic shifts through the origin of adaptive novelties in small populations experiencing drift (see Wade & Goodnight 1998). The phenotypic consequences of an allele with non-additive effects will depend on the other allele at the same locus (in the case of dominance) or the genotype at other loci (in the case of epistasis). For example, consider the task of predicting the phenotypic effects of allele A1 at a particular locus. This effect changes (i.e. cannot be predicted from knowledge of A1 alone) if the genotype at the locus is A1A1 or A1A2 (in the case of dominance). Note that under purely additive gene action we *can* make a statement predicting the phenotypic effect of the A1 allele without reference to the second allele at a locus (see Section 7.2). Similarly, the phenotypic contribution of the A1 allele can also change if the genotype at another locus is B1B1, B1B2, or B2B2 (in the case of epistasis). When population sizes are small and allele frequencies change by genetic drift, only a small number of possible genetic backgrounds will be present in a population, and sampling error will govern the distribution of genetic backgrounds, which will increase the impact of alleles with non-additive effects. Although the rate of adaptive evolution still depends upon the amount of additive genetic variance, dominance and epistatic variances play important roles in the potential for local adaptation and in structuring the adaptive landscapes in a population (Whitlock *et al.* 1995; Fenster *et al.* 1997).

Three general observations indicate that a closer look into the contribution of non-additive components to evolutionary dynamics is warranted. The first reason is purely methodological, yet has great consequences for studying adaptive evolution in wild populations. If dominance genetic effects are present, but not included in a statistical model of phenotypic variation, their exclusion can potentially bias the prediction of the additive genetic variance (Lynch & Walsh 1998; Waldmann *et al.* 2008; but see Misztal *et al.* 1997). For example, exclusion of dominance variance from analyses of phenotypic data inflated estimates of additive genetic variance in one case (e.g. Ovaskainen *et al.* 2008) and overestimated selection responses in another (e.g., Shaw *et al.* 1998). Secondly, populations in the wild often violate the conditions under which Fisher envisaged adaptive evolution to proceed. Population genetic studies over the past few decades show species are more structured than previously thought (Fenster *et al.* 1997) and effective population sizes are often quite small (e.g. for around 50% of taxa $N_e < 100$; Roff 1997, Fig. 8.3 and Table 8.2). Furthermore, many wild populations currently under study have relatively small population sizes. For example, the song sparrow (*Melospiza melodia*) population inhabiting Mandarte Island, Canada, which has been studied since 1975 (Smith *et al.* 2006), has an average of 35 ± 3 breeding pairs (Reid 2007). Similarly, the unmanaged red deer (*Cervus elaphus*) population on the Isle of Rum, UK, that has been under intensive study since 1971 has had a resident female population size of approximately 190 since 1980 (Walling *et al.* 2010). Of course, these two examples are hand-picked and are unlikely to accurately reflect the actual distribution of population sizes in nature; but they indicate that several of the most intensively studied populations are quite small. In addition to population size considerations, reported cases of adaptive evolution appear to be most often associated with colonisation events or drastic changes in the environment (Merilä *et al.* 2001; Reznick & Ghalambor 2001); although this commonality could arise simply from publication bias alone (Reznick & Ghalambor 2001). Colonisation and changes in the environment provide opportunities for population growth and genetic drift, which alter allele frequencies and therefore the non-additive variance in a population. Finally, non-additive genetic variance plays a key role in the theory behind several evolutionary phenomena: how selection shapes the additive and non-additive variances underlying different trait types (Crnokrak & Roff 1995; Merilä & Sheldon 1999), the evolution of sexual dimorphisms (Fairbairn & Roff 2006) and sexually selected traits (Merilä & Sheldon 1999), the genetic consequences and adaptive potential of bottlenecked populations (Taft & Roff 2012), and the reduction in population trait means and genetic variances as a consequence of inbreeding (Charlesworth & Charlesworth 1995; van Buskirk & Willi 2006; Charlesworth & Willis 2009; Taft & Roff 2012). Specifically, inbreeding depression, the reduction in trait means due to inbreeding, is a consequence of non-additive gene action (see Section 7.5) and both inbreeding and inbreeding depression are present in many wild populations (Crnokrak & Roff 1999; Keller 1998). Inbreeding and its consequences have long been recognised as important to the evolutionary process (e.g. Darwin 1876), and its study has the potential to yield key insights into many areas of evolutionary ecology, such as disease resistance and persistence of small populations (e.g. Altizer *et al.* 2003), distinguishing between alternative genetic causes of senescence (Charlesworth & Hughes 1996), and the evolution of polyandry (e.g. Reid *et al.* 2011).

One example in which non-additive dynamics may advance our understanding of the evolutionary process concerns the evolution of female preferences for elaborate male secondary sexual characteristics (Chapter 3, Reid). Sexual selection theory has long stated that females that express directional mating preferences for the most elaborate males will secure the best genes for their offspring, thereby increasing the female's own fitness (Andersson 1994). Paradoxically, however, continued directional selection is expected to deplete additive genetic variance in the elaborate male trait, eventually nullifying the benefits achieved by female choice (Kirkpatrick & Ryan 1991). However, if we consider that the phenotypic effect of an allele with non-additive effects will depend on the genetic background in which it is expressed, then the location of a female's offspring on the adaptive landscape depends greatly upon the female and her mate's specific combination of alleles with non-additive effects. This suggests that the optimal male for a female can vary from one individual to the next (Merilä & Sheldon 1999), leading to the maintenance of genetic variance. This process can occur if a population experiences some inbreeding and the population is structured. Inbreeding depression for the secondary sexual trait lowers fitness, while population structure will cause individuals to vary in their relatedness with potential mates, resulting in more inbred individuals being more closely related to the set of potential mates (reviewed in Reid 2007). The take-home message here is that female preference for male traits conferring non-additive benefits to offspring greatly expands our understanding of the factors driving the evolution of secondary sexual traits and mating systems (Merilä & Sheldon 1999; Reid 2007).

In this chapter we compile and summarise the main issues, previously presented elsewhere in parts, concerned with estimating dominance genetic variance. Estimating dominance variance in wild populations is often tricky and complex, but we do not feel it is impossible. The obstacles to estimating epistatic variances using wild populations include all those for dominance variance and more (for overviews see Whitlock *et al.* 1995; Fenster *et al.* 1997; Hill *et al.* 2008). Thus, to simplify what follows, while remaining within the limits of what can be achieved in studies of wild populations, we concentrate on dominance (however, we do note the importance of epistasis in evolutionary theory concerning the maintenance of sex, the evolution of selfing, phenotypic plasticity, developmental homeostasis, founder effects, and of course the 'Fisher vs Wright debate' (reviewed in Coyne *et al.* 1997; Fenster *et al.* 1997; Wade & Goodnight 1998)). We first define dominance variance and illustrate its relationship with additive variance. We then explore the practical considerations for estimating dominance variance in wild populations and, in the absence of estimates of dominance variance from wild populations (except humans), we review estimates of dominance variances from laboratory and agricultural populations to determine the extent to which dominance variance makes a substantial contribution to overall genetic and phenotypic variation. We finish the chapter discussing how inbreeding affects additive and dominance genetic variance components within a population, a situation that is likely to be common in wild populations.

7.2 Dominance variance defined

The quantitative genetic definition of dominance variance is very different from the typical notion of dominance, commonly encountered in population genetics or when discussing the physiological result of gene action. In the following, we refer to the latter as a *dominance genetic effect* (i.e. the non-additive interaction of alleles at the same locus), whereas the term *dominance deviation* refers to a population-wide characteristic that is a component of a trait's genetic variance. The distinction is analogous to that between physiological epistasis and statistical epistasis (Phillips 1998). Throughout this chapter, we carefully distinguish *dominance genetic effect* as a separate concept from *dominance deviation*. To illustrate this difference, we will start by defining additive genetic effects and additive genetic variance and then move on to include dominance. We are purposefully brief with the mathematical details below and refer readers to Chapter 7 of Falconer and Mackay (1996) for more details.

In Figure 7.1a, we illustrate the simple scenario of a single locus with purely additive gene action and two alleles at a fixed frequency in the population: A1 ($p = 0.25$) and A2 ($q = 0.75$). We assign the following arbitrary genotypic values to the three possible genotypes: the A1A1 homozygote has a genotypic value of $-a$, the A2A2 homozygote has a genotypic value of $+a$, and the A1A2 heterozygote has a genotypic value of 0 (e.g. the filled circles in Figure 7.1a). Given these genotypic values, we can calculate the mean value in the population as the sum of the genotypic values multiplied by the genotype frequencies (Eqn. 7.2 in Falconer & Mackay 1996). In our purely additive example, we calculate the mean genotypic value in this population as $0.5a$. Genotypic values allow us to make statements about mean phenotypes in the population (because environmental effects, by definition, have a mean of 0, average genotypic values equal average phenotypic values), but because they are specified for genotypes and not alleles, genotypic values are uninformative with respect to the effects of specific alleles. We are interested in the latter because in diploid organisms, only one allele per locus is transmitted from parent to offspring, not entire genotypes. Therefore, we also need a measure of the effects of alleles in a population. This so-called additive effect (Lynch & Walsh 1998, p. 72) can be determined by averaging the genotypic values produced when a particular allele is paired with all other alleles in the population and expressing this value as a deviation from the population mean genotypic value. This is achieved by summing the probability of being paired with an A1 allele, p, and with an A2 allele, q, times the genotypic values of the resultant genotypes ($-a$, 0, or $+a$), and subtracting the mean population genotypic value (in our example: $0.5a$) from this sum. The sum of the additive effects of a parent's two alleles, often called the breeding value, determines the mean genotypic values of its offspring (open circles in Figure 7.1a,b). The mean and variance of breeding values can be estimated with knowledge of only the average allelic effects and their frequencies (see Chapters 7–8 in Falconer & Mackay 1996). The important point in this example is that the mean breeding value in a population will equal the mean genotypic and phenotypic values in that population (when expressed as deviations from the population mean phenotype, the mean genotypic and breeding values equal zero). The variance in breeding values of individuals in a population therefore yields an estimate of the additive genetic variance.

When alleles interact non-additively at a locus, for example when the A2 allele is dominant to A1, the heterozygote's genotypic value no longer equals the average genotypic value of the two homozygotes (A1A2 $\neq$ 0). Instead, the genotypic value of the heterozygote deviates from the average genotypic value of the two homozygotes by the *dominance genetic effect*, d. In our example, the genotypic value of the A1A2 heterozygote is $d = a$ (Figure 7.1b). The *dominance genetic effect* is different from

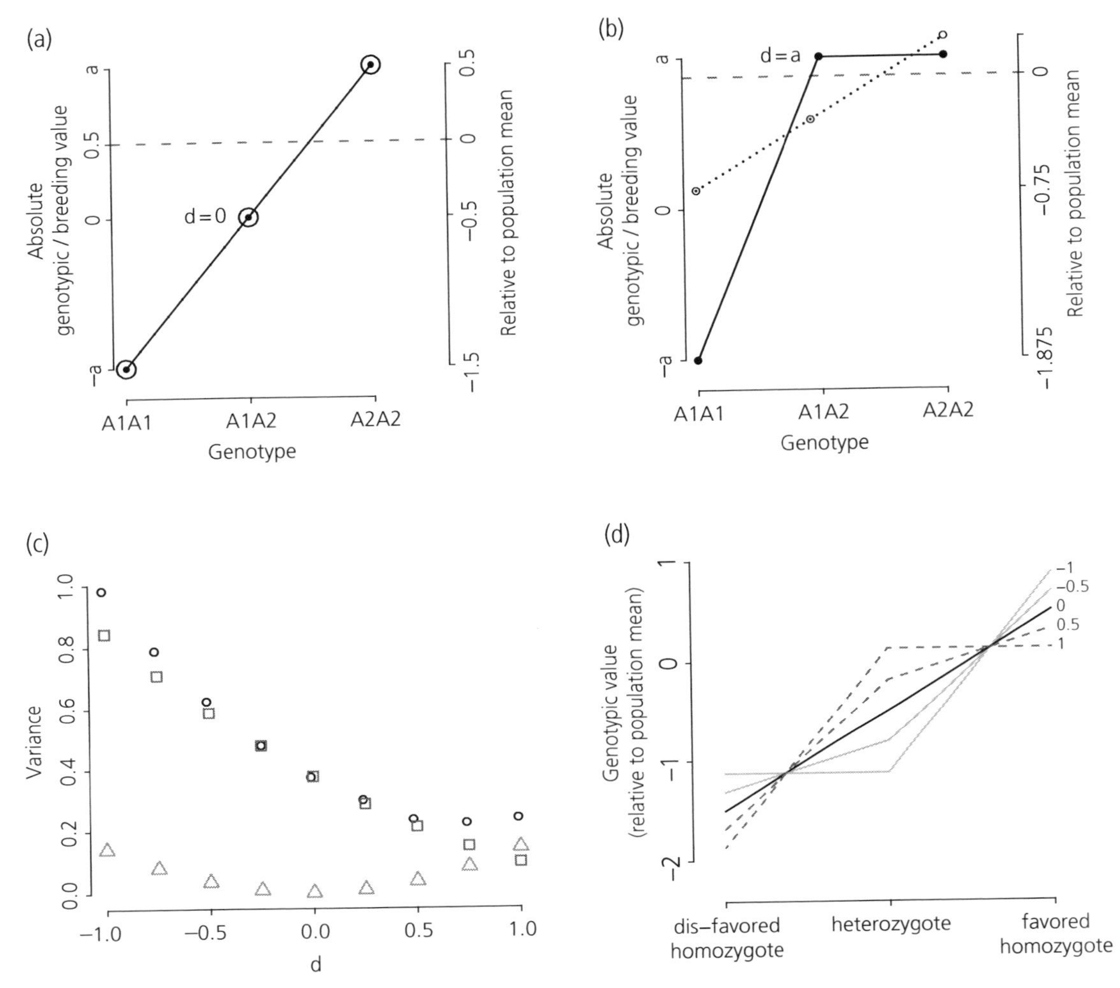

Figure 7.1 This figure defines dominance genetic effects, dominance deviations, and their relationships with additive and dominance genetic variance. For all panels, the allele frequency of the A2 allele is $q = 0.75$ and when dominance is non-zero ($d \neq 0$), the A2 allele is dominant to the A1 allele. Panels (a) and (b) depict single locus, two allele examples where the genotypic values (filled circles, solid line) or breeding values (open circles, dotted line) can either be expressed as absolute quantities (left axes) or relative to the population mean genotypic value (right axis). (a) Alleles show no *dominance genetic effects* ($d = 0$), A2A2 $= a = 1$, and the mean genotypic value in the population is 0.5 (horizontal grey dashed line). Note, the breeding values and genotypic values are equal within each genotype. (b) When the *dominance genetic effect* is 1 ($d = 1$), the mean genotypic value in the population is equal to 0.875 (horizontal grey dashed line). *Dominance deviations*, δ, are the distances between genotypic values and breeding values for each genotype. The *dominance deviations* for A1A1, A1A2, and A2A2 are, $\delta_{A1A1} = -1.875 - (-0.75) = -1.125$, $\delta_{A1A2} = 0.125 - (-0.25) = 0.375$, and $\delta_{A2A2} = 0.125 - 0.25 = -0.125$. (c) The expected genotypic variance (black circles) and the constituent additive genetic variance (dark grey squares) and dominance genetic variance (grey triangles) for different levels of the *dominance genetic effect*, *d*. (d) A range of different loci with varying levels of dominance (the *dominance genetic effect*, *d*, is represented by numbers to the right of the lines; solid grey lines, $d < 0$; solid black line, $d = 0$; and dashed grey lines, $d > 0$).

the *dominance deviation* (denoted by the symbol δ) at a particular locus. The latter are defined as the expected deviations of the genotypic value from the breeding value (Neiman & Linksvayer 2006). With dominance, we have one *dominance genetic effect* for the three possible genotypes at a locus, but three *dominance deviations* (δ, one for each possible genotype). Furthermore, because breeding values depend on gene frequencies, *dominance deviations* but not *dominance genetic effects* also depend on gene frequencies. *Dominance deviations* are thus partly measures of the degree of dominance and partly properties of populations (Falconer & Mackay 1996, p. 116). Expressed as a function of allelic effects and relative to the population mean (Falconer & Mackay 1996 pp. 116–119), the genotypic value of a particular genotype (G) equals the breeding value plus the dominance deviation δ (ignoring epistasis for simplicity, as we do throughout):

$$G = \alpha + \delta \tag{7.3}$$

Solving Eqn. 7.3 for δ emphasises that dominance deviations are the differences between genotypic value (G) and breeding value α. In Figure 7.1b, the *dominance genetic effect*, d, is portrayed by the genotypic value of the heterozygote departing from the mid-homozygote value. The *dominance deviations*, δ, of each genotype are the differences between the genotypic values (filled circles) and the breeding values (open circles). As is the case for the breeding value, the mean and variance in *dominance deviations* for a particular population are a function of the frequency of each genotype. Making assumptions similar to the additive case above, we can estimate the mean and variance of dominance deviations in a population (Chapters 7–8 in Falconer & Mackay 1996).

It is important to note here that when expressed in terms of deviations from the mean phenotype in an infinitely large population that is mating randomly, the expectation for the mean of the *dominance deviations* in a population should also equal the expected mean of the genotypic and breeding values: $E(G) = E(\alpha) = E(\delta) = 0$. We make special note of this here, because this expectation forms the basis of the statistical methodology to estimate breeding values and *dominance deviations* in wild populations (see Section 7.3.1), and it is why their estimation is problematic in light of inbreeding (see Section 7.5.1). However, assuming no inbreeding, the dominance deviation for a polygenic trait in a particular individual is the sum across all loci of *dominance deviations* at each locus (analogous to the breeding value of an individual). In the animal model discussed below, the best linear unbiased predictors (i.e. BLUPs) of the dominance random effect are the predicted values for the *dominance deviations*, δ, in each individual. Consequently, quantitative genetic dominance variance (σ^2_D) is the variance in individual *dominance deviations* in a population. Similar to Eqn. 7.3, but for the population variances, the genotypic variance is the combination of the additive variance plus the dominance variance (Figure 7.1c).

We expect selection to be efficient at removing dominant deleterious alleles from a population, leaving only those dominant alleles that increase fitness (Lynch & Walsh 1998 p. 270). In this case, we expect fitness-related traits to show directional dominance, or *dominance genetic effects* in the direction which maximise fitness. However, we point out that even though this results in all positive *dominance genetic effects*, d, across loci (e.g. all the grey dashed lines in Figure 7.1d), the mean *dominance deviations*, δ, across genotypes still equal zero. This is because the *dominance deviations*, defined as the difference between genotypic and breeding values relative to the population mean (Eqn. 7.3), are still negative for the disfavoured homozygote even when the *dominance genetic effect* is positive (e.g. Figure 7.1b).

7.3 Practical considerations

Estimating dominance variance from phenotypic data requires that we can separate it from additive genetic and non-genetic variances contributing to overall phenotypic variance in a trait. This can be achieved by contrasting the phenotypic similarity (i.e. the phenotypic covariance) of relatives to whose resemblance dominance contributes with the similarity of relatives where dominance does not contribute to the resemblance. The contribution of dominance to the phenotypic resemblance between relatives can be predicted by the probability that the two individuals share both alleles identically-by-descent (IBD) at a locus. Based on expectations of this probability in different types of relatives (e.g. Table 3.5 in Bulmer 1980), the dominance genetic variance can be separated from the additive genetic variance by contrasting the phenotypic covariance in relationships sharing both alleles at a locus IBD to the phenotypic similarity of individuals who are not expected to share two alleles IBD (Table 7.1). This

Table 7.1 Expected contribution of additive variance, dominance variance, common and genetic maternal variances, and epistatic variances to the phenotypic resemblance between an individual and various classes of relatives

Class of relative[a]	Coefficients of the proportion of total phenotypic covariance[b]							
	A	D	C	M	COV_{am}	AA	AD	DD
Father	1/2	0	0	0	1/4	1/4	0	0
Mother	1/2	0	0	1/2	5/4	1/4	0	0
FS	1/2	1/4	1	1	1	1/4	1/8	1/16
Paternal-HS	1/4	0	0	0	0	1/16	0	0
Maternal-HS	1/4	0	1	1	1	1/16	0	0
MGM	1/4	0	0	1/4	5/8	1/16	0	0
PGM	1/4	0	0	0	1/8	1/16	0	0
MGF	1/4	0	0	1/4	5/8	1/16	0	0
PGF	1/4	0	0	0	1/8	1/16	0	0
Maternal-aunt	1/4	0	0	1/2	3/4	1/16	0	0
Maternal-uncle	1/4	0	0	1/2	3/4	1/16	0	0
Paternal-aunt	1/4	0	0	0	1/4	1/16	0	0
Paternal-uncle	1/4	0	0	0	1/4	1/16	0	0
First cousin, dams FS	1/8	0	0	1/2	1/2	1/64	0	0
First cousin, sires FS	1/8	0	0	0	0	1/64	0	0
First cousin, opposite sex FS	1/8	0	0	0	1/4	1/64	0	0
DFC, opposite sex FS	1/4	1/16	0	0	1/2	1/16	1/64	1/256
DFC, sires full-brothers and dams full-sisters	1/4	1/16	0	1/2	1/2	1/16	1/64	1/256

[a] FS = full-sib; HS = half-sib; MGM = maternal grandmother; PGM = paternal grandmother; MGF = maternal grandfather; PGF = paternal grandfather; DFC = double first cousin.
[b] Contributions to: A = autosomal additive variance; D = autosomal dominance variance; C = maternal common environment variance; M = maternal genetic variance; COV_{am} = autosomal additive genetic and maternal genetic covariance; AA = autosomal additive-by-additive epistatic variance; AD = autosomal additive-by-dominance epistatic variance; DD = autosomal dominance-by-dominance epistatic variance.

occurs, for example, when comparing individuals within full-sib and among half-sib families. In Table 7.1, note the row corresponding to full-sibs has a non-zero dominance coefficient whereas this dominance coefficient is zero in the row corresponding to maternal half-sibs (siblings with the same mother, but different fathers). The sum of coefficients in a row of Table 7.1 make up a linear equation predicting the phenotypic resemblance between a certain set of relatives (e.g., $COV_{\text{Phenotypic-PHS}} = 1/4\sigma^2_A + 1/16\ \sigma^2_{AA}$). One could proceed to obtain estimates of variances by utilising methods for solving a system of linear equations. For example, if we assume only additive and dominance genetic effects, then rearranging the equation for the phenotypic covariance between paternal half-sibs would yield: $\sigma^2_A = 4COV_{\text{Phenotypic-PHS}}$. Rearranging the equation for the phenotypic covariance between full-sibs to solve for dominance variance and substitution of the additive variance for the expression in terms of the phenotypic covariance between paternal half-sibs would yield $\sigma^2_D = COV_{\text{Phenotypic-FS}} - 4COV_{\text{Phenotypic-PHS}}$. Stated another way, in our simple example, we expect the resemblance between full-sibs to be greater than that of half-sibs because of the added similarity among full-sibs due to sharing genes with dominance interactions. This approach forms the basis of the 'bio-model' developed by Cockerham & Weir (1977) to analyse reciprocal crosses.

Comparisons between heritabilities estimated from full- and half-sib subsets of data yield an estimate of the dominance variance (Roff 2007) assuming no other sources of phenotypic similarity contribute to differences between these two groups of siblings. This assumption is often violated in unmanipulated natural populations, because of common environmental or maternal genetic sources of phenotypic resemblance among full-sibs which add to the resemblance among individuals sharing mothers (Table 7.1). Therefore, we must use a method which removes phenotypic resemblance due to common maternal environment as much as possible (Kruuk & Hadfield 2007) so that what is left is the proportion of phenotypic resemblance attributable to additive and dominance variances. One way forward is to exploit systems in which full- and half-sib individuals have the same maternal environments. In such cases, dominance variance can be estimated from the difference between full- and half-sib phenotypic covariance without worry of maternal effects confounding the estimate (these maternal effects will cancel out through subtraction when they contribute equally to both full- and half-sib phenotypic resemblance). Maternal half-sibs and full-sibs could be used for this purpose, as dominance variance is not expected to contribute to phenotypic resemblance among maternal half-sibs (Table 7.1). Extra-pair copulations in many avian systems cause offspring from multiple fathers to be reared in the same nest. The result is that maternal half-sibs share the same maternal environment as full-sibs, making it easier to estimate dominance variance separately from maternal effects. Maternal half-sibs may be common in many other organisms (e.g. plants, amphibians, insects, and fish to name a few); however identifying them as different from full-sibs usually requires molecular genetic information. Alternatively, full-sibs which do not share the exact same maternal common environment could be used to disentangle dominance variance from maternal common environmental variance. This would most likely arise when full-sibs are spread out over several reproductive events separated in space or time (e.g. multiple broods/litters/clutches).

7.3.1 The animal model

Instead of comparing two classes of relatives such as full- and half-sibs, we can simultaneously compare many different pairs of relatives, varying in their degree of relatedness, using animal models (Henderson 1973, 1984; Lynch & Walsh 1998; Kruuk 2004). The animal model has been used to separate dominance genetic effects from other sources of phenotypic variation in animal and plant breeding for over 20 years (Smith & Mäki-Tanila 1990; de Boer & van Arendonk 1992; Misztal 1997). However we know of only one animal model analysis (Abney *et al.* 2000) that includes dominance genetic variance using phenotypic data from wild populations (but see breeding experiments using individuals collected from the wild in e.g. Shaw *et al.* 1998 and Waldmann 2001). The animal model can estimate dominance genetic variance by using the dominance relatedness matrix, $\boldsymbol{D}$, which contains all pairwise probabilities of IBD for both alleles at a locus. The probabilities of IBD (individual elements in $\boldsymbol{D}$) can be approximated, assuming no inbreeding, using relationships from the additive genetic relatedness matrix $\boldsymbol{A}$ (Eqn. 7.7 in Lynch & Walsh 1998, p. 140). However, in complicated, multigenerational pedigrees, this approximation often fails to produce the correct probabilities of IBD, and a simulation approach is necessary (Ovaskainen *et al.* 2008). The procedure for estimating dominance genetic variance in an animal model is analogous to estimating the additive genetic variance in an animal model by inclusion of the additive genetic relatedness matrix $\boldsymbol{A}$ (see Wolak 2012 for 'how-to' examples using the nadiv package in R to construct the dominance relatedness matrix and the MCMCglmm, ASReml, and WOMBAT software programs to run the animal model). The univariate animal model which estimates additive genetic and dominance genetic random effects is:

$$\mathbf{y} = \boldsymbol{X\beta} + \boldsymbol{Z}_{\mathrm{a}}\boldsymbol{\alpha} + \boldsymbol{Z}_{\mathrm{d}}\boldsymbol{\delta} + \mathbf{e} \qquad (7.4)$$

In the simple case where all n individuals in a population are measured once for phenotype y, then Eqn. 7.4 describes the $n \times 1$ vector of phenotypes, $\mathbf{y}$, as a function of the $p \times 1$ vector of p fixed effects in $\boldsymbol{\beta}$ (e.g. fixed effects such as year, sex, or age, in addition to the overall population mean) plus the $n \times 1$ vector of breeding values for each individual in $\boldsymbol{\alpha}$, plus the $n \times 1$ vector of dominance deviations for each individual in $\boldsymbol{\delta}$, and the $n \times 1$ vector of residual deviations in vector $\mathbf{e}$. The matrices $\boldsymbol{X}$, $\boldsymbol{Z}_{\mathrm{a}}$, and $\boldsymbol{Z}_{\mathrm{d}}$ are $n \times p$, $n \times n$, and $n \times n$ incidence matrices relating the fixed and random effects to each observation in $\mathbf{y}$. The random effects are assumed to follow independent normal distributions, where the mean and variance for additive and dominance random effects can be represented as $\boldsymbol{\alpha} \sim N(0, \sigma^2_{\mathrm{A}}\boldsymbol{A})$,

$\delta \sim N(0,\sigma_D^2 \boldsymbol{D})$. The residual deviations are assumed uncorrelated among individuals, $\mathbf{e} \sim N(0,\sigma_E^2 \mathbf{I})$, where $\mathbf{I}$ is an identity matrix ($n \times n$ with 1s along the diagonal). However, to solve the equations in the animal model and obtain estimates of the variance components (σ_A^2, σ_D^2, and σ_E^2) the inverses of the additive and dominance genetic relatedness matrices (i.e. $\boldsymbol{A}^{-1}$ and $\boldsymbol{D}^{-1}$) are necessary (Henderson 1985):

$$\begin{bmatrix} \mathbf{X}^T\mathbf{X} & \mathbf{X}^T\mathbf{Z}_\alpha & \mathbf{X}^T\mathbf{Z}_\delta \\ \mathbf{Z}_\alpha^T\mathbf{X} & \mathbf{Z}_\alpha^T\mathbf{Z}_\alpha + \lambda_A \boldsymbol{A}^{-1} & \mathbf{Z}_\alpha^T\mathbf{Z}_\delta \\ \mathbf{Z}_\delta^T\mathbf{X} & \mathbf{Z}_\delta^T\mathbf{Z}_\alpha & \mathbf{Z}_\delta^T\mathbf{Z}_\delta + \lambda_D \boldsymbol{D}^{-1} \end{bmatrix} \begin{bmatrix} \hat{\beta} \\ \hat{\alpha} \\ \hat{\delta} \end{bmatrix} = \begin{bmatrix} \mathbf{X}^T \boldsymbol{y} \\ \mathbf{Z}_\alpha^T \boldsymbol{y} \\ \mathbf{Z}_\delta^T \boldsymbol{y} \end{bmatrix} \quad (7.5)$$

where $\lambda_A = \sigma_E^2/\sigma_A^2$ and $\lambda_D = \sigma_E^2/\sigma_D^2$. Algorithms have been developed for constructing $\boldsymbol{A}^{-1}$ directly, bypassing preliminary construction of $\boldsymbol{A}$ followed by matrix inversion (e.g. Henderson 1976; Meuwissen & Luo 1992). Unfortunately, these time and computer memory saving algorithms have yet to be developed to construct $\boldsymbol{D}^{-1}$ directly for all individuals in a population (but see Hoeschele & VanRaden 1991 for estimating dominance variance from sire-dam subclasses and Smith & Mäki-Tanila 1990 for the genotypic relatedness matrix).

Interestingly, the matrix $\boldsymbol{D}^{-1}$ contains more non-zero elements than $\boldsymbol{A}^{-1}$ (in matrix parlance, $\boldsymbol{D}^{-1}$ is more dense than $\boldsymbol{A}^{-1}$). This occurs even though $\boldsymbol{A}$ is more dense than $\boldsymbol{D}$. We illustrate this in the supplementary material found online at www.oup.co.uk/companion/charmantier (Appendix S7.1 *Additive and Dominance Matrix Density*) where we depict the results from pedigrees created by simulating populations under random mating, without selection. The main point of the pedigree simulations is to demonstrate the large discrepancy in number of non-zero elements in $\boldsymbol{D}^{-1}$ as compared to $\boldsymbol{A}^{-1}$. This occurs, despite the extreme disparity between the relatively dense $\boldsymbol{A}$ matrix and relatively sparse $\boldsymbol{D}$. The matrix $\boldsymbol{D}$ is less dense than $\boldsymbol{A}$, because fewer pairs of individuals share both alleles IBD (probabilities in $\boldsymbol{D}$) than one allele IBD (probabilities in $\boldsymbol{A}$). As a consequence of matrix inversion (i.e. matrix algebra and not biology), $\boldsymbol{D}^{-1}$ is more dense than $\boldsymbol{A}^{-1}$. It is the density of $\boldsymbol{D}^{-1}$ that makes estimating dominance genetic variance in standard animal models particularly time consuming. As can be seen from Eqn. 7.5, adding dominance genetic effects as an additional random term in the animal model increases the number of rows and columns in the mixed-model equations by n (i.e. the pedigree length). Further, as $\boldsymbol{D}^{-1}$ is relatively dense (see supplementary online Figure S7.1), estimation of dominance variance greatly increases the memory and time required by the computer to solve Eqn. 7.5. This often overwhelms the animal model solving algorithms, resulting in models that fail to converge (e.g. do not return maximum likelihood estimators of all variance components or produce Markov chain Monte Carlo (MCMC) chains with poor mixing in a Bayesian animal model). An example of the added complexity can be illustrated using the pedigree length of $n = 1200$ and two offspring per mating (see supplementary online Appendix S7.1). A pedigree of this size has a $\boldsymbol{D}^{-1}$ of size $n \times n$ (i.e. 1 440 000 possible entries) with an average number of non-zero entries of 124 421 (95% quantile limits: 35 248–213 176). Thus, both the construction of $\boldsymbol{D}^{-1}$ and its implementation in an animal model to estimate the dominance variance greatly increases the computational demands as compared to the case of only estimating additive genetic variance.

7.3.2 Alternative animal model methods

Current methods exist which either work around or solve the aforementioned computational problems involved with estimating dominance genetic variance in an animal model. Henderson (1985) pointed out that dominance genetic effects can be computed indirectly after predictions of the additive genetic effects have been obtained. Although this has been suggested (Chalh & El Gazzah 2003; Schaeffer 2003), its use is not widespread, even in animal breeding studies, and this rearrangement of the mixed-model equations is not an option in common animal model software packages. Further, it is uncertain how estimates of the dominance genetic variance from these alternative methods compare to the full model using $\boldsymbol{D}^{-1}$, as the primary purpose of these proposed methods is to obtain predictions of the dominance genetic effects and not estimates of the variance components. An approach to simplifying the dimensionality of an animal model with dominance genetic effects that is potentially more useful has been proposed by Hallander *et al.* (2010) and was applied to a previously analysed dataset (e.g. Waldmann *et al.* 2008). This method reduces the time necessary for Bayesian inference of parameters in an animal model by reformulating the problem as separate conditional expressions of the multivariate normal distribution of effects to be estimated. The required code to run such analyses in the Bayesian software package WinBUGS (Lunn *et al.* 2000; freely available from http://www.mrc-bsu.cam.ac.uk/bugs/winbugs/contents.shtml) is available in the supplementary material to Hallander *et al.* (2010). However, estimates of dominance variance from the method in Hallander *et al.* (2010) were extremely similar to previous estimates of the same dataset from Waldmann *et al.* (2008), indicating only time-savings benefits and no noticeable gain in accuracy or precision from this newer method.

7.3.3 Disentangling dominance deviations

Another hurdle to estimating dominance genetic variance is the problem of identifiability, or the ability of a statistical model to correctly identify an effect, given the data. The majority of information in pedigrees comes from the nuclear family unit (i.e. parent–offspring or sibling–sibling relationships), which are often useless for separating contributions from additive genetic, dominance genetic, common environmental, and maternal genetic effects (e.g. in Table 7.1, we see that full-sibs in the same clutch/brood/litter resemble one another because they share additive, dominance, maternal genetic, and maternal common environment effects). This is particularly problematic in many datasets from wild populations in which the nuclear family accounts for many of the pairwise relationships in the population (Kruuk 2004; O'Hara *et al.* 2008). In such cases, the phenotypic deviation of an individual from the population mean can be attributed, with equal probability, to a dominance deviation, additive genetic, maternal genetic, or maternal common environment effect. In the 'bio-model' approach mentioned above, this is analogous to being unable to obtain unique solutions for the dominance, maternal genetic, and common environment variances (i.e. multiple solutions exist to the system of linear equations). Therefore, it is difficult to obtain accurate estimates from an animal model for each of these sources of phenotypic variance because they are confounded with one another (Hill 1999; Misztal & Besbes 2000; Meyer 2008). More than likely, one random effect will consistently 'steal' the effects from the other sources, resulting in overestimates of one source of variance and underestimates of the rest.

However, the ability of a particular pedigree to yield accurate estimates of dominance genetic variance is greatly improved by relationships outside of the nuclear family (e.g. you and your cousin are genetically related, but have different mothers and thus different maternal genetic and common environment effects). In pedigrees spanning many generations, relationships such as grandparent–grandoffspring, aunt/uncle–niece/nephew, and cousins will all help to separate the variance attributable to dominance deviations or maternal effects. In particular, double first cousins (individuals i and j are double first cousins if both the mothers of i and j are sisters and the fathers of i and j are brothers) are relationships which provide extra statistical power to the animal model for disentangling dominance genetic variance estimates from other variance components (Fairbairn & Roff 2006; Meyer 2008). Currently, it is unclear how many sets of double first cousins exist in wild pedigrees as this is not a commonly reported relationship (e.g. it is not reported by Morrissey & Wilson's (2010) pedantics package), but the number can be counted by the R function find DFC in the nadiv package (Wolak 2012). For example, the long-term study of song sparrows (*Melospiza melodia*) inhabiting Mandarte Island, Canada (Smith *et al.* 2006) has a pedigree size of 5756 individuals and contains 929 pairs of double first cousins (35 pairs of full-sib families whose offspring are all double first cousins to the offspring in the other full-sib family) over 26 generations. This example highlights the fact that first cousins, and hence also double first cousins, are expected to appear more often in pedigrees where the population size remains relatively small (Jacquard 1974, p. 198). This is because the probability of two siblings mating to another pair of siblings will increase as the breeding population size decreases and average full-sib size increases. The number of double first cousins from the Mandarte song sparrows is probably closer to the upper range of most wild populations with pedigree information. However, in general the ability of pedigrees from wild populations to estimate dominance genetic variance is unknown. Most likely, it is limited for many populations. Therefore, only certain systems or organisms, which tend to produce pedigree structures with many relationships supplementary to the nuclear family relationships, may be useful for testing evolutionary theories requiring estimates of dominance variance.

We encourage investigations into the ability of pedigrees from wild populations to ascertain unconfounded estimates of dominance variance through computer simulations that use phenotypes with simulated contributions of additive, dominance, and common environment effects (as well as any other sources of variation relevant to a particular population and study) based on the pedigree of a particular study (e.g. Clément *et al.* 2001; Kruuk & Hadfield 2007; Morrissey & Wilson 2010). Power of an animal model to correctly estimate dominance variance, in the presence of other confounding sources of variation (e.g. common maternal environment), is calculated by: 1) analysing the simulated data with the same model one wishes to employ on field-collected data, 2) recording the outcome of a statistical assessment of the model to correctly recover the simulation parameters (e.g. P value for likelihood ratio test between the model of simulated data and the same model structure, but with the dominance variance constrained to be equal to the simulated value), 3) resimulating phenotypes and repeating steps 1–2 many times (e.g. 1000 or 10 000), and 4) calculating the proportion of simulations for which the statistic in step 2 indicated no statistical difference between simulated values and estimated parameters. Note that this approach is slightly different than calculating the power of the animal model to yield an estimate of a variance parameter significantly different from zero (e.g. constrain the dominance variance to zero instead of the simulated value), which could be substituted as the test in step 2. Functions in the R packages MCMCglmm (Hadfield 2010) and nadiv (v2.12; Wolak 2012) can be used to simulate the component parts of the individual phenotype corresponding to breeding values, dominance deviations, common maternal environment effects, and environmental deviations, to name a few (see the rbv function in MCMCglmm and the nadiv functions grfx, makeD, and drfx). It will be interesting to see in how many natural populations researchers are able to reliably separate dominance variance from other components of variance.

7.4 Empirical estimates of dominance variance: a review of the literature

Above we noted that Hill *et al.* (2008) present theoretical models that predict that the variance in non-additive genetic effects will contribute little to phenotypic variance. They support this result with a survey of published heritability estimates. Using comparisons between different ways of estimating heritability (e.g. narrow-sense, broad-sense, and repeatability, which approximates the upper limit of heritability), Hill *et al.* note that subtraction of narrow-sense heritability (σ^2_A/σ^2_P) from broad-sense heritability ($(\sigma^2_A + \sigma^2_D + \sigma^2_I)/\sigma^2_P$) or from repeatability ($\sigma^2_{\text{Among}}/(\sigma^2_{\text{Within}} + \sigma^2_{\text{Among}})$, where additive, non-additive, and permanent environmental variances constitute σ^2_{Among}) will isolate the fraction of phenotypic variance attributable to non-additive genetic variance (e.g. $(\sigma^2_D + \sigma^2_I)/\ \sigma^2_P$; see discussion of the basis for this approach in Section 7.3). From these comparisons, they find little evidence to suggest a great discrepancy between the various ways of estimating heritability and therefore suggest the estimates they reviewed had little non-additive genetic variance contributing to overall phenotypic variation. This result disagrees, to a certain extent, with the earlier study by Crnokrak and Roff (1995), who directly quantified both the amount of dominance variance as compared to total genetic variance ($D_\alpha = \sigma^2_D/(\sigma^2_D + \sigma^2_A)$) and the contribution of dominance variation to phenotypic variance ($D_\beta = \sigma^2_D/\sigma^2_P$; note authors elsewhere have used the notation d^2 to symbolise this same quantity). Although Crnokrak and Roff were explicitly testing predictions regarding the relative amounts of dominance variance (as measured by D_α and D_β) in different trait types (i.e. life-history, behavioural, physiological, and morphological), their overall values of the two measures were substantially greater than zero.

We compiled estimates of dominance variance from the literature to specifically address two questions: 1) does additive genetic variance contribute more to phenotypic variance than the contribution from dominance variance and 2) is the contribution of dominance genetic variance negligible? Our literature search for published estimates of dominance variance yielded 86 papers (41 which were not in Crnokrak & Roff 1995) containing 559 estimates of dominance variance (331 which were not in Crnokrak & Roff) using 56 different organisms. From these studies we calculated h^2, D_α, and D_β, where possible (see supplementary online Table S7.1 for the full compilation of estimates; www.oup.co.uk/companion/charmantier). Our measures h^2, D_α, and D_β are all proportions and will most likely violate the assumptions of standard parametric statistical tests; therefore, we used non-parametric tests. To address our first question regarding the relative contributions of additive and dominance variance to phenotypic variance, we performed a paired Wilcoxon signed-rank test using our estimates of h^2 and D_β. This was a two-tailed test of the null hypothesis that the difference between these two estimates is zero. For our second question, we conducted one-tailed tests of the null hypotheses $D_\alpha = 0$ and $D_\beta = 0$, using one sample Wilcoxon signed-rank tests. Because selection is expected to shape the relative amounts of dominance and additive genetic variances resulting in differences among trait types (Crnokrak & Roff 1995), we further subdivided our dataset and assessed both questions for each trait type in either domestic or non-domestic organisms separately. We acknowledge that our analyses ignore the statistical uncertainty that accompanies estimates of the additive and dominance genetic variances, but few papers in our dataset report variance estimates accompanied by measures of uncertainty (e.g. standard errors) so we did not pursue this further. However, we note that many estimates of dominance variance contained in our dataset have very large standard errors. We encourage future analyses of this dataset, with application of more formal meta-analysis techniques, in order to answer the questions posed here and elsewhere (e.g. comparisons among trait types in Crnokrak & Roff 1995).

Overall, the difference between heritability and D_β tends to be significantly different from zero (Table 7.2), indicating that additive genetic variance contributes more to phenotypic variance than does dominance variance in these estimates from the literature. When subdividing this analysis by trait type and domestication status, we find that

Table 7.2 Parameter estimates from paired Wilcoxon signed-rank tests assessing the null hypothesis that the difference between the proportional contributions to phenotypic variance from additive genetic (heritability) and dominance genetic (D_β) variances is equal to zero

Trait type[a]	*n*	95% LCL[b]	Estimate	95% UCL[c]	P-value
Entire	291	0.154	0.184	0.214	<0.0001
All non-domestic	94	0.146	0.193	0.253	<0.0001
B non-domestic	6	−0.14	−0.045	0.01	0.062*
L non-domestic	14	−0.074	0.128	0.301	0.296*
M non-domestic	44	0.117	0.166	0.219	<0.0001
P non-domestic	30	0.215	0.35	0.47	<0.0001
All domestic	197	0.145	0.178	0.211	<0.0001
B domestic	9	−0.01	0.005	0.02	0.55*
L domestic	73	0.135	0.185	0.24	<0.0001
M domestic	81	0.056	0.106	0.165	<0.001
P domestic	32	0.32	0.395	0.453	<0.0001

[a] B = behavioural; L = life-history; M = morphological; P = physiological.
[b] LCL = lower confidence limit.
[c] UCL = upper confidence limits.
* *p*-values greater than the critical value after sequential Bonferroni correction for multiple tests, $\alpha_i = 0.05/(1 + 11 - i)$.

Table 7.3 Wilcoxon signed-rank tests assessing the null hypothesis that D_α and D_β are not significantly different from zero

Trait type[a]	Measure	n	Estimate	P-value[b]
Entire	D_α	553	0.381	<0.0001
	D_β	415	0.144	<0.0001
All non-domestic	D_α	172	0.39	<0.0001
	D_β	159	0.154	<0.0001
B non-domestic	D_α	18	0.38	<0.001
	D_β	9	0.192	<0.01
L non-domestic	D_α	52	0.632	<0.0001
	D_β	50	0.181	<0.0001
M non-domestic	D_α	71	0.257	<0.0001
	D_β	70	0.125	<0.0001
P non-domestic	D_α	31	0.452	<0.01
	D_β	30	0.361	<0.01
All domestic	D_α	381	0.379	<0.0001
	D_β	256	0.138	<0.0001
B domestic	D_α	9	0.452	<0.01
	D_β	9	0.06	<0.01
L domestic	D_α	101	0.361	<0.0001
	D_β	78	0.099	<0.0001
M domestic	D_α	205	0.405	<0.0001
	D_β	125	0.196	<0.0001
P domestic	D_α	63	0.307	<0.0001
	D_β	42	0.115	<0.0001

[a] B = behavioural; L = life-history; M = morphological; P = physiological.
[b] All p-values remain significant after sequential Bonferroni correction for multiple tests, $\alpha_i = 0.05/(1 + 22 - i)$.

life-history traits in non-domestic organisms and behavioural traits regardless of domestication do not follow the overall result and appear to have equal contributions of additive and dominance variances to phenotypic variance. Note that these particular traits have the lowest sample sizes (<14) of all trait and domestication combinations. Therefore, low power to detect a statistically significant difference could be influencing our results. We do not find support for the claim that the overall contribution of dominance variance to the total genetic variance (D_α) or the phenotypic variance (D_β) is zero (Table 7.3). This result is consistent for analyses of the entire dataset and for each trait type. Further, for both domestic and non-domestic organisms, but particularly for the former, estimates of additive and dominance variances are close to being equal.

Hill *et al.* (2008) qualitatively compared the difference between heritability in the broad sense ($(\sigma^2_A + \sigma^2_D + \sigma^2_I)/\sigma^2_P$) to narrow-sense heritability (σ^2_A/σ^2_P) and concluded that, on average, these did not differ much from one another. The difference between these two measures is equal to the ratio of total non-additive genetic variance to phenotypic variance. The D_β statistic can be thought of as a rough approximation of this ratio of non-additive genetic variance to phenotypic variance, although in reality it is an underestimate as it does not explicitly include epistatic variance. An analysis of the 415 estimates of D_β across all traits in our compiled data set revealed that D_β accounts for approximately 15% of phenotypic variation and we conclude that it is significantly greater than zero, a results that contrasts somewhat with that of Hill *et al.* (2008). A number of reasons may explain this discrepancy. First, the estimates of dominance variance we obtained from the literature may be inflated by other sources of phenotypic variance and thus bias our results towards overestimating the prevalence of dominance variance. Most information for estimating dominance variance comes from full-sibs, and so dominance variance estimates will often be completely confounded with full-sib common environmental variance (Hill 1999). For example, in laying hens, it was shown that estimates of dominance variance were inflated by the full-sib common environmental variance when models did not include a term to account for this non-genetic source of covariance between full-sibs (Misztal & Besbes 2000). Many studies in our review did use models that included terms to separately estimate full-sib common environmental variances, and many of the laboratory studies reared individuals in controlled environmental conditions, in some cases with full-sibs split across rearing containers/cages. Both practices serve to greatly reduce the full-sib common environmental variance. Therefore, it is unclear to what extent our compiled estimates from the literature may have overestimated dominance variance because of full-sib common environmental variances that were unaccounted for in the original models. Second, the difference may largely be one of perspective rather than of a major quantitative difference. Hill *et al.* (2008) concluded that typically over half of the total genetic variance is additive. We found that dominance variance rarely contributes more than half to the total genetic variance (Table 7.3). Hence, quantitatively, the two results are compatible. Overall, our estimates of D_β, combined with the results indicating that D_α is also significantly greater than zero, suggest that even though additive genetic variance may often make up most of the total genetic variance, dominance variance can make an important contribution to total phenotypic and genetic variation.

7.5 Relaxing the assumption of random mating and no genetic drift

The methods for estimating additive and dominance variances outlined above make the standard assumptions of the quantitative genetic variance partitioning framework (Eqn. 7.1), namely that populations are infinitely large and that they are mating randomly (e.g. Comstock & Robinson 1948). Most natural populations will not meet this assumption: inbreeding (mating among relatives) and genetic drift (the random changes in allele frequencies in finite populations) are common in many wild populations (e.g. Thornhill 1993; Keller & Waller 2002). Both inbreeding and genetic drift affect quantitative genetic inference because inbreeding changes genotype frequencies, and genetic drift changes allele frequencies (Chapters 3 and 4 in Falconer & Mackay 1996). If inbreeding is pronounced enough, ignoring the effects of inbreeding can lead to substantial biases in estimates of additive and dominance effects (e.g. de Boer & van Arendonk 1992; Abney *et al.* 2000). In the following, we will first give a short introduction into inbreeding and then illustrate that, in the presence of dominance, inbreeding introduces several additional (co)variance components. These additional components of phenotypic variance can be crucial to understanding the response to selection in highly inbred populations. However, in less inbred populations, the model outlined in Section 7.3.1 that ignores inbreeding may often yield reasonable estimates of additive and dominance variances, provided that a regression on individual inbreeding coefficients is added to the animal model (see Section 7.5.3).

7.5.1 Additive and dominance variance with inbreeding

Inbreeding and genetic drift are closely related but not identical concepts (e.g. Keller *et al.* 2012) that can both be measured by the probability that two alleles at a locus in an individual are identical-by-descent (the inbreeding coefficient of individual i, F_i; Falconer & Mackay 1996, p. 58). For simplicity, we will refer to inbreeding in the following and include implicitly the effects of genetic drift as measured by the inbreeding coefficient. Inbreeding reduces the frequency of heterozygotes by an amount proportional to the average inbreeding coefficient in a population, $\overline{F}$. If there is directional dominance in the direction which maximises fitness (see Section 7.2), the fitness of heterozygotes is higher than the average fitness of the two homozygotes, and hence the decline in the frequency of heterozygotes leads to decline in the mean of a trait, the well-known phenomenon of inbreeding depression (see Chapter 14 in Falconer & Mackay 1996; Chapter 10 in Lynch & Walsh 1998). Although inbreeding depression is not always observed, it is very common, especially in fitness traits (for reviews of inbreeding depression see e.g. Keller & Waller 2002 and Charlesworth & Willis 2009). Inbreeding also influences the variance of a trait (e.g. Charlesworth & Charlesworth 1995; Kristensen *et al.* 2005; van Buskirk & Willi 2006; Taft & Roff 2012).

With purely additive gene action, inbreeding only affects the genetic variance of a trait and not its mean, because inbreeding changes genotype but not allele frequencies (e.g. Falconer and Mackay 1996, p. 249 and p. 264). This change in variance can readily be incorporated in the animal model (see Section 7.3.1) through the additive genetic relationship matrix $\mathbf{A}$. The elements of $\mathbf{A}$ reflect the effects of inbreeding on the genetic covariances among relatives (for example, the diagonal elements of $\mathbf{A}$ are equal to $1 + F_i$), and this accounts for the changes in additive genetic variance in inbred populations (Kennedy and Sorensen 1988).

In the presence of dominance, the effects of inbreeding on quantitative genetic variance components become more complex. Although the basic linear variance partitioning framework can still be applied, additional variance components have to be estimated. Harris (1964) and Jacquard (1974, pp. 131–138) provide detailed derivations of the variance components under inbreeding. de Boer and Hoeschele (1993) provide an excellent technical summary and a helpful table that establishes the equivalence between the notations used by different authors. Excellent, less technical overviews can be found in Shaw *et al.* (1998) and Walsh (2005). Here we provide only a brief summary. As before, we are purposefully brief with the mathematical details and refer interested readers to the appropriate technical literature.

In the presence of dominance, two complexities arise that must be considered when partitioning the phenotypic variance of inbred populations. First, the dominance deviations (see Section 7.2), which average out to zero across all genotypes in non-inbred populations, have a non-zero expectation in inbred populations (see Table 1 in Edwards & Lamkey (2002) for the differences in expectations in inbred and outbred populations). As a consequence, inbreeding changes the mean of a trait, i.e. inbreeding depression results. If a single locus underlies a trait, inbreeding will change the mean of a trait whenever there is dominance. However, if more than one locus contributes to a trait, inbreeding will only change the mean of the trait if the dominance genetic effects underlying the trait are predominantly in one direction. In other words, directional dominance is required (Falconer & Mackay 1996, p. 250; Lynch & Walsh 1998, p. 257).

Directional dominance implies that alleles that lead to an increase of a trait are on average dominant over those that reduce a trait, or vice versa (see Section 7.2 and Figure 7.1d). Since selection is expected to remove dominant deleterious alleles quickly from a population, the majority of deleterious alleles underlying fitness-related traits are expected to be recessive (Lynch & Walsh 1998, p. 270). This is thought to be one of the reasons why fitness-related traits show more inbreeding depression than morphological traits (e.g. deRose & Roff 1999). The grey dashed lines in Figure 7.1d give a graphical representation of directional dominance, while the grey dashed and the grey solid lines together depict a situation where dominance is present but is not, on average, directional. Figure 7.1 highlights the fact that if a trait exhibits inbreeding depression, there must be dominance variance. This is true even in the unlikely case that all loci exhibit the exact same dominance genetic effect d (say, all loci show exactly the pattern depicted in Figure 7.1b), because the dominance deviations vary between genotypes. However, the reverse need not be true: if dominance is not directional, substantial dominance variance may be present in traits that show no inbreeding depression (Lynch & Walsh 1998, p. 257). In other words, the magnitude of inbreeding depression in a trait is not a good predictor of the magnitude of dominance variance in that trait.

The second complexity encountered when partitioning variance components in the presence of dominance and inbreeding is the introduction of at least three additional (co)variance components. For a population inbred to degree $\overline{F}$, instead of the familiar Eqn. 7.1, we now have (Shaw *et al.* 1998, their equation 4):

$$\sigma_P^2 = (1+\overline{F})\sigma_A^2 + (1-\overline{F})\sigma_{DR}^2 + 4\overline{F}\sigma_{ADI} + \overline{F}\sigma_{DI}^2 + \overline{F}(1-\overline{F})H^* + \sigma_E^2 \qquad (7.6)$$

where:

σ_A^2 is the additive genetic variance, as before;

σ_{DR}^2 is the variance in dominance deviations that would be expected if the allele frequencies in the inbred population were representative of a non-inbred, randomly mating population;

σ_{ADI} is the covariance between the additive effect of an allele and its dominance deviation when in a homozygous individual. In other words, this is the covariance between the average additive effect of an allele and its effects on inbreeding depression. This covariance arises, because inbreeding increases homozygosity. Therefore, even though parents only pass on a single allele, there is a greater chance that a given allele will be found in a homozygous genotype, creating a covariance between the contribution of a single allele to the breeding value of a homozygote and the dominance deviation of that genotype;

σ_{DI}^2 is the variance due to dominance deviations in completely inbred individuals;

H^* is the square of the sum over all loci of the homozygous dominance deviations. If a trait is controlled by a single locus, H^* corresponds to the square of the effects of inbreeding on the mean of a trait as calculated by conventional regression analysis (Shaw & Woolliams 1999). When many loci contribute to inbreeding depression, however, H^* can be very small even when inbreeding depression is pronounced.

In a non-inbred population $\overline{F} = 0$ and Eqn. 7.6 reduces to the familiar three-component equation $\sigma_P^2 = \sigma_A^2 + \sigma_{DR}^2 + \sigma_E^2$. Note that there are now two types of dominance variance, one due to dominance deviations in inbred individuals (σ_{DI}^2) and one due to dominance deviations in a random-mating population (σ_{DR}^2). To highlight this point, σ_D^2 has been relabelled as σ_{DR}^2 to reflect the fact that this corresponds to the dominance variance in a non-inbred, randomly mating population.

The distinction between the non-additive variance component due to dominance in a large and randomly mating population (σ_{DR}^2) on the one hand, and the additional dominance (co)variance components under inbreeding (σ_{DI}^2, H^*, and σ_{ADI}) on the other hand is very important for adaptive evolution: σ_{DR}^2 does not affect selection response, while σ_{DI}^2, H^*, and σ_{ADI} can change the selection response (Shaw *et al.* 1998; Kelly & Arathi 2003). This occurs because, in populations with inbreeding, the existence of common ancestors through both parental lines means that non-additive variance components contribute to the covariance between parents and offspring, on which selection response depends.

7.5.2 Issues related to estimating the additional dominance (co)variance components

Estimating these additional dominance (co)variance components follows the general approach outlined above for dominance variance. Because more (co)variance components need to be estimated, the estimation procedures are more complex and computationally demanding, particularly in large pedigrees (e.g. Mäki-Tanila 2007). For detailed accounts of the technical challenges see e.g. de Boer & Hoeschele (1993), Shaw & Woolliams (1999), or Abney *et al.* (2000).

One of the difficulties encountered by many of the studies is that estimates of the variance components often exhibit high sampling correlations (see Section 7.3.3 and e.g. Hoeschele and Vollema 1993; Shaw & Woolliams 1999; Edwards & Lamkey 2002). This stems from the confounding of the variance components, which is primarily a result of the fact that all genetic covariances increase with increasing relatedness. For example, assume that an individual produces two offspring, one by mating randomly, the other through a parent–offspring mating. In such a situation, relatedness will be higher between the individual and its inbred offspring than between the individual and its outbred offspring, but so will be the covariances between additive effects, between dominance effects, and between the additive and dominance effects (Table 8 in de Boer & Hoeschele 1993). This makes separating the variance components a challenging task. Carefully planned experimental designs can help a great deal (e.g. Kelly & Arathi 2003; Wardyn *et al.* 2007) but for studies of natural populations obtaining reliable, independent estimates of these variance components will likely remain a major challenge.

Given the difficulties involved in estimating all the dominance (co)variance components, it is not surprising that relatively few estimates exist. We are aware of a handful of studies that have estimated these non-additive variance components in a range of morphological, fitness, and production traits in maize (e.g. Edwards & Lamkey 2002; Wardyn *et al.* 2007 and references therein), in the plant *Nemophila menziesii* (Shaw *et al.* 1998), in *Eucalyptus globulus* (Costa e Silva *et al.* 2010), in cattle (Hoeschele & Vollema 1993), in sheep (Shaw & Woolliams 1999), in humans (Abney *et al.* 2000; 2001), and in monkey flowers (Kelly & Arathi 2003). Except for the studies in humans, we are not aware of any attempts to estimate them in an unmanipulated population. Hence, what follows is based mostly on the studies of laboratory and agricultural populations.

A qualitative survey of the results gives a varied picture. In some of the studies, many or all of the dominance (co)variance components were statistically significant and of substantial magnitude (e.g. Hoeschele & Vollema 1993; Edwards & Lamkey 2002; Wardyn *et al.* 2007). Although the results vary between studies and traits, the inbred dominance variances (σ^2_{DI}) were commonly greater than the random-mating dominance variance (σ^2_{DR}) and occasionally greater than the additive genetic variance (σ^2_{A}). Thus, the additional variance component reflecting the variance of dominance deviations in an inbred population can make an important contribution to the genetic variation observed in a number of different traits and taxa. Consequently, predicting selection response on traits in these studies requires estimates of the non-additive variance components. In other studies, substantial inbreeding depression and random-mating dominance variance (σ^2_{DR}) were found, but the additional dominance (co)variance components brought about by inbreeding were small and non-significant (e.g. Shaw & Woolliams 1999; Abney *et al.* 2001). For example, H^*, the square of the sum over all loci of the homozygous dominance deviations, was often close to zero, except in the maize studies. This result is consistent with a large number of loci of small effect contributing to inbreeding depression (Shaw *et al.* 1998). Thus, while the dominance (co)variance components were important in some of the studies, they were not important in others. An intriguing pattern emerged for the covariance between the additive effect of an allele and the dominance deviation for its corresponding homozygote (σ_{ADI}). Estimates of σ_{ADI} were sometimes small and non-significant, but when their magnitude was substantial, they were always negative. Negative values of σ_{ADI} imply that the stronger the effect of an allele on a trait, the higher its contribution to inbreeding depression. Even moderate levels of σ_{ADI} can lead to striking deviations of the selection response from that predicted with the breeder's equation (Shaw *et al.* 1998).

7.5.3 A reduced model with inbreeding depression

Shaw *et al.* (1998) highlight the need to consider the covariance between the additive effects of an allele and its dominance deviation (σ_{ADI}) when trying to predict selection response, even in the short term (but see de Boer & van Arendonk 1992). When σ_{ADI} is not zero, predictions of evolutionary response over as little as five generations are inaccurate unless all the variance components in Eqn. 7.6 are included in the model (Shaw *et al.* 1998).

However, when σ_{ADI} is assumed to be zero and/or if average inbreeding ($\bar{F}$) is low, it is not necessary to estimate all of the additional variance components to obtain reliable estimates of additive and dominance effects, and thus to make accurate predictions of the short-term selection response in inbred populations (de Boer & van Arendonk 1992; Shaw *et al.*1998). Instead, adding the effects of inbreeding on the mean of a trait to the animal model by using a regression on individual inbreeding coefficients (F_i) yields unbiased estimates of additive and dominance effects (e.g. Kennedy *et al.* 1988; Hoeschele & van Raden 1991; de Boer & van Arendonk 1992) and is sufficient to predict short-term selection response accurately (e.g. Shaw *et al.* 1998). Note that in such models, inbreeding is accounted for twice: once with the additive relationship matrix $\boldsymbol{A}$ and once with the regression on the individuals' inbreeding coefficient.

Ignoring the effects of inbreeding on the mean of a trait can lead to substantial biases in estimates of additive and dominance effects (e.g. de Boer & van Arendonk 1992; Abney *et al.* 2000). Reid & Keller (2010) simulated a natural bird population and showed that, with the levels of inbreeding and the magnitude of inbreeding depression in that population, ignoring the effects of inbreeding on the mean of a trait inflated heritability estimates by a factor of two. As outlined above, these biases can be removed by including a regression of phenotype on each individual's inbreeding coefficient in the animal models. Only recently have studies of wild populations started to include inbreeding effects in this way (e.g. Reid *et al.* 2011). The expected degree of bias depends on the levels of inbreeding and the magnitude of inbreeding depression in a population. If both are low (e.g. when $\bar{F} \leq 0.05$; Norris *et al.* 2009), the expected bias in heritability estimates will also be low, and it may not be necessary to account for inbreeding in these cases. However, given the low computational burden of including the inbreeding coefficient as a fixed effect in animal models, this should become standard practice in studies of natural populations.

There is an additional reason why inbreeding coefficients should be included as covariates in animal models: inbreeding coefficients of relatives are correlated in many populations even if they are mating randomly (Reid *et al.* 2006, Reid & Keller 2010). This correlation arises because factors such as population size, variance in reproductive success, and immigration may cause inbred individuals to be more closely related to randomly selected mates than outbred individuals are. Thus, individuals who are already themselves inbred will produce relatively inbred offspring, leading to a correlation in inbreeding between relatives. Note that this even affects parent–offspring or half-sib relationships that are not expected to have similar inbreeding coefficients under the standard theory, which assumes infinite population size (for details see Reid *et al.* 2006 and Reid & Keller 2010). One consequence of the correlation in inbreeding among relatives is that estimates of additive genetic variance are artificially inflated above and beyond the biases introduced through the mechanism outlined in the previous paragraph (Reid & Keller 2010). Given that many of the populations used to study quantitative genetic variation in the wild exhibit some inbreeding, it is important to account for inbreeding effects to avoid upwardly biased estimates of additive genetic variation.

7.6 Conclusion

The use of the quantitative genetic variance partitioning framework in natural populations is a mature field that has seen a lot of interest in the past two decades, not least because of the increased use of the animal model framework originally developed in the animal breeding community (Wilson *et al.* 2009; Chapter 2, Postma). The vast majority of applications in natural populations have ignored non-additive genetic variance components. Although additive genetic variances will continue to be the parameters of greatest interest, we believe it is time for our field to start paying more attention to non-additive genetic (co)variances, particularly to the dominance (co)variances discussed here (e.g. Eqn. 7.6).

There is no doubt that non-additive genetic effects are common in a variety of traits (e.g. Merilä & Sheldon 1999; Roff & Emerson 2006), a fact further supported by the frequent finding of inbreeding depression (Keller & Waller 2002; Charlesworth & Willis 2009). However, non-additive genetic effects at the gene level do not translate directly to non-additive genetic variance components in quantitative genetic analyses (see Section 7.2; Phillips 1998). For example, Charlesworth (1987, p. 26) showed that partially deleterious mutations are expected to create more additive than dominance variance in viability traits in *Drosophila*. Additive genetic variance therefore often accounts for much of the observed genetic variation (e.g. Hill *et al.* 2008). In fact, it seems possible that the predominance of additive genetic variance is itself a result of natural selection (Crow 2008).

However, our review of published estimates of dominance variance suggests that dominance variance nevertheless contributes substantially to overall genetic and phenotypic variation in a wide variety of traits. We do not claim to demonstrate that dominance variance is a ubiquitous feature of the quantitative genetics underlying complex traits, but we do suggest that a de facto assumption of its insignificance may not be warranted in all cases. In our review of published estimates, on average about one-third of the genetic variation ($\sigma_D^2 + \sigma_A^2$) was due to dominance variance, and in life-history traits of non-domestic species, this fraction exceeded 60% (Table 7.3). Thus, in some species and traits, dominance variance can make up a substantial component of genetic variation. The lack of dominance variance estimates from wild populations precludes conclusions for natural populations (except humans). By necessity, the empirical work reviewed in this chapter therefore stems mainly from studies of laboratory and agricultural populations. Where dominance variance has been studied explicitly, it has been studied either because it is regarded as a form of noise, which needs to be included in the statistical model to yield accurate estimates of other variance components (Mäki-Tanila 2007), or because it is of interest per se (e.g. Crnokrak & Roff 1995). Notably, dominance variance is of interest to those studying wild populations for both reasons.

Studies in animal breeding suggest that a variety of traits across many species display levels of dominance variance and inbreeding depression high enough to warrant inclusion in the statistical models used to estimate breeding values and additive genetic variances (see Section 7.1 and e.g. Misztal *et al.* 1998; Pante *et al.* 2002; Serenius *et al.* 2006; Palucci *et al.* 2007). This is not always the case, however (e.g. Miglior *et al.* 1995; Ferreira *et al.* 1999), and it will therefore be interesting to see how much estimates of breeding values and additive genetic variances in natural populations will change when dominance variance and inbreeding depression are included in animal models. Populations with many full-sibs are particularly likely to profit from the inclusion of dominance variance in the models (Uimari & Mäki-Tanila 1992). However, biased estimates of breeding values and additive genetic variances are just one issue. Perhaps more relevant to studies of evolution in wild populations, in the presence of both dominance and inbreeding the breeder's equation (Eqn. 7.2) falls short of making useful predictions of evolutionary response, even over as few as five generations (de Boer & van Arendonk 1992; Shaw *et al.* 1998). To derive useful predictions of the short-term evolutionary response under inbreeding and dominance, estimates of the additional dominance (co)variance components are essential (Eqn. 7.6). In most cases, we currently lack information on the magnitude of these components. The few empirical results available to date suggest that these components can be of substantial magnitude, particularly in highly inbred populations, e.g. of (partially) selfing plants. Thus, to accurately predict short-term evolutionary response in selfing species, estimates of dominance (co)variance components seem essential. With the population sizes and mating schemes typical of animal breeding, and in human populations, the magnitude of the dominance (co)variance components appear to be smaller. It remains to be seen whether this reflects difficulties in estimation and inadequate statistical power (e.g. Shaw & Woolliams 1999) or true biological differences. If true, then the reduced model (see Section 7.5.3) should be sufficient to predict short-term evolutionary response in these populations.

Dominance variance is, however, not only statistical noise. Perhaps more interestingly, an understanding of the dynamics of dominance variance itself can yield key insights into many areas of evolutionary theory. For example, a test of evolutionary theory has been to compare the amounts of additive and dominance variances among classes of traits generally thought to experience different intensities of selection. Evolutionary theory predicts strong selection will decrease additive genetic variance, while no change is expected for dominance variance. Thus, traits thought to be under stronger selection or closely related to fitness tend to display low additive variance, but high dominance variance (Crnokrak and Roff 1995). Similarly, the relative amounts of dominance variance, additive variance, and inbreeding depression provide a means of detecting variability maintained by selection (Charlesworth & Hughes 2000), and they lend insight into the mechanisms maintaining genetic variance. For example, analytical models postulate that if antagonistic pleiotropy acts to maintain additive genetic variance for fitness-related traits, then the amount of dominance variance relative to additive variance is expected to be around 0.5 (Curtsinger *et al.* 1994; Roff 1997; Merilä & Sheldon 1999). Kelly (1999) has shown that the contribution of rare, partially recessive, deleterious alleles can be inferred by the ratio of σ_{ADI} to σ_A. Alternatively, the amount of dominance variance relative to additive variance can be used to test the influence of mutation accumulation vs antagonistic pleiotropy in the genetic mechanisms of senescence (Charlesworth & Hughes 1996, Wilson *et al.* 2008, Escobar *et al.* 2008) or to understand evolution of sexual dimorphism in traits undergoing sexually antagonistic selection (Fairbairn & Roff 2006; Fry 2010). Further, estimates of dominance variance are necessary to understand adaptive evolutionary responses when populations undergo bottlenecks and subsequent genetic drift. In such cases non-additive genetic variance can be 'converted' to additive genetic variance, as a consequence of the change in allele frequency due to genetic drift (Cockerham & Tachida 1988; Goodnight 1988; Willis & Orr 1993; Barton & Turelli 2004). However, the dominance genetic effects themselves will also lead to a reduction in trait values for inbred populations after a bottleneck (e.g. Taft & Roff 2012), introducing further complexities as the increased potential to respond to selection may be counteracted by inbreeding depression.

For all these reasons, we encourage empiricists to place more emphasis on the estimation of dominance variance than has typically been the practice. As outlined in this chapter, there are some obstacles to overcome to obtain estimates of dominance variance from wild populations. Our intention in compiling and summarising all these issues is to inform future investigations of dominance variance in wild populations and not to impose an overwhelming sense of futility. In fact, we believe that several wild populations under study have the potential to overcome the issues noted above. This is true in particular for avian study systems with small population sizes from year to year, several offspring per reproductive effort, extra-pair copulations, more than one reproductive effort per season, and pedigree records for many generations.

In this chapter we have laid out the theory to illustrate how phenotypic variation is caused by non-additive interactions between alleles at a locus. We hope it is clear from this how the quantitative genetic framework can be used to summarise these non-additive interactions as population level contributions to phenotypic variance. Our survey of the complicating factors in practice should serve as a guide to tackle any of the key areas of evolutionary theory for which we would predict the underlying non-additive allelic interactions to result in observable dynamics within a population. Finally, our compilation of dominance variance estimates show that, though relatively scant information is available for non-domestic species, the body of previous estimates suggests an important contribution of dominance variation to phenotypic and genetic variation, which we hope will prompt more studies in the future.

Acknowledgments

We thank Ursina Tobler and Franziska Loercher for compiling estimates of dominance variance from the literature, and Philipp Becker, Timothée Bonnet, Anne Charmantier, Danny Garant, Loeske Kruuk, Pirmin Nietlisbach, and Marta Szulkin for comments that greatly improved this chapter. L.F.K. thanks Anders Christian Sørensen for teaching him how to estimate dominance variances in inbred populations. M.E.W. was supported by a graduate research fellowship from the US National Science Foundation during this work and L.F.K. by a grant from the Swiss National Science Foundation.

Literature cited

Abney, M., McPeek, M.S. & Ober, C. (2000) Estimation of variance components of quantitative traits in inbred populations. *American Journal of Human Genetics*, **66**, 629–650.

Abney, M., McPeek, M.S. & Ober, C. (2001) Broad and narrow heritabilities of quantitative traits in a founder population. *American Journal of Human Genetics*, **68**, 1302–1307.

Alitzer, S., Harvell, D. & Friedle, E. (2003). Rapid evolutionary dynamics and disease threats to biodiversity. *Trends in Ecology and Evolution*, **18**, 589–596.

Andersson, M. 1994. *Sexual selection*. Princeton University Press, Princeton.

Barton, N.H. & Turelli, M. (2004) Effects of genetic drift on variance components under a general model of epistasis. *Evolution*, **58**, 2111–2132.

Boag, P.J. & van Noordwijk, A.J. 1987. Quantitative genetics. In: *Avian genetics: a population and ecological approach* (ed. F. Cooke & P.A. Buckley), pp. 45–78. Academic Press, London.

Bulmer, M.G (1980) *The mathematical theory of quantitative genetics*. Clarendon Press, Oxford.

Chalh, A. & El Gazzah, M. (2003) Estimation of the dominance merit in non-inbred populations without recourse to its inverted relationship matrix. *Journal of Applied Genetics*, **44**, 63–69.

Charlesworth, D. & Charlesworth, B. (1987) Inbreeding depression and its evolutionary consequences. *Annual Review of Ecology and Systematics*, **18**, 237–268.

Charlesworth, D. & Charlesworth, B. (1995) Quantitative genetics in plants: the effect of the breeding system on genetic variability. *Evolution*, **49**, 911–920.

Charlesworth, B. & Charlesworth, D. (1999) The genetic basis of inbreeding depression. *Genetical Research*, **74**, 329–340.

Charlesworth, B. & Hughes, K.A. (1996) Age-specific inbreeding depression and components of genetic variance in relation to the evolution of senescence. *Proceedings of the National Academy of Sciences of the United States of America*, **93**, 6140–6145.

Charlesworth, B. & Hughes, K.A. (2000) The maintenance of genetic variation in life-history traits. In *Evolutionary genetics: from molecules to morphology* (ed. R.S. Singh & C.B. Krimbas), pp. 369–392. Cambridge University Press, Cambridge.

Charlesworth, D. & Willis, J.H. (2009) The genetics of inbreeding depression. *Nature Reviews Genetics*, **10**, 783–796.

Clément, V., Bibé, B., Verrier, E., Elsen, J., Manfredi, E., Bouix, J. & Hanocq E. (2001) Simulation analysis to test the influence of model adequacy and data structure on the estimation of genetic parameters for traits with direct and maternal effects. *Genetics, Selection, and Evolution*, **33**, 369–395.

Cockerham, C.C. & Weir, B.S. (1977). Quadratic analysis of reciprocal crosses. *Biometrics*, **33**, 187–203.

Cockerham, C.C. & Tachida, H. (1988) Permanency of response to selection for quantitative characters in finite populations. *Proceedings of the National Academy of Sciences of the United States of America*, **85**, 1563–1565

Comstock, R.E. & Robinson, H.F. (1948) The components of genetic variance in populations of biparental progenies and their use in estimating the average degree of dominance. *Biometrics*, **4**, 254–266.

Costa e Silva, J., Hardner, C. & Potts, B.M. (2010) Genetic variation and parental performance under inbreeding for growth in *Eucalyptus globulus*. *Annals of Forest Science*, **67**, 606.

Coyne, J.A., Barton, N.H. & Turelli, M. (1997) A critique of Sewall Wright's shifting balance theory of evolution. *Evolution*, **51**, 643–671.

Crnokrak, P. & Roff, D.A. (1995). Dominance variance: associations with selection and fitness. *Heredity*, **75**, 530–540.

Crnokrak, P. & Roff, D.A. (1999) Inbreeding depression in the wild. *Heredity*, **83**, 260–270.

Crow, J.F. (2008). Maintaing evolvability. *Journal of Genetics*, **87**, 349–353.

Curtsinger, J.W., Service P.M. & Prout, T. (1994) Antagonistic pleiotropy, reversal of dominance, and genetic polymorphism. *American Naturalist*, **144**, 210–228.

Darwin, C. (1876) *The effects of crossing and self-fertilization in the vegetable kingdom*. John Murray, London.

de Boer, I.J.M. & van Arendonk, J.A.M. (1992) Prediction of additive and dominance effects in selected or unselected populations with inbreeding. *Theoretical and Applied Genetics*, **84**, 451–459.

de Boer, I.J.M. & Hoeschele, I. (1993) Genetic evaluation methods for populations with dominance and inbreeding. *Theoretical and Applied Genetics*, **86**, 245–258.

DeRose, M.A. & Roff, D.A. (1999) A comparison of inbreeding depression in life-history and morphological traits in animals. *Evolution*, **53**, 1288–1292.

Edwards, J.W. & Lamkey, K.R. (2002) Quantitative genetics of inbreeding in a synthetic maize population. *Crop Science*, **42**, 1094–1104.

Escobar, J.S., Jarne, P., Charmantier, A. & David, P. (2008) Outbreeding alleviates senescence in hermaphroditic snails as expected from the mutation-accumulation theory. *Current Biology*, **18**, 906–910.

Falconer, D.S. & Mackay, T. (1996) *Introduction to quantitative genetics*, 4th edn. Longman Group Ltd., Harlow.

Fairbairn, D.J. & Roff, D.A. (2006) The quantitative genetics of sexual dimorphism: assessing the importance of sex-linkage. *Heredity*, **97**, 319–328.

Fenster, C.B., Galloway, L.F. & Chao, L. (1997) Epistasis and its consequences for the evolution of natural populations. *Trends in Ecology and Evolution*, **12**, 282–286.

Ferreira, G.B., MacNeil, M.D. & Van Vleck, L.D. (1999) Variance components and breeding values for growth traits from different statistical models. *Journal of Animal Science*, **77**, 2641–2650.

Fisher, R.A. (1918) The correlation between relatives on the supposition of Mendelian inheritance. *Transactions of the Royal Society of Edinburgh*, **52**, 399–433.

Fisher, R.A. (1958) *The genetical theory of natural selection*, 2nd edn. Dover, New York.

Fry, J.D. (2010) The genomic location of sexually antagonistic variation: some cautionary comments. *Evolution*, **64**, 1510–1516.

Goodnight, C.J. (1988) Epistasis and the effect of founder events on the additive genetic variance. *Evolution*, **42**, 441–454.

Hadfield, J.D. (2010) MCMC methods for multi-response generalized linear mixed models: The MCMCglmm R package. *Journal of Statistical Software*, **33**, 1–22.

Hallander, J., Waldmann, P., Wang, C. & Sillanpää, M.J. (2010) Bayesian inference of genetic parameters based on conditional decompositions of multivariate normal distributions. *Genetics*, **185**, 645–654.

Harris, D.L. (1964) Genotypic covariances between inbred relatives. *Genetics*, **50**, 1319–1348.

Henderson, C. R. (1973) Sire evaluation and genetic trends. *Journal of Animal Science*, **76**, 10–41.

Henderson, C.R. (1976) A simple method for computing the inverse of a numerator relationship matrix used in predicting of breeding values. *Biometrics*, **32**, 69–83.

Henderson, C.R. (1984) *Applications of linear models in animal breeding*. University of Guelph Press, Guelph.

Henderson, C.R. (1985) Best linear unbiased prediction of nonadditive genetic merits in noninbred populations. *Journal of Animal Science*, **60**, 111–117.

Hill, W.G. (1999) Advances in quantitative genetics theory. In: *From Jay L. Lush to genomics: visions for animal breeding and genetics* (ed. J. C. M. Dekkers, S. J. Lamont & M. F. Rothschild), pp. 35–46. Iowa State University, Ames.

Hill, W.G., Goddard, M.E. & Visscher P.M. (2008) Data and theory point to mainly additive genetic variance for complex traits. *PLoS Genetics*, **4**, e1000008.

Hoeschele, I. & vanRaden, P.M. (1991) Rapid inversion of dominance relationship matrices for noninbred populations by including sire by dam subclass effects. *Journal of Dairy Science*, **74**, 557–569.

Hoeschele, I. & Vollema, A.R. (1993) Estimation of variance components with dominance and inbreeding in dairy cattle. *Journal of Animal Breeding and Genetics*, **110**, 93–104.

Jacquard, A. (1974) *The Genetic structure of populations*. Springer-Verlag, New York.

Keller, L.F. (1998) Inbreeding and its fitness effects in an insular population of song sparrows (*Melospiza melodia*). *Evolution*, **52**, 240–250.

Keller, L.F. & Waller, D.M. (2002) Inbreeding effects in wild populations. *Trends in Ecology & Evolution*, **17**, 230–241.

Keller, L.F., Biebach, I., Ewing, S.R. & Hoeck, P.E.A. (2012) The genetics of reintroductions: inbreeding and genetic drift. In: *Reintroduction biology: integrating science and management*(ed. J.G. Ewen, D.P. Armstrong, K.A. Parker & P.J. Seddon), pp. 360–394. Blackwell Publishing Ltd., Chichester.

Kelly, J.K. (1999) An experimental method for evaluating the contribution of deleterious mutations to quantitative trait variation. *Genetical Research*, **73**, 263–273.

Kelly, J.K. & Arathi, H.S. (2003) Inbreeding and the genetic variance in floral traits of *Mimulus guttatus*. *Heredity*, **90**, 77–83.

Kennedy, B.W. & Sorensen, D.A. (1988) Properties of mixed-model methods for prediction of genetic merit. In: *Proceedings of the second international conference on quantitative genetics* (ed. B.S. Weir, E.J. Eisen, M.M. Goodman & G. Namkoong), pp. 91–103. Sinauer Associates, Inc., Sunderland.

Kennedy, B.W., Schaeffer, L.R. & Sorensen, D.A. (1988) Genetic properties of animal models. *Journal of Dairy Science*, **71** (Suppl. 2), 17–26.

Kirkpatrick, M. & Ryan M. (1991) The evolution of mating preferences and the paradox of the lek. *Nature*, **350**, 33–39.

Kristensen, T. N., Sørensen, A. C., Sørensen, D., Pedersen, K. S., Sørensen, J. G. & Loeschcke, V. (2005) A test of quantitative genetic theory using **Drosophila**—effects of inbreeding and rate of inbreeding on heritabilities and variance components. *Journal of Evolutionary Biology*, **18**, 763–770.

Kruuk, L.E.B. (2004) Estimating genetic parameters in natural populations using the 'animal model'. *Philosophical Transactions of the Royal Society of London B, Biological Sciences*, **359**, 873–890.

Kruuk, L.E.B. & Hadfield, J.D. (2007) How to separate genetic and environmental causes of similarity between relatives. *Journal of Evolutionary Biology*, **20**, 1890–1903.

Kruuk, L.E.B., Slate, J. & Wilson, A.J. (2008) New answers for old questions: the evolutionary quantitative genetics of wild animal populations. *Annual Review of Ecology, Evolution, and Systematics*, **39**, 525–548.

Lunn, D.J., Thomas, A., Best, N. & Spiegelhalter, D. (2000) WinBUGS—a Bayesian modeling framework: concepts, structure, and extensibility. *Statistical Computing*, **10**, 325–337.

Lynch, M. & Walsh, B. (1998) *Genetics and analysis of quantitative traits*. Sinauer, Sunderland.

Mäki-Tanila, A. (2007) An overview on quantitative and genomic tools for utilizing dominance genetic variation in improving animal production. *Agricultural and Food Science*, **16**, 188–198.

Merilä, J. & Sheldon, B.C. (1999) Genetic architecture of fitness and nonfitness traits: empirical patterns and development of ideas. *Heredity*, **83**, 103–109.

Merilä, J., Sheldon, B.C. & Kruuk, L.E.B. (2001). Explaining stasis: microevolutionary studies in natural populations. *Genetica*, **112–113**, 199–222.

Meuwissen, T.H.E. & Luo, Z. (1992) Computing inbreeding coefficients in large populations. *Genetics, Selection, and Evolution*, **24**, 305–313.

Meyer, K. (2008) Likelihood calculations to evaluate experimental designs to estimate genetic variances. *Heredity*, **101**, 212–221.

Miglior, F., Burnside, E.B. & Kennedy, B.W. (1995). Production traits of Holstein cattle: estimation of nonadditive genetic variance components and inbreeding depression. *Journal of Dairy Science*, **78**, 1174–1180.

Misztal, I. (1997) Estimation of variance components with large scale dominance models. *Journal of Dairy Science*, **80**, 965–974.

Misztal, I. & Besbes, B. (2000) Estimates of parental-dominance and full-sib permanent environment variances in laying hens. *Animal Science*, **71**, 421–426.

Misztal, I., Lawlor, T.J. & Fernando, R.L. (1997) Dominance models with method R for stature of Holsteins. *Journal of Dairy Science*, **80**, 975–978.

Misztal, I., Varona, L., Culbertson, M., Bertrand, J.K., Mabry, J., Lawlor, T.J., van Tassel, C.P. & Gengler, N. (1998). Studies on the value of incorporating the effect of dominance in genetic evaluations of dairy cattle, beef cattle and swine. *Biotechnology, Agronomy, Society, and Environment*, **2**, 227–233.

Morrissey, M.B., Kruuk, L.E.B. & Wilson, A.J. (2010) The danger of applying the breeder's equation in observational studies of natural populations. *Journal of Evolutionary Biology*, **23**, 2277–2288.

Morrissey, M.B. & Wilson, A.J. (2010) pedantics: an r package for pedigree-based genetic simulation and pedigree manipulation, characterization and viewing. *Molecular Ecology Resources*, **10**, 711–719.

Neiman, M. & Linksvayer, T.A. (2006) The conversion of variance and the evolutionary potential of restricted recombination. *Heredity*, **96**, 111–121.

Norris, D., Ngambi, W. & Mbajiorgu, C. A. (2006) The conversion of additive and non-additive genetic variances with varying levels of inbreeding. *Journal of Biological Sciences*, **9**, 254–258.

O'Hara, R.B., Cano, J.M., Ovaskainen, O., Teplitsky, C. & Alho, J.S. (2008) Bayesian approaches in evolutionary quantitative genetics. *Journal of Evolutionary Biology*, **21**, 949–957.

Ovaskainen, O., Cano, J.M. & J. Merilä (2008) A Bayesian framework for comparative quantitative genetics. *Proceedings of the Royal Society B-Biological Sciences*, **275**, 669–678.

Palucci, V., Schaeffer, L.R., Miglior, F. & Osborne, V. (2007) Non-additive genetic effects for fertility traits in Canadian Holstein cattle. *Genetics, Selection, Evolution*, **39**, 181–193.

Pante, M.J.R., Gjerde, B., McMillan, I. & Misztal, I. (2002) Estimation of additive and dominance genetic variances for body weight at harvest in rainbow trout, *Oncorhynchus mykiss*. *Aquaculture*, **204**, 383–392.

Phillips, P.C. (1998) The language of gene interaction. *Genetics*, **149**, 1167–1171.

Reid, J.M. (2007) Secondary sexual ornamentation and non-additive genetic benefits of female mate choice. *Proceedings of the Royal Society B-Biological Sciences*, **274**, 1395–1402.

Reid, J.M. & Keller, L.F. (2010) Correlated inbreeding among relatives: occurrence, magnitude, and implications. *Evolution*, **64**, 973–985.

Reid, J.M., Arcese, P. & Keller, L.F. (2006) Intrinsic parent-offspring correlation in inbreeding level in a song sparrow (*Melospiza melodia*) population open to immigration. *American Naturalist*, **168**, 1–13.

Reid, J.M., Arcese, P., Sardell, R.J. & Keller, L.F. (2011) Additive genetic variance, heritability, and inbreeding depression in male extra-pair reproductive success. *American Naturalist*, **177**, 177–187.

Reznick, D.N. & Ghalambor, C.K. (2001) The population ecology of contemporary adaptations: what empirical studies reveal about the conditions that promote adaptive evolution. *Genetica*, **112–113**, 183–198.

Roff, D.A. (1997) *Evolutionary quantitative genetics*. Chapman & Hall, New York.

Roff, D.A. (2002) Inbreeding depression: tests of the overdominance and partial dominance hypotheses. *Evolution*, **56**, 768–755.

Roff, D.A. (2006) Evolutionary quantitative genetics. *In: Evolutionary genetics: concepts and case studies*, (ed. C.W. Fox & J.B. Wolf,), pp. 267–287. Oxford University Press, New York.

Roff, D.A. (2007). Comparing sire and dam estimates of heritability: jackknife and likelihood approaches. *Heredity*, **100**, 32–38.

Roff, D.A. & Emerson, K. (2006). Epistasis and dominance: evidence for differential effects in life-history versus morphological traits. *Evolution*, **60**, 1981–1990.

Schaeffer, L.R. (2003) Computing simplifications for non-additive genetic models. *Journal of Animal Breeding and Genetics*, **120**, 394–402.

Serenius, T., Stalder, K.J. & Puonti, M. (2006) Impact of dominance effects on sow longevity. *Journal of Animal Breeding and Genetics*, **123**, 355–361.

Shaw, R.G., Byers, D.L. & Shaw, F.H. (1998) Genetic components of variation in *Nemophila menziesii* undergoing inbreeding: morphology and flowering time. *Genetics*, **150**, 1649–1661.

Shaw, F.H. & Wooliams, J.A. (1999) Variance component analysis of skin and weight data for sheep subjected to rapid inbreeding. *Genetics, Selection, Evolution*, **31**, 43–59.

Smith, J.N.M., Keller, L.F., Marr, A.B. & Arcese, P. (2006) *Conservation and biology of small populations: the song sparrows of Mandarte Island*. Oxford University Press, New York.

Smith, S.P. & Mäki-Tanila, A. (1990) Genotypic covariance matrices and their inverses for models allowing dominance and inbreeding. *Genetics, Selection, and Evolution*, **22**, 65–91.

Taft, H.R. & Roff, D.A. (2012) Do bottlenecks increase additive genetic variance? *Conservation Genetics*, **13**, 333–342.

Thornhill, N.W., ed. (1993) *The natural history of inbreeding and outbreeding: theoretical and empirical perspectives*. University of Chicago Press, Chicago.

Uimari, P. & Mäki-Tanila, A. (1992) Accuracy of genetic evaluations in dominance genetic models allowing for inbreeding. *Journal of Animal Breeding and Genetics*, **109**, 401–407.

van Buskirk, J. & Willi, Y. (2006) The change in quantitative genetic variation with inbreeding. *Evolution*, **60**, 2428–2434.

Wade, M.J. & Goodnight, C.J. (1998) The theories of Fisher and Wright in the context of metapopulations: when nature does many small experiments. *Evolution*, **52**, 1537–1553.

Waldmann, P. (2001) Additive and non-additive genetic architecture of two different-sized populations of *Scabiosa canescens*. *Heredity*, **86**, 648–657.

Waldmann, P., Hallander, J., Hoti, F. & Sillanpää, M.J. (2008) Efficient Markov Chain Monte Carlo implementation of Bayesian analysis of additive and dominance genetic variances in noninbred pedigrees. *Genetics*, **179**, 1101–1112.

Walling, C.A., Pemberton, J.M., Hadfield, J.D. & Kruuk, L.E.B. (2010) Comparing parentage inference software:

reanalysis of a red deer pedigree. *Molecular Ecology*, **19**, 1914–1928.

Walsh, B. (2005) The struggle to exploit non-additive variation. *Australian Journal of Agricultural Research*, **56**, 873–881.

Wardyn, B.M., Edwards, J.W. & Lamkey, K.R. (2007) The genetic structure of a maize population: the role of dominance. *Crop Science*, **47**, 467–476.

Whitlock, M.C., Phillips, P.C., Moore, F.B.-G. & Tonsor, S.J. (1995). Multiple fitness peaks and epistasis. *Annual Review of Ecology and Systematics*, **26**, 601–629.

Willis, J.H. & Orr, H.A. (1993) Increased heritable variation following population bottlenecks: the role of dominance. *Evolution*, **47**, 949–957.

Wilson, A.J., Charmantier, A. & Hadfield, J.D. (2008) Evolutionary genetics of ageing in the wild: empirical patterns and future perspectives. *Functional Ecology*, **22**, 431–442.

Wilson, A.J., Réale, D., Clements, M.N., Morrissey, M.M., Postma, E., Walling, C.A., Kruuk, L.E.B. & Nussey, D.H.2010. An ecologist's guide to the animal model. *Journal of Animal Ecology*, **79**, 13–26.

Wolak, M.E. (2012) nadiv: an R package to create relatedness matrices for estimating non-additive genetic variances in animal models. *Methods in Ecology and Evolution*, **3**, 792–796.

www.oup.co.uk/companion/charmantier

CHAPTER 8

Cross-pollination of plants and animals: wild quantitative genetics and plant evolutionary genetics

John R. Stinchcombe

8.1 Introduction

Several years ago, Terborgh (1988) pointed out the dramatic influence that the life-history, population density, and experimental tractability of one's focal study organism can have on the biological conclusions gained, and even on the questions posed by biologists of different stripes. Against the backdrop of the remarkable progress in wild evolutionary quantitative genetics, in this chapter I focus on how the life-history and natural history of organisms might affect the breadth and depth of application of the lessons and insights obtained from evolutionary genetic studies performed in other taxonomic groups. In particular, I examine the potential benefits of a reciprocal integration between what are now two largely disjunct literatures in evolutionary genetics: experimental studies focussed primarily on annual plants, and field studies of free-living animals. Some of the fundamental findings on the evolutionary ecology and ecological genetics of plants—fine-scale local adaptation even in the face of abundant gene flow, abundant genotype × environment ($G \times E$) interactions for traits and fitness components, ecological dependency in the strength and direction of natural selection—are likely general findings that will apply to many systems and may have important implications to studies of free-living animals. Likewise, the lessons of wild quantitative genetics in animals, and also some of its analytical tools, have largely yet to be applied or realised in experimental evolutionary ecology studies of short-lived plants.

The potential benefits of a reciprocal interaction between subfields can be illustrated with an exemplar topic: the role of environmental covariances in biasing phenotypic estimates of natural selection made with the Lande-Arnold (1983) approach. The role of the environment in influencing both traits and fitness, and thus creating a misleading impression of directional selection and its potential to generate an evolutionary response, was first pointed out in studies of bird populations, where body 'condition' could affect lay date, tarsus length, and fitness components, leading to erroneous estimates of natural selection on these traits because a correlated variable has been omitted from the analysis (Price *et al.* 1988; Schluter *et al.* 1991). Rausher and Simms (1989) and Simms and Rausher (1989) applied a solution that eliminated, or strongly reduced, the role of environmentally induced covariances: measuring selection with genetically, rather than phenotypically, based estimates of the traits and fitness (Rausher 1992). Initially, application of this approach was almost exclusively pursued in annual or short-lived perennial plants, where crossing designs could easily generate full- and half-sib families for use in experimental studies

Quantitative Genetics in the Wild. Edited by Anne Charmantier, Dany Garant, and Loeske E. B. Kruuk

(Simms & Rausher 1993; Nunez-Farfan & Dirzo 1994). The potential value of this approach was also quickly realised by those studying free-living animals (Kruuk *et al.* 2001; Merilä *et al.* 2001; Kruuk *et al.* 2002), providing evidence that environmental covariances between traits and fitness could be strong in highly mobile organisms (as opposed to just sessile plants).[1] Lately, the potential statistical problems with use of BLUPs to estimate breeding values, either to draw conclusions about temporal trends or as variables in regression-based analyses of selection, were described by workers focussing mainly on animal systems (see Postma 2006, Hadfield 2008; Hadfield *et al.* 2010 and Chapter 15, Gienapp & Brommer for a full explanation of this very technical topic), and are percolating back into experimental studies in plants (Carmona & Fornoni 2013). Further developments—such as directly estimating the genetic covariance between a trait and fitness to estimate selection—are now being applied in both plant and animal systems (Etterson & Shaw 2001; Morrissey *et al.* 2012a; 2012b). In this case, our understanding of a fundamental topic in all of evolutionary biology—natural selection—has been aided by focussing on core concepts and avoiding taxonomic blinders.

The organisation of the chapter is as follows. First, I describe a suite of fundamental findings in plant evolutionary ecology, focussing on their implications for wild animal studies. Next, I describe some of the biological, methodological, and financial challenges that are inhibiting the potential application of wild quantitative genetic approaches in free-living, unmanipulated plant populations (and other species with similar life histories). Last, I describe some underutilised or simply unutilised tools for estimating quantitative genetic parameters and natural selection that have been developed in evolutionary quantitative genetics that could yield insight on plant and other species without requiring manipulative experiments, thus potentially opening the door for future studies in wild quantitative genetics in free-living plant populations, as opposed to experimental gardens. I close by evaluating the reciprocal benefits that the lessons and outcomes of plant evolutionary biology and wild quantitative genetics, as it has been applied to animals, have for each other.

[1] The potential impact of environmental variances and covariances will be strongest when heritabilities ($h^2 = V_A/V_P = V_A/(V_A + V_E)$) are low, implying that V_E is a sizeable fraction of V_P, rather than anything to do with the taxon of the focal organism.

8.2 Lessons from plant evolutionary genetics

8.2.1 Local adaptation in the face of gene flow

Local adaptation is a ubiquitous feature of numerous biological systems. Abundant evidence exists to support local adaptation at the broad geographic scale (Cogni & Futuyma 2009; Joshi *et al.* 2001), with meta-analyses putting the frequency of local adaptation between 45% and 71% depending on the study and taxonomic focus (Leimu & Fischer 2008; Hereford 2009), and one study showing a mean fitness advantage of local populations of ~45% (Hereford 2009).

A common intuition is that local adaptive genetic differentiation occurs at broader spatial scales but is broken down by dispersal and gene flow at more local scales, and also that phenotypic plasticity is an alternative to adaptation. Despite this intuition, prominent botanical counterexamples of fine-scale adaptation include the evolution of tolerance to mine tailings (Antonovics & Bradshaw 1970), local adaptation to competitor species and genotypes in white clover (*Trifolium repens*; Turkington & Harper 1979; Aarsen & Turkington 1985), and the repeated evolution of differential growth and shade-avoidance phenotypes in touch-me-nots (*Impatiens capensis*) that grow in both forest and open habitats (Dudley & Schmitt 1995). While phenotypic plasticity is commonly posited as an alternative to adaptation, these same *Impatiens* populations exhibit adaptive differentiation in their plastic response to intraspecific competition that varies predictably between habitats (Donohue *et al.* 2000). Despite abundant gene flow between habitats inferred from molecular markers (F_{ST} of ~0.16; ~1.5 migrants per generation), the repeated evolution of differentiated quantitative traits above and beyond neutral expectations points to strong selection maintaining adaptive differentiation (von Wettberg *et al.* 2008).

What implications might these findings have for animal studies of wild quantitative genetics? The goal of many wild quantitative genetic studies is to predict the evolutionary dynamics and change (or lack of change) in the studied populations. Answering this question is aided by some knowledge of the genetic architecture of quantitative traits. For example, traits whose variance is influenced primarily by deleterious, rare, recessive alleles are less likely to show a response to selection than traits influenced by alleles that show phenotypic effects in one environment and not others ('conditional neutrality' in the QTL and association mapping literature: Fournier-Level *et al.* 2011; Hall *et al.* 2010; Anderson *et al.* 2013). A strong pattern of local adaptation leads to clear predictions about the type and amount of genetic variance that should be remaining for ecologically important traits and fitness within populations. Much of the genetic variance within populations might have been already reduced during adaptation, with the remaining variance expressed (and measured by biologists) due to deleterious alleles segregating at mutation–selection balance, or maladapted alleles that have been introduced by gene flow that are at migration–selection balance. Likewise, stabilising selection to maintain locally adapted phenotypes is expected to erode genetic variance within populations (Johnson & Barton 2005), compared to non-adapted populations. In each of these scenarios, the presence of local adaptation and/or stabilising selection is expected to affect the type and amount of genetic variation segregating within populations, leading potentially to different likelihoods of evolutionary change compared to populations that are not locally adapted.

One objection to this line of argument might be that sessile plants are more likely to show fine-scale local adaptation than mobile animals. To choose one prominent example, the potential influence of local adaptation, in concert with gene flow, has been explored in populations of the great tit (*Parus major*) on Vlieland Island (Netherlands). Female birds on the eastern half of the island were under strong selection for reduced clutch size between 1955 and 1975, and show reduced clutch sizes compared to the mainland, suggesting either adaptation or founder effects, or both (Postma & van Noordwijk 2005). In contrast, immigrant birds from the mainland have breeding values (genes) for larger clutch sizes and exhibit lower survival on Vlieland. Postma and van Noordwijk carefully show that when a female's immigrant status is included in the statistical model, selection is not acting directly on clutch size, suggesting that it does not affect survival per se, but that other characteristics of immigrants, which are correlated with clutch size, must be deleterious for fitness on Vlieland. Regardless of the specific traits under selection, these data suggest that the immigrant genes in the eastern half of Vlieland are likely to be under some form of migration–selection balance. In this case, a strong pattern of local adaptation, combined with gene flow, alters the type of genetic variance segregating in the population that is available to fuel future responses to selection, because much of the variation is due to maladaptive, immigrant-derived alleles.[2] These alleles are thus unlikely to contribute to future adaptive evolution of clutch size or other ecologically important traits, unless there is a change in the environment.

The extent of these potential problems for current and future wild quantitative genetics studies remains essentially unknown. Abundant work in plant systems suggests local adaptation, and studies of wild populations also reveal evidence for it (Postma & van Noordwijk 2005), as well as fine-scale local differentiation driven by a combination of non-random dispersal, selection, and the expression of genetic variance (Garant *et al.* 2005). Detailed ecological work, characterising population densities, habitat suitability (e.g. abundance of prey items), and associations between focal variables and ecological variables will provide clues to local adaptation, non-random habitat preferences, and ecological heterogeneity. Greater sampling intensity such as catching every animal (e.g. Reid 2012) would eliminate the possibility that the individuals in the study area are somehow a non-random

[2] The potential of maladapted alleles to slow responses to selection is at the heart of evolutionary strategies to prevent the evolution of resistance to GMO crops. By planting refuges of non-GMO crops, susceptible insects will have sufficient fitness and contribute alleles to the next generation, slowing the evolution of GMO resistance. See Carrière *et al.* (2010) for a review.

sample of a larger population. Ideally, for smaller or less mobile organisms, reciprocal transplant experiments could be performed to rigorously test for the effects of local adaptation on mean fitness and quantitative genetic variances. These studies would have to solve the challenges of preventing movement or migration back to the home site by using sufficient distances, as well as potential ethical concerns of introducing potentially maladapted individuals and alleles into new habitats. Collectively, studies such as these—in both plant and animal systems—would go a long way towards addressing whether local adaptation alters the amount and type of genetic variance segregating in populations, and therefore their potential to respond to selection.

8.2.2 Prominent $G \times E$ interactions

Plants are well known for their plastic responses to environmental conditions (Bradshaw 1965), yet one of the characteristic features of animals—behaviour—also largely involves plastic responses to environmental conditions, suggesting that both plant and animal biologists may have much to learn from each other about phenotypic plasticity. Within the evolutionary quantitative genetics context, it is important to contrast phenotypic plasticity from $G \times E$ interactions (Figure 8.1). Plasticity occurs whenever phenotypes change in response to environmental variables (Figure 8.1a), while $G \times E$ interactions occur only if genotypes differ in their plastic responses (i.e. genetic variation for plasticity, Figures 8.1b and c). The ease of planting replicate genotypes into multiple (micro)habitats has given some perspective on the geographic scale of plastic responses for life-history and morphological traits, and fitness components. The approach, sometimes referred to as the phytometer approach (Clements & Goldsmith 1924, Huber *et al.* 2004), involves using known genotypes of 'tester' plants to measure environmental gradients that can affect plant traits, performance, and fitness.

In a remarkable paper, Stratton (1994) showed how this approach could be used to measure the spatial scale of $G \times E$ interactions for fitness components. He planted replicates of three genotypes of the annual weed (*Erigeron annuus*; annual or daisy fleabane), into 630 locations in a spatially nested design, which allowed him to evaluate genotype × spatial environment interactions for survival and fecundity on spatial scales ranging from 10 cm to 100 m. When he analysed relative fitness (mean

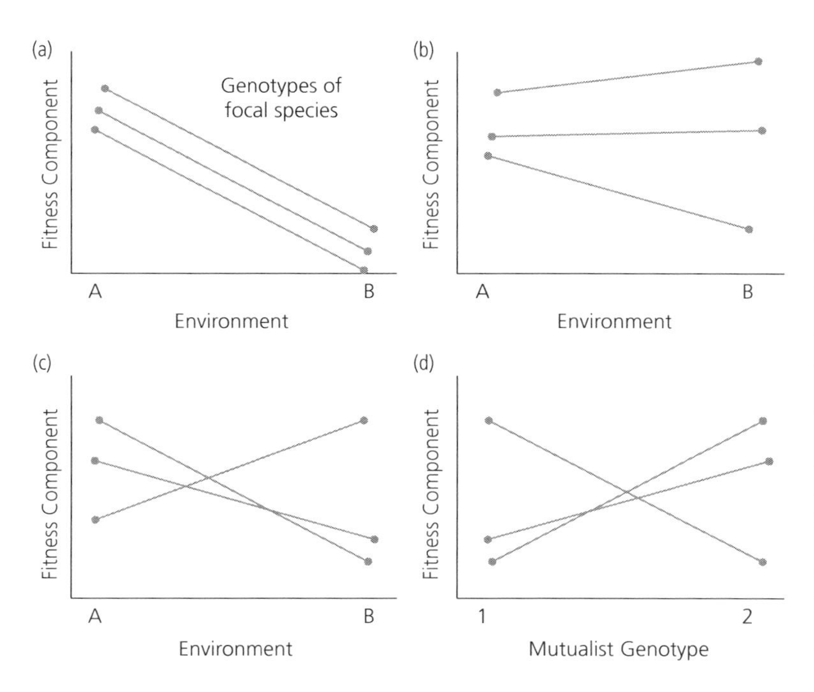

Figure 8.1 Graphical depictions of (a) phenotypic plasticity, (b) genotype × environment interactions without crossing reaction norms that will not maintain genetic variation, (c) genotype × environment interactions with crossing reaction norms that will maintain genetic variation, and (d) genotype × genotype interactions that will contribute to the maintenance of genetic variation. Shown in each panel are reaction norms for three genotypes of a focal species, with the dots at the endpoints representing genotypic means or estimated breeding values for that genotype in that environment. In panels a–c, the environment is envisioned as abiotic (high/low temperature, for example), while in panel d, the environment is the genotype of another interacting species (a mutualist in this case).

seed head number of a genotype divided by the mean seed head number of all genotypes in the same sub-plot), he found that $G \times E$ for fitness components existed on a very fine scale: there were frequently changes in the rank order of relative fitness between locations separated by as little as 10 cm. An analysis of the spatial autocorrelation of relative fitness, paired with estimates of the likely dispersal distance of seeds, revealed that relative fitness was not spatially correlated beyond 10 cm, and that most seeds were likely to be dispersed into microhabitats where their relative fitness was completely uncorrelated with the relative fitness of their maternal plant. Similar studies on the spatial and temporal grain of $G \times E$ for fitness components in animals would substantially improve our understanding of the importance of $G \times E$ for natural systems. For animals whose dispersal can be tracked and/or have home ranges or territories, more studies on the spatial and environmental autocorrelation between parental and offspring fitness would be highly useful (e.g. van der Jeugd & McCleery 2002).

The high degree of phenotypic plasticity and sensitivity to microenvironmental gradients documented by Stratton is not unique. For example, Johnson and Agrawal (2005) transplanted replicates of multiple genotypes of evening primrose (*Oenothera biennis*) into five, spatially separated habitats. They found prominent $G \times E$ interactions for a host of characters, including fruit number (a fitness estimate, Johnson 2007), biomass accumulation, and even—as a measure of 'extended phenotype'—the abundance, evenness, richness, and diversity of the arthropod community that colonised focal plants (~21 000 individual arthropods from 129 taxa; Johnson & Agrawal 2005). Genotype $\times$ environment interactions for plants are common, well known, and widely accepted (e.g. Schlichting 1986). However, the results described by Johnson and Agrawal (2005) for the arthropod community indicate remarkable plasticity (or extremely strong selection and rapid evolution) in arthropods in choosing their host plants. Subtle differences in host plants (either morphology, secondary chemistry, life-history, or some combination) lead to replicable changes in the arthropod community, depending on plant genotype and habitat.

What implications might studies like Stratton's (1994) and Johnson and Agrawal's (2005) have for studies of wild quantitative genetics? First, as is well known, the presence of genotype-by-environment interactions, even on a fine scale, complicates the ability to estimate breeding values, genetic variances and covariances, and potentially even genetically based estimates of selection: unless $G \times E$ is specifically accounted for, most estimates of these parameters will encompass or integrate over any underlying $G \times E$ interactions. Whether an estimate of genetic variance that integrates over $G \times E$ is appropriate or misleading depends on the environmental variation in natural selection (e.g. is it different within individual environments, or similar across all environments).

A growing suite of studies using free-living animal populations have started to address phenotypic plasticity and $G \times E$ interactions (e.g. Charmantier *et al.* 2004; Garant *et al.* 2004; Nussey *et al.* 2005; Wilson *et al.* 2006; Pelletier *et al.* 2007; Charmantier *et al.* 2008; Robinson *et al.* 2009; Husby *et al.* 2010; 2011; Robinson *et al.* 2012), with most studies either characterising $G \times E$ interactions (often with a focus on how environmental context affects key evolutionary parameters like genetic variances, heritabilities, and genetic correlations) or attempting to discern whether temporal trends are due to plastic or evolutionary responses. These studies, and the long history of $G \times E$ and plasticity studies in the plant literature, suggest that $G \times E$ interactions are likely to be a fundamental feature of any study system using the wild quantitative genetic framework. Given the prominent theoretical role of $G \times E$ interactions for fitness components in maintaining genetic variation (Mitchell-Olds *et al.* 2007), future work on $G \times E$ for fitness components in wild animal systems is likely to be extremely useful in evaluating competing mechanisms for the maintenance of genetic variation in the wild. If for example, $G \times E$ interactions for fitness components are of the 'crossing variety' (Figure 8.1c) where different genotypes have highest fitness in alternate environments, environmental heterogeneity can directly influence the maintenance of genetic variation in fitness-related traits. In contrast, if $G \times E$ interactions for fitness components are rare, or are due to changes in the variance expressed in each environment (Figure 8.1b), environmental heterogeneity is unlikely to contribute to the maintenance of genetic variation.

Second, the results of Johnson and Agrawal (2005) suggest that consumers—in their case herbivores and omnivores, but more generally any parasite, pathogen, or antagonist—are likely to be extremely behaviourally plastic in response to genetic variation in their hosts. In other words, the prominent differences in the arthropod community detected by Johnson and Agrawal (2005) likely indicate considerable behavioural choices in response to genetically based variation in host plants. For mobile, searching, or otherwise free-living antagonists (i.e. those able to exert any host choice), we should probably expect strong 'bottom-up' influences of host genetic variation on the distribution and abundance of antagonists. If consumers or antagonists exhibit any genetic variation in their behavioural tendencies and choices, there is strong potential for interactions between host genotype and consumer genotype. In this case, the potential for $G \times G$ interactions (the coevolutionary counterpart of $G \times E$; Figure 8.1d) between hosts and consumers is likely to be large. If $G \times G$ interactions occur for fitness components of both species, local co-adaptation is expected between hosts and consumers (Garrido *et al.* 2012) or mutualists (Heath 2010).

8.2.3 Ecological context and diffuse selection

Characterising the coevolutionary relationship between plants and their consumers has been a major goal since Ehrlich and Raven's seminal paper (1964). Early workers (Janzen 1980; Fox 1981) distinguished between 'pairwise coevolution', in which the coevolutionary dynamic between a pair of species was independent of the presence or absence of other species in the community, and 'diffuse coevolution' in which the coevolutionary dynamic between species was dependent on the presence or absence of other species. These terms were cast in the quantitative genetic framework, in which diffuse coevolution was indicated by elements of either the genetic covariance matrix (G) or the vector of selection gradients (β) leading to non-independent evolutionary responses in one of the focal species (Iwao & Rausher 1997; Stinchcombe & Rausher 2001), while pairwise coevolution was indicated by $\boldsymbol{G}$ and $\boldsymbol{\beta}$ leading to coevolution that was independent of the presence or absence of other species.

Numerous workers have attempted to evaluate whether coevolutionary dynamics are dependent on community context (e.g. Juenger & Bergelson 1998, Gomez 2003; Lau 2008; Wise 2010). A consistent finding is that elements of both $\boldsymbol{G}$ and $\boldsymbol{\beta}$ are extremely sensitive to composition of the ecological community (Pilson 1992; Simonsen & Stinchcombe 2007; Wise 2010). The evidence for natural selection in coevolutionary interactions being dependent on community context is very strong. For example, natural selection on plant defence traits against deer herbivory, and the fitness costs of these traits, depend on the presence or absence of insect herbivores (Stinchcombe & Rausher 2001; 2002; Stinchcombe 2002); selection on flower morphology depends on the presence or absence of grazing ungulates (Gomez 2003); and the presence or absence of competing plant species alters selection in annual legumes (Lau 2008); Wise and Rausher (2013) showed that the predicted evolutionary responses of resistance to nine different herbivore species were significantly affected by whether one or more of the other herbivore species was imposing selection. Strauss and Irwin (2004) and Strauss *et al.* (2005) provide thorough reviews of experimental findings. At a broader spatial scale, the interaction between the small-flowered woodland star (*Lithophragma parviflorum*) and the seed-eating pollinator, the grey moth (*Greya politella*), varies between mutualism and parasitism, depending on whether additional pollinator species are present (Thompson & Fernandez 2006), circumstantially suggesting that the presence or absence of these additional pollinator species will alter selection on seed defence and floral attraction traits. Similarly, the morphology of pine cones differs markedly between populations with cross bills and those with cross bills and squirrels, again suggesting different patterns of natural selection between habitats depending on community composition (Benkman *et al.* 2001; Siepielski & Benkman 2007).

What implications do these studies of how community composition affects coevolutionary dynamics in plant species have for studies of wild animals? The dominant result from these studies is that the strength of natural selection is usually very sensitive to the ecological context. At first glance, many might be hesitant to generalise from studies of short-lived sessile plants to mammals and birds. But existing evidence suggests that competition between bird species can have dramatic effects on natural selection, and the evolutionary response, of several traits in birds such as beak depth in Darwin's finches (Grant & Grant 2006). Bird body condition, and early growth rates and mortality, are tightly tied to the availability of insect herbivores that are the birds' food sources. Evidence to date suggests that climatic fluctuations have been associated with a mismatch between herbivore emergence, plant phenological timing, and key life stages in birds, in some but not all populations (Visser *et al.* 1998; Nussey *et al.* 2005; Both *et al.* 2006; Charmantier *et al.* 2008). The link between broad scale climatic patterns and the ecological and evolutionary dynamics of selection on feral sheep has also been established (Hallett *et al.* 2004; Ozgul *et al.* 2009). To date these early studies in birds and sheep suggest a tight coupling of ecologically based variation and evolutionary dynamics. The plant data suggest that the coupling of ecological community and the fundamentals of evolutionary change—the expression of genetic variation and the pattern of natural selection—might happen at very small spatial scales. Given these studies, it seems like a reasonable hypothesis that competing bird species, large herbivores, parasites, or pathogens could alter the expression of genetic variation and the strength of selection on focal traits in studies of free-living vertebrates.

Although demanding, and possibly requiring superhuman effort and funding, a natural next step for studies in wild animal model systems will be to continue the recent trend of examining the ecological dependence of selection and genetic variation (Pemberton 2010). While explicit experiments may not be permissible (or even advisable, in the context of long-term monitoring), natural experiments can be utilised as well by comparing 'good' and 'bad' habitats (as measured in a currency relevant to the organism, which can often be difficult; see Robinson *et al.* 2009), sites with and without competing species, or even individuals with and without various species of parasites or pathogens (e.g. Charmantier *et al.* 2004). Although these studies will be subject to the usual concerns about the use of correlative data, they would provide a first glimpse into the ecological dependence of evolutionary dynamics in wild populations.

8.2.4 The invisible fraction

Grafen (1988) coined the term the 'invisible fraction' to describe individuals that died before even being censused or measured for the traits of interest. If this process is occurring, the phenotypic variance–covariance matrix $\boldsymbol{P}$ will have already been altered by selection if mortality has been non-random. Consequently, the estimates of phenotypic and genetic variation, and selection, made at later life stages will overlook the effects of selection earlier in the life-history. Many fundamental questions in evolutionary genetics and ecology—for example, the typical strength of selection (Hereford *et al.* 2004; Kingsolver *et al.* 2012) and whether $\boldsymbol{G}$-matrices are of full rank (Hine & Blows 2006; Kirkpatrick 2009; Houle 2010)—rest on the assumption that the populations we measure (whether in the lab, field, or greenhouse) have not already experienced appreciable natural selection so as to change both the means and (co)variances of the measured traits.

Some of the best evidence of the likelihood of the invisible fraction affecting estimates of $\boldsymbol{P}$, $\boldsymbol{G}$, and β comes from studies of plants (Bennington & McGraw 1995; Donohue *et al.* 2000; Mojica & Kelly 2010; but also see Sinervo & McAdam 2008; Hadfield 2008). For instance, Bennington and McGraw (1995) experimentally thinned populations of yellow touch-me-not (*Impatiens pallida*), either selectively removing smaller plants that would normally die prior to reaching adulthood, or by allowing those plants to reach reproductive maturity. Perhaps not surprisingly, they detected significant differences between $\boldsymbol{P}$ and β between the experimental treatments. For instance, natural selection acted significantly, and in opposite directions, on early season height (at 30 days) between the two treatments. Size at this timepoint in the experiment was also ecologically relevant for understanding intraspecific competition in *Impatiens pallida*, as it was when neighbouring plants began to compete directly for light, with leaves overlapping each other. More generally, Bennington and McGraw's results suggest that typical patterns of mortality that occur prior to traits being measured and expressed will alter our understanding of $\boldsymbol{P}$ and β, and likely $\boldsymbol{G}$ (provided $\boldsymbol{G}$ and $\boldsymbol{P}$ have any relationship).

Mojica and Kelly (2010) took an alternative approach to investigating the invisible fraction: utilising artificial selection on flower size. Using the yellow monkey flower (*Mimulus guttatus*), they transplanted artificial selection lines for increased and decreased flower size (and controls) into the field over three successive years. The results they found were fundamentally important: traditional selection analysis indicated higher fitness of individuals from the large flowered selection lines, mainly through fecundity selection. However, individuals from the large flower same selection lines had significantly lower survival *prior to the flower size trait even being expressed.* But for their novel experimental design, this entire component of selection would have been missed. In this system, it appears that the mechanism is that large flowered genotypes failed to reach reproductive stage quickly enough before the end of the growing season. Consequently, mortality and fecundity selection were conflicting, with the net pattern favouring smaller flower size, despite the fecundity advantage of larger flowered genotypes, given that they survived. One advantage of the approach taken by Mojica and Kelly is that it revealed an underlying, negative genetic correlation between survival and fecundity, which the authors interpreted as evidence in support of antagonistic pleiotropy potentially contributing to the maintenance of genetic variation in flower size. These data, taken at face value, suggest that many of the reports in the plant literature of fecundity selection on flower size (or size in general) might be quantitatively biased at best, or positively misleading at worst.

Hadfield (2008) suggests several methods for dealing with the potential invisible fraction problem. Particularly promising is his suggestion of evaluating genetic covariances between survival and post survival traits, taking advantage of relatedness between surviving and non-surviving individuals (similar to estimating across-sex genetic correlations), but to my knowledge, these approaches have not been implemented. Collectively, the initial data from plant systems suggests that invisible fraction problems have the potential to substantially affect our understanding of the strength, direction, and magnitude of key evolutionary genetic parameters. Further experimentation, using genetically based data in either animals or plants, would clarify the potential role of the invisible fraction in our understanding of evolutionary processes.

8.3 Prospects and challenges in plants

The immobility of plants makes them ideal experimental subjects—in the words of the famous plant population biologist John Harper, 'they sit still and wait to be counted'. A rich literature on the quantitative genetics of annuals and short-lived perennials has developed, mainly in species that are easily crossed to produce full- and half-sib families, diallel crosses, and inbred lines. For species that are not clonal or capable of reproducing asexually, there is also an extensive literature on natural selection, frequently using fitness components (flower or fruit number, seed mass, seed set, etc.) as the estimate of fitness for the selection analysis. For species with high self-fertilisation rates, viable seed set will encompass both male and female reproductive success. Given these experimental advantages, why have studies of the quantitative genetics of wild, unmanipulated plant populations—as opposed to studies of common gardens, outplanted quantitative genetic designs, or in situ experimental manipulations—lagged behind those of difficult to study organisms like birds and mammals? In part, the answer lies with other difficult aspects of plant biology, and in part due to the reluctance of investigators to use all the tools available.

8.3.1 Challenges

Many plants exist in large, dense populations: for instance, Donohue *et al.* (2000) describe densities of *Impatiens capensis* seedlings on the order of $3000/\text{m}^2$; in the same species, Stinchcombe *et al.* (2010) noted that densities exceeding 625 plants/m^2 occur in $\sim$20% of the sites/years sampled. Many species,

Impatiens included (Schmitt *et al.* 1987), exhibit density dependent mortality (often referred to as self-thinning), such that smaller or more size suppressed individuals die prior to reproductive maturity, and as a consequence adult populations frequency have much lower densities than seedling populations. In contrast to most vertebrate and bird species, fecundity is also frequently very high: even small, short-lived plant species can produce 100s to 1000s of seeds. While seed set is a reasonable approximation of lifetime reproductive success for many species with high selfing rates, for many species a full picture of fitness species requires estimating male fitness. Outcross siring success is challenging to measure, even with allozymes or microsatellites, as pollen dispersal distances are difficult to estimate, and long-distance pollen dispersal from outside study areas can make paternity assignment difficult, even with intensive sampling of all potential local sires (see e.g. Hodgins & Barrett 2008). Finally, many plant populations have large seed banks, where a fraction of the population remains dormant and below ground, often for years at a time.

The challenges of seed banks, pollen dispersal, high fecundity, and self-thinning pose a number of unique problems for wild quantitative genetic studies, where reliable tracking of individuals, estimation of parentage, and measuring reproductive success are essential. For example, individuals can be difficult to track from germination to reproduction in dense populations, and selection on a host of life-history, morphological, and phenological traits often occurs during density dependent mortality. Estimating reproductive success requires sampling seeds or developing fruits, disrupting natural dispersal vectors and phenologies. The size of plant populations combined with the costs of genetic marker analysis often means that paternity can only be measured on a small fraction of individuals, or in small populations. Dormant seasonal stages between generations (winter and summer for summer and winter annuals, respectively) mean that both the maternal and paternal parents of individuals in the current generation have to be identified with markers (as opposed to just paternity). Finally, there is abundant opportunity for selection in the seed bank, and shortly after germination, prior to adult traits even being expressed. As a consequence, many investigators choose to focus on experimental populations or gardens where these forces can be mitigated (e.g. by collecting seeds to estimate fitness, using constant densities and spacing of plants, etc.), rather than study wild, unmanipulated plant populations. A typical approach is to follow the fate of transplanted seeds or seedlings generated from a crossing design for a single generation, rather than long-term quantitative genetic studies that have been used to such success in birds and ungulates.

8.3.2 What do we know and what tools are available?

Despite the limitations described above, several investigators have estimated quantitative genetic parameters in natural, unmanipulated plant populations not subject to transplanting seeds or seedlings of known parentage or experimental manipulations of ecological variables. Because a frequent goal of estimating genetic variances and heritabilities is to pair them with estimates of selection to predict microevolutionary responses, I consider both aspects of genetic variation and selection here. These studies have frequently examined single-generation components of variance or selection.

8.3.2.1 Heritability

The majority of studies of genetic variation in natural plant populations (see Table 8.1 for a representative list) take advantage of work by Ritland (1996a, 1996b, 2000; Lynch & Ritland 1999) that uses information on pairwise relatedness, inferred from molecular markers, to estimate heritability. Although pedigree-based approaches are preferable in obligate outbreeders (Thomas *et al.* 2002; Coltman 2005; see Garant & Kruuk 2005), many plants show some form of selfing, biparental inbreeding, and other matings between close relatives that make Ritland's approach more practical for plant systems. A prerequisite for the Ritland approach is that variation in the degree of marker-based relatedness must exist in the sample, as the method is based on a regression of phenotypic similarity on relatedness, as inferred from markers. As clearly can be seen from Table 8.1, the implementation has been taken up in three types of species: 1) long-lived trees, 2) short-lived annuals, (frequently monkey flowers

Table 8.1 Exemplar applications of Ritland's method, or similar approaches to estimating heritability in the wild[1]

Species	Traits	% Impermissible Estimates	% Significant	Median h^2	# Markers	Design	Reference
Yellow monkey flower (*Mimulus guttatus*)	Floral and life-history traits, fitness components	5/17	8/17	0.43	10 allozymes	300 naturally growing plants per transect, × 2 transects	Ritland and Ritland (1996)
Turkey oak (*Quercus laevis*)	Phenolics (defensive compounds)	.	.	.	5 microsatellites	10 naturally growing trees, from each of 20 unique multilocus allozyme genotypes	Klaper *et al.* (2001)
Yellow box eucalyptus (*Eucalyptus melliodora*)	Plant defence chemicals	1/4	3/4	0.77	6 microsatellites	259 naturally growing trees	Andrew *et al.* (2005)
Yellow monkey flower (*M. guttatus*)	Floral traits	0/7	4/7	0.18	8 allozymes	3 greenhouse grown offspring of 203 maternal seed families	van Kleunen and Ritland (2005)
White carob tree (*Prosopis alba*)	Leaf morphology traits	6/13	9/13	0.42	6 microsatellites	142 trees from 32 families in a progeny trial	Bessega *et al.* (2009)
Balsam poplar (*Populus balsmifera*)	Phenology and ecophysiology	0	9/13	0.25	100 single nucleotide polymorphisms	10 genotypes × 20 populations, grown in a common garden	Keller *et al.* (2011)
Perennial columbines (*Aquilegia vulgaris vulgaris, A. v. nevadensis, A. pyrenaica pyrenaica, A. p. cazorlensis*)	Floral and vegetative traits	79/156	24/156	0.161	10 microsatellites	33–60 naturally growing individuals × 13 populations	Castellanos *et al.* (2011)
Cazorla violet (*Viola cazorlensis*)	Fruit production over 20 years	0/1	1/1	0.5931 or 0.631	365 amplified fragment length polymorphisms	52 naturally growing plants, monitored for 20 years	Herrera and Bazaga (2009)

[1] I did not include estimates of $h^2 > 1$ or $h^2 < 0$ in my calculations of the median h^2, as they are not quantitatively comparable to traditional h^2 estimates on the same 0–1 scale; however, estimates >1 are often statistically significant, and are included in the counts of significance. For Ritland and Ritland (1996), I used data from Table 3; for Andrew *et al.* 2005, I used data from Table 4, with the 60 m distance cut-off; Klaper *et al.* (2001) found non-significant variance in relatedness, and did not present h^2 estimates; Herrera and Bazaga (2009) estimate the fraction of phenotypic variance explained, in total, by binary amplified fragment length polymorphism genotype scores of 11 loci significantly associated with fecundity (this figure will be subject to a substantial ascertainment bias).

in the genus *Mimulus*), and 3) perennials, often found in small populations or in restricted habitats, where outplanting hundreds to thousands of seeds or seedlings of known parentage would be infeasible. Investigators have also typically applied this approach to the study of life-history or plant defensive traits, where a specific goal is to estimate the heritability of the trait *in situ*, rather than a common garden experiment where $G \times E$ effects or differential adaptation to the common garden could affect the estimates. Estimates of heritability using this and other inferential approaches can yield impermissible values (<0, or >1; see Castellanos *et al.* 2011); in this case, comparison of different traits within the same population is still informative, and values >1 generally indicate some real genetic variance in the trait.

These early studies provide the first glimpse of the heritability of ecologically important traits for a suite of species that had previously been inaccessible to the quantitative genetic approach. As marker development and genotyping costs continue to fall in the next-generation sequencing era, it is likely that our ability to estimate the heritability of traits in wild populations not amenable to crossing designs will expand. One tentatively promising note from Table 8.1 is that the only study to use large numbers of genome-wide markers (Keller *et al.* 2011) also failed to detect any heritabilities outside the permissible values. These data suggest that increased information from more informative markers, and more markers, will be useful for clarifying the typical heritability in wild plant populations.

8.3.2.2 Genetic correlations

A persistent goal in evolutionary quantitative genetics has been to estimate genetic correlations among traits, either to describe their joint evolution or to test theoretical predictions about trade-offs. Remarkably, methods exist to estimate genetic correlations in natural populations, without genetic markers or controlled crosses (Lynch 1999), yet they have rarely been implemented in plant systems. Lynch showed that by measuring the phenotypic covariances between many *pairs* of individuals, genetic correlations could be reasonably approximated, provided that at least some full-sibs (~20%) existed in the sample. The Lynch method relies on comparing the covariances between traits to the covariances within traits, using pairs of individuals as sampling units, in such a way that the level of relatedness among individuals cancels out in the calculations. Lynch's method provides a cheap, easy to implement, and flexible approach for estimating genetic correlations in natural populations, requiring no more than phenotypic sampling. Sampling of groups and/or jackknifing procedures have been shown to improve estimation of standard errors of the genetic correlation (Réale & Roff 2001). Given how often investigators evaluate the relationship between $\mathbf{P}$ and $\mathbf{G}$ (frequently with the desire that $\mathbf{P}$ closely resembles $\mathbf{G}$, because of its ease of estimation, Waitt & Levin 1998), or are forced to use $\mathbf{P}$ as a substitute for $\mathbf{G}$ given a lack of data (Agrawal & Stinchcombe 2009), it is remarkable that Lynch's method appears to have been used only once for plant populations (van Kleunen & Ritland 2004), despite early calls for its investigation (Fischer & van Kleunen 2000; Conner *et al.* 2003). van Kleunen and Ritland found a high correlation ($r = 0.78$) between point estimates of the genetic correlation based on the Lynch method and those based on parent–offspring regression for a set of eight traits. While not a panacea, it would clearly be an improvement on purely phenotypic data, and would be a useful intermediate step prior to controlled crosses or marker development. With the high likelihood of environmentally induced covariances between traits and fitness from purely phenotypic selection analysis carried out in the field (Schluter *et al.* 1991; Stinchcombe *et al.* 2002; Kruuk *et al.* 2002), it is surprising that no investigators have applied the Lynch method to estimate genetic correlations between traits and fitness more reliably. Unfortunately, these data leave us in the position of not knowing much of anything about the typical genetic correlation structure of natural populations that have not been experimentally produced or created via a mating design.

8.3.2.3 Selection

There are numerous estimates of selection in the wild using female fitness, and purely phenotypic data (see meta-analyses by Kingsolver *et al.* 2012 and Geber *et al.* 2003). Unfortunately, given the high likelihood of environmentally induced bias in these estimates, their relevance is unclear; while the presence and direction of selection are unlikely to be affected (Stinchcombe *et al.* 2002), the quantitative magnitude of the selection differentials and gradients s and β from these studies are likely to be severely over- or underestimated. These problems lead not only to erroneous estimates of the strength of selection because a correlated variable, some aspect of the environment, has been omitted from the regression model, but also to erroneous predictions of evolutionary responses if the covariance between traits and fitness has no underlying genetic basis (Rausher 1992; Stinchcombe *et al.* 2002; Morrissey *et al.* 2012a, 2012b).

Studies of male fitness in plants in natural populations are much rarer, but are probably less susceptible to these problems. While the environment will clearly affect the phenotypic traits of interest (size, floral characteristics, etc.), fitness in these assays is typically measured as the fraction of seeds in the population sired, as estimated with genetic markers. In this case, a covariance between environmental conditions of the pollen donor and the number of set seeds elsewhere in the population with the appropriate multilocus genotype, is much less likely, except in the cases of geitonogamy, selfing, or highly local pollen dispersal. Under geitonogamy, selfing, and local pollen dispersal, the majority of pollen is deposited locally in the same environment where the trait was measured, potentially leading to the environment jointly influencing pollen donors and recipients, as well as the focal traits. Estimating male fitness is also essential for fully characterising fitness in outcrossing or mixed-mating plants, and for testing hypotheses about the evolution of floral form, display, and function, making it regrettable that relatively few estimates of selection using male fitness are available.

A representative list of studies estimating selection using male fitness, as determined from markers, is given in Table 8.2. In general, the dominant trend is for investigators to study selection on aspects of floral form and display, which is unsurprising given the 'male function' hypothesis for the evolution of these characters (Burd & Callahan 2000). However, there is no reason why, in principle, the same approach could not be applied to estimating selection on defensive, life-history, phenological, or architectural traits. Inspection of Table 8.2 reveals that large numbers of progeny need to be genotyped to evaluate male fitness for even a reasonable handful of parental plants. Significant statistical challenges for estimating selection through male fitness remain (e.g. significance testing via likelihood ratio tests, permutation, or bootstrapping), and many studies fail to detect any selection. A hypothesis worth testing (empirically, or with theory or simulation) is whether one would have greater power to detect selection by genotyping fewer seeds per individual, and more individuals. Larger sample sizes and more genetic markers are likely required, and forthcoming, as none of the studies in Table 8.2 and 8.1 have used next-generation technologies to generate large numbers of markers.

8.4 Prospects in plants and lessons from wild animals

The advent of next-generation sequencing technologies (see also Chapter 13, Jensen *et al.*) promises to change the study of the evolutionary genetics of wild plant populations. Recent pioneering work by Visscher and colleagues (Yang *et al.* 2010) points the way towards estimating the heritability of traits in the absence of breeding designs. By utilising the identity by descent of a large number of markers (hundreds of thousands of single nucleotide polymorphisms (SNPs)), it is possible to estimate how much of the phenotypic variance among individuals is due to shared alleles among individuals, provided one has large samples of seemingly unrelated individuals (~4000).

A handful of studies have performed association mapping using 250 000 SNPs (estimated via genotyping) to scan the genome for regions influencing ecologically important traits in *Arabidopsis thaliana* (Fournier-Level *et al.* 2011), and can successfully predict phenotypes from nothing but SNP data (Hancock *et al.* 2011). Rather than using the collective contribution of many SNPs segregating in a population to estimate heritability, these studies focus on the contribution of individual loci to ecologically important phenotypes. The advent to next-generation (and eventually third-generation) sequencing promises to move studies like this from

Table 8.2 Exemplar applications of estimates of selection gradients or differentials through male fitness.[1] In the table below, β refers to an estimate of a directional selection gradient, γ_{ii} a stabilising/disruptive selection gradient, and γ_{ij} a correlational selection gradient

Species	Traits	Significant selection?	Sample sizes	Reference
Wild radish (*Raphanus raphanistrum*)	Flower size, anther exsertion, pollen #, flower production	4 of 11 β; 0 of 6 γ_{ii}	6708 seeds, from 49–112 plants; 8 allozymes, experimental plantings	Conner *et al.* (1996)
Spotted bellflower (*Campanula punctata*)	Corolla length and width	1 of 5 β; 0 of 1 γ_{ij}	1913 seeds, from 16 plants; 5 allozymes; potted pollen donors/recipients placed in natural population	Kobayashi *et al.* (1999)
Blazing star (*Chamaelirium luteum*)	Rosette leaf number, stalk leaf number and length, raceme length	0/4 β	2294 seeds from 70 females, 273 potential males; 8 allozymes; Natural population	Smouse *et al.* (1999)
Horsenettle (*Solanum carolinense*)	Flower number and size characters (length and width of organs), % male flowers, vegetative size	4 of 21 univariate β; 0 of 21 multivariate β	43 plants per plot; 77–1202 progeny produced; 5 allozymes; 6 experimental plot–year combinations	Elle and Meagher (2000)
Wild radish (*R. raphanistrum*)	Flower size, anther exsertion, pollen #, flower production	8 of 11 β; 3 of 6 γ_{ii}	6708 seeds, from 49–112 plants; 8 allozymes, experimental plantings	Morgan and Conner (2001)
Yellow monkey flower (*M. guttatus*)	Corolla width and length traits, anther length; anther–stigma distance; spot number and fluctuating asymmetry in spot number; ovary size	1 of 8 *s*; 1 of 8 β	1794 seeds from 230 plants; 8 allozymes; natural population	van Kleunen and Ritland (2004)
White campion (*Silene latifolia*)	Corolla and calyx diameter; claw length	1997: 0 of 3 β; 1 of 3 γ_{ii} 1998: 0 of 3 β; 0 of 3 γ_{ii}	1187 and 1836 progeny; 70–263 plants; 6 allozymes; natural populations	Wright and Meagher (2004)
Yellow monkey flower (*M. guttatus*)	Corolla width and length traits, anther length; anther–stigma distance; spot number and fluctuating asymmetry in spot number; ovary size	6 of 8 β; 4 of 8 γ_{ii}	1794 seeds; from 230 plants; 8 allozymes; natural population	van Kleunen and Burczyk (2008)
Daffodil (*Narcissus triandrus*)	Flower number, flower length, corolla width	2 of 8 *s*; 2 of 8 β; 3 of 8 γ_{ii}	221–305 progeny from 41 and 50 maternal plants; 5 microsatellites; natural populations	Hodgins and Barrett (2008)
Jacob's ladder (*Polemonium brandegeei*)	Floral architecture traits (size, position of organs), tube diameter, sugar content	2 of at least 6 β	355 progeny from 94 plants, 6 microsatellites; arrays of plants exposed to pollinators	Kulbaba and Worley (2012)
Wild radish (*R. raphanistrum*)	Flower size, flower number, anther exsertion, stamen dimorphism	NA	1217 progeny from 3 arrays of 20 plants; plants were reused for different pollination treatments when new flowers developed; 4 microsatellites; selection by individual pollinators and in combination in outdoor flight cages	Sahli and Conner (2011)

[1] Morgan and Conner (2001) is a reanalysis of Conner *et al.* (1996), with updated methods. I used the 'all traits' column of Table 8.1 to ensure comparisons to Conner *et al.* (1996). Morgan and Conner (2001) also present both χ^2 tests and simulation tests; my tally of significance is based on the χ^2 tests, although results are similar. Wright and Meagher (2004) present χ^2 and permutation tests for 1997 data, but only χ^2 for 1998 data. There was a wide discrepancy in the *p*-values for the 1997 data, so I present the permutation results which are more conservative, and in line with how the authors appear to interpret these results. Likewise, Elle and Meagher (2000) present both χ^2 and permutation tests, but put more emphasis on the permutation tests. van Kleunen and Burczyk (2008) is a reanalysis of van Kleunen and Ritland (2004), but with updated methods. Kulbaba and Worley (2012) present β from models selected with backwards elimination, but it is unclear how many traits started in the initial models. Sahli and Conner (2011) do not present estimates of β for all traits at once using male fitness as the response, but instead partition selection on an individual trait into the relative contributions of different groups of pollinators.

being marker limited (e.g. 8–10 microsatellites or allozymes; see Tables 8.1 and 8.2) to being limited by phenotypes, individuals, and our ability to draw inferences about loci with very small effect sizes. The rapid drop in genome sequencing cost should soon allow much whole-genome sequencing studies of large samples of plants (provided the species has a small enough genome to make it amenable to next-generation sequencing and assembly). The coming era suggests that ecological considerations—the ability to measure, sample, and phenotype 100s to 1000s of individuals, in ecologically realistic contexts—is more likely to limit our progress than molecular, chemical, or economic challenges of marker development.

8.4.1 'But it's too hard': lame excuses shouldn't cut it anymore

Studies of wild populations of birds, deer, sheep and other free-living organisms have, in my opinion, substantially raised the bar for the study of the evolutionary genetics of short-lived plants. In short, if investigators can estimate the genetic covariance between traits and fitness in long-lived, free-living animals with overlapping generations under difficult terrain, there should be no reason why investigators cannot estimate genetic variances, heritabilities, and genetic covariances between traits and fitness in annual plants and short-lived perennials that live in easily accessible old fields, disturbed sites, forest understories. McKay *et al.* (2001) pointed out several years ago that the fear of things being 'too hard' was impeding our ability to address important questions.

The success of quantitative genetics in wild vertebrate populations only reinforces McKay *et al.*'s point: the biological, statistical, and analytical advances in free-living vertebrate populations are sufficiently compelling to demand notice. The ease of experimental manipulation, reciprocal transplant experiments, and characterisation of the ecological forces acting on plant populations suggests that wider adoption of the approaches and questions being addressed in vertebrate populations would have wide-ranging consequences for our understanding of free-living plant populations. Likewise, the findings of a long-tradition of experimental ecology and evolutionary genetics in plants suggest new ways forward and promising avenues for investigation in wild animal studies. A greater integration of the findings and approaches of the two literatures offers new prospective questions for investigation: do plants and animals show similar relationships between census size, effective population size, and heritabilities and genetic variances? Do short-lived plants and animals show patterns of genetic variation more similar to each other, or to longer-lived close relatives of the same taxonomic flavour? Is the ecological context-dependence of natural selection that is so common in plants a general finding, or specific to plant evolutionary ecology? Is the contribution of $G \times E$ interactions to the maintenance of quantitative genetic variance the same in plants and animals? These and surely many more questions suggest that further cross-pollination of the two literatures may improve our understanding of the generality of various evolutionary processes.

Acknowledgments

I thank Loeske Kruuk, Dany Garant, Anne Charmantier, Erik Postma, and Juan Fornoni for comments on the manuscript. My work on the ecological and evolutionary genetics of plants is supported by the Natural Sciences and Engineering Research Council of Canada.

References

Aarssen, L.W. & Turkington, R. (1985) Biotic specialization between neighbouring genotypes in *Lolium perenne* and *Trifolium repens* from a permanent pasture. *Journal of Ecology*, **73**, 605–614.

Agrawal, A.F. & Stinchcombe, J.R. (2009) How much do genetic covariances alter the rate of adaptation? *Proceedings of the Royal Society B-Biological Sciences*, **276**, 1183–1191.

Anderson, J.T., Lee, C.-R., Rushworth, C.A., Colautti, R.I. & Mitchell-Olds, T. (2013) Genetic trade-offs and conditional neutrality contribute to local adaptation. *Molecular Ecology*, **22**, 699–708.

Andrew, R.L., Peakall, R., Wallis, I.R., Wood, J.T., Knight, E.J. & Foley, W.J. (2005) Marker-based quantitative genetics in the wild?: The heritability and genetic correlation of chemical defenses in *Eucalyptus*. *Genetics*, **171**, 1989–1998.

Antonovics, J. & Bradshaw, A.D. (1970) Evolution in closely adjacent plant populations. VIII. Clinal patterns at a mine boundary. *Heredity*, **25**, 349–362.

Bradshaw, A.D. (1965) Evolutionary significance of phenotypic plasticity in plants. *Advances in Genetics*, **13**, 115–155.

Benkman, C.W., Holimon, W.C. & Smith, J.W. (2001) The influence of a competitor on the geographic mosaic of coevolution between crossbills and lodgepole pine. *Evolution*, **55**, 282–294.

Bennington, C.C. & McGraw, J.B. (1995) Phenotypic selection in an artificial population of *Impatiens pallida*: The importance of the invisible fraction. *Evolution*, **49**, 317–324.

Bessega, C., Saidman, B.O., Darquier, M.R., Ewens, M., Sánchez, L., Rozenberg, P. & Vilardi, J.C. (2009) Consistency between marker- and genealogy-based heritability estimates in an experimental stand of *Prosopis alba* (Leguminosae). *American Journal of Botany*, **96**, 458–465.

Both, C., Bouwhuis, S., Lessells, C. & Visser, M.E. (2006) Climate change and population declines in a long-distance migratory bird. *Nature*, **441**, 81–83.

Burd, M. & Callahan, H.S. (2000) What does the male function hypothesis claim? *Journal of Evolutionary Biology*, **13**, 735–742.

Carrière, Y., Crowder, D.W. & Tabashnik, B.E. (2010) Evolutionary ecology of insect adaptation to Bt crops. *Evolutionary Applications*, **3**, 561–573.

Carmona, D. & Fornoni, J. (2013) Herbivores can select for mixed defensive strategies in plants. *New Phytologist*, **197**, 576–585.

Castellanos, M.C., Alcantara, J.M., Rey, P.J. & Bastida, J.M. (2011) Intra-population comparison of vegetative and floral trait heritabilities estimated from molecular markers in wild *Aquilegia* populations. *Molecular Ecology*, **20**, 3513–3524.

Charmantier, A., Kruuk, L.E.B. & Lambrechts, M.M. (2004) Parasitism reduces the potential for evolution in a wild bird population. *Evolution*, **58**, 203–206.

Charmantier, A. & Garant, D. (2005) Environmental quality and evolutionary potential: lessons from wild populations. *Proceedings of the Royal Society B-Biological Sciences*, **272**, 1415–1425.

Charmantier, A., McCleery, R.H., Cole, L.R., Perrins, C., Kruuk, L.E.B. & Sheldon, B.C. (2008) Adaptive phenotypic plasticity in response to climate change in a wild bird population. *Science*, **320**, 800–803.

Clements, F.E. & Goldsmith, G.W. (1924) *The phytometer method in ecology: the plant and community as instruments*. The Carnegie Institution of Washington, Washington, D. C.

Cogni, R. & Futuyma, D.J. (2009) Local adaptation in a plant herbivore interaction depends on the spatial scale. *Biological Journal of the Linnean Society*, **97**, 494–502.

Coltman, D.W. (2005) Testing marker-based estimates of heritability in the wild. *Molecular Ecology*, **14**, 2593–2599.

Conner, J.K., Rush, S. & Jennetten, P. (1996) Measurements of natural selection on floral traits in wild radish (*Raphanus raphanistrum*). I. Selection through lifetime female fitness. *Evolution*, **50**, 1127–1136.

Conner, J.K., Franks, R. & Stewart, C. (2003) Expression of additive genetic variances and covariances for wild radish floral traits: Comparison between field and greenhouse environments. *Evolution*, **57**, 487–495.

Donohue, K., Messiqua, D., Pyle, E.H., Heschel, M.S. & Schmitt, J. (2000) Evidence of adaptive divergence in plasticity: density- and site-dependent selection on shade-avoidance responses in *Impatiens capensis*. *Evolution*, **54**, 1956–1968.

Dudley, S.A. & Schmitt, J. (1995) Genetic differentiation in morphological responses to simulated foliage shade between populations of *Impatiens capensis* from open and woodland sites. *Functional Ecology*, **9**, 655–666.

Ehrlich, P.R. & Raven, P.H. (1964) Butterflies and plants: a study in coevolution. *Evolution*, **18**, 568–608.

Elle, E. & Meagher, T.R. (2000) Sex allocation and reproductive success in the andromonoecious perennial *Solanum carolinense* (Solanaceae). II. Paternity and functional gender. *The American Naturalist*, **156**, 622–636.

Etterson, J.R. & Shaw, R.G. (2001) Constraint to adaptive evolution in response to global warming. *Science*, **294**, 151–154.

Fischer, M. & Van Kleunen, M. (2001) On the evolution of clonal plant life histories. *Evolutionary Ecology*, **15**, 565–582.

Fournier-Level, A., Korte, A., Cooper, M.D., Nordborg, M., Schmitt, J. & Wilczek, A.M. (2011) A map of local adaptation in *Arabidopsis thaliana*. *Science*, **334**, 86–89.

Fox, L.L. (1981) Defense and dynamics in plant-herbivore systems. *American Zoologist*, **21**, 853–864.

Fry, J.D. (1993) The general vigor problem: can antagonistic pleiotropy be detected when genetic covariances are positive? *Evolution*, **47**, 327–333.

Garant, D., Sheldon, B.C. & Gustafsson, L. (2004) Climatic and temporal effects on the expression of secondary sexual characters: genetic and environmental components. *Evolution*, **58**, 634–644.

Garant, D., Kruuk, L.E., Wilkin, T.A., McCleery, R.H. & Sheldon, B.C. (2005) Evolution driven by differential dispersal within a wild bird population. *Nature*, **433**, 60–65.

Garant, D. & Kruuk, L.E.B. (2005) How to use molecular marker data to measure evolutionary parameters in wild populations. *Molecular Ecology*, **14**, 1843–1859.

Garrido, E., Andraca-Gómez, G. & Fornoni, J. (2012) Local adaptation: simultaneously considering herbivores and their host plants. *New Phytologist*, **193**, 445–453.

Geber, M.A. & Griffen, L.R. (2003) Inheritance and natural selection on functional traits. *International Journal of Plant Sciences*, **164**, S21–S42.

Gómez, J.M. (2003) Herbivory reduces the strength of pollinator mediated selection in the Mediterranean herb *Erysimum mediohispanicum*: consequences for plant specialization. *The American Naturalist*, **162**, 242–256.

Grafen, A. (1988) On the uses of data on lifetime reproductive success. *Reproductive Success*(ed. T.H. Clutton-Brock), pp. 454–471. University of Chicago Press, Chicago.

Grant, P.R. & Grant, B.R. (2006) Evolution of character displacement in Darwin's finches. *Science*, **313**, 224–226.

Hadfield, J.D. (2008) Estimating evolutionary parameters when viability selection is operating. *Proceedings of the Royal Society B-Biological Sciences*, **275**, 723–734.

Hadfield, J.D., Wilson, A.J., Garant, D., Sheldon, B.C. & Kruuk, L.E.B. (2010) The misuse of BLUP in ecology and evolution. *The American Naturalist*, **175**, 116–125.

Hadfield, J.D. (2010) MCMC methods for multi-response generalized linear mixed models: the MCMCglmm R package. *Journal of Statistical Software*, **33**, 1–22.

Hall, M.C., Lowry, D.B. & Willis, J.H. (2010) Is local adaptation in *Mimulus guttatus* caused by trade-offs at individual loci? *Molecular Ecology*, **19**, 2739–2753.

Hallett, T.B., Coulson, T., Pilkington, J.G., Clutton-Brock, T.H., Pemberton, J.M. & Grenfell, B.T. (2004) Why large-scale climate indices seem to predict ecological processes better than local weather. *Nature*, **430**, 71–75.

Hancock, A.M., Brachi, B., Faure, N., Horton, M.W., Jarymowycz, L.B., Sperone, F.G., Toomajian, C., Roux, F. & Bergelson, J. (2011) Adaptation to climate across the *Arabidopsis thaliana* genome. *Science*, **334**, 83–86.

Heath, K.D. (2010) Intergenomic epistasis and coevolutionary constraints in plants and rhizobia. *Evolution*, **64**, 1446–1458.

Hereford, J., Hansen, T.F. & Houle, D. (2004) Comparing strengths of directional selection: how strong is strong? *Evolution*, **58**, 2133–2143.

Hereford, J. (2009) A quantitative survey of local adaptation and fitness trade-offs. *The American Naturalist*, **173**, 579–588.

Herrera, C.M. & Bazaga, P. (2009) Quantifying the genetic component of phenotypic variation in unpedigreed wild plants: tailoring genomic scan for within-population use. *Molecular Ecology*, **18**, 2602–2614.

Hine, E. & Blows, M.W. (2006) Determining the effective dimensionality of the genetic variance-covariance matrix. *Genetics*, **173**, 1135–1144.

Hodgins, K.A. & Barrett, S.C.H. (2008) Natural selection on floral traits through male and female function in wild populations of the heterostylous daffodil *Narcissus triandrus*. *Evolution*, **62**, 1751–1763.

Hoffmann, A.A. & Merilä, J. (1999) Heritable variation and evolution under favourable and unfavourable conditions. *Trends in Ecology & Evolution*, **14**, 96–101.

Houle, D. (2010) Numbering the hairs on our heads: The shared challenge and promise of phenomics. *Proceedings of the National Academy of Sciences of the United States of America*, **107**, 1793–1799.

Huber, H., Kane, N.C., Heschel, M.S., von Wettberg, E.J., Banta, J., Leuck, A.M. & Schmitt, J. (2004) Frequency and microenvironmental pattern of selection on plastic shade-avoidance traits in a natural population of *Impatiens capensis*. *American Naturalist*, **163**, 548–563.

Husby, A., Nussey, D.H., Visser, M.E., Wilson, A.J., Sheldon, B.C. & Kruuk, L.E.B. (2010) Contrasting patterns of phenotypic plasticity in reproductive traits in two great tit (*Parus major*) populations. *Evolution*, **64**, 2221–2237.

Husby, A., Visser, M.E. & Kruuk, L.E.B. (2011) Speeding up microevolution: The effects of increasing temperature on selection and genetic variance in a wild bird population. *PLoS Biology*, **9**, e1000585.

Iwao, K. & Rausher, M.D. (1997) Evolution of plant resistance to multiple herbivores: quantifying diffuse coevolution. *American Naturalist*, **149**, 316–335.

Janzen, D.H. (1980) When is it coevolution? *Evolution*, **34**, 611–612.

Johnson, T. & Barton, N. (2005) Theoretical models of selection and mutation on quantitative traits. *Philosophical Transactions of the Royal Society of London Series B, Biological Sciences*, **360**, 1411–1425.

Johnson, M.T.J. & Agrawal, A.A. (2005) Plant genotype and environment interact to shape a diverse arthropod community on evening primrose (*Oenothera biennis*). *Ecology*, **86**, 874–885.

Johnson, M.T.J. (2007) Genotype-by-environment interactions leads to variable selection on life-history strategy in Common Evening Primrose (Oenothera biennis). *Journal of Evolutionary Biology*, **20**, 190–200.

Joshi, J., Schmid, B., Caldeira, M.C., Dimitrakopoulos, P.G., Good, J., Harris, R., Hector, A., Huss-Danell, K., Jumpponen, A., Minns, A., Mulder, C.P.H., Pereira, J.S., Prinz, A., Scherer-Lorenzen, M., Siamantziouras, A.S.D., Terry, A.C., Troumbis, A.Y. & Lawton, J.H. (2001) Local adaptation enhances performance of common plant species. *Ecology Letters*, **4**, 536–544.

Juenger, T. & Bergelson, J. (1998) Pairwise versus diffuse natural selection and the multiple herbivores of scarlet gilia, *Ipomopsis aggregata*. *Evolution*, **52**, 1583–1592.

Keller, S.R., Soolanayakanahally, R.Y., Guy, R.D., Silim, S.N., Olson, M.S. & Tiffin, P. (2011) Climate-driven local adaptation of ecophysiology and phenology in balsam poplar, *Populus balsamifera* L. (Salicaceae). *American Journal of Botany*, **98**, 99–108.

Kingsolver, J.G., Diamond, S.E., Siepielski, A.M. & Carlson, S.M. (2012) Synthetic analyses of phenotypic selection in natural populations: lessons, limitations and future directions. *Evolutionary Ecology*, **26**, 1101–1118.

Kirkpatrick, M. (2009) Patterns of quantitative genetic variation in multiple dimensions. *Genetica*, **136**, 271–284.

Kobayashi, S., Inoue, K. & Kato, M. (1999) Mechanism of selection favoring a wide tubular corolla in *Campaula punctata*. *Evolution*, **53**, 752–757.

Lande, R. & Arnold, S.J. (1983) The measurement of selection on correlated characters. *Evolution*, **37**, 1210–1226.

Lau, J.A. (2008) Beyond the ecological: Biological invasions alter natural selection on a native plant species. *Ecology*, **89**, 1023–1031.

Leimu, R. & Fischer, M. (2008) A meta-analysis of local adaptation in plants. *PLoS ONE*, **3**, e4010.

Lynch, M. (1999) Estimating genetic correlations in natural populations. *Genetical Research*, **74**, 255–264.

Lynch, M. & Ritland, K. (1999) Estimation of pairwise relatedness with molecular markers. *Genetics*, **152**, 1753–1766.

Klaper, R., Ritland, K., Mousseau, T.A. & Hunter, M.D. (2001) Heritability of phenolics in *Quercus laevis* inferred using molecular markers. *Journal of Heredity*, **92**, 421–426.

Kruuk, L.E.B., Merila, J. & Sheldon, B.C. (2001) Phenotypic selection on a heritable size trait revisited. *The American Naturalist*, **158**, 557–571.

Kruuk, L.E.B., Slate, J., Pemberton, J.M., Brotherstone, S., Guinness, F. & Clutton-Brock, T. (2002) Antler size in red deer: heritability and selection but no evolution. *Evolution*, **56**, 1683–1695.

Kulbaba, M.W. & Worley, A.C. (2012) Selection on floral design in *Polemonium brandegeei* (Polemoniaceae): Female and male fitness under hawkmoth pollination. *Evolution*, **66**, 1344–1359.

McKay, J.K., Bishop, J.G., Lin, J.Z., Richards, J.H., Sala, A. & Mitchell—Olds, T. (2001) Local adaptation across a climatic gradient despite small effective population size in the rare sapphire rockcress. *Proceedings of the Royal Society B-Biological Sciences*, **268**, 1715–1721.

Merila, J., Sheldon, B.C. & Kruuk, L.E.B. (2001) Explaining stasis: microevolutionary studies in natural populations. *Genetica*, **112**, 199–222.

Mitchell-Olds, T., Willis, J.H. & Goldstein, D.B. (2007) Which evolutionary processes influence natural genetic variation for phenotypic traits? *Nature Reviews Genetics*, **8**, 845–856.

Mojica, J.P. & Kelly, J.K. (2010) Viability selection prior to trait expression is an essential component of natural selection. *Proceedings of the Royal Society B-Biological Sciences*, **277**, 2945–2950.

Morgan, M.T. & Conner, J.K. (2001) Using genetic markers to directly estimate male selection gradients. *Evolution*, **55**, 272–281.

Morrissey, M.B., Kruuk, L.E.B. & Wilson, A.J. (2010) The danger of applying the breeder's equation in observational studies of natural populations. *Journal of Evolutionary Biology*, **23**, 2277–2288.

Morrissey, M.B., Parker, D.J., Korsten, P., Pemberton, J.M., Kruuk, L.E.B. & Wilson, A.J. (2012a) The prediction of adaptive evolution: empirical application of the secondary theorem of selection and comparison to the breeder's equation. *Evolution*, **66**, 2399–2410.

Morrissey, M.B., Walling, C.A., Wilson, A.J., Pemberton, J.M., Clutton-Brock, T.H. & Kruuk, L.E.B. (2012b) Genetic analysis of life-history constraint and evolution in a wild ungulate population. *The American Naturalist*, **179**, E97–E114.

Nuñez—Farfan, J. & Dirzo, R. (1994) Evolutionary ecology of *Datura stramonium* L. in central Mexico: natural selection for resistance to herbivorous insects. *Evolution*, **48**, 423–436.

Nussey, D.H., Postma, E., Gienapp, P. & Visser, M.E. (2005) Selection on heritable phenotypic plasticity in a wild bird population. *Science*, **310**, 304–306.

Ozgul, A., Tuljapurkar, S., Benton, T.G., Pemberton, J.M., Clutton-Brock, T.H. & Coulson, T. (2009) The dynamics of phenotypic change and the shrinking sheep of St. Kilda. *Science*, **325**, 464–467.

Pelletier, F., Reale, D., Garant, D., Coltman, D.W. & Festa—Bianchet, M. (2007) Selection on heritable seasonal phenotypic plasticity of body mass. *Evolution*, **61**, 1969–1979.

Pemberton, J.M. (2010) Evolution of quantitative traits in the wild: mind the ecology. *Philosophical Transactions of the Royal Society of London Series B, Biological Sciences*, **365**, 2431–2438.

Postma, E. & van Noordwijk, A.J. (2005) Gene flow maintains a large genetic difference in clutch size at a small spatial scale. *Nature*, **433**, 65–68.

Postma, E. (2006) Implications of the difference between true and predicted breeding values for the study of natural selection and micro-evolution. *Journal of Evolutionary Biology*, **19**, 309–320.

Price, T., Kirkpatrick, M. & Arnold, S.J. (1988) Directional selection and the evolution of breeding date in birds. *Science*, **240**, 798–799.

Rausher, M.D. & Simms, E.L. (1989) The evolution of resistance to herbivory in *Ipomoea purpurea*. I. Attempts to detect selection. *Evolution*, **43**, 563–572.

Rausher, M.D. (1992) The measurement of selection on quantitative traits: Biases due to environmental covariances between traits and fitness. *Evolution*, **46**, 616–626.

Reale, D. & Roff, D.A. (2001) Estimating genetic correlations in natural populations in the absence of pedigree information: accuracy and precision of the Lynch method. *Evolution*, **55**, 1249–1255.

Reid, J.M. (2012) Predicting evolutionary responses to selection on polyandry in the wild: additive genetic covariances with female extra-pair reproduction. *Proceedings of the Royal Society B-Biological Sciences*.

Ritland, K. (1996) Marker-based method for inferences about quantitative inheritance in natural populations. *Evolution*, **50**, 1062–1073.

Ritland, K. (1996) Estimators for pairwise relatedness and individual inbreeding coefficients. *Genetical Research*, **67**, 175–185.

Ritland, K. (2000) Marker-inferred relatedness as a tool for detecting heritability in nature. *Molecular Ecology*, **9**, 1195–1204.

Ritland, K. & Ritland, C. (1996) Inferences about quantitative inheritance based on natural population structure in the yellow monkeyflower, *Mimulus guttatus*. *Evolution*, 1074–1082.

Robinson, M.R., Wilson, A.J., Pilkington, J.G., Clutton-Brock, T.H., Pemberton, J.M. & Kruuk, L.E.B. (2009) The impact of environmental heterogeneity on genetic architecture in a wild population of Soay Sheep. *Genetics*, **181**, 1639–1648.

Robinson, M.R., Sander van Doorn, G., Gustafsson, L. & Qvarnström, A. (2012) Environment-dependent selection on mate choice in a natural population of birds. *Ecology Letters*, **15**, 611–618.

Sahli, H.F. & Conner, J.K. (2011) Testing for conflicting and nonadditive selection: floral adaptation to multiple pollinators through male and female fitness. *Evolution*, **65**, 1457–1473.

Schluter, D., Price, T.D. & Rowe, L. (1991) Conflicting selection pressures and life-history trade-offs. *Proceedings of the Royal Society B-Biological Sciences*, **246**, 11–17.

Schmitt, J., Eccleston, J. & Ehrhardt, D.W. (1987) Dominance and suppression, size-dependent growth, and self-thinning in a natural *Impatiens capensis* population. *Journal of Ecology*, **75**, 651–665.

Sgrò, C.M. & Hoffmann, A.A. (2004) Genetic correlations, tradeoffs and environmental variation. *Heredity*, **93**, 241–248.

Simms, E.L. & Rausher, M.D. (1989) The evolution of resistance to herbivory in *Ipomoea purpurea*. II. Natural selection by insects and costs of resistance. *Evolution*, **43**, 573–585.

Simms, E.L. & Rausher, M.D. (1993) Patterns of selection on phytophage resistance in *Ipomoea purpurea*. *Evolution*, **47**, 970–976.

Simonsen, A.K. & Stinchcombe, J.R. (2007) Induced responses in *Ipomoea hederacea*: simulated mammalian herbivory induces resistance and susceptibility to insect herbivores. *Arthropod-Plant Interactions*, **1**, 129–136.

Sinervo, B. & McAdam, A.G. (2008) Maturational costs of reproduction due to clutch size and ontogenetic conflict as revealed in the invisible fraction. *Proceedings of the Royal Society B-Biological Sciences*, **275**, 629–638.

Smouse, P.E., Meagher, T.R. & Kobak, C.J. (1999) Parentage analysis in *Chamaelirium luteum* (L.) Gray (Liliaceae): why do some males have higher reproductive contributions? *Journal of Evolutionary Biology*, **12**, 1069–1077.

Stratton, D.A. (1994) Genotype-by-environment interactions for fitness of *Erigeron annuus* show fine-scale selective heterogeneity. *Evolution*, **48**, 1607–1618.

Stinchcombe, J.R. & Rausher, M.D. (2001) Diffuse selection on resistance to deer herbivory in the ivyleaf morning glory, *Ipomoea hederacea*. *American Naturalist*, **158**, 376–388.

Stinchcombe, J.R. & Rausher, M.D. (2002) The evolution of tolerance to deer herbivory: modifications caused by the abundance of insect herbivores. *Proceedings of the Royal Society B-Biological Sciences*, **269**, 1241–1246.

Stinchcombe, J.R. (2002) Environmental dependency in the expression of costs of tolerance to deer herbivory. *Evolution*, **56**, 1063–1067.

Stinchcombe, J.R., Rutter, M.T., Burdick, D.S., Tiffin, P., Rausher, M.D. & Mauricio, R. (2002) Testing for environmentally induced bias in phenotypic estimates of natural selection: theory and practice. *The American Naturalist*, **160**, 511–523.

Stinchcombe, J.R., Izem, R., Heschel, M.S., McGoey, B.V. & Schmitt, J. (2010) Across-environment genetic correlations and the frequency of selective environments shape the evolutionary dynamics of growth rate in *Impatiens capensis*. *Evolution*, **64**, 2887–2903.

Strauss, S.Y. & Irwin, R.E. (2004) Ecological and evolutionary consequences of multispecies plant-animal interactions. *Annual Review of Ecology and Systematics*, **35**, 435–466.

Strauss, S.Y., Sahli, H. & Conner, J.K. (2005) Toward a more trait-centered approach to diffuse (co)evolution. *New Phytologist*, **165**, 81–90.

Terborgh, J. (1988) The big things that run the world—a sequel to EO Wilson. *Conservation Biology*, **2**, 402–403.

Thomas, S.C., Coltman, D.W. & Pemberton, J.M. (2002) The use of marker-based relationship information to estimate the heritability of body weight in a natural population: a cautionary tale. *Journal of Evolutionary Biology*, **15**, 92–99.

Thompson, J.N. & Fernandez, C.C. (2006) Temporal dynamics of antagonism and mutualism in a geographically variable plant-insect interaction. *Ecology*, **87**, 103–112.

Turkington, R. & Harper, J.L. (1979) The growth, distribution and neighbor relationships of *Trifolium repens* in a permanent pasture. IV. Fine-scale biotic differentiation. *Journal of Ecology*, **67**, 245–254.

van der Jeugd, H.P. & McCleery, R. (2002) Effects of spatial autocorrelation, natal philopatry and phenotypic plasticity on the heritability of laying date. *Journal of Evolutionary Biology*, **15**, 380–387.

van Kleunen, M. & Ritland, K. (2004) Predicting evolution of floral traits associated with mating system in a natural plant population. *Journal of Evolutionary Biology*, **17**, 1389–1399.

van Kleunen, M. & Ritland, K. (2005) Estimating heritabilities and genetic correlations with marker-based methods: an experimental test in *Mimulus guttatus*. *Journal of Heredity*, **96**, 368–375.

van Kleunen, M. & Burczyk, J. (2008) Selection on floral traits through male fertility in a natural plant population. *Evolutionary Ecology*, **22**, 39–54.

Visser, M.E., Noordwijk, A.J.v., Tinbergen, J.M. & Lessells, C.M. (1998) Warmer springs lead to mistimed reproduction in great tits (*Parus major*). *Proceedings of the Royal Society B-Biological Sciences*, **265**, 1867–1870.

Waitt, D.E. & Levin, D.A. (1998) Genetic and phenotypic correlations in plants: a botanical test of Cheverud's conjecture. *Heredity*, **80**, 310–319.

Wilson, A.J., Pemberton, J.M., Pilkington, J.G., Coltman, D.W., Mifsud, D.V., Clutton-Brock, T.H. & Kruuk, L.E.B. (2006) Environmental coupling of selection and heritability limits evolution. *PLoS Biology*, **4**, e216.

Wise, M.J. (2010) Diffuse interactions between two herbivores and constraints on the evolution of resistance in horsenettle (*Solanum carolinense*). *Arthropod-Plant Interactions*, **4**, 159–164.

Wise, M.J. & Rausher, M.D. (2013) Evolution of resistance to a multiple-herbivore community: genetic correlations, diffuse coevolution, and constraints on the plant's response to selection. *Evolution*, **67**, 1767–1779.

Wright, J.W. & Meagher, T.R. (2004) Selection on floral characters in natural Spanish populations of *Silene latifolia*. *Journal of Evolutionary Biology*, **17**, 382–395.

Yang, J., Benyamin, B., McEvoy, B.P., Gordon, S., Henders, A.K., Nyholt, D.R., Madden, P.A., Heath, A.C., Martin, N.G., Montgomery, G.W., Goddard, M.E. & Visscher, P.M. (2010) Common SNPs explain a large proportion of the heritability for human height. *Nature Genetics*, **42**, 565–569.

CHAPTER 9

Quantitative genetics of wild populations of arthropods

Felix Zajitschek and Russell Bonduriansky

9.1 Why study invertebrates in the field?

This book discusses how to study quantitative genetics of traits measured in wild populations, and in this chapter, we focus on the potential of arthropod species in this respect. The study of the genetic basis of quantitative traits is fundamental to gaining an understanding of how organisms' phenotypes evolve and, more specifically, how trait variation is maintained, how traits are correlated genetically, and what implications these genetic parameters have for the evolution of traits under selection (Lande 1982; Barton & Turelli 1989). As we make clear below, our ability to answer these questions comprehensively depends on our understanding of the quantitative genetics of natural populations across a wide range of biological diversity. Extension of this research to arthropods (and, ultimately, other invertebrates) will be an important step in revealing how these parameters vary across the tree of life, and in relation to life-history strategies and ecological niches.

Field studies are enormously important in evolutionary research because they make it possible to investigate variation in fitness-related traits under natural conditions. Fitness, and traits closely associated with fitness, are extremely sensitive to the ambient environment (physical, ecological, social) that an organism experiences, and this plasticity can lead to confounding effects in laboratory studies: if phenotypic variation interacts with environmental factors affecting fitness, laboratory estimates of fitness variation and natural selection can be misleading (Irschick 2003). These considerations apply equally to the study of genetic variation in fitness. A genetic variant associated with high fitness in the lab may have no such effect, or even the opposite effect, in a natural environment (i.e. there may be extensive genotype × environment interaction, or $G \times E$, for fitness). A response to selection measured in laboratory cages may not be representative of processes in natural populations of the same species. For this reason, it is imperative to estimate fitness in natural populations, and to determine the importance of $G \times E$ for fitness as a check on laboratory studies. A less ambitious goal of field studies is to estimate 'true' heritabilities or 'true' genetic correlations for phenotypic traits. When obtained in natural populations, such estimates are based on realistic amounts and sources of environmental variance, and can therefore allow us to predict the strength and direction of evolutionary change in the natural population. Ideally, it should be possible to re-create the relevant environmental conditions in a precisely controlled laboratory environment. In practice, however, we simply do not know what factors or combinations of factors are important, and only by going into the field and tracking wild animals can we hope to find out.

While field estimates of fitness are clearly of considerable value, they are also very challenging to obtain. Fitness is best estimated from long-term (ideally, lifetime) performance data on individuals. Heritabilities and genetic correlation structure can often only be estimated by measuring phenotypes of individuals of known degrees of relatedness,

Quantitative Genetics in the Wild. Edited by Anne Charmantier, Dany Garant, and Loeske E. B. Kruuk

such as parents and their offspring, or members of a more extensive pedigree. To obtain such data for individuals in natural populations, it is necessary to follow individuals over their lifetimes and, ideally, to be able to identify and follow their offspring as well. This is very difficult to do even for large-bodied, long-lived vertebrates such as mammals and birds. The difficulties have long seemed insurmountable for small-bodied, short-lived invertebrates such as arthropods. However, as we detail below, there already exists a suite of old and new methodologies that can be combined into a toolbox for future research in this field.

Until recently, quantitative genetic approaches have mainly been applied to study traits under situations of low environmental noise compared to natural environments, for example in laboratories or livestock breeding facilities. In recent years, computing power has opened up new analytical possibilities, and molecular analysis has become more sophisticated and affordable for large samples. High and steadily increasing computational power is available today. This has made it possible to employ statistical methods that allow for inclusion of random effects in models, the use of better likelihood maximisation procedures and the use of Bayesian statistics. These advances make it possible to account for environmental variation, and to use pedigree data from wild populations to estimate quantitative genetic parameters (see other chapters in this book). The analysis of some well-established long-term data sets has thereby produced some important insights into the evolution of traits in natural populations (e.g. Sheldon *et al.* 2003; Charmantier *et al.* 2007; Morrissey *et al.* 2012) by quantifying the strength of selection and evolutionary potential for some traits, under ecologically more relevant circumstances than under laboratory or farming conditions.

9.1.1 Suitability of model systems

Despite great advances in analytical power, the practical difficulties of establishing relatedness and obtaining longitudinal data on individuals in natural populations of any species remain. In this chapter, we discuss some approaches and study systems that minimise these difficulties. Study systems that are most amenable to these kinds of analyses are characterised by having moderate population sizes, large-bodied individuals that can be individually marked and tracked, for example with the help of electronic devices (e.g. using harmonic radar, Boiteau *et al.* 2010) or simple re-sighting approaches, and moderate offspring numbers. Low offspring numbers make it possible to genotype offspring, assign parentage and potentially create a deep, well-resolved pedigree over many generations of sampling. Species that have high site-fidelity, or form stable mating aggregations, are most useful for such research. If the ability and possibility to disperse is not very high, as is the case for some plant species and terrestrial animals, and especially for populations located in geographically contained locations, data on relatedness and fitness traits can be very complete. To date, these practical considerations have unfortunately led to a bias in longitudinal and quantitative genetic studies of wild populations whereby a considerable amount of work has been done on natural populations of birds and mammals, but arthropods and other invertebrate taxa are very poorly represented. In this chapter, we suggest arthropod systems and approaches that could help to redress this bias.

Selection of the most suitable arthropod systems for field research can, of course, lead to biases as well. Since the most suitable arthropod species share some of the characteristics of vertebrate models, it could be argued that these arthropods still represent a narrow range of life-history parameters, compared to the whole of arthropod diversity. This is a valid concern that applies to vertebrate systems as well. Nonetheless, even those arthropod taxa that are most suitable for this type of research (see below) span an enormous range of body sizes, life-histories and ecological niches, and will certainly represent a great improvement in our coverage of biological diversity, relative to studies on birds and mammals alone. Being able to quantify the relationship between fundamentally different life-histories and quantitative genetic parameters (e.g. what effect does the form of metamorphosis have on adult genetic architecture?) will be an important task. This scientific goal can be achieved by comparisons across the whole realised life-history parameter space. As more powerful techniques are developed, further extension to taxa with more divergent life-histories may become possible.

9.1.2 Why arthropods?

Why go to the trouble of studying natural populations of invertebrates when data are already available for many species of mammals and birds? We believe that studies of invertebrates are important from a theoretical perspective, in that the development of robust evolutionary theory requires understanding how life-history strategies vary and evolve in organisms spanning a broad spectrum of variation in body size, life expectancy, reproductive strategies, ecological roles and life cycle complexity. Most invertebrates are small-bodied and short-lived, and therefore cluster towards the opposite end of the ecological, physiological and life-history spectrum from relatively large-bodied and long-lived mammalian and avian field models. It is therefore risky to extrapolate and generalise from studies on large-bodied vertebrates to other taxonomic groups that differ in important life-history parameters, and are biologically more diverse. Longitudinal studies of phenotypic variation and fitness in natural populations of arthropods and other invertebrates, particularly when combined with quantitative genetic studies, can provide a theoretically informative contrast to studies on natural populations of mammals and birds. From the more applied perspective of conservation biology, certain invertebrate species with short life-cycles might be suitable as systems to monitor how genetic variation in critical traits affects the potential of populations to adapt to a changing climate (Sgro *et al*. 2011; Hansen *et al*. 2012). While the bias towards lab studies and, for studies in wild populations, towards vertebrates has been recognised and discussed to a certain degree before (Moore & Kukuk 2002; McGuigan 2006; Kruuk & Hill 2008; Hill 2012), we can still only refer to a few arthropod field studies that have examined genetic (co)variation of quantitative traits (see below for notable exceptions). For a discussion of the current state of quantitative genetic research in natural populations of plants, we refer the reader to Chapter 8, Stinchcombe.

In regard to evolutionary questions, the most relevant information to collect includes the proportion of a phenotypic character that is due to additive genetic variation (i.e., trait heritability), how that character is genetically correlated with other traits and influenced by environmental factors (genetic architecture), and the nature and strength of selection on this character, i.e. how it is related to fitness. Fundamental quantitative genetic parameters can be estimated in a variety of ways that always involve measuring variation in quantitative traits among individuals of a known degree of relatedness. Suitable sample sizes depend on the quantitative genetic parameter of interest, on the quality (repeatability) of measurement, and the environmental variance associated with the studied trait(s). Typically, much of this information is unavailable and difficult to guess before the study is conducted. Generally, sample sizes needed to estimate quantitative genetic parameters are quite high compared to purely phenotypic studies. This disparity is even more pronounced when traits of interest are measured in natural populations, as compared to more benign laboratory environments. In the wild, the statistically unaccounted for environmental variation in a trait is likely to be higher, which in turn inflates the total phenotypic variance and decreases the ratio of additive genetic variance to the total phenotypic variance of the trait—its heritability. A good review on this topic can be found in Moore and Kukuk (2002).

9.1.3 Non-genetic effects

It is also very important, whenever possible, to estimate the contribution of non-genetic maternal and paternal effects to the total phenotypic variance in fitness-related traits (see also Chapter 6, McAdam *et al*.). Laboratory studies have shown that such effects, mediated by mechanisms such as transgenerational epigenetic inheritance, or the transfer of resources, environments, hormones, or pathogens from parents to offspring, can have large effects on a great variety of phenotypic traits (reviewed in Jablonka & Lamb 1995; Bonduriansky & Day 2009; Jablonka & Lamb 2010; Danchin *et al*. 2011). Experimental evidence in mice (Cropley *et al*. 2012) and *Drosophila* (Sollars *et al*. 2002) shows that heritable epigenetic variation can respond to selection. Importantly, these mechanisms also have the potential to transmit effects of environment across generations, and theoretical analyses suggest that such effects can have important consequences for the dynamics and direction of evolution (reviewed in Bonduriansky *et al.* 2012). Although epigenetic variation is clearly present in natural populations (Vaughn *et al.* 2007; Herrera & Bazaga 2011), and maternal effects have been studied in natural populations of a few birds and mammals (Badyaev *et al.* 2002; McAdam *et al.* 2002; Chapter 6, McAdam *et al.*), very little is presently known about the contribution of such effects to total phenotypic variance in natural populations of invertebrates (but see Agrawal *et al.* 1999). Understanding the role of such effects in natural populations will require a combination of field studies that partition variance among genes, parental environment and offspring environment, as well as laboratory experiments that verify the potential for environmental factors to influence future generations.

9.1.4 Selection

Individuals in wild populations are often affected by whole suites of biotic and abiotic selective pressures—the ecology of a specific population—whose combined effect drives evolutionary change, given enough standing genetic variation in the traits and trait combinations under selection. The strength and direction of multivariate selection in the wild can be very complex, variable and episodic, i.e. it might change over small geographic and temporal scales. Nevertheless, strong and consistent patterns of selection may exist, and it is possible to estimate them. For example, in a study on sexual selection in a wild population of ambush bugs (*Phymata americana*, Punzalan *et al.* 2010), a total of ten cross-sectional samplings were performed over two generations. The authors showed that the linear selection gradient for dark lateral coloration in males was positive and significant in four cases, negative and significant in one case and not significant during five sampling periods. This case illustrates that selective regimes in the wild can be successfully studied, although detecting overall patterns can arguably be very challenging.

It has been argued that studies on wild populations might not shed much light on adaptive evolution because of the ubiquitous and confounding effects of genotype by environment ($G \times E$) interactions on genetic correlations and, therefore, on trade-offs (in this specific case the authors refer to *Drosophila melanogaster*; Prasad & Joshi 2003). Identifying and quantifying the most important selective agents can be difficult even in a stable, homogeneous environment. With $G \times E$ present, the magnitude and sign of selection might change dramatically from one environment to another. Population differences in genetic architecture, due to drift, different levels of inbreeding, or slightly different selective regimes in the populations' evolutionary past, are likely to be another source of variation that affects phenotypes and that is rather unpredictable. Yet, $G \times E$ and fluctuations over space and time are precisely the reason why studies on natural populations are essential. After all, these are factors that populations experience under natural and realistic conditions, and only field studies can reveal their nature and importance.

Moreover, it should be possible to identify populations that have a shared evolutionary history and that have only recently (on an evolutionary timescale) begun to evolve in different ecological backgrounds. Under this comparative approach, effects of different ecological factors can potentially be partitioned out, even if genetic background differs across populations. More details on comparative quantitative genetics can be found in a review by Steppan *et al.* (2002).

9.1.5 Lab vs wild

Under laboratory conditions, the full spectrum of selection pressures operating in natural populations can never be comprehensively known or simulated, but scenarios of fluctuating environments, and therefore fluctuating selection pressures, can be experimentally tested to a certain degree (e.g. seed beetles under fluctuating thermal environment, Hallsson & Björklund 2012). The problem of potentially rapid change of environmental selection pressures for the estimation of phenotypic trait values and their genetic basis in wild populations is not unique to invertebrate systems and affects populations of birds and large mammals in similar ways. Of course there are likely to be differences of scale, mainly geographic and temporal in nature, that might lead to different suitabilities of methods of how to measure trait values and assess their corresponding environmental variables. For example, whereas climatic conditions can potentially be gathered from local government-run weather stations that are located in proximity to populations of larger herbivores or birds, microclimatic measurements for an insect population might require customised measurement solutions on a much more localised scale, using, for example, data loggers for local temperature and humidity (Gibbs *et al.* 2003).

Arthropod species that are suitable for longitudinal and possibly quantitative genetic field studies might have confined, largely stable populations, relatively low population sizes, life-cycles that are not too long (some cicada species, for example, might not be very suitable as they spend several years in the larval stage; Saulich 2010), sufficiently large body size for marking and possibly for taking DNA samples, and a non-cryptic lifestyle. One of the biggest advantages of studying arthropod species, compared to much larger sized vertebrates, is the opportunity to complement studies in the wild with studies in the laboratory, using the same source populations. Many studies in the wild are not experimental in so far as they are based solely on observations and measurements of the traits of interest, without manipulating any of the variables that might affect these traits and drive their evolution. Single snapshots of quantitative genetic (co)variation fall under this category, irrespective of whether they are taken in the lab or in natural populations. Ideally, more evolutionary quantitative genetic studies will consist of longer term laboratory and field studies to utilise the advantages of arthropod over vertebrate systems and to tap their full outstanding potential as model organisms. If we want to find out whether laboratory studies provide relevant quantitative genetic estimates, studying arthropods in both lab and wild conditions is one of the most promising approaches. Moreover, at present, we do not know to what extent we might be misled by lab results, nor how to go about designing more suitable conditions for lab studies. Lab–field comparisons will inform the design of experimental studies in the lab, allowing for the creation of more appropriate environments within which to test evolutionary questions.

One potential approach is to collect individuals from the wild, transfer them into a stable lab environment and rear them for one or two generations in a common garden environment, before setting up breeding experiments (e.g. Hendrickx *et al.* 2008). In a study on *D. melanogaster*, for example, Robinson *et al.* (2012) collected mating pairs of a natural fly population. After sampling, females were allowed to lay eggs on standardised medium, and males and females were genotyped at 23 microsatellite marker loci. Several pairwise relatedness scores were calculated to evaluate the effect of relatedness on mate choice. A more ambitious approach is to estimate fitness and quantitative genetic parameters entirely within a natural population, while also establishing a laboratory population from the same genetic stock in order to investigate mechanisms or test specific hypotheses generated by the field data. In either case, the most powerful approach is likely to be a combination of lab and field studies—an approach to which invertebrate systems are particularly well suited, and indispensable, when it comes to lab-based evolutionary experiments, incorporating genotype by environment effects, that need very large sample sizes.

9.1.6 Estimation of relatedness

Establishing relatedness among individuals in natural populations is challenging. This is especially the case in comparatively small and short-lived arthropods for which it is often impossible to infer relatedness from direct observations in the field. However, new (next-generation) molecular genomic methods offer a potential solution to this problem. While this topic is discussed in more detail by Jensen *et al.* in Chapter 13, we feel that its enormous potential for studying quantitative genetic variation in natural populations, especially of arthropods, warrants a general outline here. As DNA sequencing becomes ever faster and more affordable, it is now feasible to create dense enough genetic marker maps for relatedness analyses in wild populations of some suitable arthropod species. For example, restriction site associated

DNA sequencing (RADSeq) can identify and score an immense number of markers (Davey & Blaxter 2011), which can be used to estimate relatedness as well as genetic (co)variances of traits. There exist different statistical methods for the use of molecular marker-based methods in heritability estimation (Ritland 2000; Sillanpäa 2011; Hill 2012) and we refer the reader to Chapter 2 (Postma) for a critical review on these methodological issues. Another, it seems yet underexplored method to estimate genetic correlations, was proposed by Lynch (1999) and later refined (Reale & Roff 2001). Here, relatedness between sampled individuals does not need to be known, and reliable estimates of genetic correlations can be achieved with a couple of hundred sampled individuals, which is a very feasible sample size for many arthropod species. But the critical condition for this method to provide reliable estimates is the percentage of related pairs of individuals in the sample, with a minimum of at least 20–30% (given a sample size of 400–500 pairs), which might be hard to assess for any given natural population under many circumstances.

9.2 Examples of promising arthropod model systems, and groundbreaking studies

Despite the considerable challenges of conducting longitudinal field studies on most invertebrate groups, the enormous phenotypic and ecological diversity they offer presents a rich resource. The challenge is to find the right model species to use. In the following section, we discuss in more detail some experimental designs and animal systems that have the potential to be employed to address quantitative genetic questions in wild invertebrate populations. We also outline several groundbreaking studies that have taken advantage of these systems to investigate the quantitative genetics of wild arthropod populations. These studies illustrate the potential value of such work, and point the way forward. Importantly, these studies provide examples of methodologies that can be used to estimate fitness of one or both sexes in natural populations, as well as techniques for the estimation of relatedness. The combination of these techniques thus makes it possible to investigate the quantitative genetics of fitness in natural populations of small-bodied animals.

We focus on arthropods, as they include a large number of well-studied and diverse taxa. However, the potential for longitudinal field research exists in other invertebrate groups as well. For example, some species of molluscs are large enough to lend themselves to such work (e.g. Henry *et al*. 2003).

9.2.1 Crickets

Various species of crickets have been used in a wide range of studies on animals in or from natural populations (e.g. Zuk *et al*. 1998; Gray & Cade 1999; Simmons *et al*. 2005; Bentsen *et al*. 2006). However, longitudinal studies of captured, marked and released crickets in the wild are not often undertaken (but see e.g. Holzer *et al*. 2003; Zajitschek *et al*. 2009; Rodriguez-Munoz *et al*. 2010). For a few cricket species with large enough body size, it is possible to radio-tag individuals in order to track and recapture them in the wild. This method was used with Mormon crickets (*Anabrus simplex*, Lorch *et al*. 2005) and giant wetas (*Deinacrida rugosa*; Kelly *et al*. 2008), but no quantitative genetic data were gathered in these studies, probably because of the difficulty of assessing genetic relatedness between tracked individuals. In general, while small and light radio tags are available, only large arthropod species are suitable for radio-tagging (but see Boiteau *et al*. 2010). In addition, the risk of disturbing natural behaviour patterns through the weight and size of the tag (for example, mounting during mating or avoiding predation by using an underground shelter) still seems to be quite high.

We would like to highlight a recent landmark study by Tom Tregenza and his group on a wild population of the European field cricket (*Gryllus campestris*) located in northern Spain. This groundbreaking study combined longitudinal observations and relatedness estimates that allow the estimation of quantitative genetic parameters. Rodriguez-Munoz *et al*. (2010) showed that it is possible to monitor the behaviour of individually marked adult crickets of a single population throughout the season, in fact, 24 hours a day, over several years, by videotaping individual borrows, in or

Figure 9.1 *Gryllus campestris* study by Rodriguez-Muñoz *et al.* (2010). (a) The study site. (b) Video camera filming an individual burrow. (c) Close-up of two marked crickets in front of a burrow. © www.wildcrickets.org

near which adult crickets spend most of their time (Figure 9.1). The size of the population was estimated as being between 150 and 200 adult crickets in two seasons, living on a 40m × 20m meadow. The population seemed to be very stable, with low adult migration rates (estimated 4–5% adult immigration for two consecutive seasons), and newly emerged adults could be captured, marked and released again. Individual marker tags were attached to the pronotum with cyanoacrylate (super glue), after the pronotum had been slightly abraded. By taking a tissue or hemolymph sample, every captured cricket was genotyped and assessed at 11 microsatellite loci for parentage analyses. This allowed the researchers to assess relatedness between individuals within and across generations, and to build a pedigree of the entire adult population. This species is univoltine, and individuals overwinter as late instar nymphs. Consequently, lifetime reproductive success for males and females can be estimated as the number of offspring captured as newly eclosed adults. As expected in a stable population with F2 recruitment replacing F1 mortality, this number was not very high in absolute terms (on average within the study sample, females had 1.79 ± 2.46 and males had 1.92 ± 3.66 offspring that survived to adulthood).

With the estimate of lifetime reproductive success (LRS) as a good measure of fitness in hand, the authors were able to test and support a fundamental prediction in sexual selection theory: the variance in LRS was larger in males than in females. In both males and females, larger body size, longer life span or having more mating partners was beneficial in terms of increased LRS. But the study also showed that male mating success was actually not a good predictor for male reproductive success: although dominance positively influenced mating success, it had no effect on LRS. Because of the extensive video monitoring and the fact that adult crickets spend most of their time near borrows, it was possible to evaluate spatial distances between all male–female pairs for most of the observation time. This information, together with relatedness of paired males and females, made it possible to test for (and exclude) inbreeding avoidance in this population (Bretman *et al.* 2011).

While total sample sizes per season, given as captured adult individuals, are not especially high for quantitative genetic estimation purposes, the available data over several generations together with the pedigree make this system amenable for animal model analyses (Wilson *et al.* 2010; see also this book, e.g. Chapter 2, Postma; Chapter 3, Reid; and Chapter 10, Kruuk *et al.*). Quantitative genetic analyses will shed light on the genetic underpinnings of some of these results. For example, the sex-independent positive selection on lifespan and number of mating partners on the phenotypic level might be corroborated by underlying genetic correlations between these traits and fitness. This would furnish a strong case against sexual conflict over these traits in this system. The *Gryllus campestris* model system therefore offers a valuable example of the potential for quantitative genetic research on natural populations of arthropods.

9.2.2 Damselflies

Like field crickets, damselflies have proven to be a very suitable taxon for ecological and evolutionary field studies (e.g. Cordero 1995; Svensson *et al.* 2006; Steele, Siepielski & McPeek 2011; Thompson *et al.* 2011). Recently, a research team has shown that it is feasible to combine classic capture-mark-recapture methodology with genotyping over multiple field seasons. In total, 1042 individuals of the damselfly *Coenagrion puella*, including parents and F1 offspring, were marked over two consecutive seasons (Lowe *et al.* 2009; Figure 9.2a). The distance of the sampled population to other populations and water bodies was sufficient to avoid immigration into the studied population, so that all adult individuals could be captured, marked and genotyped, using DNA samples from one of the middle legs that was severed after the first capture. All individuals were genotyped at 12 microsatellite loci and parentage of 74% of offspring could be assigned at the 95% level. The heritability estimates for maturation date and hindwing length of about 0.05 were very low and not significantly different from zero. For maturation date only, the maternal effect was high and significant. Behavioural traits are expected to be more variable than morphometric traits, both because of generally higher measurement error, and because morphometric traits are often largely fixed in adult individuals, whereas behaviours can vary at the time of measurement due to undetected environmental effects, resulting in low measurement repeatability (see also Chapter 3, Reid). Accordingly, measurement error and low repeatability (caused by environmental effects) can lead to lower estimates of heritability (Hoffmann 2000). Given the sample size and a highly variable environment, the power to detect significant additive genetic variation for maturation date was low in this study. Individual development time, which was inferred from observed matings and the fact that a female lays all her eggs shortly after a mating, was estimated as well, but the sample sizes were considerably lower than for date of maturity, and therefore heritability could not be estimated for this trait. The lack of significant heritability for hind wing length is more surprising.

Several damselfly species already rank highly among established arthropod systems for ecological research in natural populations, which means that there may be fewer methodological issues compared to less studied taxa. It therefore seems only a comparatively small step to also establish damselflies for studying quantitative genetic

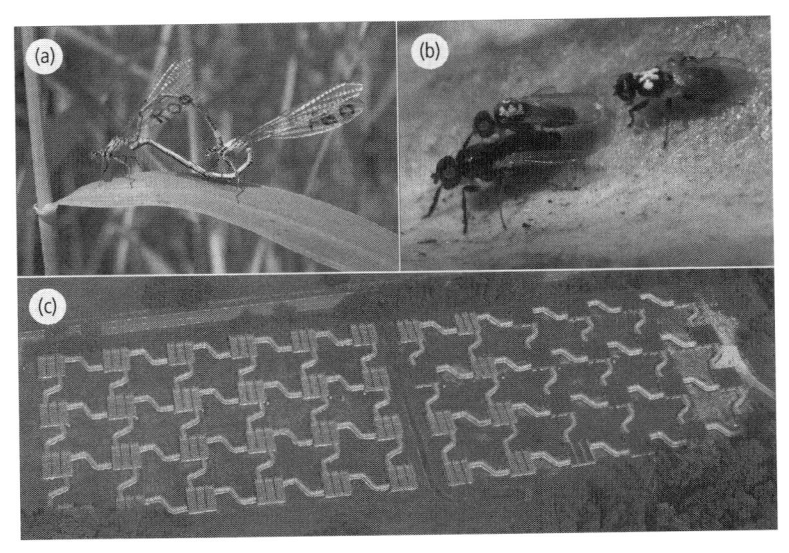

Figure 9.2 (a) Two mating, individually marked *C. puella* damselflies in the wild. © Phill Watts (b) Wild, individually marked antler flies (*P. litigata*) on a moose antler. A single male pursues a female guarded by another male following mating. © R. Bonduriansky (c) The 48 experimental cages (each 10 × 10 m) of the Metatron, linked with corridors. © Quentin Bénard, SEEM, CNRS.

variation in the wild. One problem common to all research on wild invertebrates is lack of funding. The study by Lowe *et al.* (2009) was conducted over two field seasons and then aborted due to insufficient funding. But, at least in theory, there is no reason why more long-term studies on arthropods, similar to the well-established flagship vertebrate mammal and avian systems, should not be funded. This might happen once a threshold of high-quality experimental findings, gained directly from natural invertebrate populations, is reached.

9.2.3 Antler flies

Although only 2–3 mm in length, the antler fly (*Protopiophila litigata*), a dipteran in the family Piophilidae, is a valuable field model because of its unusual site-fidelity and tolerance for close observation. This species breeds exclusively on the discarded antlers of cervids such as moose and deer in eastern Canada: adults form complex aggregations of up to several hundred individuals on an antler's surface, females oviposit into cracks and pores in the antler, and larvae develop in the porous bone inside the antler and emerge to pupate in the surrounding soil (Bonduriansky 1995; Bonduriansky & Brooks 1998; Bonduriansky & Brooks 1999). Remarkably, male antler flies typically spend each day of their lives on the same antler, leaving only to spend the night in nearby vegetation. Adults can be marked with individual codes using a simple restraining device and enamel paint (Bonduriansky & Brooks 1997), and these codes can be read in the field (Figure 9.2b), making it possible to construct detailed biographies for these tiny insects in a fully natural environment. Studies on this species in Algonquin Provincial Park (Ontario, Canada) have examined ageing (Bonduriansky & Brassil 2002; Bonduriansky & Brassil 2005) and sexual selection (Bonduriansky & Brooks 1998; Bonduriansky & Brooks 1999) in the wild. With the advent of efficient and inexpensive techniques for parentage analysis, it should now be possible to estimate reproductive success for individuals that have been marked and observed throughout their lives in their natural habitat. DNA samples can be obtained from wild-collected males by taking sperm samples from mated females, and a sample of individuals from the following generation can be subjected to paternity analysis. Offspring can be collected as final-instar larvae that emerge during rain from cracks and pores in antlers on their way to pupation sites in the surrounding soil, or as adults that return to their natal antler. Most individuals emerging near high-quality antlers (i.e. antlers that are five years old or less) return to their natal antler to breed. This species offers a potentially valuable contrast with other arthropod field models, such as crickets, damselflies and spiders, because of its much smaller body size, holometabolous life-history, and relatively close relatedness to *Drosophila melanogaster*.

9.2.4 Spiders

Many species of web-building spiders represent potential models for quantitative genetic research in natural populations because they are largely sessile as adults, allowing for easy identification and observation of individuals throughout their lives. For example, field studies based on longitudinal data have investigated variation in male and female fitness in several species of spiders (e.g. Kasumovic *et al.* 2007; Rabaneda-Bueno *et al.* 2008; Kasumovic & Jordan 2013). In one study, diet manipulation (supplementary feeding) was applied in a natural spider population to examine the interaction between resource availability and age (Moya-Laraño 2002). A study has also employed microsatellite-based genetic analysis of spiderlings collected from webs, along with their mothers, to investigate patterns of polyandry and genetic variation (Tuni *et al.* 2012). There is far more scope for this type of research. The contrasting biology of spiders offers a valuable point of comparison with insect models. In particular their predatory lifestyle and sessile habits would extend the range of biological variation.

9.3 Semi-natural: in between the lab and the wild

Animals can be observed and measured in a variety of different environments. On one side of the spectrum, lineages of experimental animals are kept for many generations under standardised rearing conditions in the lab, with low variation in environmental variables and often with a very constrained reproductive schedule. For example, *D. melanogaster* are normally kept on one type of food, at a constant temperature, humidity and light–dark cycle. Generations are often kept separate: a specific number of male–female pairs per vial are allowed to mate for a few days after eclosion and are then discarded. Finally, eggs laid in vials are reduced in number and left to develop into the next generation of adults. A variation to this protocol is to keep flies in mixed sex groups in population cages with an effective population size of several thousand. New food, which is also the egg-laying medium, is provided weekly, but old food containers that hold the developing eggs and larvae are left in the cage until most adults have eclosed from them. This results in an age-structured cage population and allows for more of the interactions and behaviours that flies in nature would engage in.

The value of populations of larger-bodied animals kept under artificial conditions in zoos, with existing pedigree information, has been pointed out before (Pelletier *et al.* 2009). Not only does a wealth of data from systematic breeding programs in zoos exist, but the data are also often on threatened species that are difficult to study in the wild, due to, for example, low population sizes or sensitivity to research-related disturbance. Animal model approaches make it possible to study quantitative genetic variation for certain life-history traits in captive populations and to assess genetic adaptations to captivity, which have a high information value for reintroduction and genetic rescue programs (Pelletier *et al.* 2009).

We want to emphasise the potential of semi-natural enclosures in the study of arthropods. Enclosures cannot provide a completely realistic model of the natural environment, as, by definition, they enclose populations into a restricted space and thereby hinder some abiotic and biotic interactions and migration processes that would take place in the population's natural environment. For the purpose of estimating quantitative genetic parameters, it would be advantageous to keep populations in enclosures for more than one generation. In experimental populations of some species, strongly restricted space availability will keep offspring from dispersing away from their natal site and radically alter the naturally occurring population density, for example. The specifics depend to a large extent on the effective population size of the founder generation, the number of offspring produced and the mobility of juveniles. But some of these factors are problems of scale. By using an experimental structure large enough to sustain populations of several thousand individuals and to allow for offspring dispersal, invertebrates' small body size compared to vertebrates can be turned to advantage.

One enclosure system that could fulfil these requirements is the Metatron in South-Western France, a system that comprises 48 population cages, each covering a ground area of 10 × 10 metres, with a height of 2 metres (Legrand *et al.* 2012; Figure 9.2c). Cages are connected by corridors that can be closed off. The upper parts of the cages and corridors are made out of fine, insect-proof mesh, the lower parts, which are built like a trough, enclosing 0.5 metres of soil, are made from plastic tarp. Setting up a system of several cages with a suitable locally occurring insect species and running it for several years could be a way to make a semi-natural enclosure system much more realistic than most cage systems used to date. Researchers can also manipulate humidity and temperature levels of each Metatron cage, which opens up the possibility of experimental approaches to test, for example, the effects of changing temperature regimes on quantitative genetic parameters. The choice of suitable experimental species, again, is paramount. Even in the absence of larger predators in the enclosures, some mortality hazards, such as climatic hazards (e.g. rainfall) or invertebrate predators (e.g. spiders, mites) in the cages can mimic mortality risks in completely natural environments quite well, especially for species whose larval and adult mortality, and population growth depends to a large extent on these factors. This system has already been used successfully with an arthropod species, the large white butterfly (*Pieris brassicae*; Legrand *et al.* 2012). Individuals of this species can easily be marked and recaptured, and have a comparatively short generation time (about one month) and adult lifespan (average adult lifespan in females was 7.35 days, and 6.28 days in males; Legrand *et al.* 2012). Additionally, life-history traits of individuals with known parents can be measured (see for example Ducatez *et al.* 2012). Together with the possibility to manipulate not only major climatic parameters in the Metatron cages, but also, for example, food availability and quality (Romeis & Wäckers 2002) or predation pressure (for example through lizards, kept in the same cages; Legrand *et al.* 2012), this opens up approaches to tackle elusive and hard-to-measure $G \times E$ interactions and their effects on quantitative genetic parameters in this semi-natural system.

9.4 Conclusion

Understanding the quantitative genetics of natural populations of invertebrates is of considerable theoretical importance, but this research has been neglected for practical reasons. The greatest challenges involved in this kind of work are to obtain longitudinal (preferably lifetime) field data, allowing for high-quality estimates of fitness under natural conditions, and to determine relatedness of focal individuals. However, less ambitious studies involving semi-natural enclosures can also yield valuable data. We have outlined recent advances in quantitative genetic analysis based on next-generation sequencing technology that will facilitate research on natural populations of invertebrates, and reviewed pioneering studies that have taken advantage of these new techniques. We have also discussed several systems (including crickets, odonates, antler flies and spiders) that have the potential to be used in quantitative genetic studies on natural populations. Although arthropods are currently most amenable to such studies, the four taxa highlighted here represent an enormous range of life-history strategies, body sizes and ecological niches. The tools discussed here will ultimately allow researchers to fill a major gap in our understanding of the evolutionary ecology of natural populations—the nature and magnitude of genetic variation in fitness in natural populations of small, short-lived organisms.

References

Agrawal, A.A., Laforsch, C. & Tollrian, R. (1999) Transgenerational induction of defences in animals and plants. *Nature*, **401**, 60–63.

Badyaev, A.V., Hill, G.E. & Whittingham, L.A. (2002) Population consequences of maternal effects: sex-bias in egg-laying order facilitates divergence in sexual dimorphism between bird populations. *Journal of Evolutionary Biology*, **15**, 997–1003.

Barton, N.H. & Turelli, M. (1989) Evolutionary quantitative genetics: how little do we know? *Annual Review of Genetics*, **23**, 337–370.

Bentsen, C.L., Hunt, J., Jennions, M.D. & Brooks, R. (2006) Complex multivariate sexual selection on male acoustic signaling in a wild population of *Teleogryllus commodus*. *The American Naturalist*, **167**, 102–116.

Boiteau, G., Vincent, C., Meloche, F. & Leskey, T.C. (2010) Harmonic radar: Assessing the impact of tag weight on walking activity of colorado potato beetle, plum curculio, and western corn rootworm. *Journal of Economic Entomology*, **103**, 63–69.

Bonduriansky, R. (1995) A new Nearctic species of *Protopiophila* Duda (Diptera: Piophilidae), with notes on its behaviour and comparison with *P. latipes* (Meigen). *Canadian Entomologist*, **127**, 859–863.

Bonduriansky, R. & Brassil, C.E. (2002) Rapid and costly ageing in wild male flies. *Nature*, **420**, 377.

Bonduriansky, R. & Brassil, C.E. (2005) Reproductive ageing and sexual selection on male body size in a wild population of antler flies (*Protopiophila litigata*). *Journal of Evolutionary Biology*, **18**, 1332–1340.

Bonduriansky, R. & Brooks, R.J. (1997) A technique for measuring and marking live flies. *Canadian Entomologist*, **129**, 827–830.

Bonduriansky, R. & Brooks, R.J. (1998) Male antler flies (*Protopiophila litigata*; Diptera: Piophilidae) are more selective than females in mate choice. *Canadian Journal of Zoology*, **76**, 1277–1285.

Bonduriansky, R. & Brooks, R.J. (1999) Why do male antler flies (*Protopiophila litigata*) fight? The role of male combat in the structure of mating aggregations on moose antlers. *Ethology Ecology & Evolution*, **11**, 287–301.

Bonduriansky, R., Crean, A.J. & Day, T. (2012) The implications of nongenetic inheritance for evolution in changing environments. *Evolutionary Applications*, **5**, 192–201.

Bonduriansky, R. & Day, T. (2009) Nongenetic inheritance and its evolutionary implications. *Annual Review of Ecology, Evolution and Systematics*, **40**, 103–125.

Bretman, A., Rodriguez-Munoz, R., Walling, C., Slate, J. & Tregenza, T. (2011) Fine-scale population structure, inbreeding risk and avoidance in a wild insect population. *Molecular Ecology*, **20**, 3045–3055.

Charmantier, A., Keyser, A.J. & Promislow, D.E.L. (2007) First evidence for heritable variation in cooperative breeding behaviour. *Proceedings of the Royal Society B-Biological Sciences*, **274**, 1757–1761.

Cordero, A. (1995) Correlates of male mating success in two natural-populations of the damselfly *Ischnura graellsii* (Odonata, Coenagrionidae). *Ecological Entomology*, **20**, 213–222.

Cropley, J.E., Dang, T.H.Y., Martin, D.I.K. & Suter, C.M. (2012) The penetrance of an epigenetic trait in mice is progressively yet reversibly increased by selection and environment. *Proceedings of the Royal Society B-Biological Sciences*, **279**, 2347–2353.

Danchin, E., Charmantier, A., Champagne, F.A., Mesoudi, A., Pujol, B. & Blanchet, S. (2011) Beyond DNA: integrating inclusive inheritance into an extended theory of evolution. *Nature Reviews Genetics*, **12**, 475–486.

Davey, J.L. & Blaxter, M.L. (2011) RADSeq: next-generation population genetics. *Briefings in Functional Genomics*, **9**, 416–423.

Ducatez, S., Baguette, M., Stevens, V.M., Legrand, D. & Freville, H. (2012) Complex interactions between paternal and maternal effects: Parental experience and age at reproduction affect fecundity and offspring performance in a butterfly. *Evolution*, **66**, 3558–3569.

Gibbs, A.G., Perkins, M.C. & Markow, T.A. (2003) No place to hide: microclimates of Sonoran Desert *Drosophila*. *Journal of Thermal Biology*, **28**, 353–362.

Gray, D.A. & Cade, W.H. (1999) Sex, death, and genetic variation: natural and sexual selection on cricket song. *Proceedings of the Royal Society B-Biological Sciences*, **266**, 707–709.

Hallsson, L.R. & Björklund, M. (2012) Selection in a fluctuating environment leads to decreased genetic variation and facilitates the evolution of phenotypic plasticity. *Journal of Evolutionary Biology*, **25**, 1275–1290.

Hansen, M.M., Olivieri, I., Waller, D.M., Nielsen, E.E. & Ge, M.W.G. (2012) Monitoring adaptive genetic responses to environmental change. *Molecular Ecology*, **21**, 1311–1329.

Hendrickx, F., Maelfait, J.P. & Lens, L. (2008) Effect of metal stress on life history divergence and quantitative genetic architecture in a wolf spider. *Journal of Evolutionary Biology*, **21**, 183–193.

Henry, P.Y., Pradel, R. & Jarne, P. (2003) Environment-dependent inbreeding depression in a hermaphroditic freshwater snail. *Journal of Evolutionary Biology*, **16**, 1211–1222.

Herrera, C.M. & Bazaga, P. (2011) Untangling individual variation in natural populations: ecological, genetic and epigenetic correlates of long-term inequality in herbivory. *Molecular Ecology*, **20**, 1675–1688.

Hill, W.G. (2012) Quantitative genetics in the genomics era. *Current Genomics*, **13**, 196–206.

Hoffmann, A.A. (2000) Laboratory and field heritabilities: some lessons from *Drosophila*. In: *Adaptive genetic variation in the wild* (ed. T.A. Mousseau, B. Sinervo & J.A. Endler), pp. 265. Oxford University Press, New York.

Holzer, B., Jacot, A. & Brinkhof, M.W.G. (2003) Condition-dependent signaling affects male sexual attractiveness in field crickets, *Gryllus campestris*. *Behavioral Ecology*, **14**, 353–359.

Irschick, D.J. (2003) Measuring performance in nature: implications for studies of fitness within populations. *Integrative and Comparative Biology*, **43**, 396–407.

Jablonka, E. & Lamb, M.J. (1995) *Epigenetic inheritance and evolution*. Oxford University Press, Oxford.

Jablonka, E. & Lamb, M.J. (2010) Transgenerational epigenetic inheritance. In: *Evolution—the extended synthesis* (ed. M. Pigliucci & G.B. Müller), pp. 137–174. The MIT Press, Cambridge.

Kasumovic, M.M., Bruce, M.J., Herberstein, M.E. & Andrade, M.C.B. (2007) Risky mate search and mate preference in the golden orb-web spider (*Nephila plumipes*). *Behavioral Ecology*, **18**, 189–195.

Kasumovic, M.M. & Jordan, L.A. (2013) The social factors driving settlement and relocation decisions in a solitary and aggregative spider. *American Naturalist*, **182**, 532–541.

Kelly, C.D., Bussiere, L.F. & Gwynne, D.T. (2008) Sexual selection for male mobility in a giant insect with female-biased size dimorphism. *American Naturalist*, **172**, 417–423.

Kruuk, L.E.B. & Hill, W.G. (2008) Introduction. Evolutionary dynamics of wild populations: the use of long-term pedigree data. *Proceedings of the Royal Society B-Biological Sciences*, **275**, 593–596.

Lande, R. (1982) A quantitative genetic theory of life-history evolution. *Ecology*, **63**, 607–615.

Legrand, D., Guillaume, O., Baguette, M., Cote, J., Trochez, A., Calvez, O., Zajitschek, S., Zajitschek, F., Lecomte, J., Bénard, Q., Le Galliard, J.F. & Clobert, J. (2012) The Metatron: an experimental system to study dispersal and metaecosystem dynamics for terrestrial vertebrates. *Nature Methods*, **9**, 828–833.

Lorch, P.D., Sword, G.A., Gwynne, D.T. & Anderson, G.L. (2005) Radiotelemetry reveals differences in individual movement patterns between outbreak and non-outbreak Mormon cricket populations. *Ecological Entomology*, **30**, 548–555.

Lowe, C.D., Harvey, I.F., Watts, P.C. & Thompson, D.J. (2009) Reproductive timing and patterns of development for the damselfly *Coenagrion puella* in the field. *Ecology*, **90**, 2202–2212.

Lynch, M. (1999) Estimating genetic correlations in natural populations. *Genetical Research*, **74**, 255–264.

McAdam, A.G., Boutin, S., Réale, D. & Berteaux, D. (2002) Maternal effects and the potential for evolution in a natural population of animals. *Evolution*, **56**, 846–851.

McGuigan, K. (2006) Studying phenotypic evolution using multivariate quantitative genetics. *Molecular Ecology*, **15**, 883–896.

Moore, A.J. & Kukuk, P.F. (2002) Quantitative genetic analysis of natural populations. *Nature Reviews Genetics*, **3**, 971–978.

Morrissey, M.B., Walling, C.A., Wilson, A.J., Pemberton, J.M., Clutton-Brock, T.H. & Kruuk, L.E.B. (2012) Genetic analysis of life-history constraint and evolution in a wild ungulate population. *American Naturalist*, **179**, E97–E114.

Moya-Laraño, J. (2002) Senescence and food limitation in a slowly ageing spider. *Functional Ecology*, **16**, 734–741.

Pelletier, F., Reale, D., Watters, J., Boakes, E.H. & Garant, D. (2009) Value of captive populations for quantitative genetics research. *Trends in Ecology & Evolution*, **24**, 263–270.

Prasad, N.G. & Joshi, A. (2003) What have two decades of laboratory life-history evolution studies on *Drosophila melanogaster* taught us? *Journal of Genetics*, **82**, 45–76.

Punzalan, D., Rodd, F.H. & Rowe, L. (2010) Temporally variable multivariate sexual selection on sexually dimorphic traits in a wild insect population. *American Naturalist*, **175**, 401–414.

Rabaneda-Bueno, R., Rodríguez-Gironés, M.Á., Aguado-de-la-Paz, S., Fernández-Montraveta, C., De Mas, E., Wise, D.H. & Moya-Laraño, J. (2008) Sexual cannibalism: high incidence in a natural population with benefits to females. *PLoS ONE*, **3**, e3484.

Reale, D. & Roff, D.A. (2001) Estimating genetic correlations in natural populations in the absence of pedigree information: accuracy and precision of the Lynch method. *Evolution*, **55**, 1249–1255.

Ritland, K. (2000) Marker-inferred relatedness as a tool for detecting heritability in nature. *Molecular Ecology*, **9**, 1195–1204.

Robinson, S.P., Kennington, W.J. & Simmons, L.W. (2012) Assortative mating for relatedness in a large naturally occurring population of *Drosophila melanogaster*. *Journal of Evolutionary Biology*, **25**, 716–725.

Rodriguez-Munoz, R., Bretman, A., Slate, J., Walling, C.A. & Tregenza, T. (2010) Natural and sexual selection in a wild insect population. *Science*, **328**, 1269–1272.

Romeis, J. & Wäckers, F.L. (2002) Nutritional suitability of individual carbohydrates and amino acids for adult *Pieris brassicae*. *Physiological Entomology*, **27**, 148–156.

Saulich, A. (2010) Long life cycles in insects. *Entomological Review*, **90**, 1127–1152.

Sgro, C.M., Lowe, A.J. & Hoffmann, A.A. (2011) Building evolutionary resilience for conserving biodiversity under climate change. *Evolutionary Applications*, **4**, 326–337.

Sheldon, B.C., Kruuk, L.E.B. & Merila, J. (2003) Natural selection and inheritance of breeding time and clutch size in the collared flycatcher. *Evolution*, **57**, 406–420.

Sillanpäa, M.J. (2011) On statistical methods for estimating heritability in wild populations. *Molecular Ecology*, **20**, 1324–1332.

Simmons, L.W., Zuk, M. & Rotenberry, J.T. (2005) Immune function reflected in calling song characteristics in a natural population of the cricket *Teleogryllus commodus*. *Animal Behaviour*, **69**, 1235–1241.

Sollars, V., Lu, X., Xiao, L., Wang, X., Garfinkel, M.D. & Ruden, D.M. (2002) Evidence for an epigenetic mechanism by which Hsp90 acts as a capacitor for morphological evolution. *Nature Genetics*, **33**, 70–74.

Steele, D.B., Siepielski, A.M. & McPeek, M.A. (2011) Sexual selection and temporal phenotypic variation in a damselfly population. *Journal of Evolutionary Biology*, **24**, 1517–1532.

Steppan, S.J., Phillips, P.C. & Houle, D. (2002) Comparative quantitative genetics: evolution of the G matrix. *Trends in Ecology & Evolution*, **17**, 320–327.

Svensson, E.I., Eroukhmanoff, F., Friberg, M. & Benkman, C. (2006) Effects of natural and sexual selection on adaptive population divergence and premating isolation in a damselfly. *Evolution*, **60**, 1242–1253.

Thompson, D.J., Hassall, C., Lowe, C.D. & Watts, P.C. (2011) Field estimates of reproductive success in a model insect: behavioural surrogates are poor predictors of fitness. *Ecology Letters*, **14**, 905–913.

Tuni, C., Goodacre, S., Bechsgaard, J. & Bilde, T. (2012) Moderate multiple parentage and low genetic variation reduces the potential for genetic incompatibility avoidance despite high risk of inbreeding. *PLoS ONE*, **7**, e29636.

Vaughn, M.W., Tanurdžić, M., Lippman, Z., Jiang, H., Carrasquillo, R., Rabinowicz, P.D., Dedhia, N., McCombie, R., Agier, N., Bulski, A., Colot, V., Doerge, R.W. & Martienssen, R.A. (2007) Epigenetic natural variation in *Arabidopsis thaliana*. *PLoS Biology*, **5**, e174.

Wilson, A.J., Reale, D., Clements, M.N., Morrissey, M.M., Postma, E., Walling, C.A., Kruuk, L.E.B. & Nussey, D.H. (2010) An ecologist's guide to the animal model. *Journal of Animal Ecology*, **79**, 13–26.

Zajitschek, F., Brassil, C.E., Bonduriansky, R. & Brooks, R.C. (2009) Sex effects on life span and senescence in the wild when dates of birth and death are unknown. *Ecology*, **90**, 1698–1707.

Zuk, M., Rotenberry, J.T. & Simmons, L.W. (1998) Calling songs of field crickets. *Evolution*, **166**, 1–8.

CHAPTER 10

Case study: quantitative genetics and sexual selection of weaponry in a wild ungulate

Loeske E. B. Kruuk, Tim Clutton-Brock
and Josephine M. Pemberton

10.1 Introduction

A central problem in current evolutionary quantitative genetics is how to explain the maintenance of abundant genetic variance in the face of the eroding effects of selection (Roff 1997; Walsh & Blows 2009). A related problem is that we also do not understand why observed evolutionary responses to selection frequently fail to match predictions (Merilä *et al.* 2001; Walsh & Blows 2009). These two issues may have common explanations: for example, it may be that genetic variation is not aligned with the multivariate direction of selection (Blows & Hoffmann 2005), or alternatively, that the phenotypic traits of interest do not causally affect fitness but only appear to be under selection because of associations of both trait and fitness with environmentally induced aspects of condition (Price *et al.* 1988; Rausher 1992). Under such scenarios, no evolutionary response to selection will occur, thus preventing the loss of genetic variation underlying phenotypic traits. Testing these hypotheses involves exploration of the genetic architecture underlying phenotypic variation. The statistical quantitative genetic tools of animal breeding have been critical for this and have facilitated much recent activity in studies of natural populations, as evidenced by the chapters in this book. However this work has ultimately served to underline the above paradoxes, repeatedly providing evidence of abundant genetic variation, strong directional selection and yet apparent microevolutionary stasis in natural populations (Merilä *et al.* 2001; Kruuk *et al.* 2008). To reconcile these inconsistencies, we now need to combine analyses of the process of selection with quantitative genetic models. The ability to do so has been motivated by several recent developments in the field of evolutionary quantitative genetics, which we outline below; we then illustrate these with an analysis of sexually selected weaponry in a wild ungulate population.

The first motivation is the increasing appreciation of the relevance of quantitative genetic analyses to studies of selection, and in particular of the need to consider the genetic rather than just the phenotypic causes of variation and covariation. When considering relationships amongst traits, phenotypic associations are not the same as genetic ones (Hadfield 2008; Kruuk *et al.* 2008). The difference between the two may be especially apparent in wild populations in natural environments, experiencing greater environmental variation than artificial populations in controlled environments, and also when considering the association between two traits where one is an estimate of fitness, i.e. when considering selection pressures. A clear illustration of the implication of this difference comes

Quantitative Genetics in the Wild. Edited by Anne Charmantier, Dany Garant, and Loeske E. B. Kruuk

from considering predictions for microevolutionary responses to selection. The classic expectation is that directional selection for a heritable trait should generate an evolutionary response (the 'breeder's equation', Lande 1979; Falconer & Mackay 1996). However if the apparent selection (estimated from the phenotypic association between trait and fitness) is only driven by covariances of both trait and fitness with external environmental factors, there will be no cross-generational or evolutionary response in the trait (Price *et al.* 1988; Rausher 1992); the breeder's equation thus fails to accurately predict change if the focal trait(s) in the model are not causal determinants of fitness (Hadfield 2008; Morrissey *et al.* 2010). The assumption of causation can be tested by comparing the environmental relationship between trait and fitness with the genetically induced relationship: specifically, the conditions of the breeder's equation involve (amongst other things) equality of the genetic and environmental regression gradients on a trait (Rausher 1992; Hadfield 2008; Morrissey *et al.* 2010), so evidence of inequality implies a trait will not evolve in the manner predicted by its phenotypic association with fitness.

An alternative approach to predicting a selection response that is not dependent on whether associations with fitness are causal is to estimate the genetic covariance between trait and relative fitness (the Robertson-Price covariance, Robertson 1966; Price 1970). This approach remains valid whatever the patterns of environmental covariance and whether or not all traits with causal effects on fitness are included in the model (Hadfield 2008; Morrissey *et al.* 2010). A lack of genetic covariance with fitness thus implies that a trait cannot evolve in response to selection. The obvious corollary to this point is that genetic variance for fitness is a prerequisite for evolutionary change; where a component of fitness is being considered, genetic variance for that component is required for selection via it to have any effect. These arguments imply that if we are considering the potential for selection for a trait to generate an evolutionary response, knowledge of both the genetic variance in fitness (or its components) and the genetic basis of the relationship between traits and fitness is important: in other words, a quantitative genetic analysis can inform our understanding of the evolutionary dynamics of selection. However to date, estimates of the heritability of fitness are extremely rare (see review in Chapter 2, Postma), especially when compared to the thousands of estimates of selection available in the literature (Kingsolver *et al.* 2001), as are empirical estimates of genetic covariances or correlations between trait and fitness (Etterson & Shaw 2001; Kruuk *et al.* 2002; Bolund *et al.* 2011; Teplitsky *et al.* 2011; Morrissey *et al.* 2012a; Stinchcombe *et al.* 2014).

A second point that has recently received fuller appreciation is the need to consider multivariate associations between multiple traits, and hence to move beyond the 'flatland' (Blows & Walsh 2009) of analysis of single traits. Whilst it has always been accepted that adaptation is a multivariate process, the extent to which studies of higher dimensions can reveal otherwise cryptic constraints on evolutionary responses has become increasingly apparent (Blows 2007; Walsh & Blows 2009; Chapter 12, Teplitsky *et al.*). For example, evolutionary dynamics will be constrained if the direction in multivariate space of useful genetic variance is different from the direction of selection (Schluter 1996; Blows *et al.* 2004). In the simplest case, a negative genetic correlation between two positively selected traits will constrain the evolution of both traits, but algebraic examples show that piecewise consideration of pairwise covariances may not reveal the full extent of constraints within variance–covariance matrices (Dickerson 1955). However, multivariate analyses make proportionately much greater demands on data (analysis of k traits requires estimation of $k(k+1)/2$ parameters for each component of variance), and these demands can cause problems for studies of natural populations in which sample sizes and statistical power are often limiting. This inevitably generates a dilemma of wishing to explore higher-order multivariate patterns but at the same time facing limitations of what is realistically feasible with a given dataset. Some studies have dealt with the problem by fitting multiple analyses of subsets of the full set of traits (e.g. Jensen *et al.* 2003; Coltman *et al.* 2005). Alternatively, eigenstructure analyses such as factor analysis can be employed to reduce the dimensionality by focussing on the dimensions of multivariate space with most variation (Kirkpatrick & Meyer 2004;

Hine & Blows 2006). Other lower-dimension metrics that summarise contributions from multiple traits may also be useful: for example, a range of different metrics of evolutionary constraint have recently been developed and used to test the effects of multitrait associations (see reviews in Hansen & Houle 2008; Walsh & Blows 2009; Simonsen & Stinchcombe 2010; Chapter 12, Teplitsky *et al.*). In summary, multivariate analyses may be challenging, but they can also provide information that is critical for an understanding of the evolutionary dynamics of a system.

Finally, a third recent development that has been important for quantitative genetic and selection analyses is the availability of statistical tools to deal with non-standard distributions typical of measures of fitness or its components, within a Bayesian Markov chain Monte Carlo (MCMC) framework (e.g. Sorensen & Gianola 2002; O'Hara *et al.* 2008; Ovaskainen *et al.* 2008; Hadfield 2009; Chapter 14, Morrissey *et al.*). The frequentist (for example, restricted maximum likelihood (REML)-based) approaches initially used in evolutionary biology typically make an assumption that residuals follow a normal distribution. This assumption may be justified for the majority of traits in which, for example, animal or plant breeders are interested, but it is probably less robust for many of the traits of interest to evolutionary biologists, in particular for estimates of fitness components such as survival or number of offspring (Bolker *et al.* 2009). Although there is increasing interest in generalised linear models in quantitative genetic analyses, the implications of using a more appropriate statistical distribution for estimates of key parameters such as the heritability of a trait are not yet clear. A Bayesian MCMC framework also provides a very effective means of assessing statistical uncertainty around derived parameters by examining their posterior distributions—for example for metrics of multivariate evolutionary constraint or estimates of selection pressures (see for example Ovaskainen *et al.* 2008; Bolund *et al.* 2011; Morrissey *et al.* 2012b; Stinchcombe *et al.* 2014).

To illustrate the implications of these developments for studies of quantitative genetics of wild populations, we present here analysis of data from a 40-year study of an unmanaged red deer (*Cervus elaphus*) population on the Isle of Rum, UK (Clutton-Brock *et al.* 1982). The construction of a multigenerational pedigree (Walling *et al.* 2010) has made possible a series of quantitative genetic analyses on data from this population (e.g. Kruuk *et al.* 2000; Wilson *et al.* 2007; Stopher *et al.* 2012), including application of some of the multivariate approaches discussed above, specifically factor analysis and estimates of evolutionary constraint (Morrissey *et al.* 2012b; Walling *et al.* unpubl.). We focus here on the secondary sexual trait of male weaponry.

Red deer are polygynous, sexually dimorphic ungulates in which males develop antlers (Figure 10.1). During the breeding season, males defend harems of females against rivals; antlers are used in fights during this intra-sexual competition for mates (Clutton-Brock *et al.* 1982), and their experimental removal has been shown to reduce dominance rank and fighting success (Suttie 1979). In previous analyses of data from the Rum deer population, we have shown strong directional

Figure 10.1 Twelve-year-old stag Ivy96 during the breeding season on Rum, in a year in which he fathered three offspring. Ivy96 died before casting these antlers, but from previous year's measurements, each antler is likely to have weighed >1 kg. (Photo Loeske Kruuk)

selection for antler mass: antler mass is positively associated with both annual and lifetime breeding success, including after correcting for body size (Kruuk *et al.* 2002). This previous study also found significant heritability for antler mass but no genetic correlation with breeding success, suggesting that the phenotypic association was driven by indirect correlations with environmental variables (Kruuk *et al.* 2002). Here, we revisit this issue with reference to the developments described above, using a much-extended and improved dataset: we estimate heritability of the relevant fitness component—male annual breeding success—using a more appropriate statistical distribution, and use a multivariate analysis to explore the alignment of multivariate genetic variation with sexual selection. We focus on two measures of antler morphology, and distinguish between 'total' selection on a given trait and 'direct' selection accounting for the effects of the other trait; finally, we compare the genetic and environmental associations between traits and fitness component. From the previous results (Kruuk *et al.* 2002), we predict a lack of significant genetic covariance between antler traits and annual breeding success, but it is not clear if this may be due to a lack of genetic variance in annual breeding success, or a lack of causal effect of antler morphology on fitness.

10.2 Methods

10.2.1 Study population

The individual-level study of the unmanaged red deer population in the North Block of the Isle of Rum, Scotland has been running in its present form since 1972; it currently contains records of more than 4000 individuals (Clutton-Brock *et al.* 1982). Each deer in the study population can be recognised individually, and has been monitored throughout life, from birth—when calves are caught and marked—through all breeding attempts, to death. Tissue samples are collected from newborn calves and deer found soon after death for genetic analyses, and DNA is also extracted from cast antlers to recover genotypes of any males, for example immigrants, not sampled at birth. Daily censuses during the autumn rut provide information on male rutting activity and in particular which females a male is holding in his harem. We considered here males born from 1970 onwards, and measures of antlers and annual breeding success from 1973 to 2011. Because we used annual measures in all analyses, the dataset included records from individuals who are still alive, as well as from those who had not died a natural death but had been shot in culls in other parts of the island.

10.2.1.1 Pedigree construction

We constructed a multigenerational pedigree to provide estimates of individual breeding success and for quantitative genetic analyses. Maternity can be assigned reliably based on field observations of associations of mothers and calves. Paternity was assigned using both genetic (genotypes at up to 15 microsatellite loci) and informative behavioural or phenotypic data, using a combination of two parentage-assignment programmes: MasterBayes (Hadfield *et al.* 2006) and COLONY2 (Wang 2004; Wang & Santure 2009). Further details are given in the online Supplementary Information for this chapter, or for a fuller description, see Walling *et al.* (2010).

10.2.1.2 Male Annual Breeding Success

We used annual breeding success (*ABS*) of males as our metric of a component of fitness. *ABS* was defined as the number of calves for which a given male in a given year was assigned paternity by the above procedures, under the condition that he was seen in the study area during the mating season (rut) in which those calves were conceived; a score of 0 was assigned if the male met these criteria but was not assigned paternity of any calves that year. We considered only individuals aged three years or more as this was the youngest age at which an individual achieved non-zero breeding success (so mean *ABS* within each age category is >0). This resulted in a total of 2408 records of *ABS* for 570 individual males, which we could then compare with the size of antlers grown (see below) and used in the rut in the corresponding year. Where relative *ABS* was required, we divided by mean *ABS*. Use of an annual measure enabled us to fit models of covariance with annually expressed antler traits using the Bayesian MCMC package MCMCglmm (Hadfield 2009; see below); it also enabled a far larger sample size that did not need to be restricted to individuals known to have died a natural death.

10.2.1.2 Antler traits

Male red deer grow a new set of antlers (Figure 10.1) each year, casting the old set between March and May. We focussed on data for two antler traits:

(i) mass: total dry mass (g) of cast antler. Cast antlers are recovered in the field, and those cast by individuals aged three years or more can be reliably assigned to known individuals given variation in shape (Kruuk *et al*. 2002).

(ii) form: the number of points on an antler, including all branches (or tines) and the top points. Estimates of form are available from field observations of stags each year, and so are not dependent on a cast antler being found in the field; sample sizes are correspondingly much higher for form than for mass (note that previous analyses of form in this population (e.g. Nussey *et al*. 2009) have been restricted to measures from cast antlers, whereas the current analysis involves a much-extended data-set).

Where measures on both antlers for an individual in a given year were available, we used the average. Summary statistics and sample sizes are given in Table 10.1; further details of antler measures are given in the online Supplementary Information. To avoid scale effects causing particular traits to dominate analyses of variance–covariance matrices,

Table 10.1 Summary statistics, estimates of selection via annual breeding success (*ABS*) and heritability and genetic correlations for antler mass and form. Phenotypic models were run in MCMCglmm in order to estimate 95% confidence intervals for selection gradients, using models assuming (a) an overdispersed Poisson distribution for *ABS*; and, for comparison with published results, (b) a Gaussian error distribution for relative *ABS*. Selection differentials $\mathbf{s}_p$ are the covariances between standardised trait values and *ABS*; standardised selection gradients $\boldsymbol{\beta}_p$ estimate direct selection on the focal trait accounting for selection via the other trait and are estimated from $\boldsymbol{\beta}_p = \mathbf{P}^{-1}\mathbf{s}_p$ (see Section 10.2). Estimates of heritability and the genetic correlation between antler mass and form are from the trivariate animal model presented in more detail in Table 10.2 (note that although analyses used antler trait values standardised to unit variance, heritability is not equal to V_a because of the inclusion of fixed effects of age.).

Trait	Mass (g)	Form (# points)
Summary statistics		
n observations; *n* individuals	706; 263	3555; 762
Mean (SD)	636.71 (240.08)	4.35 (1.19)
Selection estimates from phenotypic models		
(a) Assuming overdispersed Poisson distribution for ABS		
Selection differential $\mathbf{s}_p$	0.238 (0.137, 0.349)*	0.140 (0.073, 0.226)*
Selection gradient $\boldsymbol{\beta}_p$	0.520 (0.041, 0.827)*	0.041 (–0.162, 0.166)
(b) Assuming Gaussian distribution for relative ABS		
Selection differential $\mathbf{s}_p$	0.241 (0.146, 0.328)*	0.129 (0.062, 0.202)*
Selection gradient $\boldsymbol{\beta}_p$	0.517 (0.104, 0.904)*	–0.036 (–0.175, 0.138)
Correlations and heritability for antler traits		
Phenotypic correlation mass-form	0.495 (0.412, 0.568)*	
Heritability	0.317 (0.163, 0.432)*	0.265 (0.154, 0.334)*
Genetic correlation mass-form	0.568 (0.379, 0.839)*	

* indicates a parameter for which the 95% CI excludes zero.

both antler variables were transformed to unit variance prior to analysis.

In addition to the measures of antler mass and form, data were also available on three other antler traits: the total length of the antler (cm); the length of the brow tine (the lowest and first grown of the branches; cm); and the circumference at the base of the antler (cm). However, the dataset was not sufficient to generate a six-trait quantitative genetic model of all five antler traits and *ABS* (see below), so we focussed on multivariate analyses for mass and form only; we chose mass for comparison with previous analyses (Kruuk *et al.* 2002), and form because of its substantially larger sample size. Details of analyses of phenotypic selection for all antler five traits and their (co)variance components are given in the online supplementary information (Tables S10.1 & S10.2; www.oup.co.uk/companion/charmantier).

10.2.1.2 Age

Both *ABS* and antler traits showed substantial variation with age, reaching a peak at prime age of approximately ten years (Kruuk *et al.* 2002; Nussey *et al.* 2009). We therefore included an individual's age as a multi-level category as a fixed effect in all models described below. Ages of 14 years and upwards (up to a maximum age of 17) were combined into a single level because of smaller sample sizes (a total of 31 observations). Figure 10.2 shows the change in mean *ABS*, antler mass and antler form across ages.

10.2.2 Statistical analyses

Models presented in the main text were fitted using a Bayesian approach in the R-package MCMCglmm (Hadfield 2009). For comparison, in some places we also present results from REML models fitted in ASReml (Gilmour *et al.* 2009).

10.2.2.1 Mixed model analyses in MCMCglmm

We fitted trivariate mixed models in MCMCglmm, first at the phenotypic level to provide estimates of uncertainty on selection parameters (see below) and second as animal models (Lynch and Walsh 1998) using the pedigree information to partition (co)variances. All models included a fixed effect of age category. We used parameter expanded priors throughout (V = diag (3), $\nu = 2$,

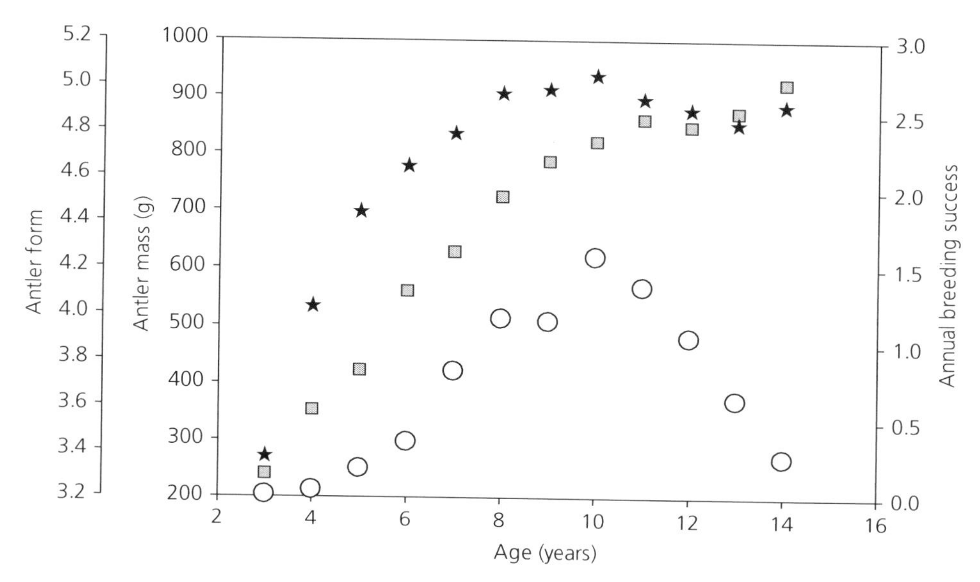

Figure 10.2 Age-related changes in mean *ABS* (open circles), antler mass (filled squares) and antler form (stars) from age three onwards, for male red deer. Individuals aged 14+ years are combined into a single category. (Note that the increase in average of antler traits right at the end of life represents selective disappearance rather than within-individual change; Nussey *et al.* 2009.)

alpha.mu = rep (0,3), alpha.V = diag (3) × 1000); as residual multivariate variance structures cannot have parameter expansion in MCMCglmm, the prior for the residual covariance structure was defined by (V = diag (3), $\nu = 0.002$). Choice of an appropriate prior distribution for a Bayesian model is a difficult subject (see for example discussion in Chapter 14, Morrissey *et al.*), and although parameter expansion has many advantages, e.g. in terms of improved rates of convergence, understanding of its properties is still developing (Hadfield 2012). Here, we compared different priors, varying V and ν, and found little effect on the resulting parameter estimates: we report estimates for additive genetic variance and covariances from a range of models in Supplementary Information Table S10.4. Models were run with a burn-in period of 6×10^5 iterations followed by a further 20×10^5 iterations with a thinning interval of 2000, resulting in a total sample size of 1000. Convergence of the MCMC sampling was assessed by visual inspection of the variance component chains; autocorrelation between successive observations was <0.12 across all analyses. Gaussian error distributions were assumed for the two antler traits, but for *ABS* we compared models with different distributions (see below). Conclusions are based on the modes and 95% credible intervals of the posterior distributions of respective parameter estimates and any derived metrics.

ABS is a non-negative integer with a very high proportion of zeroes (see below), and so the assumption of normality typically used in selection analyses is not valid. We therefore analysed *ABS* assuming an overdispersed Poisson distribution: this model uses a log-link and estimates (co)variance components for a latent trait θ, where *ABS* follows a Poisson distribution with mean exp (θ). MCMCglmm fits an additive overdispersion model in which overdispersion is represented in the residual variance V_{res} of θ (Hadfield 2012); see Chapter 14, Morrissey *et al.* for further details, and Reid *et al.* (2011) for another example. Posterior predictive simulation (MCMCglmm course notes, Section 5.3.1 Hadfield 2012) confirmed that this overdispersed Poisson distribution was sufficient to represent the distribution of *ABS* (for details see Supplementary Information).

10.2.2.2 Estimates of selection differentials and gradients

We first estimated phenotypic selection differentials and gradients for the antler traits, using a three-trait phenotypic-level MCMCglmm mixed model for *ABS* and the two standardised antler traits. Phenotypic selection *differentials* are given by the vector of covariances between trait and *ABS*, $\mathbf{s}_P$; these estimate the 'total' selection for the trait resulting both from its own direct effects on ABS and from indirect effects of correlated traits. From these differentials, we then calculated a vector of selection *gradients* $\boldsymbol{\beta}_P = \boldsymbol{P}_2^{-1}\mathbf{s}_P$, where $\boldsymbol{P}_2$ is the phenotypic variance–covariance matrix for the two antler traits (the subscript 2 is to differentiate from the 3 × 3 matrix $\boldsymbol{P}$ estimated in the trivariate model, of which $\boldsymbol{P}_2$ is a subset); selection gradients are equivalent to the 'direct' selection removing any indirect effects of other traits in the model (Lande & Arnold 1983), in this case the one other antler trait. The selection gradient for antler mass therefore represents the associations of mass with ABS whilst effectively holding form constant, and vice versa, as in a multiple regression of both traits. Selection differentials and gradients were derived for each sample of the MCMC posterior distribution of the model, generating posterior distributions for $\mathbf{s}_P$ and $\boldsymbol{\beta}_P$ from which modes and 95% credible intervals could be taken.

Selection differentials and gradients from the Poisson model are not equivalent to typical selection parameters (e.g. as reviewed in Kingsolver *et al.* 2001). This is because the generalised linear model of *ABS* estimates variances and covariances of the latent trait θ rather than of the observed data and, furthermore, because Poisson models of *ABS* necessarily use the observed values (a discrete count), rather than the measure of relative *ABS* typically used in estimates of selection differentials (Lande & Arnold 1983). The latent scale estimates can still be used to give an indication of the direction and magnitude of selection, but for comparison with published standardised selection estimates (e.g. Kingsolver *et al.* 2001), we also report selection parameters from a MCMCglmm phenotypic model using relative *ABS* and assuming Gaussian errors.

10.2.2.3 Multivariate animal models

We next extended the phenotypic models to estimate components of variance and covariance using a three-trait animal model (Lynch & Walsh 1998), with age category fitted as a fixed effect, and random effects for each trait of additive genetic effects (variance V_a), year effects (V_{yr}), 'permanent environment' (Kruuk and Hadfield 2007) effects (V_{pe}) and residual effects (V_{res}).

As mentioned above, the Poisson model of *ABS* estimates the variance (and covariance) components for the latent trait θ (i.e. on a transformed scale). As is discussed in Chapter 14, (Morrissey *et al.*) discuss, heritability can therefore be reported on different scales. Following their terminology used in Chapter 14, we estimate here i) heritability of *ABS* on the *latent* scale θ, defined as: $h^2_{latent} = V_a/(V_a + V_{pe} + V_{yr} + V_{res})$; and ii) heritability of *ABS* on the *link* scale (i.e. exp (θ)) defined as $h^2 = V_a/(V_a + V_{pe} + V_{yr} + V_{res} + \ln(1 + 1 / \exp(\alpha_0)))$, where α_0 is the intercept on the latent scale (Nakagawa & Schielzeth 2010), and can be estimated by setting exp (α_0) equal to mean *ABS*. (Note that Morrissey *et al.*'s terminology, which we follow here, differs from that of Nakagawa & Schielzeth 2010.)

As an aside, because previous analyses of male *ABS* in this population have assumed Gaussian errors within a REML framework (Kruuk *et al.* 2000; Foerster *et al.* 2007; Walling *et al.* unpubl.), and estimates of the heritability of fitness components assuming any other distribution are still relatively rare for any system, we also ran univariate animal models of *ABS* to compare the overdispersed Poisson distribution with a model assuming Gaussian errors, the latter run both in MCMCglmm and in ASReml. These are reported in detail in the supplementary information Table S10.3.

As a test for whether the direction of maximum genetic variance (highest evolvability) in antler morphology differed from the direction of selection (Schluter 1996; Blows *et al.* 2004), we estimated the angle between the phenotypic selection gradient $\boldsymbol{\beta}_p$ and the first principal component of the ***G***-matrix for mass and form, **gmax**. This was done for each sample of the posterior distribution to give a distribution of angles; note that considering only two traits makes estimation of a signed angle between two vectors, and hence significance testing for any difference from zero, more straightforward than for higher dimensions.

We repeated the procedure described above to estimate differentials and gradients at the genetic and non-genetic level. Thus we estimated a *genetic differential*[1] vector $\mathbf{s}_g$ equal to the additive genetic covariances between each trait and ABS (Rausher 1992), and an equivalent non-genetic, or *environmental differential* vector, $\mathbf{s}_e$ equal to the sum of the other covariances (permanent environment, year and residual). From these estimates of differentials, we then derived estimates of gradients. We first estimated *genetic gradients*, defined as $\boldsymbol{\beta}_g = \boldsymbol{G}_2^{-1}\mathbf{s}_g$, where $\boldsymbol{G}_2$ is the genetic variance–covariance matrix for the two antler traits. This is analogous to a multiple regression of fitness measure on breeding values for mass and form (Rausher 1992); Stinchcombe *et al.* (2014) present an example of equivalent calculations from multivariate animal models. We also estimated an equivalent *environmental gradient* for the two traits, $\boldsymbol{\beta}_e = \boldsymbol{E}_2^{-1}\mathbf{s}_e$, where $\boldsymbol{E}_2$ is the two-dimensional variance–covariance matrix defined as the sum of the matrices of the three non-genetic effects; this is analogous to a multiple regression of the fitness measure on the environmental components of mass and form. To test the assumption of the breeder's equation, that the genetically induced relationship between traits and fitness should be equivalent to the environmentally induced relationship (Rausher 1992; Hadfield 2008; Morrissey *et al.* 2010), we then estimated the difference between β_e and β_g for each trait.

The MCMC approach allowed us to examine posterior distributions for the various parameters derived above, specifically the angle between vectors, the different estimates of gradients, and the differences between them, by running the relevant

[1] The vector $\mathbf{s}_g$ is sometimes referred to as a 'genetic selection differential'. However to emphasise the fact that selection is a phenotypic process (Morrissey *et al.* 2010), we refer to this parameter as simply a 'genetic differential': algebraically, the 'phenotypic selection differential' and the 'genetic differential' are analogous, but they represent different processes. The same argument applies regarding 'environmental differentials', which are sometimes referred to as 'environmental selection differentials', and also by extension to gradients, both genetic and environmental.

calculations for each sample of the posterior distribution and then extracting the modes and 95% confidence interval (CI) for the posterior distributions of the derived metrics. Note that due to sampling covariances the mode of the posterior distribution of a function of parameters is not necessarily equal to the same function applied to the posterior modes of those parameters: for example, the mode of the posterior distribution of the heritability of a trait is not exactly the ratio of the posterior modes of the additive and phenotypic variance components.

10.3 Results

10.3.1 Selection for antler mass and form

There was evidence of positive selection via ABS for both antler mass and form (Table 10.1). The selection differentials $\mathbf{s}_\mathrm{P}$, representing the overall covariance of each trait with ABS (i.e. both direct and indirect selection), were all significantly positive. However the selection gradients, estimated from $\boldsymbol{\beta}_p = \boldsymbol{P}^{-1}\mathbf{s}_\mathrm{P}$, indicated that only antler mass was under direct selection after taking the other trait into account; in other words, for a given antler mass, there was no selection for antler form. These conclusions held whether *ABS* was fitted in a model with either an overdispersed Poisson or—for comparison with published estimates—a Gaussian error structure (Table 10.1a vs b). Results from the overdispersed Poisson model are also shown in Figure 10.3.

10.3.2 Components of variance and covariance of antler traits

Table 10.2 contains estimates of the contributions from additive genetic, permanent environment and year effects to variance and covariance of the two antler traits (and also *ABS*). Overall, the fixed and random effects of the model jointly explained 90% of the total variance in antler mass, and 75% of that in antler form, with substantial contributions from

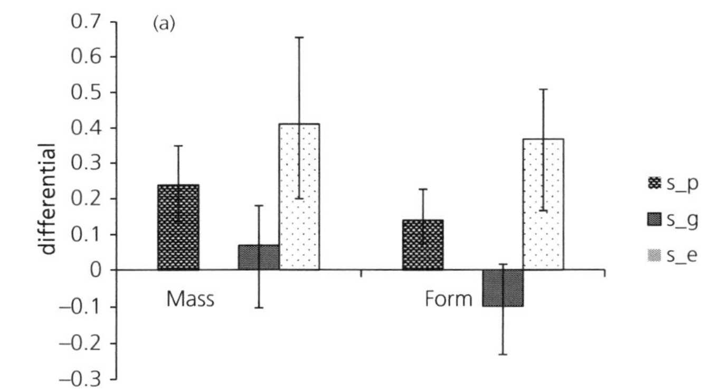

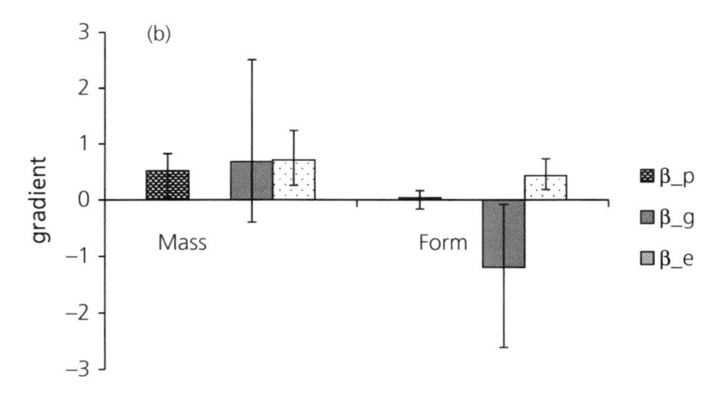

Figure 10.3 Selection via annual breeding success for antler mass and form, and genetic and environmental components. Values are modes of posterior distributions from a trivariate MCMCglmm model, assuming overdispersed Poisson errors for *ABS* and Gaussian errors for antler mass and form. Error bars are 95% CI. (a) *Differentials* (i.e. covariances) incorporating effects of all correlated traits and so representing 'total' associations between trait and *ABS*, at phenotypic (s_p), genetic (s_g) and environmental (s_e) levels; (b) *Gradients*, representing 'direct' associations corrected for effect of the other trait, at phenotypic (β_p), genetic (β_g) and environmental levels (β_e).

Table 10.2 Three-trait model of annual breeding success (*ABS*), antler mass and antler form, fitted in MCMCglmm with an overdispersed Poisson distribution for *ABS* and Gaussian errors for the antler traits. (a) Variances (on diagonal) and covariances (below diagonal), with 95% confidence interval (CI) taken from posterior distribution (note that correlations are not defined given the latent scale for *ABS*). (b) Estimates of differentials, taken from covariances (either genetic or environmental) between antler traits and log (*ABS*). (c) Estimates of gradients, equivalent to multiple regression (either genetic or environmental) of (log) *ABS* on antler traits.

(a) Variance and covariance components			
	ABS	**Mass**	**Form**
Additive genetic			
ABS	0.691 (0.224, 1.284)*		
Mass	0.069 (−0.103, 0.180)	0.182 (0.082, 0.235)*	
Form	−0.098 (−0.231, 0.016)	0.105 (0.044, 0.162)*	0.164 (0.104, 0.241)*
Permanent environment			
ABS	1.072 (0.634, 1.594)*		
Mass	0.186 (0.103, 0.370)*	0.078 (0.022, 0.143)*	
Form	0.288 (0.174, 0.420)*	0.076 (0.027, 0.123)*	0.165 (0.103, 0.227)*
Year			
ABS	0.263 (0.101, 0.692)*		
Mass	0.104 (−0.035, 0.321)	0.182 (0.106, 0.354)*	
Form	0.087 (−0.023, 0.226)	0.097 (0.041, 0.173)*	0.093 (0.057, 0.173)*
Residual			
ABS	0.606 (0.492, 0.856)*		
Mass	0.029 (−0.017, 0.085)	0.095 (0.086, 0.112)*	
Form	0.017 (−0.041, 0.061)	0.020 (0.006, 0.038)*	0.253 (0.242, 0.268)*

(b) Differentials	Mass	Form
Genetic: s_g	0.069 (−0.103, 0.180)	−0.098 (−0.231, 0.016)
Environmental: s_e	0.411 (0.169, 0.623)*	0.368 (0.228, 0.570)*

(c) Gradients	Mass	Form
Genetic: β_g	0.683 (−0.396, 2.508)	−1.196 (−2.604, −0.081)*
Environmental: β_e	0.713 (0.255, 1.236)*	0.435 (0.187, 0.735)*
Difference: $\beta_e - \beta_g$	−0.129 (−1.793, 1.485) **	1.756 (0.361, 3.155)*

* Indicates a CI that does not include zero.
** Although the difference in posterior modes of β_e and β_g is positive, the posterior mode of the difference is actually negative, but with large CIs.

year of antler growth and permanent environment effects for both traits. There was also evidence of significant additive genetic variance, with a heritability of 0.317 (95% CI 0.163, 0.432) for mass and 0.265 (0.154, 0.334) for form, and a positive genetic correlation between the two traits of 0.568 (0.379, 0.839; Table 10.1). Both traits loaded positively on the first principal component (**gmax**) of the two-dimensional *G*-matrix for mass and form; **gmax** accounted for 80.4% of their total genetic variance. There was a moderate discrepancy between the direction of **gmax** and that of β_p, with a posterior mode of the angle between the two vectors of 37.62° (95% CI 6.43, 62.34).

For comparison, a REML animal model analysis of all five antler traits (fitted in ASReml, without *ABS*) similarly indicated significant levels of additive genetic variance and positive covariances across the other antler traits, and positive loadings on the major axis of genetic variance and on major axes for other random effects (Supplementary Information, Table S10.2).

10.3.2 Heritability of *ABS*

Male *ABS* showed substantial age-related variation (Figure 10.2) and a highly skewed distribution, with a modal value of 0 and a maximum value of 14. Analysis of *ABS* assuming an overdispersed Poisson distribution in the three-trait model indicated additive genetic variance V_a of 0.691 (95% CI 0.224, 1.284) on the latent scale (Table 10.2a). This corresponds to an estimate of heritability on the latent scale of 0.250 (95% CI 0.097, 0.444), and of heritability on the link scale (i.e. $\log^{-1}(\theta)$) of 0.190 (95% CI 0.070, 0.333).

For comparison, a univariate MCMCglmm model assuming Gaussian errors for *ABS* resulted in poor convergence, as assessed by the trace and posterior distribution of the estimate of V_a. The 95% CI for estimates of V_a from this model were substantially lower (heritability $= 3.75 \times 10^{-4}$ (95% CI 0.010×10^{-4}, 0.063), as were the non-significant estimates from an ASReml model assuming Gaussian errors (heritability $= 0.029 \pm 0.023$ standard error (SE)); see Supplementary Information Table S10.3 for details.

10.3.3 Multivariate quantitative genetic analysis

The three-trait animal model provided estimates of the genetic and non-genetic covariances between *ABS* and the antler traits, analogous to selection differentials at a genetic, $\mathbf{s}_g$, and an environmental, $\mathbf{s}_e$, level. Neither antler trait showed evidence of genetic covariance with *ABS* (Table 10.2b; Figure 10.3a). In contrast, the environmental covariances $\mathbf{s}_e$ were significantly positive for both traits (Table 10.2b; Figure 10.3a).

We then assessed 'direct' associations corrected for any indirect effects of the other trait, using the regression gradients $\boldsymbol{\beta}_g = \boldsymbol{G}_2^{-1}\mathbf{s}_g$ and $\boldsymbol{\beta}_e = \boldsymbol{E}_2^{-1}\mathbf{s}_e$. For antler mass, the genetic gradient was positive but with a 95% CI that encompassed zero, whereas the environmental gradient was significantly positive (Table 10.2c; Figure 10.3b). There was no evidence of a difference between the environmental and the genetic selection gradients, $\boldsymbol{\beta}_e - \boldsymbol{\beta}_g$, for mass (Table 10.2c). Somewhat surprisingly, the estimate of the genetic gradient for antler form was negative, with a 95% CI that did not encompass zero (Table 10.2c; Figure 10.3b). The environmental gradient of *ABS* on form was positive, and significantly different from the genetic gradient (Table 10.2c; $\boldsymbol{\beta}_e - \boldsymbol{\beta}_g$ was positive in 99.1% of samples of the posterior distribution).

Figure 10.4 shows the vectors of **gmax** and the gradients $\boldsymbol{\beta}_p$, $\boldsymbol{\beta}_g$, and $\boldsymbol{\beta}_e$ for mass and form (estimated as the modes of the posterior distributions of

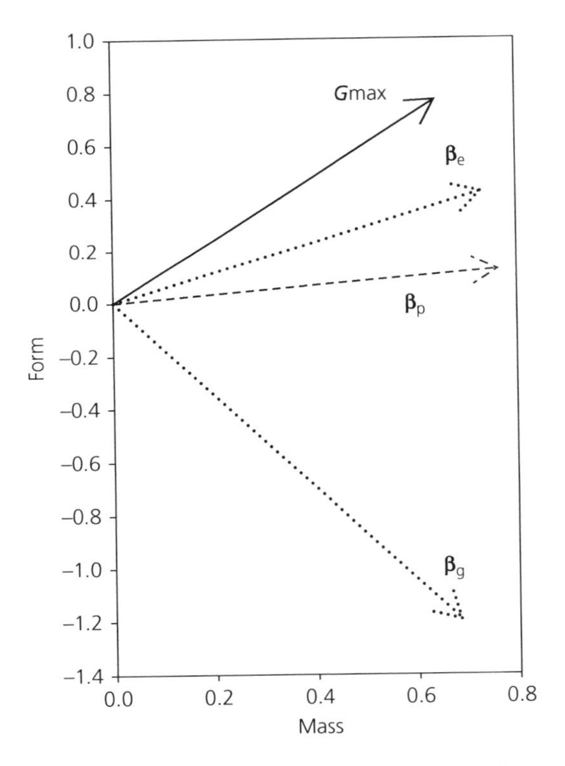

Figure 10.4 Comparison of directions of genetic variance and selection for antler mass and form. Arrows show the direction of **gmax** (solid line; the major axis of genetic variation from the two-dimensional **G**-matrix of mass and form, which accounts for 80.4% of genetic variation in mass and form) and gradients for antler mass and form: phenotypic selection gradient $\boldsymbol{\beta}_p$, (dashed line), genetic gradient $\boldsymbol{\beta}_g$ and environmental gradient $\boldsymbol{\beta}_e$ (dotted lines). Values shown are modes of the posterior distributions. The posterior mode of the angle between **gmax** and $\boldsymbol{\beta}_p$ is 37.62° (95% CI 6.43.–62.34.).

the respective derived parameters). The phenotypic associations between ABS and the two traits reflected in β_p are determined by both the genetic and environmental components, represented by β_g, and β_e respectively. Figure 10.4 illustrates the difference between β_g and β_e, and also suggests that the effect of β_g has been to deflect selection away from the principal axis of genetic variance.

10.4 Discussion

We have presented here a case study to illustrate some topical issues in evolutionary quantitative genetics as applied to studies of natural populations, many of which are reviewed more generally in other chapters of this book. In particular, we hope to have demonstrated the role that a quantitative genetic analysis can play in informing investigations of selection and evolutionary dynamics of phenotypic traits in the wild. Below, we first consider the main results and then discuss some of the limitations of the study, many of which may be typical of studies of wild populations.

Our analyses of red deer antler morphology revealed additive genetic variance for all traits considered, positive covariances between almost all pairwise combinations, and positive phenotypic selection via male *ABS* (see Supplementary Information for results for all five traits). There was also evidence of additive genetic variance for *ABS* when modelled assuming an overdispersed Poisson distribution. Computational constraints meant that we were unable to run a six-trait animal model including all five of the antler traits and *ABS* in MCMCglmm, so we focussed on quantitative genetic models of the relationships between antler mass, antler form, and *ABS*.

Considering the genetic and non-genetic components of associations between antler traits and *ABS* revealed strong positive environmental covariance for both antler size (mass) and shape (form). Biologically, positive environmental associations make sense: breeding success in red deer is determined by a stag's ability to defend a harem of females from other males through roaring contests and fighting (Clutton-Brock *et al.* 1979), whereas secondary sexual weaponry typically shows a strong condition-dependent element (Andersson 1994). Both contest success and antler size may be heavily dependent on an individual's nutritional state and hence condition. Consequently, it seems plausible that spatial and temporal variation in the environmental conditions experienced by individual males could affect both their breeding success and their antler morphology in a given year (Kruuk *et al.* 2002). This would generate a scenario in which the phenotypic covariance between a trait and fitness is not due to causal effects of the trait on fitness, but rather reflects the joint association of both trait and fitness component with external, environmental conditions. Here, by way of a formal test of this scenario, there was a significant difference between the genetic and environmental gradients of *ABS* on form. Under such a scenario, the joint occurrence of heritability and directional selection is not sufficient for the trait's evolution (Price *et al.* 1988; Rausher 1992).

This conclusion is in agreement with previous analyses in the Rum red deer showing a lack of genetic correlation between antler mass and male breeding success (Kruuk *et al.* 2002).[2] The current results suggest that the absence of a genetically induced association between ABS and antler mass (and hence an apparent constraint on the evolution of mass) is not merely a lack of genetic variance in the relevant fitness component, as we found evidence for heritability of ABS. The genetic covariance (or differential, s_g) reflects the combined impact of direct effects of genetic associations between antler mass and ABS as well as indirect effects of other correlated traits with which it is genetically correlated. We cannot identify and measure all of the latter, but we did find interesting effects of another antler trait, form. Mass was genetically correlated with antler form (r_g = +0.57 (0.38, 0.84)), and the estimate of the genetic gradient β_g of male ABS on form was negative (–1.20 (–2.60, –0.08), Table 10.2c). The CIs on this estimate of an antagonistic association between *ABS* and form are wide. However if it

[2] The previous analyses (Kruuk *et al.* 2002) had also quantified genetic and environmental associations between antler mass and male breeding success, but had focussed on BLUP estimates of individual breeding and environmental values, which may have generated biased estimates (Hadfield *et al.* 2010); the aim here was to focus on the direct estimates of genetic covariance from a multivariate model.

is truly negative, we potentially have two traits, one (form) negatively genetically associated with *ABS*, the other (mass) with a genetic association that is not different from zero, and a genetic correlation between them meaning that the overall covariance of either with *ABS* is not different from zero—effectively a constraint on any evolutionary increase of either trait. We can only speculate about the possible genomic basis for these relationships: one possible scenario would be that the genetic variants responsible for the positive association between mass and form also have antagonistic associations with *ABS* (potentially because those with positive associations have gone to fixation, Roff 1996), but that there is also other genetic variation generating positive associations between just mass and *ABS*. Further investigation of the molecular basis of this variation would be extremely interesting.

Further indication of some constraint on potential evolution was also apparent in the difference between the direction of selection and the direction of maximum genetic variance (Schluter 1996). Presumably as a result of the antagonistic associations between male fecundity fitness and antler form at the genetic level, phenotypic selection was aligned in a moderately (nearly midway between a value of no difference versus a value of 90°) different direction to that of maximum genetic variance (Figure 10.4; 37.62°). Considering a principal axis of genetic variance from the analysis of all five antler traits (Supplementary Information), the angle between **gmax** (Table S10.2) and the corresponding selection gradients β_{5p} was 65.02°; we have no estimates of the statistical uncertainty of this value (as **gmax** was derived from an ASReml rather than MCMCglmm model), but it is also in accordance with some discrepancy. However these values are lower than an equivalent value of 116° observed for **gmax** in a multivariate analysis of multiple traits in bighorn sheep (*Ovis canadensi;*, Coltman *et al.* 2005), or for example than the values reported for sexual selection of *Drosophila serrata* cuticular hydrocarbons (84.3° for **gmax**, Blows *et al.* 2004). Other analyses of evolutionary constraints in the same red deer population, using different metrics of multivariate constraint, have also suggested low or moderate levels of constraint across multiple life history traits, sometimes with insufficient statistical power to distinguish them from zero (Morrissey *et al.* 2012b; Walling *et al.* unpubl.), implying that in general evolutionary constraints in this population may not be strong. One of our aims in this chapter was to illustrate the effect that including an additional phenotypic trait—antler form—has on our knowledge of the system (Figure 10.4), and hopefully the increased application of multivariate analyses to other systems will broaden our knowledge of the extent of such constraints in natural populations.

Analyses such as those presented here have typically focussed on combining estimates of selection and genetic variance to predict a response to selection (Merilä *et al.* 2001; Kruuk *et al.* 2002; Morrissey *et al.* 2012b), including comparison of predictions for cross-generation response from the breeder's equation vs the Robertson–Price covariance (Morrissey *et al.* 2012a; Stinchcombe *et al.* 2014). In this system, our analyses showed that the underlying assumptions of the breeder's equation are not met, suggesting that the Robertson–Price covariance would provide a better tool for estimating change given observed patterns of selection and inheritance (Morrissey *et al.* 2010). We also now have evidence that average antler mass in the Rum study population has increased slightly since 1980, at a rate of +3.90 ± 1.30 SE g per year (Moyes *et al.* 2011). However we do not attempt to compare predicted and observed rates of phenotypic change, for two reasons. First, it has become increasingly clear that predicting cross-generation responses to natural selection in wild populations is overly ambitious (Merilä *et al.* 2001; Kruuk *et al.* 2008). This is not least because natural environments are currently experiencing a large-scale inadvertent experiment on the effects of changing climate which immediately violates any assumption of a constant environment. The extent to which the observed change in antler mass in the Rum red deer is due to phenotypic plasticity in response to warmer temperatures remains to be determined, but we expect it to be high (see for example Teplitsky *et al.* 2008). On a more mundane level, use of a Poisson model meant that variances and covariances are estimated on the latent scale due to the log-link function inherent in a generalised linear model. Jansen and Stern (1998) outline a transformation for coefficients from a logistic regression (e.g. for binary survival data) to estimate an average linear effect of trait on fitness, which they suggest can then be used as selection gradients, for example in breeder's equation predictions. Equivalent possibilities for log-link functions such as these will be worth exploring, but we have focussed here on dissecting the observed phenotypic associations between traits and components of fitness, and exploring the extent to which these are shaped by genetic associations, rather than aiming to use estimates for linear predictions. Most importantly, the comparison of the genetic vs environmental gradients gives insights into the causal effects of traits on fitness variation.

The Bayesian MCMC framework of MCMCglmm (Hadfield 2009) allowed appropriate analysis of our highly skewed component of fitness, male ABS. Conclusions as to the extent of additive genetic variance underlying variation in *ABS* changed with the use of an overdispersed Poisson error distribution rather than assuming normality (either within MCMCglmm or REML; although note that Walling *et al.* (unpubl.) found evidence for a low but significant heritability (0.053 ± 0.019SE) of (square-root transformed) male *ABS* in a REML analysis of the red deer data). It will be useful to explore the effect of statistical distribution for other components of fitness similarly better suited to non-normal distributions which may in the past have been analysed assuming normality. The MCMC approach is also highly valuable in providing estimates of uncertainty on derived parameters (e.g. Stinchcombe *et al.* 2014). However a REML framework does have the benefit of orders of magnitude difference in run time, and for this reason our analyses of covariances amongst all five antler traits (Supplementary Information) were taken from ASReml runs. Furthermore, the possible dependence of the posterior distributions of parameter estimates on how the prior is specified in a Bayesian analyses is an area requiring much more exploration, especially for complex models fitted to small data sets (Hadfield 2012; Chapter 14, Morrissey *et al.*).

There are several limitations to the current study which it is worth emphasising, some of which are generally relevant to other studies of quantitative genetics of wild populations. First, our analyses use a single component of fitness, *ABS*, rather than a lifetime measure. *ABS* provides an assessment of the effects of sexual selection (competition for access to mates, Andersson 1994), and thereby useful insights into the processes by which it may affect a trait. An annual measure was also chosen for convenience of modelling associations with an annually expressed trait in MCMCglmm (and see other points in Section 10.2). However the overall evolutionary dynamics will also be determined by components of selection at other times in the life history. Ideally we would therefore also estimate selection via other components of fitness such as survival, and ultimately lifetime traits such as lifetime breeding success (*LBS*), although previous analyses of the same study population suggest that patterns of selection in relation to antler size are very similar whether annual or *LBS* is considered (Kruuk *et al.* 2002), and also that both the phenotypic and genetic covariances between male *ABS* and longevity are positive (Walling *et al.* unpubl.).

Second, it would be informative to explore the effect on male sexual selection of other traits such as body size, in addition to weaponry. As for *LBS*, we did not do so here because of the focus on annually expressed traits (our measures of body size are largely restricted to at birth or death), but previous analyses indicate positive selection on antler size even after correcting for body size at death (Kruuk *et al.* 2002). Third, all traits under consideration are expressed only in adults, so we necessarily miss any component of genetic covariance carried by individuals who did not survive to adulthood (Hadfield 2008). Fourth, we have estimated single values for **G** and selection pressures across our study period and across different ages, making no allowance for any variation in any parameters due to a changing environment or due to ageing effects. Determining the extent to which the multivariate expression of genetic variation and covariance with fitness changes with environmental conditions (Sgro & Hoffmann 2004) or with age (Chapter 5, Charmantier *et al.*) will provide a future substantial challenge.

Finally, the 40-year Rum red deer study has generated arguably one of the best datasets available for a wild mammal population, and within it measures of antler morphology comprise one of the most extensive sets of phenotypic records. With repeated measures on individuals and a relatively low likelihood of additive genetic variance being confounded with common environment effects given male dispersal patterns (Kruuk & Hadfield 2007), the data offer a relatively good opportunity to dissect phenotypic variance and covariance into genetic and non-genetic components. It is therefore notable that the parameters estimated are associated with wide CIs, indicating a lack of statistical power—and reflecting the fact that our sample sizes are not large by standards of any typical animal breeding design or laboratory experiment. Multivariate analyses will be important for exploring evolutionary constraints, but they make greater demands than may be feasible for data from natural (or even experimental; see Bolund *et al.* 2011) populations. Extensions to test for $G \times E$ or $G \times Age$ interactions (see previous point) obviously do so even more. This issue is a persistent limitation for studies of natural populations and one which necessarily has to be borne in mind in interpretation of results; however, as estimates of key parameters hopefully accumulate, they can still provide valuable contributions to meta-analyses of general patterns.

10.4.1 Conclusions

Our analyses illustrate three key themes of current evolutionary quantitative genetics that are relevant to studies of wild populations: the potential contribution of environmental covariance to estimates of selection, the insights gained from multivariate models, and the utility of a Bayesian MCMC approach. However they also demonstrate many of the limitations of such studies. The results illustrate environmentally driven associations between traits and components of fitness which can generate the appearance of selection, but with no evolutionary relevance because of the lack of appropriate genetic covariance between trait and fitness component—despite the underlying genetic variance in the fitness component. The conclusions drawn from a multivariate quantitative genetic analysis may therefore differ from those that would arise from a univariate phenotypic analysis. Overall, we hope to have demonstrated the insights that a quantitative genetic analysis can bring to an understanding of selection and its evolutionary implications.

Acknowledgments

We are very grateful to the many people who have contributed to field data collection, study logistics and data management over the years, in particular Fiona Guinness, Ali Morris, Sean Morris, Martyn Baker, Steve Albon, Tim Coulson, Mick Crawley, Craig Walling, Katie Stopher, Dan Nussey and Ian Stevenson. We also thank Michael Morrissey, Craig Walling, Jarrod Hadfield, Anne Charmantier and Dany Garant for very useful comments on this chapter, and John Stinchcombe for access to a (at-the-time) unpublished manuscript using similar methodology. The research was funded by the Natural Environment Research Council (UK); we are also grateful to Scottish Natural Heritage for permission to work on the Isle of Rum and to their local staff for help and support.

References

Andersson, M. (1994) *Sexual selection*. Princeton University Press, Princeton.

Blows, M. & Walsh, B. (2009) *Spherical cows grazing in Flatland: constraints to selection and adaptation*. In *Adaption and Fitness in Animal Populations: Evolutionary and Breeding Perspectives on Genetic Resource Management*, ed. J. Van derWerk, H.U. Graser, R. Frankham, C. Gondro, pp. 82 –102. Springer, Dordrecht.

Blows, M.W. (2007) A tale of two matrices: multivariate approaches in evolutionary biology. *Journal of Evolutionary Biology*, **20**, 1–8.

Blows, M.W., Chenoweth, S.F. & Hine, E. (2004) Orientation of the genetic variance-covariance matrix and the fitness surface for multiple male sexually selected traits. *American Naturalist*, **163**, E329–E340.

Blows, M.W. & Hoffmann, A.A. (2005) A reassessment of genetic limits to evolutionary change. *Ecology*, **86**, 1371–1384.

Bolker, B.M., Brooks, M.E., Clark, C.J., Geange, S.W., Poulsen, J.R., Stevens, M.H.H. & White, J.S.S. (2009) Generalized linear mixed models: a practical guide for ecology and evolution. *Trends in Ecology & Evolution*, **24**, 127–135.

Bolund, E., Schielzeth, H. & Forstmeier, W. (2011) Correlates of male fitness in captive zebra finches—a comparison of methods to disentangle genetic and environmental effects. *BMC Evolutionary Biology*, **11**.

Clutton-Brock, T.H., Albon, S.D., Gibson, R.M. & Guinness, F.E. (1979) The logical stag—adaptive aspects of fighting in red deer (*Cervus elaphus*). *Animal Behavior*, **27**, 211–225.

Clutton-Brock, T.H., Guinness, F.E. & Albon, S.D. (1982) *Red deer—Behaviour and ecology of two sexes*. University of Chicago Press, Chicago.

Coltman, D.W., O'Donoghue, P., Hogg, J.T. & Festa-Bianchet, M. (2005) Selection and genetic (co)variance in bighorn sheep. *Evolution*, **59**, 1372–1382.

Dickerson, G.E. (1955) Genetic slippage in response to selection for multiple objectives. *Cold Spring Harbor Symposia on Quantitative Biology*, **20**, 213–224.

Etterson, J.R. & Shaw, R.G. (2001) Constraint to adaptive evolution in response to global warming. *Science*, **294**, 151–154.

Falconer, D.S. & Mackay, T.F.C. (1996) *Introduction to quantitative genetics*. Longman, Essex.

Foerster, K., Coulson, T., Sheldon, B.C., Pemberton, J.M., Clutton-Brock, T.H. & Kruuk, L.E.B. (2007) Sexually antagonistic genetic variation for fitness in red deer. *Nature*, **447**, 1107–1119.

Gilmour, A.R., Gogel, B.J., Cullis, B.R. & Thompson, R. (2009) *ASReml user guide release 3.0*. VSN International Ltd, Hemel Hempstead. http://www.vsni.co.uk.

Hadfield, J. (2008) Estimating evolutionary parameters when viability selection is operating *Proceedings of the Royal Society B-Biological*, **275**, 723–734.

Hadfield, J.D. (2009) MCMC methods for multi-response generalised linear mixed models: the MCMCglmm R package. *Journal of Statistical Software*, **33**, 1–22.

Hadfield, J.D. (2012) *MCMCglmm course notes. http://cran.r-project.org/web/packages/MCMCglmm/vignettes/CourseNotes.pdf*.

Hadfield, J.D., Richardson, D.S. & Burke, T. (2006) Towards unbiased parentage assignment: combining genetic, behavioural and spatial data in a Bayesian framework. *Molecular Ecology*, **15**, 3715–3730.

Hadfield, J.D., Wilson, A.J., Garant, D., Sheldon, B.C. & Kruuk, L.E.B. (2010) The misuse of BLUP in ecology and evolution. *American Naturalist*, **175**, 116–125.

Hansen, T.F. & Houle, D. (2008) Measuring and comparing evolvability and constraint in multivariate characters. *Journal of Evolutionary Biology*, **21**, 1201–1219.

Hine, E. & Blows, M.W. (2006) Determining the effective dimensionality of the genetic variance-covariance matrix. *Genetics*, **173**, 1135–1144.

Janzen, F.J. & Stern, H.S. (1998) Logistic regression for empirical studies of multivariate selection. *Evolution*, **52**, 1564–1571.

Jensen, H., Sæther, B.E., Ringsby, T.H., Tufto, J., Griffith, S.C. & Ellegren, H. (2003) Sexual variation in heritability and genetic correlations of morphological traits in house sparrow (*Passer domesticus*). *Journal of Evolutionary Biology*, **16**, 1296–1307.

King, E.G., Roff, D.A. & Fairbairn, D.J. (2011) The evolutionary genetics of acquisition and allocation in the wing dimorphic cricket, *Gryllus firmus*. *Evolution*, **65**, 2273–2285.

Kingsolver, J.G., Hoekstra, H.E., Hoekstra, J.M., Berrigan, D., Vignieri, S.N., Hill, C.E., Hoang, A., Gibert, P. & Beerli, P. (2001) The strength of phenotypic selection in natural populations. *American Naturalist*, **157**, 245–261.

Kirkpatrick, M. & Meyer, K. (2004) Direct estimation of genetic principal components: simplified analysis of complex phenotypes. *Genetics*, **168**, 2295–2306.

Kruuk, L.E.B., Clutton-Brock, T.H., Slate, J., Pemberton, J.M., Brotherstone, S. & Guinness, F.E. (2000) Heritability of fitness in a wild mammal population. *Proceedings of the National Academy of Sciences of the United States of America*, **97**, 698–703.

Kruuk, L.E.B. & Hadfield, J.D. (2007) How to separate genetic and environmental causes of similarity between relatives. *Journal of Evolutionary Biology*, **20**, 1890–1903.

Kruuk, L.E.B., Slate, J., Pemberton, J.M., Brotherstone, S., Guinness, F.E. & Clutton-Brock, T.H. (2002) Antler size in red deer: heritability and selection but no evolution. *Evolution*, **56**, 1683–1695.

Kruuk, L.E.B., Slate, J. & Wilson, A.J. (2008) New answers for old questions: the evolutionary quantitative genetics of wild animal populations. *Annual Review of Ecology, Evolution and Systematics*, **39**, 525–548.

Lande, R. (1979) Quantitative analysis of multivariate evolution, applied to brain:body size allometry. *Evolution*, **33**, 402–416.

Lande, R. & Arnold, S.J. (1983) The measurement of selection on correlated characters. *Evolution*, **37**, 1210–1226.

Lynch, M. & Walsh, B. (1998) *Genetics and analysis of quantitative traits*. Sinauer, Sunderland.

Merilä, J., Sheldon, B.C. & Kruuk, L.E.B. (2001) Explaining stasis: microevolutionary studies of natural populations. *Genetica*, **112**, 119–222.

Morrissey, M.B., Kruuk, L.E.B. & Wilson, A.J. (2010) The danger of applying the breeder's equation in observational studies of natural populations. *Journal of Evolutionary Biology*, **23**, 2277–2288.

Morrissey, M.B., Parker, D.J., Korsten, P., Pemberton, J.M., Kruuk, L.E.B. & Wilson, A.J. (2012a) The prediction of adaptive evolution: empirical application of the secondary theorem of selection and comparison to the breeder's equation. *Evolution*, **66**, 2399–2410.

Morrissey, M.B., Walling, C.A., Wilson, A.J., Pemberton, J.M., Clutton-Brock, T.H. & Kruuk, L.E.B. (2012b) Genetic analysis of life-history constraint and evolution in a wild ungulate population. *American Naturalist*, **179**, E97–E114.

Moyes, K., Nussey, D.H., Clements, M.N., Guinness, F.E., Morris, A., Morris, S., Pemberton, J.M., Kruuk, L.E.B. & Clutton-Brock, T.H. (2011) Climate change and breeding phenology in a wild red deer population. *Global Change Biology*, **17**, 2455–2469.

Nakagawa, S. & Schielzeth, H. (2010) Repeatability for Gaussian and non-Gaussian data: a practical guide for biologists. *Biological Reviews*, **85**, 935–956.

Nussey, D.H., Kruuk, L.E.B., Morris, A., Clements, M.N., Pemberton, J.M. & Clutton-Brock, T.H. (2009) Inter- and intrasexual variation in aging patterns across reproductive traits in a wild red deer population. *American Naturalist*, **174**, 342–357.

O'Hara, R.B., Cano, J.M., Ovaskainen, O., Teplitsky, C. & Alho, J.S. (2008) Bayesian approaches in evolutionary quantitative genetics. *Journal of Evolutionary Biology*, **21**, 949–957.

Ovaskainen, O., Cano, J.M. & Merilä, J. (2008) A Bayesian framework for comparative quantitative genetics. *Proceedings of the Royal Society B-Biological Sciences*, **275**, 669–678.

Price, G. (1970) Selection and covariance. *Nature*, **227**, 520–521.

Price, T., Kirkpatrick, M. & Arnold, S.J. (1988) Directional selection and the evolution of breeding date in birds. *Science*, **240**, 798–799.

Rausher, M.D. (1992) The measurement of selection on quantitative traits: biases due to environmental covariances between traits and fitness. *Evolution*, **46**, 616–626.

Reid, J.M., Arcese, P., Sardell, R.J. & Keller, L.F. (2011) Additive genetic variance, heritability, and inbreeding depression in male extra-pair reproductive success. *American Naturalist*, **177**, 177–187.

Robertson, A. (1966) A mathematical model of the culling process in dairy cattle. *Animal Production*, **8**, 95–108.

Roff, D.A. (1996) The evolution of genetic correlations: an analysis of patterns. *Evolution*, **50**, 1392–1403.

Roff, D.A. (1997) *Evolutionary quantitative genetics*. Chapman & Hall, New York.

Schluter, D. (1996) Ecological speciation in postglacial fishes. *Philosophical Transactions Of The Royal Society Of London Series B, Biological Sciences*, **351**, 807–814.

Sgro, C.M. & Hoffmann, A.A. (2004) Genetic correlations, tradeoffs and environmental variation. *Heredity*, **93**, 241–248.

Simonsen, A.K. & Stinchcombe, J.R. (2010) Quantifying evolutionary genetic constraints in the ivyleaf morning glory, *Ipomoea hederacea*. *International Journal of Plant Sciences*, **171**, 972–986.

Sorensen, D.A. & Gianola, D. (2002) *Likelihood, Bayesian and MCMC methods in quantitative genetics*. Springer-Verlag, New York.

Stinchcombe, J.R., Simonsen, A.K. & Blows, M.W. (2014) Estimating uncertainty in multivariate responses to selection. *Evolution*, in press.

Stopher, K.V., Walling, C.A., Morris, A., Guinness, F.E., Clutton-Brock, T.H., Pemberton, J.M. & Nussey, D.H. (2012) Shared spatial effects on quantitative genetic parameters: accounting for spatial autocorrelation and home range overlap reduces estimates of heritability in wild red deer. *Evolution*, **66**, 2411–2426.

Suttie, J.M. (1979) The effect of antler removal on dominance and fighting behaviour in farmed red deer stags. *Journal of Zoology*, **190**, 217–224.

Teplitsky, C., Mills, J.A., Alho, J.S., Yarrall, J.W. & Merila, J. (2008) Bergmann's rule and climate change revisited: disentangling environmental and genetic responses in a wild bird population. *Proceedings of the National Academy of Sciences of the United States of America*, **105**, 13492–13496.

Teplitsky, C., Mouawad, N.G., Balbontin, J., de Lope, F. & Moller, A.P. (2011) Quantitative genetics of migration syndromes: a study of two barn swallow populations. *Journal of Evolutionary Biology*, **24**, 2025–2039.

Walling, C.A., Morrissey, M.B., Foerster, K., Pemberton, J.M., Clutton-Brock, T.H. & Kruuk, L.E.B. (unpubl.) A multivariate analysis of genetic constraints to life history evolution in a wild population of red deer.

Walling, C.A., Pemberton, J.M., Hadfield, J.D. & Kruuk, L.E.B. (2010) Comparing parentage inference software: reanalysis of a red deer pedigree. *Molecular Ecology*, **19**, 1914–1928.

Walsh, B. & Blows, M.W. (2009) Abundant genetic variation plus strong selection = multivariate genetic constraints: a geometric view of adaptation. *Annual Review of Ecology, Evolution and Systematics*, **40**, 41–59.

Wang, J. (2004) Sibship reconstruction from genetic data with typing errors. *Genetics*, **166**, 1963–1979.

Wang, J. & Santure, A.W. (2009) Parentage and sibship inference from multilocus genotype data under polygamy. *Genetics*, **181**, 1579–1594.

Wilson, A.J., Nussey, D.H., Pemberton, J.M., Pilkington, J.G., Morris, A., Pelletier, F., Clutton-Brock, T.H. & Kruuk, L.E.B. (2007) Evidence for a genetic basis of aging in two wild vertebrate populations. *Current Biology*, **17**, 2136–2142.

www.oup.co.uk/companion/charmantier

CHAPTER 11

Epigenetic processes and genetic architecture in character origination and evolution

Alexander V. Badyaev and J. Bruce Walsh

11.1 Origination of the relationship between genetic, developmental, and functional dimensions of a phenotype

How do genes get associated with a particular developmental or functional roles and contexts in evolution? How and why do they lose it? Why is there a general lack of determinism in the genotype–phenotype map and how does this map evolve? Few questions are more important to evolutionary theory, and few have been more elusive. The field of epigenetics studies properties of emergent, self-regulatory, and compensatory interactions that arise above the level of the gene, but are not directly predictable from the intrinsic properties of either phenotype or genotype (e.g. Hallgrimsson & Hall 2011). Here we suggest that explicit consideration of these mechanisms in the evolution of integration between genomic and phenotypic dimensions of organisms provides an important insight into these problems.

The evolution of genetic architecture reflects historical correspondence among genomic, genetic and phenotypic dimensions of an organism (Turelli 1988; Falconer & MacKay 1996; Hansen 2006; Arnold *et al.* 2008). Within each of these dimensions, evolutionary dynamics are dictated by patterns of connectivity among its elements (e.g. among genes, alleles, proteins, or traits) and the rate and time at which a population can sample various combinations of such elements (Fisher 1930; Lynch & Walsh 1998). Once a trait combination with significant fitness consequences is found, natural selection can either stabilise or delete it, and genetic architecture (summarized by the ***G***-matrix) then evolves to approach the shape of the fitness landscape (Cheverud 1996; Wagner & Altenberg 1996; Lynch & Walsh 1998)—that is to mimic the most consistently beneficial functional and developmental relationships of the organism (Cheverud 1988). Genetic architecture is further modified by random drift and by non-additive effects introduced by epistatic interactions of alleles across changing genetic backgrounds (Wright 1931; Rice 2001; Hansen & Houle 2008).

Because of its relationship to the fitness landscape, the ***G***-matrix, by definition, is considered to be agnostic to specific proximate processes that generate phenotypes and integration of its components; instead it is shaped by the effects of individual alleles/genotypes on the traits of interest. Part of this view reflects an approach where character identity is either given outright or is defined by independent contributions to fitness (e.g. Wagner 2001). For instance, the theory of morphological integration (Olson & Miller 1958; Berg 1960; Cheverud 1982; Riska 1989) posits that

Quantitative Genetics in the Wild. Edited by Anne Charmantier, Dany Garant, and Loeske E. B. Kruuk

stabilising selection leads to the evolution of strong genetic correlations among functionally compatible traits, ultimately resulting in the modularization of the phenotype into relatively independent, functionally interacting groups of characters (see also Badyaev & Foresman 2004)—in other words, the formation of developmental *maps* that partly canalises the *input* values (that may range from timing and levels of transcripts to intermediate developmental stages of emerging traits). It is assumed that developmental processes can produce and transmit the required genotype–phenotype modularity for such outcomes; but direct evidence is lacking and some empirical and conceptual results suggest that evolution of such developmental modularity is unlikely from both population–genetic and developmental perspectives (Bulmer 1971; Slatkin & Frank 1990; Rice 2001; Beldade *et al.* 2002; Welch & Waxman 2003; Wagner *et al.* 2007; Conner 2012; Salazar-Ciudad & Marin-Riera 2013).

A central element of development is the divergent goals of canalisation (buffering genetic and environmental perturbations) on one hand and phenotypic plasticity on the other. Both of these processes can have genetic components, but the extent to which these processes are captured by a ***G***-matrix is unclear (see Section 11.5 for an empirical illustration). In general, the ***G***-matrix can be strongly shaped by modification of the underlying developmental processes (i.e. *maps*) on which segregating variation is translated into traits. A one-dimensional example of this is the mapping of some underlying score (a liability) into a trait. In a simple threshold model (where a trait only appears when the liability score of an individual exceeds a critical value), either evolution to increase the mean liability score, or evolution to decrease the threshold value (or potentially both), results in an increase in the population frequency of the trait (see also Wright 1934; Lande 1977). On such a one-dimensional scale, the difference between changes in the developmental map by moving the threshold vs changes in the liability given a set threshold is rather trivial. In a more complex morphospace, however, this distinction is of fundamental importance, especially if one envisions a set of loosely interconnected modular developmental maps acting on different suites of traits (see Section 11.5). We can think of this distinction as variation/evolution in the inputs (e.g. the state of early developmental structures or molecular features that interact with the entire developmental sequence to produce the final phenotype) that developmental maps transform into trait values vs variation/evolution in the shape/form of the maps and the connectivity between the various maps that together form the suite of traits under consideration. Small underlying genetic changes in one or more of these maps could result in significant morphological changes that are then further canalised from genetic variance in the input variables (liability in the simple case). This allows genetic variation to be shielded from selection whilst at the same time generating covariance structures that appear to be held under stabilising selection.

Proximately, the consistency and coordination of genomic and functional elements—which are captured by the ***G***-matrix—are at least partially produced by epigenetic processes of development (that is, emergent interactions that arise above (i.e. *epi-*) the level of the gene (in the sense of DNA sequence) and those properties are not predictable from intrinsic properties of either phenotype or genotype). Therefore, the patterns and prevalence of such processes within a population have significant consequences for the distribution of variances among characters that form integrated complexes (Rice 2002, 2004ab; Hansen 2011). Explicit consideration of the correspondence between the genomic and phenotypic dimensions of an organism (its developmental dynamics) can resolve apparent discrepancies between predicted and observed evolutionary patterns. For example, when micro-evolutionary response in the direction of abundant genetic variation is inhibited by developmental interactions (Maynard Smith *et al.* 1985) or when significant response to selection is produced in the direction orthogonal to absolute genetic correlation among traits, or, more generally, when strong selection coexists with abundant genetic variation, and both coexist with a preponderance of evolutionary stasis (Arnold 1992; Hansen 2006, Walsh and Blows 2009). Overall, however, these discrepancies show that whatever the ***G***-matrix indicates, it does not fully cover the mechanisms that actually produce an evolutionary response. Ignorance of the mechanisms by which the phenotype is produced translates, by necessity, into ignorance of the mechanisms by which it changes.

11.2 A developmental view of the *G*-matrix

Whilst most individual traits harbour significant additive genetic variation, most of the additive genetic variance in multivariate space is often concentrated on just a few axes of variation (the first few eigenvalues of ***G*** account for the vast majority of its total variation, e.g. Figure 11.1; Björklund 1996; Schluter 1996; Arnold *et al.* 2008; Kirkpatrick 2009; Walsh & Blows 2009). Thus, the presence of additive genetic variation (i.e. non-zero heritability) for each element in a suite of traits under selection is no guarantee that the population will evolve, as the direction favoured by directional selection may be largely orthogonal to these major axes of variation (reviewed by Teplitsky *et al.* in Chapter 12). The population–genetic explanation for the observation of low effective dimensionality for the *G*-matrices is that persistent selection has eroded genetic variation in that direction (Falconer & MacKay 1996). This leaves significant variation in other directions, and the projection of this variation onto single traits yields significant amounts of additive genetic variance. This argument is independent of any underlying developmental processes.

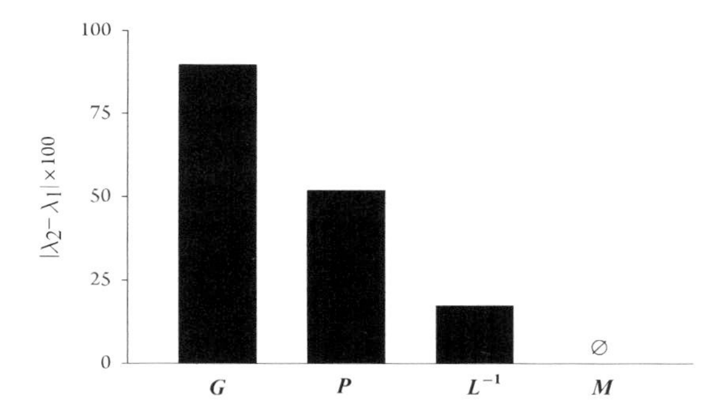

Figure 11.1 Relative difference between the first (λ_1) and the second (λ_2) eigenvalues of additive genetic covariance matrix (***G***), phenotypic covariance matrix (***P***), matrix of ontogenetic variation ($\boldsymbol{L}^{-1}$), and matrix describing adaptive landscape (***M***) in Montana house finches over 19 generations after population establishment. Smaller values indicate greater dimensionality of variability. The third eigenvalues were not significant in any of the matrices. From Badyaev 2010.

An alternative explanation is developmental, as in our model of variation in input variables and variation in the structure of developmental maps transforming these input variables into an integrated multivariate phenotype (see Section 11.1). As is the case for a threshold trait, there is significant variation in a population in the underlying liability score, much of which may be additive genetic. When mapped into the observed presence/absence trait, the resulting variation is much lower. In multivariate morphospace, one can imagine a set of input variables showing considerable variation that is transformed by development maps into a more limited space. The average tendency for these maps over a large number of offspring is a measure of their genetic value, whilst specific individuals may show deviations from this average trend due to environmental and developmental noise. Hence, the distribution of phenotypes can be rather different from the underlying set of distributions of average developmental trajectories (the genetic values of the maps averaged over input variables). Therefore, the structure of the *G*-matrix may be more reflective of these development maps, and hence show canalised multivariate directions. Conversely, if an underlying single trait is more strongly influenced by input variables than is the entire integrated multivariate phenotype, then the trait would show more variation; but this is constrained by the map into a more limited range. Importantly, as with the input variables, these developmental maps can also respond to selection, potentially selecting for more favourable maps (i.e. more variation in constraints along the directions favoured by selection).

Obviously, the current structure of the *G*-matrix could be a result of both of these features. However, if some *G*-matrices are indeed more strongly shaped by changes in developmental maps, then the potential exists for rapid transformation of one covariance structure into another, whilst still keeping most of the variation in the underlying traits.

11.3 Does interchangeability of genotype–phenotype relationships matter for stability of the *G*-matrix, and what role does development play in this process?

Discoveries from genomics bring additional considerations to our understanding of the relationship between genotypes and phenotypes and to the overall realisation that the evolutionary forces that shape the evolution of genomes and their architecture are likely to be distinct from those that shape phenotypic diversity (Müller & Newman 2003; Badyaev 2011b; Koonin 2011). Five insights are particularly relevant here. The first is the tremendous evolutionary conservation of most genes (i.e. their orthologous lineages) that persist unchanged over billions of years of exceptional phenotypic diversifications (Tatusov *et al.* 2003; Shubin *et al.* 2009; Wolf *et al.* 2009). The second is the relative fluidity of genomes, with gene orders and genome architecture often showing little conservation even among closely related species (reviewed in Koonin 2011). The third is the interchangeability of the relationship between genes and their functions at all levels of organismal organisation, from non-orthologous gene replacements, where unrelated sets of genes fulfil identical sets of cell functions (Wolf *et al.* 2006) to the 'many to many' mapping of genes on developmental and functional aspects of phenotypes (Wilkins 2001; Wainwright 2007). The fourth, and related to it, is a new appreciation of the high dimensionality of genotypes in relation to phenotypic diversity and therefore of exceptional redundancy of the genotype-to-fitness relationship (van Nimwegen & Crutchfield 2000; Gavrilets 2004; Wagner 2011). The fifth is the empirical finding that, despite such dimensionality and redundancy, only a small portion of theoretically possible pathways are assessable to evolution, a constraint that emphasises the overwhelming importance of epistasis in maintaining the evolutionary cohesiveness of evolving genomes despite their high dimensionality and ubiquitous redundancy of relationships with phenotypes (Weinreich *et al.* 2005; Gravner *et al.* 2007; Poelwijk *et al.* 2007; Breen *et al.* 2012). This echoes a persistent finding of quantitative genetics that genetic variance tends to be concentrated along only a few axes of morphospace (Kirkpatrick 2009; Walsh & Blows 2009), such that most empirically derived *G*-matrices often have far fewer dimensions than their phenotypic equivalents (although the issue might be due to limited statistical power in some cases), whilst patterns of pleiotropy tend to be highly restrictive, reflecting variational modularity of phenotypes (Hansen 2011; Wagner & Zhang 2011). Such channelling of variation to a few dimensions has significant implications for evolution because it makes it more likely that random mutational input will affect beneficial combination of traits. The effect is further amplified if these combinations of traits can actually be produced by developmental processes (Alberch 1991; West-Eberhard 2003).

These insights bring a new urgency to understanding the rules by which ubiquitous interchangeability of genotype–phenotype relationships is translated into the evolutionary stability of the *G*-matrix and the role developmental processes play in this process. Can non-linear and emergent developmental interactions be realistically taken into account when phenotypic evolution is modelled? A recent shift in focus from concerns about availability of genetic variance (that is rarely limited by mutational input) to the patterns of its distribution among trait combinations (Larsen 2005; Blows 2007; Blows & Walsh 2009) needs to be augmented by the realisation that evolution of complex structures depends on coordinated variability, which often arises through epigenetic developmental processes (e.g. Section 11.2; see also Riedl 1978; Hallgrimsson & Hall 2011).

11.4 How do emergent processes influence genetic dimensionality of the phenotype?

What is the relationship between basic elements of morphogenesis and its genetic architecture? At what point in organic evolution do the genes ('building blocks of basic processes') get associated with the consistent developmental programs that the *G*-matrix describes?

One suggestion is that the origination and evolution of structures are based on different mechanisms: emergent processes of morphogenesis form a template (developmental maps) for the subsequent accumulation of genetic networks that generate variation in the developed phenotype (Newman & Müller 2000; Newman 2005; Newman & Bhat 2008). Under this framework, the initial prevalence of plastic and environmentally contingent developmental processes is gradually replaced, over evolutionary time, by genetic networks that assure reliability of developmental outcomes in particular environmental conditions (Reid 2007; Newman 2011). In terms of our model, this amounts to genetic assimilation of a new developmental map (Waddington 1961). Thus, more recurrent organism–environment associations are expected to accumulate greater redundancy and determinism in their genetic architecture—these associations are then most likely to form additive genotype–phenotype associations (Nowak *et al.* 1997; Badyaev 2007; 2011b). One striking observation on this point comes from isogenic inbred lines in mice and *Drosophila* (Lemos *et al.* 2005; Flint & Mackay 2009). When these lines are used to introgress specific combinations of alleles from different loci into an otherwise identical genetic background, they tend to show strong epistasis. However, when these alleles are scored in outbred populations, they are largely additive. The strong epistatic effects that appear when scored over a single genetic background are randomised into additive effects when scored over a diverse background of random genotypes. Overall, according to this scenario, correspondence of one genotype to one phenotype is a highly derived condition in which an overdetermining genetic circuitry ensures that changes in external and internal environments have less impact on phenotypic outcomes (Salazar-Ciudad, Newman & Sole 2001; Newman 2005; Badyaev 2011b). In less recurrent associations, there is a greater role of environmental plasticity and contingent developmental interactions (West-Eberhard 2003).

Such a scenario can play out not just over evolutionary time, but also at different levels of organismal organisation where reciprocal interactions between traits and tissue types provide each other 'environments' during development; that is form linkages and associations that ensure the development of a structure, but which themselves do not experience selection acting on the final structures (and thus do not deplete variance among their elements; Kirschner & Gerhart 2005; Gerhart & Kirschner 2007). Such hierarchically arranged variability harbours genetic variance and can contribute to variational pleiotropy of complex structures (Hansen 2011; Wagner & Zhang 2011; Salazar-Ciudad & Marin-Riera 2013).

Further, selection efficacy negatively correlates with the complexity of organisms, including the complexity of genetic architecture (Lynch 2010), although this model assumes weak or absent indirect selection on the elements of genetic architecture. If the model is correct, then in complex organisms occurring in small populations, the selection ability to streamline and stabilise the most fit configurations of a structure is increasingly determined by homeostatic, entropy-reducing mechanisms (Badyaev 2013) because such elements of selection are most consistent across environmental contexts (Schmalhausen 1938). Accumulations of complexity and associated channelling effects of development on genetic and phenotypic variance further emphasise geometric considerations in studies of genetic variance; developmental modularity becomes an increasingly powerful force in shaping possible functional and, ultimately, additive genetic associations among traits.

11.5 Reconciling precise adaptation and evolutionary change: lessons from the long-term study of avian beak evolution

Avian beaks are some of the best examples of precise adaptation and extreme evolutionary diversification (Lack 1947; Grant 1986; Smith 1990; Benkman 1993). These traits typically experience strong selection on precise functional integration needed for food handling and bite force, and such functional integration is accomplished by coordinated changes of many developmental components that are often under distinct genetic control (Eames & Helm 2004; Grant *et al.* 2006; Mallarino *et al.* 2011, 2012). Thus, it is a particularly suitable structure in which to study the evolutionary changes in integration of genetic, developmental, and functional dimensions of the phenotype. A particular paradox is how to reconcile adaptation and diversification—evolution of local adaptations requires close genetic integration of beak components and high heritability of their development for incremental fine-tuning of beak morphology. However, such consistent reduction of developmental variability should, at the same time, prevent the evolutionary diversifications in beaks routinely observed in birds.

The proximate resolution of this paradox could come from significant redundancy of the developmental pathways that produce beaks and the ubiquitous reuse of conserved regulatory elements in developmental and functional integration of beak components (Badyaev 2011a)— in other words, changes in the structure of the development *map* rather than in the *input* variables. The evolutionary significance of such modular organisation and conserved signalling (Figure 11.2) is that only a few genetic changes in regulatory elements are needed for rapid evolution of local adaptation and extensive evolutionary diversification without depletion of genetic variance in beak morphology (see Section 11.1). Under this scenario, the role of natural selection is limited to eliminating developmental abnormalities and to stabilising developmental configurations most adaptive under prevalent conditions. Genetic fixation of mutations in regulatory elements can enable the evolutionary persistence of the most favoured configurations, but redundancy of the regulatory network, compensatory interactions among its elements, and the overall highly modular organisation assures short-term evolutionary retention of many functional configurations of beaks (Badyaev 2011a).

What consequences does such developmental organisation have on the evolution of the **G**-matrix? To answer this question empirically, we undertook a long-term study of multivariate coevolution of genetic, developmental, and functional integration in beaks of house finches (*Carpodacus mexicanus*),

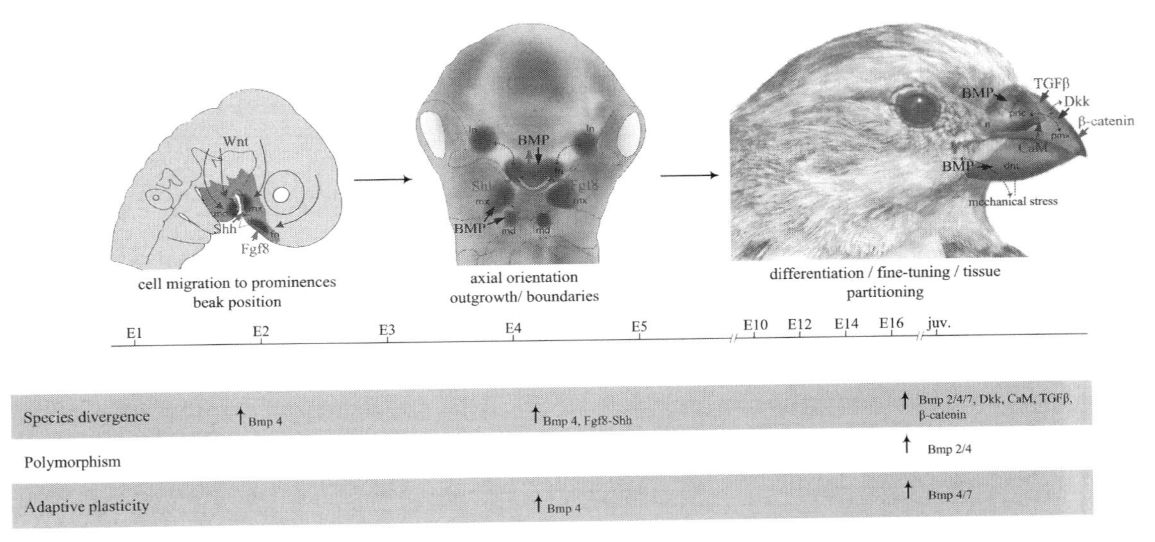

Figure 11.2 Modular core processes are regulated by conserved growth factors in beak development. Main growth factors are wingless type (*Wnt*), fibroblast growth factor 8 (*Fgf8*), sonic hedgehog (*Shh*), bone morphogenetic proteins (BMP), transforming growth factor beta (*TGFβ*), calmodulin (*CaM*), Dickkopf (*Dkk*), and *β-catenin*. Facial prominences (shown in dark), formed by proliferation of neural crest cells, are frontonasal (fn), lateral nasal (ln), mandibular (md), and maxillary (mx). Cartilage and bone areas arising in late development are prenasal cartilage (pnc), premaxillary bone (pmx), nasal bone (n), and dentary bone (dnt). Double-headed arrows show interactions between neighbouring prominences during growth and expansion (left and middle figure), tissue partitioning between pnc and pmx and effects of mechanical stress conductance in late developmental stages (right figure). Table shows developmental stage and growth factors that were shown to regulate species divergence, polymorphism, and adaptive plasticity in beak size and shape. Based on references in Badyaev 2011a and Mallarino *et al*. 2012.

tracing the evolution of novel beak configurations across 20 consecutive generations during the origin of local adaptation (Badyaev 2010). We were particularly interested in how abundant genetic variance in beak ontogeny is modulated to produce both precise local adaptations (that require significant reduction of variance; Badyaev *et al.* 2000; Badyaev & Martin 2000) and continuing diversification of beak morphology among different populations as this species expands its ecological and geographic range (Badyaev, Belloni & Hill 2012).

We derived an overall genetic variance–covariance matrix and genetic correlations among three beak components (width, length, and depth) from a fully resolved pedigree consisting of full- and half-sib groups across 11 generations (Badyaev 2005, 2010). Additive genetic (***G***) and common environmental (***E***; e.g. nest effects) variance–covariance matrices were derived by fitting phenotypic data to a multivariate animal model of the general form: $\mathbf{y} = \boldsymbol{X}\mathbf{b} + \boldsymbol{Z}\mathbf{a} + E\mathbf{c} + \mathbf{e}$, where **y** is a vector of trait values, **b** is a vector of fixed effects, **a** is a vector of additive genetic effects, **c** is a vector of common environmental effects, **e** is a vector of residual variation, and ***X***, *Z*, and *E* are incidence matrices for the fixed, additive genetic, and common environmental effects, respectively (Lynch & Walsh 1998). Analysis was carried out using restricted maximum likelihood implemented in ASReml (2.0) software. We used univariate general linear models (PROC GLM, SAS Inc.) to identify significant fixed effects, and any fixed-effect term that was significant in at least one trait was included in the final multivariate model. This resulted in the inclusion of year, offspring sex and age. Nest identity (nested within dam identity) was included as a random effect representing variance due to common environment (e.g. nest environment, parental effects).

In each generation, we measured multivariate selection on beak morphology by fitting the full second-order polynomial equation (Lande & Arnold 1983) $w = \alpha + z^{T}\beta + z^{T}\gamma z$, where z are the three original traits, w is juvenile survival associated with the onset of independent foraging (from 40 to 80 days post-fledging), $\boldsymbol{\beta}$ is the vector of standardised directional selection gradients, and γ is the matrix of quadratic and cross-product terms among the traits. We also performed canonical rotation of the γ matrix (Box & Draper 1987; Blows & Brooks 2003) to create the ***M***-matrix, in which the eigenvectors m_i describe the shape of the response surface and the direction of its principal orientation. The largest eigenvalues, λ_i, are associated with the greatest curvatures in the response surface; positive eigenvalues indicate upward curvature, and negative eigenvalues indicate downward curvature (Phillips & Arnold 1989).

For each generation, we also measured developmental variability by constructing age-specific correlation matrices for the growth sequence of all beak components from age 1–16 days and calculated the overall correlational matrix ***L*** and associated eigenvalues and eigenvectors for the entire growth sequence (e.g. Badyaev & Martin 2000). To assess multivariate directions of greatest independent developmental variation (i.e. lowest ontogenetic integration) of the three beak components, we calculated $\boldsymbol{L}^{-1}$ for each generation, whose vectors are projections of traits with greatest independent variation during growth.

Although we measured 'end products'—beak dimensions during different times of ontogeny—our measure of the ***G***-matrix and multivariate selection nevertheless accurately described patterns of microevolutionary change (Figure 11.3). The direction of long-term stabilising selection on beak depth was indistinguishable from the direction of maximum variance of the ***G***-matrix whereas fluctuating selection on relative expression of beak length–beak width dimensions was strongly concordant with the second eigenvalue of the *G*-matrix. The role of directional selection was largely confined to elimination of phenotypic extremes formed by compensatory developmental interactions (Figure 11.4) and these diverse developmental interactions during initial stages of population establishment did not have a significant imprint in either genetic variance–covariance structure or patterns of long-term stabilising selection (Figure 11.4).

We also found high genetic correlations between beak components that experienced variable and frequently antagonistic developmental interactions,

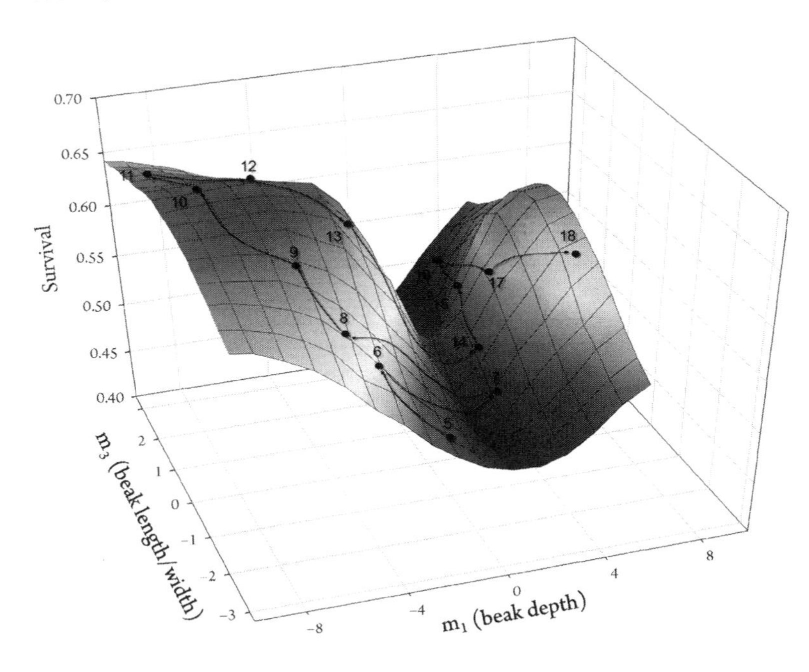

Figure 11.3 Overall fitness surface (juvenile post-fledging survival) defined by variation in beak depth (m_1) and relative expression of beak length and width (m_3). The ***G***-matrix orients congruently with the ***M***-matrix (85% along m_1 and 8.7% along m_3). Dots are average coordinates of m_1, m_3, and survival for each generation (shown by consecutive numbers), starting with generation 5. Arrows connect subsequent generations. Modified from Badyaev (2010).

had low additive genetic variance, and were likely regulated by non-overlapping gene cascades (Abzhanov *et al.* 2006; Badyaev 2010). This further emphasises that genetic and developmental integration are rarely congruent, even for frequently co-selected traits (Badyaev 2004; Frankino *et al.* 2007). At the same time, this finding confirmed the expectation that when variation in partitioning of developmental precursors among traits is consistently greater than variation in fitness consequences of their end products—as is expected in adaptively equivalent combinations of traits–the developmentally interacting traits will evolve genetic correlations even when their relative expression is antagonistic (Bulmer 1971; Houle 1991).

Compensatory adjustments of beak components during development and the adaptive equivalence of distinct configurations can have important consequences for the structure of the *G*-matrix that can reconcile adaptation and evolutionary change. First, such developmental interactions extended the phenotypic range of locally adaptive morphologies converting overall disruptive selection during colonisation of a novel environment to overall stabilising selection where several adaptive phenotypes could be maintained—a condition that favours long-term stability in the *G*-matrix (Lande 1976; Kopp & Hermisson 2006). Second, compensatory adjustments among beak components shielded genetic variance in individual traits, such that a small number of conserved developmental modules produced both locally adaptive morphology and evolutionary diversifications. Third, compensatory development and functional equivalence (when distinct configurations of traits produce the same functional outcome) might explain the puzzling result whereby dimensions of the strongest and most persistent stabilising selection coincided with dimensions of greatest additive genetic variance, whilst traits under fluctuating selection had lower additive genetic variance. Thus, diverse developmental adjustments among beak components can either replenish genetic variance along the axis of most consistent selection or shield variance in the trait that is most consistently subject to selection (Walsh & Blows 2009). The finding that only the most recurrent developmental and functional interactions were represented in *G*-matrix structure provides an important insight into the hierarchical arrangement of epigenetic processes in development in relation to their genetic stabilisation.

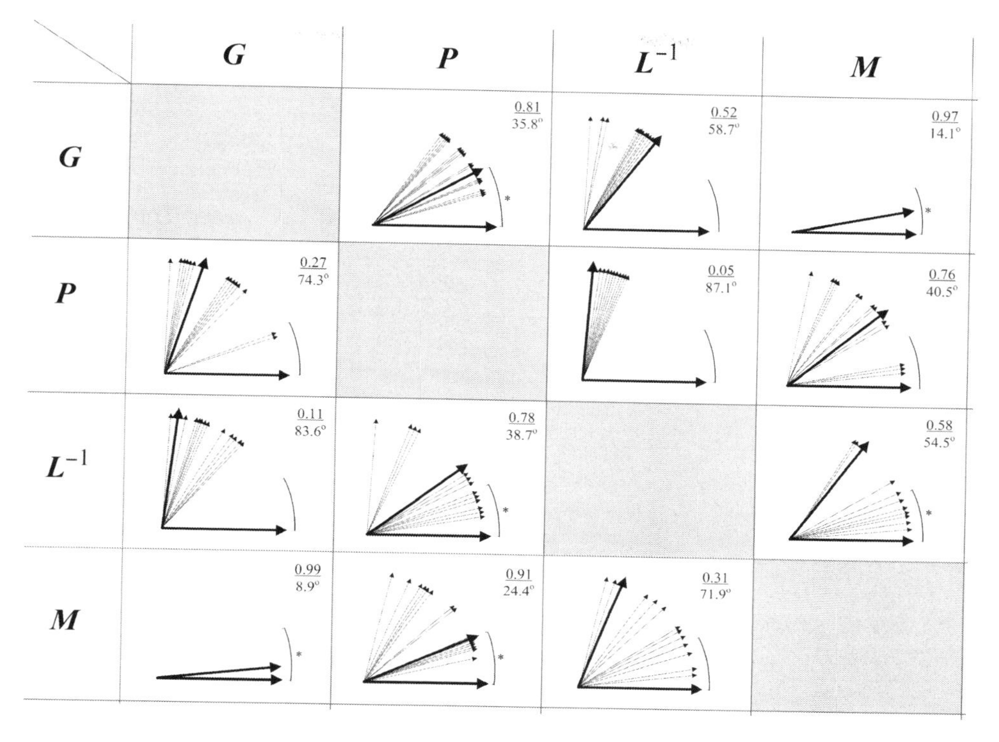

Figure 11.4 Concordance of $\boldsymbol{G}$ (additive genetic covariance matrix), $\boldsymbol{P}$ (phenotypic covariance matrix), $\boldsymbol{L}^{-1}$ (negative inverse covariance matrix of ontogenetic integration), and $\boldsymbol{M}$ (matrix of multivariate selection) of beak morphology in Montana house finches during 19 generations following population establishment. Shown are the first (above diagonal) and the second (below diagonal) eigenvectors and the corresponding vector correlations and angles (in degrees) between them. Thick lines show overall vectors calculated across all generations, thin lines show within-generation vectors. For $\boldsymbol{L}^{-1}$ vs $\boldsymbol{G}$ and $\boldsymbol{P}$ vs $\boldsymbol{G}$ comparisons, within-generation vectors of $\boldsymbol{L}^{-1}$ and $\boldsymbol{P}$ are compared with the vectors of overall $\boldsymbol{G}$. For all other comparisons ($\boldsymbol{L}^{-1}$ vs $\boldsymbol{P}$, $\boldsymbol{L}^{-1}$ vs $\boldsymbol{M}$, and $\boldsymbol{P}$ vs $\boldsymbol{M}$) vectors are compared within each generation. Brackets show bootstrapped limits ($n = 100$) within which the vectors are not distinct. Asterisks indicate similarity of overall vector correlations (if thick vectors are within the bracket) or similarity of within-generation vectors (if most of the within-generation vectors are within the bracket). From Badyaev (2010).

11.6 Conclusions

One of the central questions in evolutionary quantitative genetics is, what forces shape the current *G*-matrix for a given population? Are these largely population–genetic, namely the removal of additive variation along persistent directions of natural selection or, as we suggest here, could emergent processes of development also play a major role? Either of these forces can generate a *G*-matrix that is highly constrained (the first few eigenvalues accounting for the vast majority of variation). In such cases, evolution along the lines of genetic least resistance (in directions of maximal variation in the *G*-matrix) is inevitable. If trait evolution is largely governed by drift (potentially in the face of weak stabilising selection), then the maximal divergence occurs along axes with maximal variation. Conversely, under directional selection, with a very constrained *G*-matrix the projection of any random selection gradient onto the first eigenvector likely dominates the projection of that vector onto the whole of the *G*-matrix, and hence the response closely aligns to the first eigenvalue, i.e. along the line of least resistance. In our empirical example, it seems that the lines of least genetic resistance are, proximately speaking, the lines of greatest recurrence of organism–environment associations that reflect the evolved correspondence between epigenetic processes of developmental and functional integrations stabilised by increasingly redundant additive genetic linkages. This and other empirical studies show that the genetic dimensionality of a structure is often far smaller than is expected from dimensionality of its phenotype.

Acknowledgments

Empirical work reported here was supported by the National Science Foundation (DEB-0075388, DEB-0077804, IBN-0218313) and the David and Lucille Packard Foundation Fellowship. We are grateful to Anne Charmantier, Loeske Kruuk, and Dany Garant for invitation to contribute. We thank the editors, Renee Duckworth, Badyaev Lab, and anonymous reviewers for insightful suggestions and comments. AVB thanks the Distinguished Visiting Professor program of the University of Miami for their hospitality in Coral Gables during writing of this chapter.

References

Abzhanov, A., Kuo, W.P., Hartmann, C., Grant, B.R., Grant, P.R. & Tabin, C.J. (2006) The calmodulin pathway and evolution of elongated beak morphology in Darwin's finches. *Nature*, **442**, 563–567.

Alberch, P. (1991) From genes to phenotype: dynamical systems and evolvability. *Genetica*, **84**, 5–11.

Arnold, S.J. (1992) Constraints on phenotypic evolution. *American Naturalist*, **140**, S85–S107.

Arnold, S.J., Bürger, R., Hohenlohe, P.A., Ajie, B.C. & Jones, A.G. (2008) Understanding the evolution and stability of the G-matrix. *Evolution*, **62**, 2451–2461.

Badyaev, A.V. (2004) Integration and modularity in the evolution of sexual ornaments: an overlooked perspective. In: *Phenotypic integration: the evolutionary biology of complex phenotypes*. (ed. M. Pigliucci & K. Preston), pp. 50–79. Oxford University Press, Oxford.

Badyaev, A.V. (2005) Maternal inheritance and rapid evolution of sexual size dimorphism: passive effects or active strategies? *American Naturalist*, **166**, S17–S30.

Badyaev, A.V. (2007) Evolvability and robustness in color displays: bridging the gap between theory and data. *Evolutionary Biology*, **34**, 61–71.

Badyaev, A.V. (2010) The beak of the other finch: coevolution of genetic covariance structure and developmental modularity during adaptive evolution. *Philosophical Transactions of The Royal Society of London Series B, Biological Sciences*, **365**, 1111–1126.

Badyaev, A.V. (2011a) How do precise adaptive features arise in development? Examples with evolution of context-specific sex-ratios and perfect beaks. *Auk*, **128**, 467–474.

Badyaev, A.V. (2011b) Origin of the fittest: link between emergent variation and evolutionary change as a critical question in evolutionary biology. *Proceedings of the Royal Society B-Biological Sciences*, **278**, 1921–1929.

Badyaev, A.V. (2013) "Homeostatic hitchhiking": a mechanism for the evolutionary retention of complex adaptations *Integrative and Comparative Biology*, **53**, 913–922.

Badyaev, A.V., Belloni, V. & Hill, G.E. (2012) House Finch (*Haemorhous mexicanus*). *The birds of North America online* (ed. A. Poole), pp. doi:10.2173/bna.2146. Cornell Lab of Ornithology, Ithaca.

Badyaev, A.V. & Foresman, K.R. (2004) Evolution of morphological integration: I. Functional units channel stress-induced variation in shrew mandibles. *American Naturalist*, **163**, 868–879.

Badyaev, A.V., Hill, G.E., Stoehr, A.M., Nolan, P.M. & McGraw, K.J. (2000) The evolution of sexual dimorphism in the house finch: II. Population divergence in relation to local selection. *Evolution*, **54**, 2134–2144.

Badyaev, A.V. & Martin, T.E. (2000) Individual variation in growth trajectories: phenotypic and genetic correlations in ontogeny of the house finch (*Carpodacus mexicanus*). *Journal of Evolutionary Biology*, **13**, 290–301.

Beldade, P., Koops, K. & Brakefield, P.M. (2002) Developmental constraints versus flexibility in morphological evolution. *Nature*, **416**, 844–846.

Benkman, C.W. (1993) Adaptation to single resources and the evolution of crossbill (*Loxia*) diversity. *Ecological Monographs*, **63**, 305–325.

Berg, R.L. (1960) The ecological significance of correlation pleiades. *Evolution*, **14**, 171–180.

Björklund, M. (1996) The importance of evolutionary constraints in ecological time scales. *Evolutionary Ecology*, **10**, 423–431.

Blows, M.W. (2007) A tale of two matrices: multivariate approaches in evolutionary biology. *Journal of Evolutionary Biology*, **20**, 1–8.

Blows, M.W. & Brooks, R. (2003) Measuring nonlinear selection. *American Naturalist*, **162**, 815–820.

Blows, M.W. & Walsh, B. (2009) Spherical cows grazing in Flatland: constraints to selection and adaptation. In: *Adaptation and fitness in animal populations: evolutionary and breeding perspectives on genetic resource management* (ed. J. van der Werf, H.-U. Graser, R. Frankham & C. Gondro), pp. 83–101. Springer Netherlands, Dordrecht.

Box, G.E.P. & Draper, N.R. (1987) *Empirical model-building and response surfaces*. Wiley Series in Probability and Mathematical Statistics, New York.

Breen, M.S., Kemena, C., Vlasov, P.K., Notredame, C. & Kondrashov, F.A. (2012) Epistasis as the primary factor in molecular evolution. *Nature*, **490**, 535–538.

Bulmer, M.G. (1971) The effect of selection on genetic variability. *American Naturalist*, **105**, 201–211.

Cheverud, J.M. (1982) Relationships among ontogenetic, static, and evolutionary allometry. *American Journal of Physical Anthropology*, **59**, 139–149.

Cheverud, J.M. (1988) A comparison of genetic and phenotypic correlations. *Evolution*, **42**, 958–968.

Cheverud, J.M. (1996) Developmental integration and the evolution of pleiotropy. *American Zoologist*, **36**, 44–50.

Conner, J.K. (2012) Quantitative genetic approaches to evolutionary constraint: how useful? *Evolution*, **66**, 3313–3320.

Eames, B.F. & Helm, J.W. (2004) Conserved molecular program regulating cranial and appendicular skeletogenesis. *Developmental Dynamics*, **231**, 4–13.

Falconer, D.S. & MacKay, T.D. (1996) *Quantitative Genetics*. Longman, Essex, UK.

Fisher, R.A. (1930) *The genetical theory of natural selection*. Clarendon Press, Oxford.

Flint, J. & Mackay, T.F.C. (2009) Genetic architecture of quantitative traits in mice, flies, and humans. *Genome Research*, **19**, 723–733.

Frankino, W.A., Zwaan, B.J., Stern, D.L. & Brakefield, P.M. (2007) Internal and external constraints in the evolution of morphological allometries in a butterfly. *Evolution*, **61**, 2958–2970.

Gavrilets, S. (2004) *Fitness landscapes and the origin of species*. Princeton University Press, Princeton, NJ.

Gerhart, J. & Kirschner, M. (2007) The theory of facilitated variation. *Proceedings of the National Academy of Sciences of the United States of America*, **104**, 8582–8589.

Grant, P.R. (1986) *Ecology and evolution of Darwin's finches*. Princeton University Press, Princeton.

Grant, P.R., Grant, B.R. & Abzhanov, A. (2006) A developing paradigm for the development of bird beaks. *Biological Journal of Linnean Society*, **88**, 17–22.

Gravner, J., Pitman, D. & Gavrilets, S. (2007) Percolation on fitness landscapes: Effect of correlation, phenotype, and incompatibilities. *Journal of Theoretical Biology*, **248**, 627–645.

Hallgrimsson, B. & Hall, B.K. (2011) Epigenetics: linking genotype and phenotype in development and evolution. Univertsity of California Press, Berkeley.

Hansen, T.F. (2006) The evolution of genetic architecture. *Annual Review of Ecology, Evolution and Systematics*, **37**, 123–157.

Hansen, T.F. (2011) Epigenetics: adaptation or contingency? In: *Epigenetics: linking phenotype and genotype in development and evolution* (ed. B. Hallgrimsson & B.K. Hall), pp. 357–376. University of California Press, Berkeley.

Hansen, T.F. & Houle, D. (2008) Measuring and comparing evolvability and constraint in multivariate characters. *Journal of Evolutionary Biology*, **21**, 1201–1219.

Houle, D. (1991) Genetic covariance of fitness correlates: what genetic correlations are made of and why it matters. *Evolution*, **45**, 630–648.

Kirkpatrick, M. (2009) Patterns of quantitative genetic variation in multiple dimensions. *Genetica*, **136**, 271–284.

Kirschner, M. & Gerhart, J.C. (2005) *The plausibility of life: resolving Darwin's Dilemma*. Yale University Press, New Haven.

Koonin, E.V. (2011) *The logic of chance: the nature and origin of biological evolution*. FT Press Science, Upper Saddle River.

Kopp, M. & Hermisson, J. (2006) The evolution of genetic architecture under frequency-dependent disruptive selection. *Evolution*, **60**, 1537–1550.

Lack, A.J. (1947) *Darwin's finches*. Cambridge University Press, Cambridge.

Lande, R. (1976) Natural selection and random genetic drift in phenotypic evolution. *Evolution*, **30**, 314–334.

Lande, R. (1977) Evolutionary mechanisms of limb loss in tetrapods. *Evolution*, **32**, 73–92.

Lande, R. & Arnold, S.J. (1983) The measurement of selection on correlated characters. *Evolution*, **37**, 1210–1226.

Larsen, E. (2005) Developmental origins of variation. In: *Variation* (ed. B. Hallgrímsson & B.K. Hall), pp. 113–129. Elsevier Academic Press, San Diego.

Lemos, B., Meiklejohn, C.D., Cáceres, M. & Hartl, D.L. (2005) Rates of divergence in gene expression profiles of primates, mice, and flies: stabilizing selection and variability among functional catagories. *Evolution*, **59**, 126–137.

Lynch, M. (2010) Scaling expectations for the time to establishment of complex adaptations. *Proceedings of the National Academy of Sciences of the United States of America*, **107**, 16577–16582.

Lynch, M. & Walsh, B. (1998) *Genetics and analysis of quantitative traits*. Sinauer Associates, Sunderland.

Mallarino, R., Campàs, O., Fritz, J.A., Burns, K.J., Weeks, O.G., Brenner, M.I. & Abzhanov, A. (2012) Closely related bird species demonstrate flexibility between beak morphology and underlying developmental programs. *Proceedings of the National Academy of Sciences of the United States of America*, **109**, 16222–16227.

Mallarino, R., Grant, P.R., Grant, B.R., Herrel, A., Kuoa, W.P. & Abzhanov, A. (2011) Two developmental modules establish 3D beak-shape variation in Darwin's finches. *Proceedings of the National Academy of Sciences of the United States of America*, **108**, 4057–4062.

Maynard Smith, J., Burian, R., Kauffman, S., Alberch, P., Campbell, J., Goodwin, B., Lande, R., Raup, D. & Wolpert, L. (1985) Developmental constraints and evolution. *The Quarterly Review of Biology*, **60**, 266–287.

Müller, G.B. & Newman, S. (2003) *Origination of organismal form: beyond the gene in developmental and evolutionary biology*. The MIT Press, Cambridge.

Newman, S.A. (2005) The pre-Mendelian, pre-Darwinian world: shifting relations between genetic and epigenetic mechanisms in early multicellular evolution. *Journal of Biosciences*, **30**, 75–85.

Newman, S.A. (2011) The evolution of evolutionary mechanisms: a new perspective. In: *Biological evolution: facts and theories* (ed. G. Auletta, M. Leclerc & R.A. Martinez), pp. 169–191. Gregorian & Biblical Press, Rome.

Newman, S.A. & Bhat, R. (2008) Dynamical patterning modules: physico-genetic determinants of morphological development and evolution. *Physical Biology*, **5**, 1–14.

Newman, S.A. & Müller, G.B. (2000) Epigenetic mechanisms of character origination. *Journal of Experimental Zoology*, **288**, 304–314.

Nowak, M.A., Boerlijst, M.C., Cooke, J. & Smith, J.M. (1997) Evolution of genetic redundancy. *Nature*, **388**, 167–171.

Olson, E.C. & Miller, R.L. (1958) *Morphological integration*. University of Chicago Press, Chicago.

Phillips, P.C. & Arnold, S.J. (1989) Visualizing multivariate selection. *Evolution*, **43**, 1209–1222.

Poelwijk, F.J., Kiviet, D.J., Weinreich, D.M. & Tans, S.J. (2007) Empirical fitness landscapes reveal accessible evolutionary paths. *Nature*, **445**, 383–386.

Reid, R.G.B. (2007) *Biological emergences: evolution by natural experiment*. MIT Press, Cambridge.

Rice, S.H. (2001) The evolution of developmental interactions; epistasis, canalization, and integration. In: *Epistasis and the evolutionary process* (ed. J.B. Wolf, I. Brodie Edmund D & M.J. Wade), pp. 82–98. Oxford University Press, New York.

Rice, S.H. (2002) A general population genetic theory for the evolution of developmental interactions. *Proceedings of the National Academy of Sciences of the United States of America*, **99**, 15518–15523.

Rice, S.H. (2004a) Developmental associations between traits: covariance and beyond. *Genetics*, **166**, 513–526.

Rice, S.H. (2004b) *Evolutionary theory: mathematical and conceptual foundations*. Sinauer Associates, Sunderland.

Riedl, R. (1978) *Order in living organisms: a systems analysis of evolution*. Wiley, New York.

Riska, B. (1989) Composite traits, selection response, and evolution. *Evolution*, **43**, 1172–1191.

Salazar-Ciudad, I. & Marin-Riera, M. (2013) Adaptive dynamics under dvelopment-based genotype-phenotype maps. *Nature*, **497**, 361–364.

Salazar-Ciudad, I., Newman, S.A. & Sole, R.V. (2001) Phenotypic and dynamical transitions in model genetic networks I. Emergence of patterns and genotype-phenotype relationships. *Evolution and Development*, **3**, 84–94.

Schluter, D. (1996) Adaptive radiation along genetic lines of least resistance. *Evolution*, **50**, 1766–1774.

Schmalhausen, I.I. (1938) *Organism as a whole in individual development and history*. Academy of Sciences, Leningrad.

Shubin, N., Tabin, C.J. & Carroll, S. (2009) Deep homology and the origins of evolutionary novelty. *Nature*, **457**, 818–823.

Slatkin, M. & Frank, S.A. (1990) The quantitative genetic consequences of pleiotropy under stabilizing and directional selection. *Genetics*, **125**, 207–213.

Smith, T.B. (1990) Natural selection on bill characters in the two bill morphs of the African finch *Pyrenestes ostrinus*. *Evolution*, **44**, 832–842.

Tatusov, R.L., Fedorova, N.D., Jackson, J.D., Jacobs, A.R., Kiryutin, B., Koonin, E.V., Krylov, D.M., Mazumder, R., Mekhedov, S.L., Nikolskaya, A.N., Rao, B.S., Smirnov, S., Sverdlov, A.V., Vasudevan, S., Wolf, Y.I., Yin, J.J. & Natale, D.A. (2003) The COG database: an updated version includes eukaryotes. *BMC Bioinformatics*, **4**, 41.

Turelli, M. (1988) Phenotypic evolution, constant covaraince, and the maintenance of additive genetic variance. *Evolution*, **42**, 1342–1347.

van Nimwegen, E. & Crutchfield, J.P. (2000) Metastable evolutionary dynamics: crossing fitness barriers or escaping via neutral paths. *Bulletin of Mathematical Biology*, **62**, 799–848.

Waddington, C.H. (1961) Genetic assimilation. *Advances in Genetics*, **10**, 257–290.

Wagner, A. (2011) *The origins of evolutionary innovations: a theory of transformative change in living systems*. Oxford University Press, Oxford.

Wagner, G.P. (2001) *The character concept in evolutionary biology*. Academic Press, San Diego.

Wagner, G.P. & Altenberg, L. (1996) Complex adaptation and the evolution of evolvability. *Evolution*, **50**, 967–976.

Wagner, G.P., Pavlicev, M. & Cheverud, J.M. (2007) The road to modularity. *Nature Reviews Genetics*, **8**, 921–931.

Wagner, G.P. & Zhang, J. (2011) The pleotropic structure of the genotype-phenotype map: the evolvability of complex organisms. *Nature Reviews Genetics*, **12**, 204–213.

Wainwright, P.C. (2007) Functional versus morphological diversity in macroevolution. *Annual Review of Ecology, Evolution and Systematics*, **38**, 381–401.

Walsh, B. & Blows, M.W. (2009) Abundant genetic variation + strong selection = multivariate genetic constraints: a geometric view of adaptation. *Annual Review of Ecology, Evolution, and Systematics*, **40**, 41–59.

Weinreich, D.M., Watson, R.A. & Chao, L. (2005) Perspective: sign epistasis and genetic constraint on evolutionary trajectories. *Evolution*, **59**, 1165–1174.

Welch, J.J. & Waxman, D. (2003) Modularity and the cost of complexity. *Evolution*, **57**, 1723–1734.

West-Eberhard, M.J. (2003) *Developmental plasticity and evolution*. Oxford University Press, Oxford.

Wilkins, A.S. (2001) *The evolution of developmental pathways*. Sinauer Associates, Sunderland.

Wolf, Y.I., Carmel, L. & Koonin, E.V. (2006) Unifying measures of gene function and evolution. *Proceedings of the Royal Society B–Biological Sciences*, **273**, 1507–1515.

Wolf, Y.I., Novichkov, P.S., Karev, G.P., Koonin, E.V. & Lipman, D.J. (2009) The universal distribution of evolutionary rates of genes and distinct characteristics of eukaryotic genes of different apparent ages. *Proceedings of the National Academy of Sciences of the United States of America*, **106**, 7273–7280.

Wright, S. (1931) Evolution in Mendelian populations. *Genetics*, **16**, 97–159.

Wright, S. (1934) An analysis of variability in number of digits in an inbred strain of guinea pigs. *Genetics*, **19**, 506–536.

CHAPTER 12

Evolutionary potential and constraints in wild populations

Céline Teplitsky, Matthew R. Robinson and Juha Merilä

12.1 Introduction

The question of how the evolutionary forces of natural selection, genetic drift and mutation shape the evolutionary potential of traits (i.e. properties of the genetic variance–covariance matrix, **G**), and translate to microevolutionary changes and population differentiation, forms one of the core areas of research in contemporary evolutionary biology. Understanding these issues is key to reconciling the causes of underlying evolutionary stasis (Hansen & Houle 2004) as well as to our understanding of whether and which populations will adapt or become extinct in the face of environmental changes (Futuyma 2010). For instance, concerns about the ability of populations to adapt to ongoing global changes are as pressing as ever, not only in case of currently endangered populations, but for organisms in general (e.g. Hoffmann & Sgro 2011; Merilä 2012), and understanding the evolutionary potential of populations at the edge of their range distribution can give insights into the factors shaping species range limits (Sexton *et al.* 2009). In addition to its theoretical interest, such knowledge could also allow us to predict the rate of adaptation to ongoing climate change and to estimate sustainable rates of environmental change (Chevin *et al.* 2010), information needed to evaluate and implement appropriate management and conservation actions. In the case of invasive species, an accurate evaluation of their evolutionary potential could improve our understanding of their success and potential to spread to new areas (Lee & Gelembiuk 2008; Bacigalupe 2009).

A crucial step towards resolving these issues depends on understanding the genetic architecture, expression and dynamics of standing genetic variation in ecologically important traits, and thus, the evolutionary potential of wild populations. The amount of additive genetic variance in a single trait is often used as a first approximation of a trait's and population's evolutionary potential (Falconer & Mackay 1996). However, this may be an oversimplification in the sense that evolutionary potential can be impacted by various environmental factors (e.g. genotype–environment interactions, $G \times E$) and by genetic correlations among traits. Hence, an "inclusive" measure of evolutionary potential would require the integration of several other aspects—such as developmental effects (ontogeny and individual history; see Chapter 5, Charmantier *et al.*), the influence of the biotic environment ($G \times E$ and indirect genetic effects (IGEs); see Chapter 6, McAdam *et al.*) and the genetic associations, or integration, among traits—into our models.

As vividly pointed out by Gould and Lewontin (1979), traits within an organism are not free to evolve independently, but form integrated units. From an adaptive point of view, this ensures that traits can function together. In this perspective,

Quantitative Genetics in the Wild. Edited by Anne Charmantier, Dany Garant, and Loeske E. B. Kruuk

integration can be viewed as a consequence of correlative selection acting simultaneously on traits and shaping them together (Merilä & Björklund 2004). Integration can also be a by-product of a shared genetic basis and common developmental pathways (Dobzhansky 1956), and the role of integration as an adaptation or a constraint requires further investigation. Thus, integration is also likely to be reflected in the organismal genetic architecture, and thereby in genetic correlations among traits (Merilä & Björklund 2004). As selection often acts on combinations of traits rather than on single traits (Phillips & Arnold 1989), the study of integration is likely to be critical to our understanding of the evolutionary potential of ecologically important traits. Indeed, genetic correlations among traits can impose constraints on responses to selection, for example if the correlated response in a trait (i.e. due to the correlation with another trait under selection) pushes it away from its optimal value.

Here, we provide an overview of our current understanding of evolutionary potential in wild populations and then examine what is known about the stability of genetic architecture across different ecological timescales. In particular, we review current knowledge about: 1) multivariate evolutionary potential, i.e. the ability to respond to a potential selective challenge (Hansen & Houle 2008); 2) genetic constraints both in terms of i) absolute genetic constraints when some traits combinations simply cannot be achieved, and ii) relative constraints when genetic correlations affect the pace and direction of evolution; and 3) the effect of evolutionary forces on the structure of *G*-matrices with specific reference to natural populations. In addition, we test the ability of a range of proposed matrix comparison statistics to detect differences in *G*-matrices with the aid of numerical simulations.

12.2 Evolutionary potential and constraints in natural populations

12.2.1 Estimates of multivariate evolutionary potential

A univariate eye-view of additive genetic variance may be overly narrow: a multivariate perspective shows that additive genetic variance can be low in some directions of phenotypic space, although individual traits may still harbour significant amount of genetic variance (Box 12.1; Figure 12.1). Hence, the amount of additive genetic variance in traits is not often sufficient to describe their evolutionary potential (Walsh & Blows 2009).

As a first approach, it could be tempting to equate evolutionary potential with the amount of additive genetic variance accounted for by the vector containing the maximum genetic variance in a set of traits, known as $\mathbf{g}_{max}$ (Schluter 1996). However, whilst $\mathbf{g}_{max}$ is the dimension that contains most of the shared additive genetic variance in a suite of traits, it does not automatically mean it explains most of the genetic variance in the data: empirical estimates of the variance accounted for range from 37 to 97% (Table S12.1 in the supplementary online material at [insert website address when confirmed]). Hence, taking into account other dimensions than $\mathbf{g}_{max}$ is also warranted and several metrics have been designed for this effect (see Hansen & Houle 2008; Walsh & Blows 2009). The 'multivariate evolvability' as defined by Hansen and Houle (2008) is an integrative measure with a straightforward and meaningful biological interpretation (Figure 12.2). It estimates the amount of the predicted response to selection that occurs exactly in the direction of actual selection ($e(\beta)$). If selection cannot be quantified—or if an overall assessment of evolutionary potential is needed—multivariate evolvability can also be estimated utilising random selection gradients to get an average evolvability in all directions of phenotypic space (mean of the eigenvalues; Hansen & Houle 2008).

These metrics have not been yet much used in studies of natural populations. Björklund *et al.* (2013) found an average evolvability of 0.07% for morphological traits in the collared flycatcher (*Ficedula albicollis*). Hansen and Voje (2011) reanalysed data from Berner *et al.* (2010) on 18 populations of sticklebacks that had diverged from an ancestral marine population and found an average evolvability of 1.5% in the direction of divergence among populations, which is twice the evolvability in random phenotypic directions

Box 12.1 *G*, $\mathbf{g}_{max}$, and constraints on response to selection

The additive genetic variance–covariance matrix (***G***) can be visualised as an ellipse that contains the breeding values of the population (Arnold *et al.* 2001). The major axis of ***G***, which represents the highest percentage of additive genetic variance contained in ***G***, is called $\mathbf{g}_{max}$ (Figure 12.1a). As is apparent from Figure 12.1a, whilst individual traits can harbour a lot of additive genetic variance, some directions of phenotypic space can lack genetic variance when genetic correlations are high. ***G*** is predicted to bias evolutionary trajectories because evolution occurs most easily along the lines of least genetic resistance ($\mathbf{g}_{max}$; Figure 12.1b). If $\mathbf{g}_{max}$ is aligned with selection, the response to selection is unconstrained and should be fast (Figure 12.1b—along β_1), whilst it is expected to be slower if dimensions of lower variance are aligned with selection (moderate constraint,Figure 12.1b—along β_2). Finally, if ***G*** and selection are not aligned, evolutionary response should deviate from selection, because of the biasing effect of $\mathbf{g}_{max}$ (Figure 12.1b—along β_3; Schluter 1996), so that evolution is predicted to occur first along $\mathbf{g}_{max}$ before reaching the higher fitness peak, resulting in a curvilinear evolutionary trajectory. The population can eventually reach the adaptive peak, but with reduced rate (high constraint). Figure 12.1b highlights the importance of the shape and position of ***G*** relative to selection. Microevolutionary constraints thus impose slower response to selection and/or response not exactly in the direction of selection.

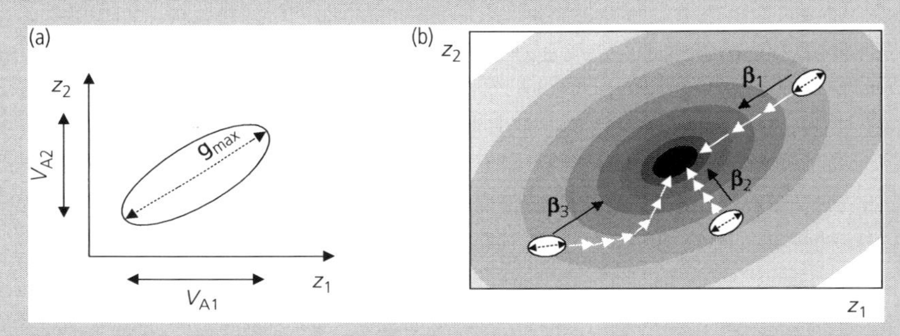

Figure 12.1 (a) Variances (V_A) and covariances of two traits z_1 and z_2. The ellipse represents ***G***, the matrix of additive genetic variances and covariances, and $\mathbf{g}_{max}$ is the first eigenvector of ***G***, i.e. the direction containing the highest percentage of genetic variance. (b) Interplay between ***G*** (white ellipses, dotted arrow denotes g_{max}) and directional selection (β, black arrows). The adaptive landscape is represented by the grey gradient, where darker means higher fitness, with the adaptive peak as the black region in the centre. Three ***G***-matrices are represented: (1) selection and $\mathbf{g}_{max}$ aligned, (2) selection orthogonal to $\mathbf{g}_{max}$ but aligned with $\mathbf{g}_{min}$, and (3) selection and $\mathbf{g}_{max}$ not aligned. Modified from Arnold *et al.* (2001) with kind permission from Springer Science + Business Media B.V.

(0.70%). Whilst nicely illustrating the utility of the method, Hansen's and Voje's (2011) inference was based on purely phenotypic data. Hence, little is known about the range of multivariate evolvabilities in the wild. In general, multivariate quantitative genetic studies have been more focussed on investigating potential evolutionary constraints than evolvability as such. Hence, in the next sections, we review evidence for various forms of constraints which may inhibit an adaptive response to selection.

12.2.2 Absolute constraints: lack of additive genetic variance in some directions (impossible combinations)

Lack of additive genetic variance in some directions of multivariate space represents an absolute constraint, as some trait combinations are basically unreachable. Put in technical terms, this corresponds to a situation where *G* has fewer dimensions than the number of traits measured, and *G* is termed ill conditioned. Ill-conditioning is notoriously difficult to demonstrate, as estimating the dimensionality of G is challenging (Hill & Thompson 1978; Blows 2007). However, it is worth noting that the importance of G being ill conditioned in terms of constraints depends on the direction of selection (i.e. is selection occurring in the direction of no genetic variance?). To our knowledge, no information is available yet about the prevalence of these constraints in wild populations.

Advances in modelling, computer algorithms, and computer hardware have enabled large-scale analyses comprising of numerous records on a number of traits to be conducted (animal models, Box 12.2). However, there are still a number of issues which require further consideration. For instance, estimating G for p traits requires $p(p+1)/2$ parameters, and modelling environmental and measurement error requires at least as many additional parameters. Such large numbers of parameters can lead to unstable parameter estimates. If G is poorly estimated (i.e. the confidence intervals of the covariance terms are large), then it is likely that only a limited number of eigenvectors will be detected (Hill & Thompson 1978). A second issue is that the estimation of eigenvalues is a variance maximising process, where the first eigenvalues

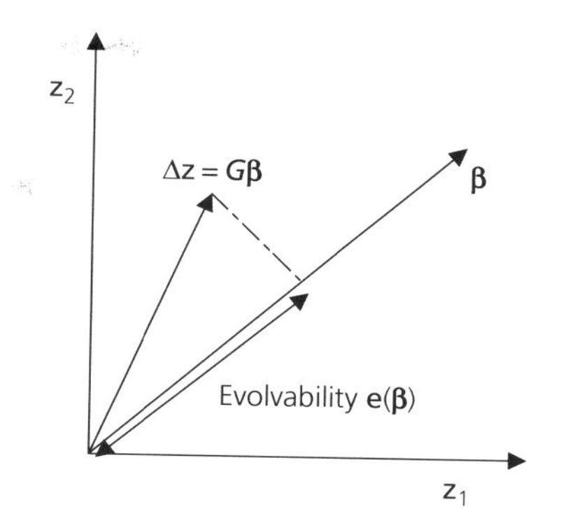

Figure 12.2 Evolvability along a selection gradient $\boldsymbol{\beta}$. Traits $\mathbf{z}_1$ and $\mathbf{z}_2$ are genetically correlated, and Δz is their predicted response to a selection gradient $\boldsymbol{\beta}$ given $\mathbf{G}$. Evolvability ($e(\boldsymbol{\beta})$)is the projection of $\Delta \mathbf{z}$ on $\boldsymbol{\beta}$. Redrawn from Hansen and Houle (2008). Reproduced with permission from John Wiley & Sons.

Box 12.2 Estimating G-matrices: animal model and estimation methods

$\boldsymbol{G}$-matrices can be estimated with multivariate animal models (Kruuk 2004):

$$\boldsymbol{y} = \boldsymbol{u} + \boldsymbol{X}\beta + \boldsymbol{Z}a + \mathbf{e} \qquad \text{(B12.2.1)}$$

where in this case $\boldsymbol{y}$ is a matrix of observations for N traits measured on individuals within a population, $\boldsymbol{Z}$ is a design matrix relating individuals to the vector of additive genetic effects $\mathbf{a}$, which estimates an $N \times N$ genetic variance–covariance matrix ($\boldsymbol{G}$). All random effects are assumed to be normally distributed, and elements of $\mathbf{a}$ are assumed to be drawn from $\mathbf{a} \sim N(0, \boldsymbol{G} \otimes \boldsymbol{A})$ where $\boldsymbol{A}$ is the relatedness matrix derived from the pedigree.

Eq. B.12.2.1 can be estimated within a restricted maximum likelihood framework which provides (co)variance estimates along with approximate standard errors, or alternatively, within a Bayesian MCMC framework which provides a joint posterior distribution for the parameters. The strong advantages of the latter are the possibility to propagate error through to eigen decomposition statistics and any calculation involving the estimated $\boldsymbol{G}$ such as matrix comparisons, and also to predicted responses to selection using the multivariate version of the breeder's equation (Chapter 14, Morrissey *et al.* 2012).

The applicability of these models to pedigreed populations from the wild is often limited by data: to estimate the additive genetic variance–covariance of N traits requires $N(N+1)/2$ genetic variances and covariances and typically an equal or larger number of parameters that describe environmental sources of variation and measurement error to be estimated. Data limitation and trait correlations cause precision of estimates to decrease rapidly as the number of measured traits increases. This issue is a particular concern in studies of wild populations, where the numbers of individuals measured are usually relatively modest and the number of different traits measured can be numerous. As a result, it is often difficult to fit a multivariate model which combines all traits of interest. Failing to integrate some traits is likely to decrease the number of detected constraints, as constraints can emerge even with low genetic covariances if the number of traits is large (Kirkpatrick & Lofsvold 1992).

are systematically biased upwards and the smallest values downwards, so that a high proportion of variance will always be attributed to $\mathbf{g}_{max}$ (Hayes & Hill 1981; Meyer & Kirkpatrick 2010). This limitation suggests that use of eigenvalues may actually lead to an overestimation of the amount of absolute multivariate constraints, and this problem may become more pronounced with decreasing size of datasets and increasing number of traits. Reducing this sampling variance is the focus of many current avenues of research (Mezey & Houle 2005; Meyer & Kirkpatrick 2010) with the aim of gaining better estimates of the covariance structure of quantitative traits. Some promising methods may have emerged, for example estimating a constrained ***G***-matrix that is low rank, and thus, specified by fewer parameters (e.g. the factorial analyses of Kirkpatrick and Meyer (2004), Hine and Blows (2006)). Nevertheless, the question about the dimensionality of ***G*** in natural populations is still very much an open question requiring both theoretically and empirically orientated further investigations (Blows & Hoffmann 2005).

12.2.3 Relative constraints: rate and direction of evolution

In the absence of absolute genetic constraints, a response to selection can occur, but both the rate and direction of evolution can be influenced by the genetic covariances among the traits (Box 12.1).

12.2.3.1 Rate of adaptation

Depending on the interplay between $\mathbf{g}_{max}$ and selection, genetic correlations can either speed up or slow down responses to selection as compared to the predicted responses from univariate evolutionary models. A common example is the expectation that negative genetic correlations between life-history traits—which are expected to be positively selected—are going to slow down a response to selection. In contrast, integration through genetic correlations can also facilitate adaptation as it can preserve the functionality of traits acting together. For example, the two recently diverged *Nicotiana* species *N. alata* and *N. forgetiana* present very different floral morphology, related to differences in their main pollinators. *N. alata* is pollinated by hawkmoth and has pale white or greenish flowers with long narrow corolla tube. In turn, *N. forgetiana* is pollinated by hummingbirds and has purple-red flowers with shorter and wider tubes. The existence of separate modules for traits involving pollinator attraction (attractive features from the limb) and traits involved in pollen transfer (length and width of corolla tube) in these two species may have facilitated specialisation to different pollinators (Bissell & Diggle 2010).

One way to quantify the effect of genetic correlations on the rate of adaptation is to compare predicted responses to selection using a multivariate (i.e. integrating ***G***) vs. univariate (i.e. genetic correlations fixed to zero) framework. For example Etterson and Shaw (2001) concluded that genetic correlations antagonistic to the direction of selection slow down adaptation of a North American prairie plant *Chamaecrista fasciculata* to global warming as multivariate responses were systematically lower than responses predicted from a univariate equation. Agrawal and Stinchcombe (2009) suggested a metric to quantitatively assess the amount of constraint that genetic correlations impose on the rate of a selection response. It is estimated by the relative rate of adaptation, which considers the gain or loss in fitness due to genetic correlations by contrasting the expected rate of adaptation in the presence of genetic correlation to that expected with genetic correlations fixed to zero. As it evaluates the change in fitness along a directional selection gradient, when non-linear selection is weak, this metric is actually equivalent to the ratio between the multivariate evolvabilities along a selection gradient ($\mathbf{e}(\boldsymbol{\beta})$; Hansen & Houle 2008) when genetic correlations are set to zero and when they are not (Agrawal & Stinchcombe 2009, Morrissey *et al.* 2012).

Agrawal and Stinchcombe (2009) reviewed the literature and found that across all studies available so far, the average impact of genetic correlations on the relative rate of adaptation was zero. However, in 27% of studies, the rate of adaptation was decreased by 30% by genetic correlations whilst it increased by the same amount in 16% of studies. Although, note that in a vast majority of the studies reported (29 out of 45), ***P*** (the phenotypic variance–covariance matrix) was used as a surrogate for ***G***, rather than examining *G* directly. Two recent studies of natural populations have found that genetic correlations could be constraining the rate of adaptation in life-history traits by 48% in barn swallows (*Hirundo rustica*; Teplitsky *et al.* 2011) and 40% in red deer (*Cervus elaphus*; Morrissey *et al.* 2012). In the case of the barn swallow, this constraint arises because there is a negative genetic correlation between arrival date and the delay between arrival and laying date (i.e. individuals arriving earlier also wait longer) whilst selection favours individuals arriving earlier and waiting the least time before commencing to breed.

One conceptual issue to note about the study of constraint vs. facilitation is that constraints might actually be detected more readily than facilitation. This makes the results of null average effect from Agrawal and Stinchcombe (2009) all the more surprising, as a more skewed distribution could have been expected. Nevertheless, assuming that genetic correlations facilitate the response to selection, populations should adapt and be submitted to less selection. Hence, facilitation could be a transient state whilst constraints would represent a more stable state. If this is the case, facilitation may be observed more readily in populations submitted to novel selection pressures such as invasive species or populations submitted to novel environments. Following real time evolution of the interplay between selection and *G* in such populations could provide valuable information about the dynamics of evolutionary facilitation and constraints, and possibly, about the transition from one to another.

12.2.3.2 Direction of evolutionary response

Beyond their impact on the pace of adaptation, genetic correlations may also bias the direction of the evolutionary trajectory away from selection and towards $\mathbf{g}_{max}$ (Box 12.1). The prevalence and/or importance of such constraints are under active study because several key questions remain as yet unanswered. For instance, the extent to which $\mathbf{g}_{max}$ biases evolutionary trajectories is still unclear. To test for this, studies generally compare the orientation of $\mathbf{g}_{max}$ to the orientation of divergence among populations or closely related species. The results concerning the constraints imposed by $\mathbf{g}_{max}$ are contrasting: whilst some studies have found that populations have diverged along $\mathbf{g}_{max}$ (Schluter 1996; Begin & Roff 2003; Chenoweth *et al.* 2010; Colautti & Barrett 2011), others have not found such a possible constraining effect of $\mathbf{g}_{max}$ on divergence (McGuigan *et al.* 2005; Berner *et al.* 2008; Berner *et al.* 2010; Kimmel *et al.* 2012). It is also worth noting that all these listed studies have used designs where individuals from the wild were brought to the laboratory to estimate the *G*-matrix. To our knowledge, such comparisons have not been performed yet for wild populations. Apart from the logistic constraints of collecting pedigree data from multiple populations, such studies are challenging in the view that it is practically impossible to control for environmental influences on population mean phenotypes in data collected in the wild.

It should be noted that divergence along $\mathbf{g}_{max}$ is not synonymous with constraints on the response to selection for two reasons. First, divergence along $\mathbf{g}_{max}$ is not synonymous with constraint because selection can be aligned with $\mathbf{g}_{max}$, allowing an even faster evolutionary response (Arnold *et al.* 2001). Second, population divergence along $\mathbf{g}_{max}$ can also result from drift (Lande 1979; Phillips *et al.* 2001), meaning that approaches such as Q_{ST}/F_{ST} analyses are needed to distinguish between these alternatives (Merilä & Björklund 2004). The method of Ovaskainen *et al.* (2011)—as implemented in the program driftsel (Karhunen *et al.* 2013)—provides one readily usable approach to this effect.

Whilst alignment of divergence among populations and $\mathbf{g}_{max}$ is not a proof of evolutionary constraint, neither are deviations from the line of least resistance proof of the absence of genetic constraints (Hansen & Voje 2011). This is because the existence of additive genetic variance along eigenvectors other than $\mathbf{g}_{max}$ can also promote evolution in directions other than those dictated by $\mathbf{g}_{max}$ (Blows & Higgie 2003). In other words, if evolution occurs not along $\mathbf{g}_{max}$ but in directions where evolvability is greater than on average (i.e. the evolutionary trajectory would not follow a line of least resistance, but a plane or a volume of least resistance), the role for genetic constraints cannot be excluded. The example from Kimmel *et al.* (2012) on divergence in opercle shape (described by ten principal component (PC) analysis scores) among 22 stickleback populations illustrates this point nicely: there was no association between the direction of divergence and $\mathbf{g}_{max}$, but the amount of genetic variance along the axis of divergence was higher than that in all other possible directions.

Several methods have recently been developed to include dimensions of ***G*** other than $\mathbf{g}_{max}$ to study paths of evolutionary divergence among populations. These include: comparison of the matrix of population divergence ***D*** with the ***G***-matrix (Blows & Higgie 2003; Colautti & Barrett 2011), projecting ***D*** into subspaces of ***G*** to assess the association between divergence and subspaces of ***G*** (McGuigan *et al.* 2005) and evaluating whether ***D*** is more similar to selection or to the predicted response to selection when including genetic constraints (Chenoweth *et al.* 2010). Finally, Hansen and Houle (2008; see also Hansen & Voje 2011) also suggested comparing the expected evolvability in the direction of selection to the expected evolvability in random directions. These approaches should better describe the potential for evolutionary divergence among populations, but as yet they have not often been applied to studies of wild populations.

Although the biasing effect of $\mathbf{g}_{max}$ can have a long-term impact on population divergence, its influence is supposed to decay with time. However, to our knowledge, the only study which has focussed on the decaying effects of $\mathbf{g}_{max}$ with time was the seminal study by Schluter (1996) on populations recently derived from a common ancestor. The biasing effect of $\mathbf{g}_{max}$ was found to decay with time, but effects were nonetheless estimated to last on a scale of million years.

Ultimately, the rate of decay in the influence of $\mathbf{g}_{max}$ on population divergence must depend on the stability of the orientation of the ***G***-matrix. The stability of ***G*** is thus of fundamental importance to understand the time scale on which genetic correlations affect rate and direction of evolution. For example, genetic correlations will represent insurmountable genetic constraints only if they remain stable over long periods of time. Thus, to make general predictions about the effect of multivariate genetic constraints on evolutionary change, we need to understand the evolutionary and ecological factors shaping ***G***-matrices, and especially the orientation of ***G*** relative to selection pressures. Thus in the next section, we will outline what is known about the interplay between the stability of ***G***, selection and genetic drift in wild populations.

12.2.4 Stability of G-matrices

The predictive power of quantitative genetics and the importance of genetic constraints in dictating patterns and pace of population differentiation depend on the stability of ***G***. However, with recent theoretical and empirical developments, the question has switched from asking whether ***G*** is stable (Arnold *et al.* 2001) to assessing the conditions favouring its stability or influencing its evolution (Steppan *et al.* 2002; Jones *et al.* 2003; Begin & Roff 2004; Arnold *et al.* 2008). Arnold *et al.* (2008) have recently reviewed the evolutionary models put forth to elucidate the evolution of ***G***-matrices. In what follows, our aim is to give an overview of the testable hypotheses in this context using data from the wild.

12.2.5 G-matrix stability under selection, migration and drift

Simulation studies (Jones *et al.* 2003; Jones *et al.* 2004; Guillaume & Whitlock 2007) can provide predictions about the stability of ***G***, many of which have yet to be tested or discussed in the context of natural populations. The time scale at which the stability of ***G*** is investigated matters since ***G*** can fluctuate even over very short time scales (within a generation), but it also evolves in response to selection on longer time scales (Arnold *et al.* 2008). Hence the time scale used to assess the stability of ***G*** within populations could have a very strong impact on the conclusions of a given study.

Selection can both shape and stabilise ***G***: theory predicts that ***G*** should become aligned with the prevailing adaptive landscape (Lande 1980; Arnold *et al.* 2001). Correlative selection and alignment of directional selection and $\mathbf{g}_{max}$ should promote stability of ***G*** (Jones *et al.* 2003; 2004). Under stabilising selection, large effective population sizes should promote stability of the ***G***-matrix (Jones *et al.* 2003). However, the limited number of attempts to assess the orientation of ***G*** relative to adaptive landscape

(Blows *et al.* 2004; Simonsen & Stinchcombe 2010) along with the existence of predicted constraints, suggest that **G** may not always be aligned with the prevailing adaptive landscape. One hypothesis is that fluctuating selection may continually shift the adaptive landscape, but to date the stability of selection is still debated (Siepielski *et al.* 2009; Morrissey & Hadfield 2012). Many of these hypotheses and/or predictions could now be tested using available methods as applied to datasets from the wild, and knowledge on the stability of **G** will benefit from further investigation of the stability of selection.

Concerning other evolutionary forces, theory predicts that drift should lead to proportional changes in **G**-matrices (Lande 1979) but it must be noted that matrix proportionality is predicted in terms of expectation, not for a particular realisation. As to the effects of migration, theoretical work shows it can affect the orientation, shape and size of **G**-matrices, depending on the migration–selection balance (Guillaume & Whitlock 2007). Notably, migration among populations could lead to additional additive genetic variance in the direction of divergence, creating an alignment between divergence and $\mathbf{g}_{max}$ (Guillaume & Whitlock 2007).

12.2.6 Matrix comparison methods

12.2.6.1 Overview of methods

Many methods for comparing **G**-matrices have been developed during the recent years (supplementary online Table S12.2, and Box 12.3). These methods differ in their philosophy: whilst some take a geometrical view of **G**-matrices and aim to assess differences in their size, shape and orientation (Krzanowski 1979; Flury 1988; Ovaskainen *et al.* 2008), others aim to compare **G**-matrices by evaluating how similar their effects are on predicted evolutionary trajectories (Hansen & Houle 2008; Calsbeek & Goodnight 2009). So far, the most commonly used method is the Flury hierarchy (Flury 1988), which tests for differences among matrices in a hierarchical way, from complete similarity (all axes and eigenvalues similar among matrices) to matrices being unrelated (no proportionality among eigenvalues and no eigenvector in common). Whilst it is difficult to implement this method for complex breeding designs (Roff *et al.* 2012), the ideas underlying this analysis have been carried over to several new tests (Box 12.3).

In the ten years since the influential review of Steppan *et al.* (2002), the range of available methods to compare matrices has considerably expanded, and several major breakthroughs have opened the way for more meaningful and powerful matrix comparisons. First, methods based on overall matrix similarity comparisons offer a much more powerful approach than element-by-element comparisons (Blows 2007). The methods of tensors (Hine *et al.* 2009; Robinson & Beckerman 2013) and matrix distances (Mitteroecker & Bookstein 2009) are particularly interesting in this respect, as they allow an investigation of changes in the structure of **G** along a gradient instead of only comparing matrices two by two (e.g. Robinson & Beckerman 2013). A further advantage of a tensor approach is that it allows assessment of which trait combinations have contributed to **G**-matrix divergence, in terms of their relative contributions to the changes in additive genetic (co)variance along an environmental gradient.

Second, the implementation of Bayesian methods to quantitative genetics (Chapter 14, Morrissey *et al.*; Ovaskainen *et al.* 2008; Hadfield 2010) has removed the need for randomisation in order to assess significance of comparison metrics, as the use of the full posterior distribution both of **G** and of derived metrics allows an easy estimation of their confidence intervals (Ovaskainen *et al.* 2008). In the case of wild populations, this approach is of particular interest as it provides confidence intervals around comparison metrics easily from animal model outputs.

One limitation of these methods is that they do not assess the relative importance of drift and selection in shaping **G**-matrices. Briefly, two methods can be used to evaluate this: first, a multivariate version of the traditional Q_{ST}/F_{ST} approach (Chenoweth & Blows 2008; Martin *et al.* 2008) that requires the use of neutral markers; and second, the method of Ovaskainen *et al.* (2011), which also makes use of the neutral markers information but appears to outperform multivariate Q_{ST}/F_{ST} approaches especially for small datasets. In addition, this method accounts for

Box 12.3 Matrix comparison methods

A number of different methods to compare the similarity of $\boldsymbol{G}$-matrices have been developed. For the purposes of our methodological comparisons, we focussed on a subset of methods that are the most widely used and implementable using Bayesian methods. These include the following approaches to test for differences between two matrices ($\boldsymbol{G}_1$ and $\boldsymbol{G}_2$). For further details of each method, see the respective references.

1) Similarity estimates
   - *Ovaskainen's* D (Ovaskainen *et al.* 2008): gives an overall distance between the underlying distributions of $\boldsymbol{G}_1$ and $\boldsymbol{G}_2$. If $f(x)$ and $g(x)$ are estimates of the multivariate distributions from which $\boldsymbol{G}_1$ and $\boldsymbol{G}_2$ are drawn, then $D = \left(\frac{1}{2} \int_{R^d} \frac{[f(x)-g(x)]^2}{f(x)+g(x)} dx \right)^{\frac{1}{2}}$
   - *Random skewers* (Calsbeek & Goodnight 2009): The matrices are multiplied by some random selection vectors ($\boldsymbol{\beta}$), as if to predict response to selection, so that $R_1 = \boldsymbol{G}_1\boldsymbol{\beta}$ and $R_2 = \boldsymbol{G}_2\boldsymbol{\beta}$. The correlation between R_1 and R_2 is then calculated, and the process reiterated over a thousand random vectors to give an average correlation between response vectors, that is an estimate of similarity among matrices.
   - *Krzanowski test* (Krzanowski 1979; Krzanowski 2000): gives an overall index of matrix similarity, which is the sum of the eigenvalues of matrix $\boldsymbol{S}$ (with $\boldsymbol{S} = \boldsymbol{G}_1^T\boldsymbol{G}_2\boldsymbol{G}_2^T\boldsymbol{G}_1$) finding the minimum angle between subspaces of $\boldsymbol{G}_1$ and $\boldsymbol{G}_2$.
   - *Difference in average response* $\bar{d}$ (Hansen & Houle 2008). Similarly to random skewers, this method uses predicted responses to selection and how they diverge to assess differences among matrices. The difference of average response is the expectation of difference over random selection gradients $\boldsymbol{E}\left[\sqrt{\beta'(\boldsymbol{G}_1 - \boldsymbol{G}_2)^2\beta}\right]$
2) Angle between subspaces of matrices
   - The *angle between subspaces* of $\boldsymbol{G}_1$ and $\boldsymbol{G}_2$ is estimated by $\cos^{-1}\sqrt{\lambda_1}$ where λ_1 is the largest eigenvalues of the matrix $\boldsymbol{S}$ defined above (Krzanowski 1979; Krzanowski 2000).
3) Size and shape of matrices
   - *Volume* of matrices expressed as $0.5\pi^2 \prod(\sqrt{\lambda_n})$ or $\sum\lambda$ (Ovaskainen *et al.* 2008).
   - *Eccentricity*: the ratio of the first to the second eigenvalue (Jones *et al.* 2003).
   - *Proportion of variance along* $\boldsymbol{g}_{max}$ calculated as $\lambda_1/\sum\lambda$ (Kirkpatrick 2009).
   - *Trait change index* (Robinson & Beckerman 2013): estimates the proportion of variance of $\boldsymbol{G}_2$ explained by the first three eigenvectors of $\boldsymbol{G}_1$, tests whether different eigenvectors explain the (co) variance in $\boldsymbol{G}_1$ and $\boldsymbol{G}_2$.

When comparing $\boldsymbol{G}$-matrices estimated from real data, significance testing depends on the metric used. For metrics that are estimated within populations and then compared between populations (e.g. volume), significance is simply given by whether the 95% confidence interval overlaps zero. In turn for metrics that are directly built on a between-matrix comparison (e.g. angles, similarity indices), in order to account for sampling error, significance testing should be done by comparing the intra-population differences to the inter-population differences (Ovaskainen *et al.* 2008; Robinson & Beckerman 2013), the null hypothesis being that difference among populations are similar to differences within populations due to sampling error, i.e.:

$$P = 1 - \text{Count}\,([(M_{w1} + M_{w2}) - 2 \times M_{b1,2}] < 0)/N \quad \text{(B12.3.1)}$$

where M_{w1} and M_{w2} are metrics estimated within populations 1 and 2 (based on sampling from posterior distributions within population), and $M_{b1,2}$ is the metric estimated between populations. N is the sample size of the posterior distribution.

shared evolutionary history among analysed populations (Ovaskainen *et al.* 2011).

The diversity of methods used in contemporary literature makes it possible to fine-tune hypothesis testing in regard to which aspect of G one is most interested in. However, some caution is required. A recent comparison of different methods (Roff *et al.* 2012) showed that they each have different levels of power to detect differences among G-matrices. However, Roff *et al.* (2012) did not explore the behaviour of some of the newest methods. To fill this gap, we assessed the performance of different methods using simulated datasets in contrasting situations.

12.2.6.2 Comparison of methods to test for differences in G-matrices

We compared how the results obtained from different methods of assessing the similarity of two G-matrices relate to each other, and compared the power and sensitivity of each method. We focussed on four estimates of overall matrix similarity (viz. Ovaskainen's *D*, Krzanowski's *S*, random skewers, and average response difference $\bar{d}$), and on a selection of metrics focussed on specific geometric aspects: angle between $\mathbf{g}_{max}$, size (volume), shape and orientation (eccentricity, trait change index) of matrices. These methods and their rationale are summarised in Box 12.3.

The first and striking—yet comforting—result was that all four different overall similarity indices gave similar information despite their very different

Box 12.4 Comparison of methods: simulated scenarios

Simulated data: For simplicity, the **G**-matrices of two populations (or environments) were simulated and compared for four traits. Five scenarios (Figure 12.3) were devised: 1) similar matrices, 2) proportional matrices, 3) matrices with one shared component (common principal component or CPC) and 4) unrelated matrices. Two unrelated scenarios were used: Unrelated (I) with an angle of 24° and Unrelated (II) with an angle of 35° between sample $\mathbf{g}_{max}$-vectors.

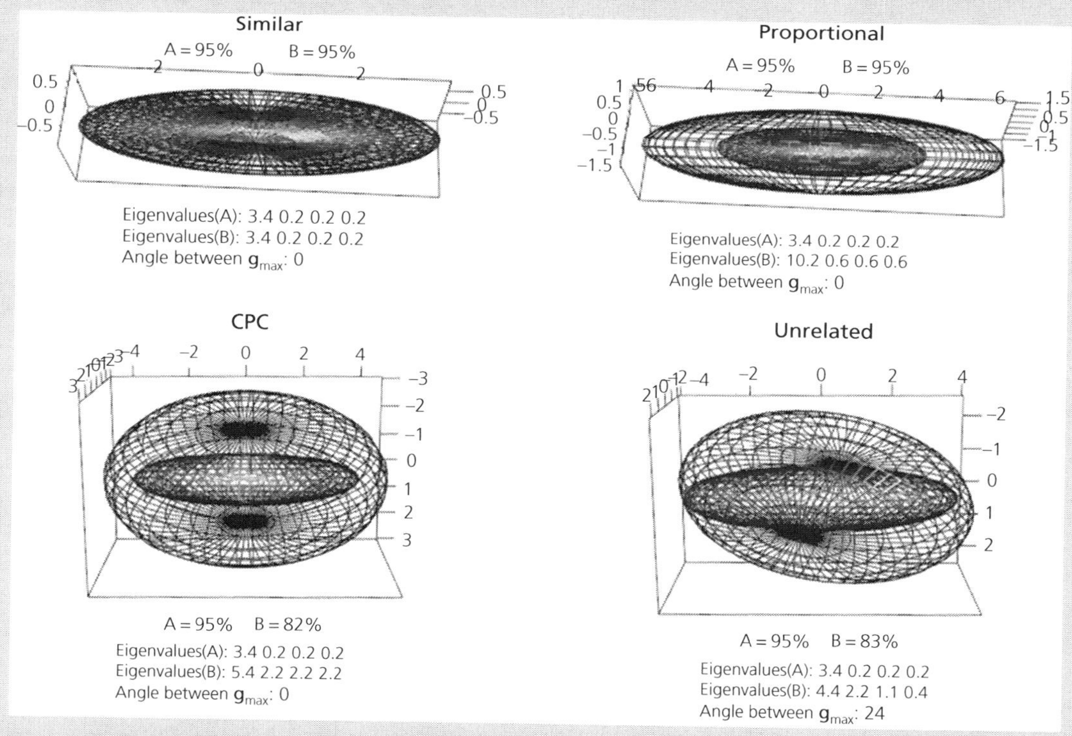

Figure 12.3 Three dimensional representations of two **G**-matrices (three first dimensions) under the four different scenarios. A and B in each plot represent the two different matrices; except in the 'Similar' scenario, B is the largest matrix (see Table 12.1 for the detailed variance–covariance structure). Associated numbers include eigenvalues and the angle between $\mathbf{g}_{max}$ of the original **G**-matrices used in the simulations. The percentage represents the proportion of variance from each matrix that is represented in the subspace defined by the three first eigenvectors of A.

continued

Box 12.4 *Continued*

Both populations were founded by 100 males and 100 females. The maximum number of offspring produced in any one year was two, mating was random, sex ratios equal, and each generation 40% of males and 20% of females died, with the probability of death increasing with age. The full pedigree and data included 1200 individuals measured over six generations, with 200 individuals each generation. Phenotypic values for each of the four traits were simulated as a combination of additive genetic effects estimated from the pedigree, and residual error drawn from a Gaussian distribution. The phenotypic variance was 1, with a heritability of 0.4 for each trait. The simulated **G**-matrices under the different scenarios are presented in Table 12.1. To evaluate power issues, datasets were also restricted to 600, 300 and 150 individuals by randomly deleting complete individuals or records; in this last case missing values were created so that some individuals had complete records, some had only missing values and some individuals had no records and were deleted.

G-matrices from these simulated data were estimated in MCMCglmm (Hadfield 2010), using informative priors ($V =$ diag $(n) \times V_p / r$, nu $= n$), where V_p is the phenotypic variance, n the number of traits and r the number of random factors. These priors have a direct biological interpretation: the prior specification implies that (1) variance is distributed evenly across the random terms and (2) traits are independent (Hadfield 2010). For each analysis 1,200,000 iterations were run with a burn-in phase of 200,000 and thinning of 1,000. Hence for each estimate, we had a sample of 1000 values.

Synopsis of results. The results of the five contrasted scenarios are presented in Table 12.2, emphasising the main differences for each comparison, following the Flury hierarchy. In the 'Similar' scenario no significant difference was detected whatever the method used. In the 'Proportional' scenario matrices, overall differences were indicated by Ovaskainen's D and the difference in average response $\bar{d}$. The origin of these differences is a difference in volume but not in any other shape or direction estimate. This contrasts with the 'CPC' scenario where additional differences in percentage of genetic variance along $\mathbf{g}_{max}$, and eccentricity were found, along with a significant trait change index. If matrices are unrelated, this is further shown by a significant angle between $\mathbf{g}_{max}$.

mathematical constructions (Box 12.4; Figure 12.3; Figure 12.4). However, their sensitivities in terms of significance were different: Krzanowski's S and the random skewers approach systematically failed to detect significant differences among matrices when Ovaskainen's D and mean average response $\bar{d}$ uncovered such (Table 12.2). This difference in sensitivity can be explained by the fact that the random skewer method does not take into account the length of the response vector, only its orientation, so that matrices of very different volume can appear similar (Hansen & Houle 2008). In line with the results of Roff *et al.* (2012), all other methods were positioned relatively independently (Figure 12.4), emphasising that they convey different information about differences among matrices.

In terms of statistical power, it seems that the risk of sampling error leading to the conclusion that matrices are different when they are not (type I error) is not an issue. Using Ovaskainen's D across datasets of different size, significant differences were never found when matrices were generated from the 'Similar' scenario (Figure 12.5). In contrast, the risk of concluding that the matrices are similar when they are not (type II error) seems higher: under the 'Unrelated (I)' scenario, with 300 individuals, and a fully complete pedigree, the differences between the two matrices (using Ovaskainen's D) were not significant (Figure 12.5). The sample size below which differences are detected or not depends on the difference among matrices: the threshold was $n = 300$ for the scenario 'Unrelated I' with an angle of 24° and $n = 150$ for the scenario 'Unrelated II' with an angle of 35°. Whilst our simulations do not explicitly account for missing pedigree links and measurement error, they suggest that sample sizes common to many studies in evolutionary biology are insufficient to accurately estimate *G*-matrix differentiation in the case of weak divergence. Hence, more detailed power studies are needed before firm conclusions about absence of divergence among *G*-matrices can be

Table 12.1 Matrices of variance–covariances used in the different simulation scenarios. Each matrix include four traits; variances are on the diagonal and covariances are the off-diagonal elements.

	Scenario 'Similar'	Scenario 'Proportional'	Scenario 'CPC'	Scenario 'Unrelated I'	Scenario 'Unrelated II'
A	$\begin{bmatrix} 1 & 0.8 & 0.8 & 0.8 \\ 0.8 & 1 & 0.8 & 0.8 \\ 0.8 & 0.8 & 1 & 0.8 \\ 0.8 & 0.8 & 0.8 & 1 \end{bmatrix}$	$\begin{bmatrix} 1 & 0.8 & 0.8 & 0.8 \\ 0.8 & 1 & 0.8 & 0.8 \\ 0.8 & 0.8 & 1 & 0.8 \\ 0.8 & 0.8 & 0.8 & 1 \end{bmatrix}$	$\begin{bmatrix} 1 & 0.8 & 0.8 & 0.8 \\ 0.8 & 1 & 0.8 & 0.8 \\ 0.8 & 0.8 & 1 & 0.8 \\ 0.8 & 0.8 & 0.8 & 1 \end{bmatrix}$	$\begin{bmatrix} 1 & 0.8 & 0.8 & 0.8 \\ 0.8 & 1 & 0.8 & 0.8 \\ 0.8 & 0.8 & 1 & 0.8 \\ 0.8 & 0.8 & 0.8 & 1 \end{bmatrix}$	$\begin{bmatrix} 1 & 0.8 & 0.8 & 0.8 \\ 0.8 & 1 & 0.8 & 0.8 \\ 0.8 & 0.8 & 1 & 0.8 \\ 0.8 & 0.8 & 0.8 & 1 \end{bmatrix}$
B	$\begin{bmatrix} 1 & 0.8 & 0.8 & 0.8 \\ 0.8 & 1 & 0.8 & 0.8 \\ 0.8 & 0.8 & 1 & 0.8 \\ 0.8 & 0.8 & 0.8 & 1 \end{bmatrix}$	$\begin{bmatrix} 3 & 2.4 & 2.4 & 2.4 \\ 2.4 & 3 & 2.4 & 2.4 \\ 2.4 & 2.4 & 3 & 2.4 \\ 2.4 & 2.4 & 2.4 & 3 \end{bmatrix}$	$\begin{bmatrix} 3 & 0.8 & 0.8 & 0.8 \\ 0.8 & 3 & 0.8 & 0.8 \\ 0.8 & 0.8 & 3 & 0.8 \\ 0.8 & 0.8 & 0.8 & 3 \end{bmatrix}$	$\begin{bmatrix} 1 & 0.8 & 0.4 & 0 \\ 0.8 & 1.5 & 1 & 0.4 \\ 0.4 & 1 & 2.5 & 1.2 \\ 0 & 0.4 & 1.2 & 3 \end{bmatrix}$	$\begin{bmatrix} 1 & 0.4 & 0.2 & 0 \\ 0.4 & 1 & 0.2 & 0 \\ 0.2 & 0.2 & 1 & 0 \\ 0 & 0 & 0 & 1 \end{bmatrix}$

Table 12.2 Assessment of differences among **G**-matrices using some of the methods described in Box 12.3. Bold numbers indicate that the method detected a significant difference between populations. For methods comparing values within populations (e.g. volume), significance is assessed by evaluating whether the 95% confidence interval (CI; given in square brackets) includes 0. For other methods, followed by a '*', significance is not evaluated by whether the CI includes zero, but by a comparison of differences within and among populations (see Box 12.3). The methods cited in the boxes under the brackets are the ones estimating the relevant changes when comparing scenarios.

Method	'Similar'		'Proportional'		'CPC'		'Unrelated (I)'		'Unrelated (II)'	
	Estimate [95% CI]	*p*-value	Estimate [95% CI]	*p*-value	Estimate [95% CI]	p-value	Estimate [95% CI]	*p*-value	Estimate [95% CI]	*p*-value
Krzanowski's S^*	1.29 [1; 1.75]	0.64	1.46 [1; 1.74]	0.67	1.04 [1; 1.69]	0.735	1.44 [0.91; 1.68]	0.75	1.32 [0.98; 1.59]	1
Ovaskainen's D^*	0.02 [0; 0.10]	0.38	**0.19 [0.14;0.23]**	**0**	**0.19 [0.15;0.24]**	**0**	**0.17 [0.12;0.24]**	**0**	**0.15 [0.12;0.20]**	0
Mean response difference $\bar{d}$*	0.33 [0.2; 0.5]	0.16	**3.05 [2.60; 3.62]**	**0**	**2.11 [1.90; 2.33]**	**0**	**1.24 [1.06;1.41]**	**0**	**1.68 [1.48; 1.87]**	**0**
Random skewers*	0.99 [0.87; 1]	1	1 [0.99;1]	0.89	0.96 [0.74; 0.99]	1	0.83 [0.53; 0.96]	1	0.88 [0.80; 0.94]	1
Difference in volume	–0.08 [–0.5; 0.4]	>0.05	**–5.88 [–7.83; –4.16]**	**<0.05**	**–39.03 [–49.95; –34.49]**	**<0.05**	**–4.32 [–6.03; –3.08]**	**<0.05**	**–10.43 [–13.3; –7.72]**	**<0.05**
Difference in eigenvolume	–0.60 [–1.44; 0.20]	>0.05	**–8.55 [–10; –6.47]**	**<0.05**	**–8.04 [–9.64; –6.91]**	**<0.05**	–0.23 [–1.17; 0.65]	>0.05	**–4.58 [–5.8; –3.6]**	**<0.05**
Difference in % variance in **g**max	–0.03 [–0.09; 0.04]	>0.05	–0.004 [–0.07; 0.04]	>0.05	**0.40 [0.34; 0.46]**	**<0.05**	**0.47 [0.41; 0.54]**	**<0.05**	**0.21 [0.15; 0.28]**	**<0.05**
Trait change index	1 [–1; 5]	>0.05	2 [0; 4]	>0.05	**15 [11; 19]**	**<0.05**	**17 [9; 27]**	**<0.05**	**12 [5; 24]**	**<0.05**
Eccentricity difference	–0.45 [–5.89; 3.81]	0.7	–2.75 [–8.23; 4.33]	0.81	**8.69 [6.40; 15.50]**	**<0.05**	**13.11 [7.89; 18.17]**	**<0.05**	**3.96 [2.44; 7.82]**	**<0.05**
Angle between **g**max*	3.76 [2.12; 7.47]	0.39	5.05 [2.75; 7.79]	0.067	8.33 [3.90; 13.29]	0.22	23.33 [12.4; 55.2]	0.06	**35.34 [29.25; 39.9]**	**0**

–similarity indices
–volume difference

–% V_A along $\mathbf{g}_{max}$
–Trait change index
–Eccentricity

Angle between $\mathbf{g}_{max}$

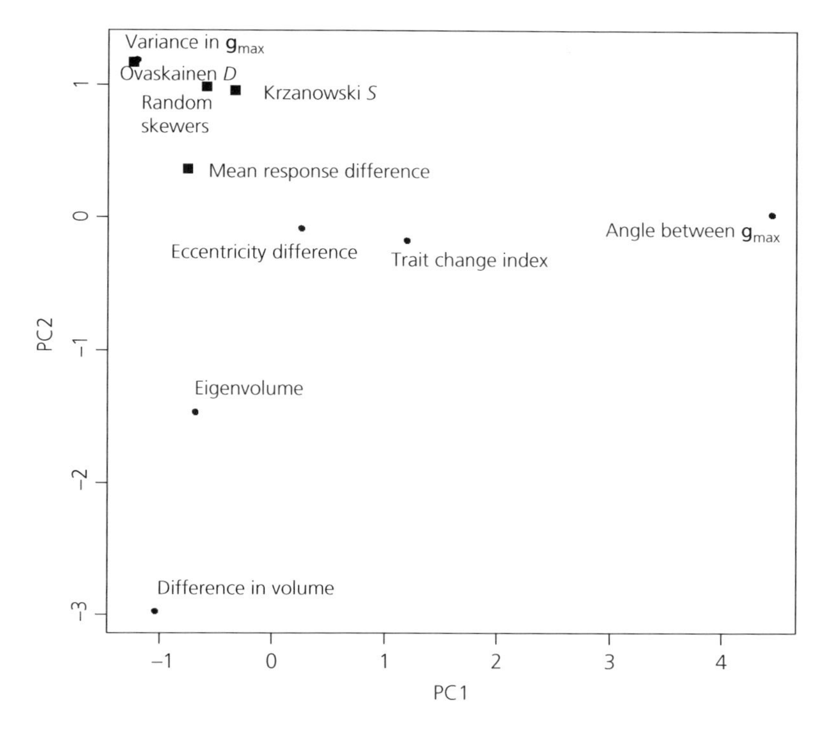

Figure 12.4 Illustration of similarity of results yielded by ten different methods devised to compare ***G***-matrices. To investigate similarity of the information given by the different methods, we performed a PC analysis on their output scores under the five scenarios explained in Box 12.4. The two first axes sum up 85% of the variance in results among methods (PC1: 53 % and PC2: 32%). Proximity of different methods on these two axes thus reveals the extent to which they provide similar information. Similarity indices are represented by square symbols, other methods by circles.

made, especially in comparisons involving recently and/or weakly divergent populations or species.

12.3 Results from natural populations

Investigation of the stability of *G* has been undertaken both at the intra- and interspecific levels, hence allowing evaluation of the impact of ecology and phylogeny on the structure of *G*. Comparison of *G*-matrices for different populations and closely related species has revealed a surprising degree of conservation of *G*-matrices among populations (Roff 2000) but also among species (Arnold *et al.* 2008). Reviewing 31 studies that used the Flury hierarchy to compare *G*-matrices among experimental treatments, sexes, populations or species, Arnold *et al.* (2008) found that in 75% of the cases, matrices shared at least one eigenvector, suggesting frequent conservatism in *G*. However, in the light of the results of the sensitivity analyses presented in Figure 12.3, caution is needed as the constancy of *G*-matrices noted by Roff (2000) and Arnold *et al.* (2008) could also be partly due to low statistical power to detect differences.

In natural populations, several studies contrasting replicated populations occupying different habitats (supplementary online Table S12.2) have indicated that *G*-matrices were conserved within habitats, but differed among them. Such studies are probably so far the best support in favour of the view that *G*-matrix differentiation due to selection is occurring. For example, in reed canary grass (*Phalaris arundinacea*), invasive populations originate from a mixture of European populations: 150 years after introduction, *G*-matrices of invasive populations in America are more similar to European populations developing at a similar latitude than they are to each other (Calsbeek *et al.* 2011). Similarly, in Swedish lakes, ecotypes of an aquatic isopod (*Asellus aquaticus*) have ***G***-matrices

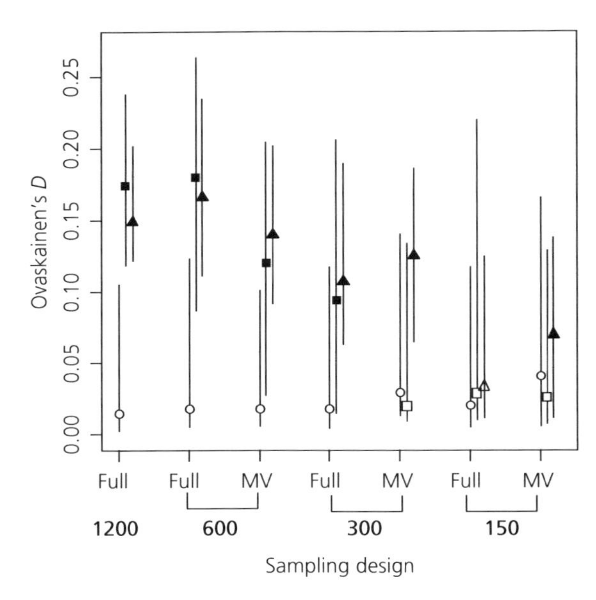

Figure 12.5 Power analysis results of matrix comparison method Ovaskainen's *D* for three scenarios ('Similar' = circles, 'Unrelated (I)' = squares and 'Unrelated (II)' = triangles) with varying sample sizes from 150 to 1,200 individuals. 'Full' indicates that only full records were kept, 'MV' that missing values were created within individual records. White dots represent non-significant results whilst black dots represent detection of significant differences between populations.

that are more similar among similar ecotypes from different lakes, than among different ecotypes from the same lake (Eroukhmanoff & Svensson 2011). Hence, results from both of these studies suggest that changes in the structure of **G** among populations can be fast.

At the intra-population level, Björklund *et al.* (2013) found that **G** was evolving extremely fast in collared flycatchers (*Ficedula albicollis*) as **G**-matrices for five different five-year periods were extremely different, likely because of selection or genotype–environment interactions ($G \times E$), but probably not due to genetic drift. In contrast, it appears that **G** can also be very stable. Garant *et al.* (2008) showed that the **G** for life-history traits (viz. egg size, egg number and laying date) in great tits (*Parus major*) had not changed over the 40 years of study despite the climatic changes that had occurred over the same time period.

Disentangling the effect of $G \times E$ from changes in the structure of **G** resulting from changing selection pressures across environments prior to measurement is difficult. When comparing several populations, assessing $G \times E$ requires a common garden experiment. This was the protocol used to collect data for the two examples of divergence among habitats cited above (*P. arundinacea* in Calsbeek *et al.* 2011; *A. aquaticus* in Eroukhmanoff & Svensson 2011), supporting the concept that differences among **G** among habitats were due to selection and not $G \times E$. Within populations, initial studies examining genetic variance across environmental gradients have used a 'random regression' approach (Wilson *et al.* 2006; Robinson *et al.* 2009; Husby *et al.* 2011). A study in Soay sheep (*Ovis aries*; Robinson *et al.* 2009) suggests that $G \times E$ are important in shaping **G**: in males, genetic correlations between horn length, body weight and parasite load were weaker in more favourable environment as compared to more stressful environments as estimated by lamb survival rate. Random regressions make strong assumptions that variance changes as a quadratic function of a predictor variable, and thus may provide a poorer fit to the data compared to other techniques, meaning that further testing of different approaches is required to assess intra-population **G**-matrix stability in the wild (Pletcher & Geyer 1999). Taken together, this argues for method development and common garden experiments at the intra-population level if we are to understand $G \times E$ in the wild.

12.4 Conclusions and future directions

When using a multivariate framework, questions are framed more often in terms of evolutionary constraints than evolutionary potential. Although they represent two sides of the same coin, focus on constraints may lead to an oversight of the role of genetic correlations in facilitating responses to selection. This focus also suggests that we do not know much about multivariate evolvabilities in wild populations yet. This is surprising as characterising evolutionary potential using a univariate framework has been a long-standing aim in evolutionary biology. We thus suggest that future studies should also focus more on evaluating multivariate evolvabilities, especially in the context of rapid global changes. However, we do not mean to imply that further work about genetic constraints is not needed. Whilst there is strong evidence that genetic constraints are likely to influence adaptive responses both in terms of their rate and direction, our knowledge of the prevalence of these constraints, as well as their stability and the time scale on which they affect the evolutionary trajectories, is still limited.

To clarify the role of constraints in multivariate evolution, several issues regarding the stability of **G** need to be empirically addressed further. First, under which conditions is **G** changing and at which pace? Analysing whether changes in **G** are due to selection (changes in allele frequencies) or due to $G \times E$ will be fundamental to understanding the pace and evolutionary consequences of such changes. Changes in **G** due to selection could be less transient than changes due to $G \times E$ and thus impose stronger evolutionary constraints. Changes due to $G \times E$ would be much faster and understanding when evolutionary potential is coupled to selection positively (Husby *et al.* 2011) or negatively (Wilson *et al.* 2006) would improve our understanding of mechanisms shaping evolvability. Furthermore, how is **G** changing? Theory predicts alignment of **G** with selection, but little if any empirical evidence exists to this effect. The issue of the orientation of **G** relative to the adaptive landscape has been a reoccurring question throughout this review (see also Merilä & Björklund 2004), and it is likely to hold one of the keys to understanding the role of genetic constraints in evolution, both in terms of the evolution of **G** and the constraints it imposes on evolutionary trajectories.

We have also outlined methods to statistically compare **G** both within and across populations using a Bayesian Monte Carlo Markov Chain (MCMC) framework, which provides a framework to reach accurate and reliable insights into the evolutionary potential of wild populations. Our simulation study revealed that with sample sizes representative of those usually analysed in studies of wild populations, low power might be an issue to detect differences among recently diverged populations. This emphasises the value of long-term population studies based on large numbers of individuals, as well development of new statistical approaches and meta-analyses capable of making more refined inference on multivariate inheritance and evolution in the wild.

Acknowledgments

CT was funded by the Agence Nationale de la Recherche (grant ANR-12-ADAP-0006 - PEPS), MRR by an NERC research fellowship, and JM by Academy of Finland (grants: 250435 & 265211). We thank Anne Charmantier, Dany Garant, Loeske Kruuk, Erik Postma and an anonymous reviewer for their help in improving this chapter.

References

Agrawal, A.F. & Stinchcombe, J.R. (2009) How much do genetic covariances alter the rate of adaptation? *Proceedings of the Royal Society B-Biological Sciences*, **276**, 1183–1191.

Arnold, S.J., Burger, R., Hohenlohe, P.A., Ajie, B.C. & Jones, A.G. (2008) Understanding the evolution and stability of the **G**-matrix. *Evolution*, **62**, 2451–2461.

Arnold, S.J., Pfrender, M.E. & Jones, A.G. (2001) The adaptive landscape as a conceptual bridge between micro- and macroevolution. *Genetica*, **112**, 9–32.

Bacigalupe, L.D. (2009) Biological invasions and phenotypic evolution: a quantitative genetic perspective. *Biological Invasions*, **11**, 2243–2250.

Begin, M. & Roff, D.A. (2003) The constancy of the **G**-matrix through species divergence and the effects of quantitative genetic constraints on phenotypic evolution: a case study in crickets. *Evolution*, **57**, 1107–1120.

Begin, M. & Roff, D.A. (2004) From micro- to macroevolution through quantitative genetic variation: positive evidence from field crickets. *Evolution*, **58**, 2287–2304.

Berner, D., Adams, D.C., Grandchamp, A.C. & Hendry, A.P. (2008) Natural selection drives patterns of lake-stream divergence in stickleback foraging morphology. *Journal of Evolutionary Biology*, **21**, 1653–1665.

Berner, D., Stutz, W.E. & Bolnick, D.I. (2010) Foraging trait (co)variances in stickleback evolve deterministically and do not predict trajectories of adaptive diversification. *Evolution*, **64**, 2265–2277.

Bissell, E.K. & Diggle, P.K. (2010) Modular architecture of floral morphology in *Nicotiana*: quantitative genetic and comparative phenotypic approaches to floral integration. *Journal of Evolutionary Biology*, **23**, 1744–1758.

Björklund, M., Husby, A. & Gustafsson, L. (2013) Rapid and unpredictable changes of the **G**-matrix in a natural bird population over 25 years. *Journal of Evolutionary Biology*, **26**, 1–13.

Blows, M.W. (2007) A tale of two matrices: multivariate approaches in evolutionary biology. *Journal of Evolutionary Biology*, **20**, 1–8.

Blows, M.W., Chenoweth, S.F. & Hine, E. (2004) Orientation of the genetic variance-covariance matrix and the fitness surface for multiple male sexually selected traits. *American Naturalist*, **163**, 329–340.

Blows, M.W. & Higgie, M. (2003) Genetic constraints on the evolution of mate recognition under natural selection. *American Naturalist*, **161**, 240–253.

Blows, M.W. & Hoffmann, A.A. (2005) A reassessment of genetic limits to evolutionary change. *Ecology*, **86**, 1371–1384.

Calsbeek, B. & Goodnight, C.J. (2009) Empirical comparison of **G** matrix test statistics: finding biologically relevant change. *Evolution*, **63**, 2627–2635.

Calsbeek, B., Lavergne, S., Patel, M. & Molofsky, J. (2011) Comparing the genetic architecture and potential response to selection of invasive and native populations of reed canary grass. *Evolutionary Applications*, **4**, 726–735.

Chenoweth, S.F. & Blows, M.W. (2008) Qst meets the **G** matrix: The dimensionality of adaptive divergence in multiple correlated quantitative traits. *Evolution*, **62**, 1437–1449.

Chenoweth, S.F., Rundle, H.D. & Blows, M.W. (2010) The contribution of selection and genetic constraints to phenotypic divergence. *American Naturalist*, **175**, 186–196.

Chevin, L.-M., Lande, R. & Mace, G.M. (2010) Adaptation, plasticity, and extinction in a changing environment: towards a predictive theory. *Plos Biology*, **8**, e1000357.

Colautti, R.I. & Barrett, S.C.H. (2011) Population divergence along lines of genetic variance and covariance in the invasive plant *Lythrum salicaria* in Eastern North America. *Evolution*, **65**, 2514–2529.

Dobzhansky, T. (1956) What is an adaptive trait? *American Naturalist*, **90**, 337–347.

Eroukhmanoff, F. & Svensson, E.I. (2011) Evolution and stability of the **G**-matrix during the colonization of a novel environment. *Journal of Evolutionary Biology*, **24**, 1363–1373.

Etterson, J.R. & Shaw, R.G. (2001) Constraint to adaptive evolution in response to global warming. *Science*, **294**, 151–154.

Falconer, D.S. & Mackay, T.F.C. (1996) *Introduction to quantitative genetics*. Longman, New York.

Flury, B. (1988) *Common principal component and related multivariate models*. John Wiley & Sons, New York.

Futuyma, D.J. (2010) Evolutionary constraints and ecological consequences. *Evolution*, **64**, 1865–1884.

Garant, D., Hadfield, J.D., Kruuk, L.E.B. & Sheldon, B.C. (2008) Stability of genetic variance and covariance for reproductive characters in the face of climate change in a wild bird population. *Molecular Ecology*, **17**, 179–188.

Gould, S.J. & Lewontin, R.C. (1979) The spandrels of San Marco and the Panglossian paradigm: a critique of the adaptationist programme. *Proceedings of the Royal Society B-Biological Sciences*, **205**, 581–598.

Guillaume, F. & Whitlock, M.C. (2007) Effects of migration on the genetic covariance matrix. *Evolution*, **61**, 2398–2409.

Hadfield, J.D. (2010) MCMC methods for multi-response generalised linear mixed models: the MCMCglmm R package. *Journal of Statistical Software*, **33**, 1–22.

Hansen, T.F. & Houle, D. (2004) Evolvability, stabilizing selection and the problem of stasis. In: *Phenotypic integration: studying the ecology and evolution of complex phenotypes* (ed. M. Pigliucci & K. Preston), pp. 130–150. Oxford University Press, Oxford.

Hansen, T.F. & Houle, D. (2008) Measuring and comparing evolvability and constraint in multivariate characters. *Journal of Evolutionary Biology*, **21**, 1201–1219.

Hansen, T.F. & Voje, K.L. (2011) Deviation from the line of least resistance does not exclude genetic constraints: a comment on Berner *et al.* (2010). *Evolution*, **65**, 1821–1822.

Hayes, J.F. & Hill, W.G. (1981) Modifications of estimates of parameters in the construction of genetic selection indices ('bending'). *Biometrics*, **37**, 483–493.

Hill, W.G. & Thompson, R. (1978) Probabilities of non-positive definite between-group or genetic covariance matrices. *Biometrics*, **34**, 429–439.

Hine, E. & Blows, M.W. (2006) Determining the effective dimensionality of the genetic variance-covariance matrix. *Genetics*, **173**, 1135–1144.

Hine, E., Chenoweth, S.F., Rundle, H.D. & Blows, M.W. (2009) Characterizing the evolution of genetic variance using genetic covariance tensors. *Philosophical Transactions of the Royal Society of London Series B,Biological Sciences*, **364**, 1567–1578.

Hoffmann, A.A. & Sgro, C.M. (2011) Climate change and evolutionary adaptation. *Nature*, **470**, 479–485.

Husby, A., Visser, M.E. & Kruuk, L.E.B. (2011) Speeding up microevolution: the effects of increasing temperature on selection and genetic variance in a wild bird population. *Plos Biology*, **9**, e1000585.

Jones, A.G., Arnold, S.J. & Burger, R. (2004) Evolution and stability of the **G**-matrix on a landscape with a moving optimum. *Evolution*, **58**, 1639–1654.

Jones, A.G., Arnold, S.J. & Bürger, R. (2003) Stability of the G-matrix in a population experiencing pleiotropic mutation, stabilizing selection, and genetic drift. *Evolution*, **57**, 1747–1760.

Karhunen, M., Merilä, J., Leinonen, T., Cano, J.M. & Ovaskainen, O. (2013) driftsel: an R package for detecting signals of natural selection in quantitative traits. *Molecular Ecology Resources*, 13, 746–754.

Kimmel, C.B., Cresko, W.A., Phillips, P.C., Ullmann, B., Currey, M., von Hippel, F., Kristjansson, B.K., Gelmond, O. & McGuigan, K. (2012) Independent axes of genetic variation and parallel evolutionary divergence of opercle bone shape in threespine stickleback. *Evolution*, **66**, 419–434.

Kirkpatrick, M. (2009) Patterns of quantitative genetic variation in multiple dimensions. *Genetica*, **136**, 271–284.

Kirkpatrick, M. & Lofsvold, D. (1992) Selection and constraint in the evolution of growth. *Evolution*, **46**, 954–971.

Kirkpatrick, M. & Meyer, K. (2004) Direct estimation of genetic principal components: simplified analysis of complex phenotypes. *Genetics*, **168**, 2295–2306.

Kruuk, L.E.B. (2004) Estimating genetic parameters in natural populations using the 'animal model'. *Philosophical Transactions of the Royal Society of London Series B, Biological Sciences*, **359**, 873–890.

Krzanowski, W.J. (1979) Between-groups comparison of principal components. *Journal of The American Statistical Association*, **74**, 703–707.

Krzanowski, W.J. (2000) *Principles of multivariate analysis, revised edn*. Oxford University Press, Oxford.

Lande, R. (1979) Quantitative genetic analysis of multivariate evolution, applied to brain: body size allometry. *Evolution*, **33**, 402–416.

Lande, R. (1980) The genetic covariance between characters maintained by pleiotropic mutation. *Genetics*, **94**, 203–215.

Lee, C.E. & Gelembiuk, G.W. (2008) Evolutionary origins of invasive populations. *Evolutionary Applications*, **1**, 427–448.

Martin, G., Chapuis, E. & Goudet, J. (2008) Multivariate Qst–Fst comparisons: a neutrality test for the evolution of the **G** matrix in structured populations. *Genetics*, **180**, 2135–2149.

McGuigan, K., Chenoweth, S.F. & Blows, M.W. (2005) Phenotypic divergence along lines of genetic variance. *American Naturalist*, **165**, 32–43.

Merilä, J. (2012) Evolution in response to climate change: in pursuit of the missing evidence. *Bioessays*, **34**, 811–818.

Merilä, J. & Björklund, M. (2004) Phenotypic integration as a constraint and adaptation. In: *Phenotypic integration: studying the ecology and evolution of complex phenotypes* (ed. M. Pigliucci & K. Preston), pp. 170–129. Oxford University Press, Oxford.

Meyer, K. & Kirkpatrick, M. (2010) Better estimates of genetic covariance matrices by 'bending' using penalized maximum likelihood. *Genetics*, **185**, 1097–1110.

Mezey, J.G. & Houle, D. (2005) The dimensionality of genetic variation for wing shape in *Drosophila melanogaster*. *Evolution*, **59**, 1027–1038.

Mitteroecker, P. & Bookstein, F. (2009) The ontogenetic trajectory of the phenotypic covariance matrix, with examples from cranofacial shape in rats and humans. *Evolution*, **63**, 727–737.

Morrissey, M.B. & Hadfield, J.D. (2012) Directional selection in temporally replicated studies is remarkably consistent. *Evolution*, **66**, 435–442.

Morrissey, M.B., Walling, C.A., Wilson, A.J., Pemberton, J.M., Clutton-Brock, T.H. & Kruuk, L.E.B. (2012) Genetic analysis of life history constraint and evolution in a wild ungulate population. *American Naturalist*, **179**, E97–E114.

Ovaskainen, O., Cano, J.M. & Merilä, J. (2008) A Bayesian framework for comparative quantitative genetics. *Proceedings of the Royal Society B-Biological Sciences*, **275**, 669–678.

Ovaskainen, O., Karhunen, M., Zheng, C., Arias, J.M.C. & Merilä, J. (2011) A new method to uncover signatures of divergent and stabilizing selection in quantitative traits. *Genetics*, **189**, 621–632.

Phillips, P.C. & Arnold, S.J. (1989) Visualizing multivariate selection. *Evolution*, **43**, 1209–1222.

Phillips, P.C., Whitlock, M.C. & Fowler, K. (2001) Inbreeding changes the shape of the genetic covariance matrix in *Drosophila melanogaster*. *Genetics*, **158**, 1137–1145.

Pletcher, S.D. & Geyer, C.J. (1999) The genetic analysis of age-dependent traits: modeling the character process. *Genetics*, **151**, 825–835.

Robinson, M.R. & Beckerman, A.P. (2013) Genetic variation in resource acquisition drives plasticity in resource allocation to components of life history. *Ecology Letters*, **16**, 281–290.

Robinson, M.R., Wilson, A.J., Pilkington, J.G., Clutton-Brock, T.H., Pemberton, J. & Kruuk, L.E.B. (2009) The impact of environmental heterogeneity on genetic architecture in a wild population of Soay sheep. *Genetics*, **181**, 1639–1648.

Roff, D. (2000) The evolution of the **G** matrix: selection or drift? *Heredity*, **84**, 135–142.

Roff, D.A., Prokkola, J.M., Krams, I. & Rantala, M.J. (2012) There is more than one way to skin a **G** matrix. *Journal of Evolutionary Biology*, **25**, 1113–1126.

Schluter, D. (1996) Adaptive radiation along genetic lines of least resistance. *Evolution*, **50**, 1766–1774.

Sexton, J.P., McIntyre, P.J., Angert, A.L. & Rice, K.J. (2009) Evolution and ecology of species range limits. *Annual Review of Ecology and Systematics*, **40**, 415–436.

Siepielski, A.M., DiBattista, J.D. & Carlson, S.M. (2009) It's about time: the temporal dynamics of phenotypic selection in the wild. *Ecology Letters*, **12**, 1261–1276.

Simonsen, A.K. & Stinchcombe, J.R. (2010) Quantifying evolutionary genetic constraints in the Ivyleaf Morning Glory, *Ipomoea Hederacea*. *International Journal of Plant Sciences*, **171**, 972–986.

Steppan, S.J., Phillips, P.C. & Houle, D. (2002) Comparative quantitative genetics: evolution of the **G** matrix. *Trends in Ecology and Evolution*, **17**, 320–327.

Teplitsky, C., Mouawad, N.G., Balbontín, J., de Lope, F. & Møller, A.P. (2011) Quantitative genetics of migration syndromes: a study of two barn swallow populations. *Journal of Evolutionary Biology*, **24**, 2025–2038.

Walsh, B. & Blows, M.W. (2009) Abundant genetic variation plus strong selection = multivariate genetic constraints: a geometric view of adaptation. *Annual Review of Ecology Evolution and Systematics*, **40**, 41–59.

Wilson, A.J., Pemberton, J.M., Pilkington, J.G., Coltman, D.W., Mifsud, D.V., Clutton–Brock, T.H. & Kruuk, L.E.B. (2006) Environmental coupling of selection and heritability limits evolution. *Plos Biology*, **4**, 1270–1275.

www.oup.co.uk/companion/charmantier

CHAPTER 13

Molecular quantitative genetics

Henrik Jensen, Marta Szulkin and Jon Slate

13.1 Introduction

During the last decade, the rapid development, and decreasing cost, of high-throughput genomics techniques has seen them being employed in non-model species (Ekblom & Galindo 2011; Ellegren & Sheldon 2008; Slate *et al.* 2010). In particular, highly multiplexed single nucleotide polymorphism (SNP) platforms and various methods for genotyping-by-sequencing (Lank *et al.* 2010; Peterson *et al.* 2012), including restriction site associated DNA (RAD) tags (Miller *et al.* 2007; Baird *et al.* 2008), mean that hundreds of thousands of genetically variable sites can be typed for large numbers of individuals. Today even whole genome sequencing and resequencing are applicable and promising tools in studies of organisms that lack a tradition of being the focus of genomics research (Ellegren *et al.* 2012). Here, we explore how this development may complement, contribute to, and strengthen quantitative genetics analyses in studies of natural populations.

We do not take the view that 'genomics' is the solution to all questions in quantitative genetics and evolutionary ecology/biology; of course many questions can be answered without high-throughput genomics tools, especially when phenotypic measures on individuals in a pedigree are available (Hill 2012). For example, predicted responses to selection and adaptive evolutionary change can (with some assumptions/reservations) be estimated when one knows the strength and direction of selection acting on a suite of traits and the ***G***-matrix (additive genetic variance–covariance matrix) for the traits (Falconer & Mackay 1996; Hill & Kirkpatrick 2010; Lynch & Walsh 1998; see also other chapters in this book). However, evolutionary quantitative genetics can certainly be enhanced by a detailed knowledge of the number and identity of individual genes underlying trait variation, as well as their physical location in the genome, and their interaction with each other. Indeed, a fundamental assumption of quantitative genetics is that variation in quantitative traits is determined by many genes of small effect; an assumption that can only be tested with molecular markers. Importantly, knowledge of a trait's genetic architecture will also lead to a better understanding of evolutionary mechanisms, and of the causes of observable phenotypic and genetic variation, than was previously possible to achieve (Ellegren & Sheldon 2008; Slate *et al.* 2010). The goal of this chapter is to give an overview of how genomics tools are increasingly being applied to evolutionary quantitative genetics research.

The chapter is divided into two main sections. In the first, we describe how molecular tools are providing new insight into *quantitative* genetic problems, such as the genetic architecture of traits. In the second section, we examine how genomic data are enabling *population* genetic analyses of natural populations. A distinction between quantitative and population genetics is that the former tends to focus on the mean and variance of traits whereas the latter examines the effects and frequencies of alleles at loci responsible for trait variation. In the context of the work described

Quantitative Genetics in the Wild. Edited by Anne Charmantier, Dany Garant, and Loeske E. B. Kruuk

here, population genetics complements quantitative genetics, because it includes approaches that are well suited to identify genes or genomic regions under selection. We also give some guidance to sample collection and preservation for those ecologists considering applying genomics techniques to their study populations (see Table S13.1 and the supplementary material in Sections online at www.oup.co.uk/companion/charmantier).

13.2 Molecular quantitative genetics

13.2.1 Quantitative trait locus mapping

Identifying the genes that influence variation in fitness-related traits, measuring their relative importance (effect sizes), and being able to track changes in their allele frequencies across time and space are major goals in evolutionary biology (Ellegren & Sheldon 2008; Slate *et al.* 2010). In natural populations, where controlled crosses or the creation of recombinant inbred lines (e.g. Lynch & Walsh 1998) are not possible, there are two main procedures for mapping genes: either 1) linkage analysis or 2) a genome-wide association study (GWAS) (Mackay *et al.* 2009; Slate *et al.* 2010). These methods and their use in natural populations have previously been discussed in detail elsewhere (Erickson *et al.* 2004; Goddard & Hayes 2009; Slate 2005; Slate *et al.* 2010), and therefore the underlying mechanics will only briefly described here. Both approaches depend on the existence of a genetic marker map for the species under study. The marker map can be generated either by assigning marker positions based on their co-segregation across generations in a pedigree, where individuals have been genotyped (Slate 2005), or by obtaining exact physical positions from whole genome sequencing (Ellegren *et al.* 2012). The main distinction between linkage mapping and GWAS is that the former examines the co-segregation of marker alleles and quantitative trait locus (QTL) alleles within the genotyped individuals and therefore exploits recent recombination, whilst the latter detects statistical associations between marker alleles and QTL alleles that are sufficiently closely linked for ancestral recombination events to have not broken them down. As a result, linkage mapping tends to have lower precision for mapping QTL but is less prone to type 1 error. A GWAS requires much higher marker densities, but ultimately will map QTL with greater precision.

In humans, various linkage mapping approaches have been very successful in identifying disease genes for many Mendelian (and to a lesser extent, more complex) disorders (Altshuler *et al.* 2008). Furthermore, there have been numerous studies of humans, crop plants, livestock and model organisms where QTL underlying quantitative traits have been mapped using either linkage mapping or GWAS (Altshuler *et al.* 2008; Barton & Keightley 2002; Flint & Eskin 2012; Goddard & Hayes 2009). However, because QTL mapping in natural populations is more challenging than in humans, livestock and model organisms (Slate 2005; Slate *et al.* 2010), there are still very few descriptions of the genetic architecture of quantitative fitness-related traits in wild vertebrates (see Table 13.1). Furthermore, in most of these studies, the power to detect QTL was low due to low-density linkage maps and relatively few genotyped individuals (see below). The advent of relatively affordable next-generation genomic techniques will however allow detection and individual genotyping of a much larger number of SNPs, which will substantially increase the power in QTL studies in pedigreed natural populations (Van Bers *et al.* 2010; 2012). More importantly, higher marker densities open the way for GWASs, which are far less restrictive than linkage mapping in terms of the requirement of a pedigreed population.

13.2.2 Lessons from linkage mapping: are quantitative traits polygenic?

One of the principal aims of QTL mapping in wild populations is to understand the genetic architecture of quantitative traits; that is, the number and effect size of loci that contribute to genetic variation. Classical quantitative genetics assumes that continuous traits are highly polygenic with (infinitely) many loci of small effect contributing to trait variation (Falconer & Mackay 1996; Lynch & Walsh

Table 13.1 Wild or recently domesticated lab populations in which variance component QTL mapping has been attempted

Organism	Trait	Sample size	No. QTL detected‡	QTL effect size†	Reference
Red deer	Birth weight	295	1 suggestive	29	(Slate *et al.* 2002)
Soay sheep	Morphological traits + birth date	396–526	2 suggestive, 1 significant	21–36	(Beraldi *et al.* 2007b)
Soay sheep	Horn traits	217	1 significant	34–40	(Johnston *et al.* 2010)
Soay sheep	Parasite burdens	155–310	2 suggestive	Not reported	(Beraldi *et al.* 2007a)
Bighorn sheep	Horn traits	86–242	4 suggestive	33–73	(Poissant *et al.* 2012)
Great reed warblers	Wing length	333	1 significant	36	(Tarka *et al.* 2010)
Zebra finches	Beak colour	1019	4 suggestive	7–11	(Schielzeth *et al.* 2012b)
Zebra finches	Wing length	1066	5 suggestive, 1 significant	5–8	(Schielzeth *et al.* 2012a)
Zebra finches	Beak morphology	992	10 suggestive, 2 significant	5–12	(Knief *et al.* 2012)

† Expressed as percentage of phenotypic variance explained.
‡ Suggestive threshold expected once by chance in a genome scan; significant threshold expected with probability 0.05 in genome scan.

1998). If this is true, then perhaps QTL mapping studies will not detect any QTL, or at least not any real ones (i.e. significant peaks could simply represent type 1 errors). However, the aim of understanding the genetic architecture of a trait is not simply to test whether a long-held assumption is correct. The number and magnitude of individual loci can play an important role in determining how a trait will respond to directional selection (traits with large effect loci can evolve faster) and can give insight into how selection has acted on that trait (the existence of large effect loci may imply some form of balancing selection helping to maintain genetic variation; Orr 2005).

Genetic mapping and GWASs of humans, domestic and model organisms strongly support the idea that most quantitative traits are highly polygenic (Flint & Mackay 2009; Yang *et al.* 2011). It is thus surprising that for QTL mapping in natural vertebrate populations every study conducted to date has identified QTL that individually explain a large proportion of phenotypic variation (Table 13.1). However, besides the fact that only five species have been studied so far, this conclusion is almost certainly an artefact of relatively small sample sizes being used in these studies. It is well known that QTL effect sizes can be overestimated, especially when sample sizes are modest (say, less than 500 individuals). This bias is known as the Beavis Effect (Beavis 1994), and arises because QTL mapping typically involves stringent corrections for multiple testing. Due to the correlation between statistical significance and estimated effect size, it tends to be only those regions where effect sizes are (grossly) overestimated that reach statistical significance, whilst those regions where effect sizes are underestimated do not. In QTL mapping crosses it has been estimated that 500 individuals are sufficient for the Beavis Effect to be minimal. However, all QTL mapping studies conducted in wild populations have used a different statistical approach to that simulated by Beavis. In crosses, marker alleles are treated as fixed effects in a linear regression-based framework. In wild populations, complex multigenerational pedigrees are the norm, and a mixed-effects model that is an extension of the animal model has been used to map QTL (George *et al.* 2000; Slate 2005; Slate *et al.* 2010). Simulations

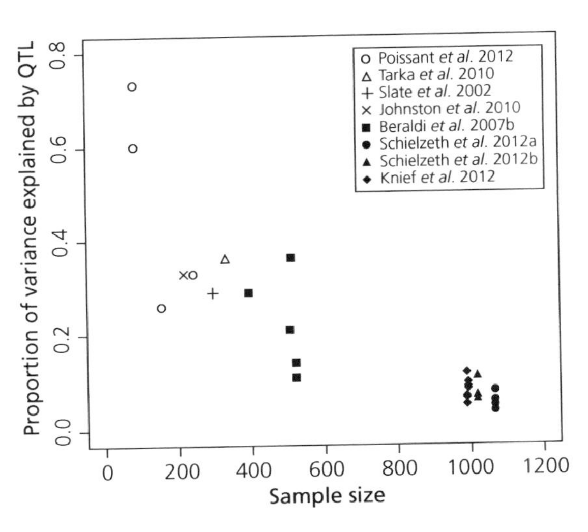

Figure 13.1 The proportion of variance explained by detected QTL as a function of sample size in linkage mapping studies of wild populations.

show that in the mixed-effects framework QTL of modest effect size (~3% of trait variation), are seriously overestimated when statistically significant, even in mapping populations of around 1000 individuals (Slate 2013). Furthermore, among the studies listed in Table 13.1, there is a very strong negative relationship between sample size and estimated QTL effect size (Figure 13.1); a pattern expected if the Beavis Effect is causing QTL effect sizes to be overestimated. In conclusion, it is too early to say whether quantitative traits in natural populations show departures from a polygenic model, but it is probable that there has been a serious problem of QTL effect size overestimation. Addressing this question requires replication of significant QTL in independent datasets, something which has seldom been achieved. One notable exception, where replication has been performed, seems to be horn size in a feral population of an ancient breed of sheep, Soay sheep (*Ovis aries*). Genetic variation is largely determined by variation in the RXFP2 gene, which not only affects horn type (presence or absence of horns) in both Soay (Johnston *et al.* 2011) and domestic sheep (Dominik *et al.* 2012; Kijas *et al.* 2012), but also explains most of the variation in horn length among Soay males with normal-shaped horns (Johnston *et al.* 2010, 2011).

13.2.3 Pedigree-free quantitative genetics

Given the difficulties in reliably estimating QTL effect size, one could question the usefulness of collecting high-density molecular marker data in natural populations. In fact, SNP data can be used to perform complementary analyses that, when conducted in tandem with QTL mapping, can provide a more complete picture of a trait's genetic architecture. In particular, if marker densities are sufficiently high, SNP-based relatedness estimators can be used to accurately perform quantitative genetic analyses, even in the absence of a pedigree, and to identify regions of the genome that have undergone selection (Section 13.3). The idea of using markers, in the absence of a pedigree, to estimate relatedness and then perform quantitative genetic analyses is not new (Ritland 1996; reviewed in Ritland 2000; Visscher *et al.* 2008). However, early attempts to perform this type of analysis were hampered by low marker densities which made accurate measurement of relatedness impossible. The relative ease with which large numbers (by which we mean ~5000 – ~500000) of markers can now be typed, has made 'molecular quantitative genetic' studies possible; in some cases these may be advantageous compared to traditional quantitative genetic approaches. Furthermore, tremendous progress in the field has been made by human geneticists and animal breeders, and there is now a strong framework and a large number of analytical tools that evolutionary geneticists can take advantage of. This potential should increase the number of natural systems in which quantitative genetics are performed, because it may now be possible to sample and measure a single cohort of individuals rather than rely on many generations of life history data.

13.2.3.1 Marker-based relationship matrices

High-density genetic data generated by next-generation genomics methods can assist and advance quantitative genetic analyses by enabling the estimation of $\mathbf{G}$ and other quantitative genetic parameters with higher accuracy and less bias than pedigree-based methods. The resemblance between relatives depends on the number of alleles they share that are identical by descent (IBD) at loci influencing the trait of interest, and this can be calculated based on the expected probability defined by their pedigree relationships (Lynch & Walsh 1998). The main distinction between pedigree and marker-based approaches to quantitative genetics is that the pedigree approach estimates the proportion of the genome IBD that two individuals are *expected* to share whereas the latter estimates the *actual* proportion of the genome that is IBD. The actual genetic relationship between all types of relatives, except monozygotic twins and parent–offspring pairs, shows stochastic variation around the expected value due to linkage and Mendelian sampling during gamete formation (Hill & Weir 2011; Visscher *et al.* 2006; see Figure 13.2). In other words, because genes are located on a relatively small number of chromosomes and recombination events are relatively rare (*c.* 1–3 per 100 centimorgan (cM)), large chromosomal segments will segregate from parents to progeny and hence generate variation around the expected value in the proportion of the genome which is shared between pairs of relatives (Visscher 2009). By using large amounts of individual genetic information (such as high-density SNP data) estimates of the actual or realised relationship matrix between the individuals in a pedigree may be more accurate than if the pedigree was used (Oliehoek *et al.* 2006; Visscher *et al.* 2006). Presumably, pedigree-based methods outperform low-density marker sets, hence the rather limited uptake of the Ritland (1996) approach at the time. The realised relationship matrix accounts for the variation among relatives in their actual relationship across the genome. When the actual IBD between all pairs of relatives is known, this information can be used to estimate additive genetic variance with higher accuracy and less bias, as well as facilitate estimation of genetic variance due to dominance and epistatic genetic effects (Hill & Weir 2011; Visscher *et al.* 2006; Visscher 2009). For example, a regression of the phenotypic similarity of full sibs on the proportion of the genome they actually share will provide estimates of additive and dominance variances for which environmental (i.e. non-genetic) covariances between family members have been removed. Such studies have recently been performed on height in humans, where genome-wide data on >400 microsatellite

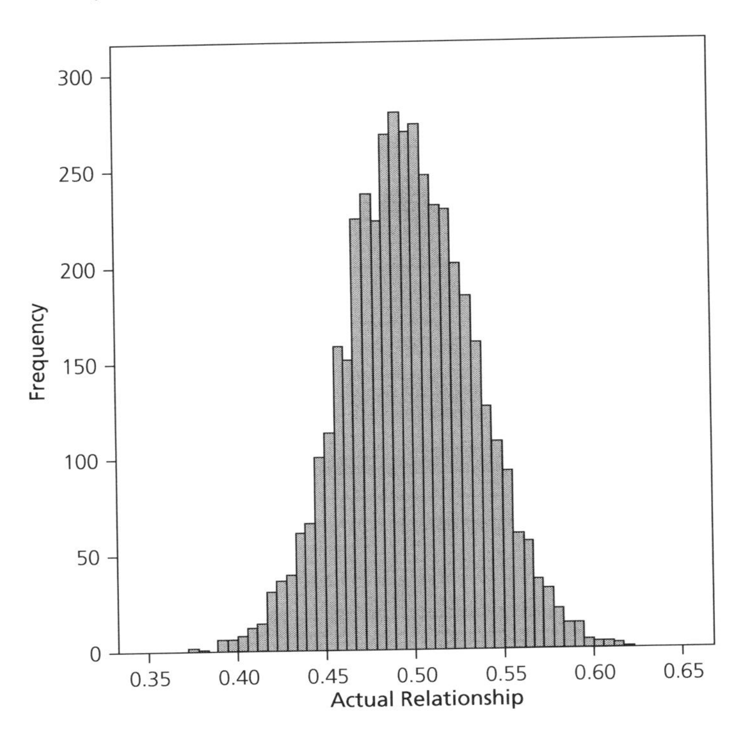

Figure 13.2 Histogram showing the actual or realised genome-wide additive genetic relationships of 4401 full-sib pairs (the expected relationship is equal to 0.5) estimated from genetic markers. Figure from Visscher *et al*. (2006).

markers typed in 11200 sibling pairs were used to demonstrate that there was no evidence for either dominance variance or between-chromosomal epistatic effects, and hence that phenotypic variation in this trait can be explained by additive effects of many loci spread across all autosomes (Visscher *et al*. 2006, 2007). Furthermore, in a study of approximately 2200 mice typed on 11700 SNPs, Lee *et al*. (2010) used the realised relationships between individuals in an animal model to obtain more accurate and less biased estimates of genetic and non-genetic components of variation compared to estimates based on the pedigree information only.

When traditional quantitative genetics methods are applied to wild populations, it is sometimes a problem that close relatives share a common environment, and partitioning out genetic and environmental variance components can be difficult. Marker-based approaches may be able to provide a solution to the problem. With high marker densities it is possible to estimate realised IBD coefficients between very distant pairs of relatives. Provided there is some variation in IBD coefficients (and in nearly all populations there will be), heritabilities can be estimated by examining how much phenotypic similarity is explained by an IBD matrix (Powell *et al*. 2010; Visscher *et al*. 2008; Yang *et al*. 2010). If distant relatives are used in such an analysis there is little, if any, covariance between environmental and genetic similarity, and so heritability estimates should be unbiased. Therefore, quantitative genetic studies should be possible in large populations of individuals who are not closely related and whose relationship to each other is unknown.

13.2.3.2 Exploration of genetic architecture without the need to map QTL

In addition to providing estimates of additive genetic variance and heritability, molecular quantitative genetics can be used to explore the genetic architecture of a trait. For example, if IBD or identical-by-state (IBS) matrices are estimated from all of the SNPs on a particular chromosome, then the contribution of that chromosome to phenotypic variation can be estimated, either by fitting IBD matrices for each chromosome simultaneously in one model, or by building a series of models with two IBD matrices, one based on the chromosome

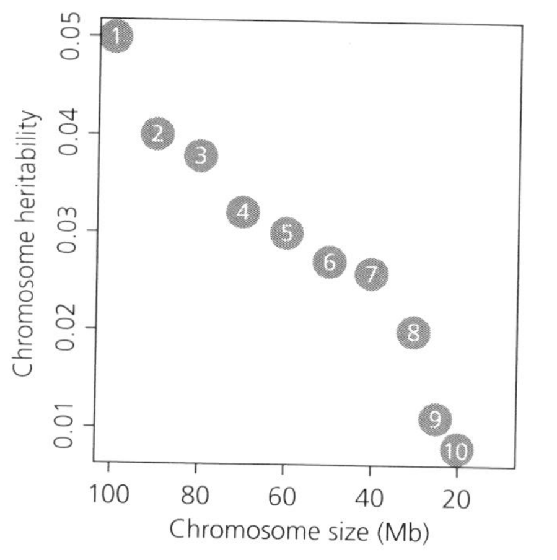

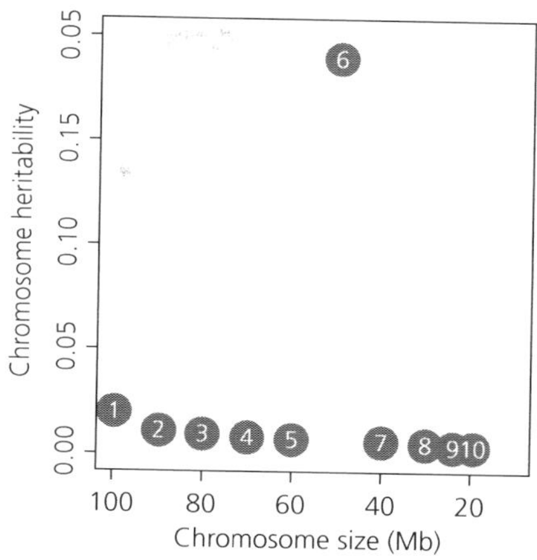

Figure 13.3 Idealised results of chromosome partitioning analysis on a polygenic trait (left panel) and a trait largely explained by a single QTL (right panel). Ten chromosomes of variable size (measured in Mb) are shown, with the proportion of phenotypic variation explained by each chromosome on the y-axis. Under a polygenic model, each chromosome will contribute to trait variation in proportion to its size. Under the major gene model, a QTL on chromosome 6 explains most of the variation and the remaining chromosomes contribute relatively little.

of interest and the other with all other SNPs used to estimate IBD (Yang *et al.* 2011; Robinson *et al.* 2013). If a trait is highly polygenic, then the proportion of variance explained by each chromosome should be proportional to the size of the chromosome (measured in million base pairs or as the number of genes), whereas a trait with a simpler genetic basis may be characterised by one or a few chromosomes, not necessarily the largest, contributing most of the variation (see Figure 13.3). Therefore, genomic partitioning i) provides an estimate of the maximal amount of additive genetic variance attributable to a given chromosome; and ii) through the strength of the regression of effect size on chromosome length, provides an indication of the degree to which the trait is polygenic. Thus, genomic partitioning provides a clear complement to QTL mapping; if chromosomes that contain suggestive or significant QTL of large effect also explain most of the variance in a chromosome partitioning analysis, the evidence supporting a simple genetic basis becomes much stronger. Conversely, if genomic partitioning supports a polygenic model, then the apparent QTL could be a sign of a false-positive peak or a true QTL with seriously inflated effect sizes.

This approach of partitioning genetic variation across chromosomes has been successfully applied to various complex traits in humans (Yang *et al.* 2011) using large samples of unrelated individuals. In contrast, most wild populations where phenotypic data have been collected contain close relatives as well as unrelated individuals. In theory, marker-based heritabilities can be estimated in populations with or without close relatives (Hayes *et al.* 2009; Robinson *et al.* 2013; VanRaden 2008; Yang *et al.* 2010). In all scenarios, the typed SNPs will not be in perfect linkage disequilibrium (LD) with all of the causal variants (which are unknown) (Powell *et al.* 2010; Yang *et al.* 2010), meaning that they will not provide exact IBS relatedness estimates, and thus that the proportion of variance explained by the SNPs is likely to be less than the 'true' heritability. A shrinkage estimator can be used to minimise the error around the marker-relatedness estimates, yielding more accurate heritability estimates that are 'unbiased' of the LD among markers (Yang *et al.* 2010; Powell *et al.* 2010; Goddard *et al.* 2011; Robinson *et al.* 2013). The method used to 'shrink' the variance of the marker-based relatedness depends upon the relatedness structure of the population, but empirical studies have shown that these methods provide increased accuracy in populations of 'unrelated' individuals (Yang *et al.* 2010; Powell *et al.* 2010); for half-sibling livestock populations (Goddard *et al.* 2011); and importantly for wild populations with complex patterns of relatedness with/without pedigrees (Robinson *et al.* 2013). In populations of 'unrelated' individuals, SNP estimates of heritability will still be less than the 'true' heritability ('missing heritability') as the SNP markers are unlikely to tag all of the causal variants (Powell *et al.* 2010; Yang *et al.* 2010). However, in populations containing close relatives, if SNP estimates are used as a substitute for the pedigree then they will provide an accurate estimate of the 'true' heritability and thus accurate partitioning of this variance across genomic regions (Robinson *et al.* 2013).

An extension of the chromosome partitioning approach, that has yet to be applied to wild populations, is to further partition additive genetic variation into different segments of a chromosome. For example, by estimating IBD matrices from blocks of consecutive markers, it would then be possible to estimate the amount of variance explained by each block. In the animal breeding literature, the approach has been termed 'regional heritability' mapping (Nagamine *et al.* 2012; Riggio *et al.* 2013). It would be of interest to ask the question whether chromosomal blocks that are gene rich tend to explain more variance than regions with a lower density of genes; under a polygenic model, a positive relationship between local gene density and proportion of variance explained would be predicted.

13.2.4 Genomic prediction

Genomic prediction was originally developed in animal breeding (Goddard & Hayes 2009). A sample of individuals with records of phenotypic values is genotyped on a genome-wide set of dense markers (called the training population). The relationship between the genome-wide marker data and phenotypic values is then used to derive a prediction equation that predicts individual genomic breeding values (GEBVs) for any individual (regardless of whether it is phenotyped) based on their marker genotypes (Goddard & Hayes 2009; Meuwissen *et al.* 2001). Knowing and using information on the relationship between the individuals in the training population is strictly speaking, not necessary, but seems advantageous (Lee *et al.* 2010). For example, the combination of pedigree information and dense SNP data on individuals in a pedigree can be used in a single-step procedure that eliminates several assumptions and parameters and hence provides more accurate GEBVs (Aguilar *et al.* 2010). When the prediction equation has been obtained, it can be applied to predict the GEBV of other individuals without a phenotype (Goddard & Hayes 2009).

Using GEBVs has recently been shown to be very advantageous in animal breeding (Goddard & Hayes 2009). For example, recent studies in cattle (e.g. Hayes *et al.* 2010), mice (Lee *et al.* 2008) and chicken (Gonzales-Recio *et al.* 2009) have shown that this approach is a powerful tool in artificial selection (Goddard & Hayes 2009). Despite the rapid uptake of genomic prediction in animal breeding, we are unaware of it being applied to wild populations, perhaps because studies of wild populations usually have smaller sample sizes and may lack power. Simulations suggest that thousands of individuals may be necessary to attain an acceptable accuracy, but that fewer individuals are needed if the heritability of the trait, the density of markers, and the extent of LD in the population are high (i.e. effective population size is low). Importantly, the accuracy of genomic prediction also increases when individuals in the training population are related to the individuals for which GEBVs are to be estimated (Makowsky *et al.* 2011). Currently it is not known how well a genomic prediction equation obtained in one population will predict phenotypes in another population because patterns of LD (see Section 13.3) between markers and causal loci may change across populations.

If the challenges associated with genomic prediction in wild species can be overcome, then it has a number of attractive properties. First, studies in animal and plant breeding have demonstrated that GEBVs are clearly superior in terms of predictive ability to breeding values estimated solely from pedigree data (Makowsky *et al.* 2011). Second, GEBVs may complement traditional pedigree-based analyses when one wishes to account for non-randomly missing data to obtain estimates of quantitative genetic parameters unbiased by the missing observations, or when the aim is to estimate the direction and strength of selection generating the missing observations (Hadfield 2008; Steinsland *et al.* 2014). Furthermore, because all markers are jointly modelled, GEBVs account for population structure, admixture, actual relatedness between relatives, and relatedness between presumed unrelated individuals better than pedigree-based breeding values do. Consequently, GEBVs estimated by genomic prediction have the potential to, for instance, track genetic change at the genome level across space or time whilst at the same time avoiding some of the problems associated with pedigree-based breeding values (Hadfield *et al.* 2010; Kruuk *et al.* 2008; other chapters in book).

13.3 Population genetics of natural populations

Of key interest to evolutionary geneticists studying wild animal populations is to understand how genetic data can be merged with long-term, individual-based field studies, ultimately bridging i) signatures of selection in genetic data detected with population genetics statistics, with ii) evidence of selection at loci explaining variation in individual fitness. One of the most anticipated opportunities opened up by high-throughput genomics is the opportunity to identify which parts of the genome are, or have been in the past, responding to selection. Importantly, different types of analyses detect selection that has operated on different evolutionary timescales (Table 13.2). In this section, we outline how quantitative genetics approaches (e.g. QTL mapping, and the estimation of heritabilities and breeding values) can complement high-throughput genotyping and whole genome sequencing-based population genetic approaches, which tend to focus on the frequencies and mean phenotypic values of different alleles at individual loci.

13.3.1 GWAS and LD

The use of GWASs is an approach used to identify QTL that has largely replaced linkage mapping in studies of humans, model organisms and agricultural organisms. A GWAS relies on the fact that recombination breaks down associations (known as LD) between linked sites. Therefore, only those SNPs physically close to causal loci will be associated with trait variation. Consequently, a high density of markers must be typed to ensure QTL are detected, something that was impractical in wild populations until recently. A frequently used graphical method to visualise a GWAS is to create a Manhattan Plot, where genomic coordinates (markers usually grouped by chromosome location) are displayed along the x-axis, and their statistical association with variation at a given trait is presented on the y-axis (see for example, Figure 13.4).

Unlike linkage mapping in wild populations, in which QTL effects are estimated as variance components, a GWAS fits each SNP as a fixed effect. This makes a fundamental difference to how results are interpreted, because the mean and frequency of each allele (or genotype) at the QTL is estimated. As a result it becomes possible to determine the relative contributions of additive and dominance genetic components at the locus. It also becomes possible to measure the relative fitness of each genotype (if fitness data are available) and to examine whether allele frequencies are temporally stable or not. Another advantage of GWAS over linkage mapping is that pedigree information is not required, and so populations where

Table 13.2 Selective forces leave different signatures in the genetic data, and are dependent on the evolutionary time scale considered. Statistical methods used to detect selection are developed accordingly

Difference between:	Species	Populations	Individuals
Time frame (in number of generations):	Millions, hundreds of thousands	Thousands, hundreds	Tens (long-term field study duration)
Statistic:	dN/dS ratios	Neutrality tests, genome scans, GWASs, QTLs	GWASs, Microevolution studies, QTLs
Reference:	(Yang & Bielawski 2000)	(Mackay 2001; Nielsen 2005)	(Barrett & Hoekstra 2011; Slate *et al.* 2010)

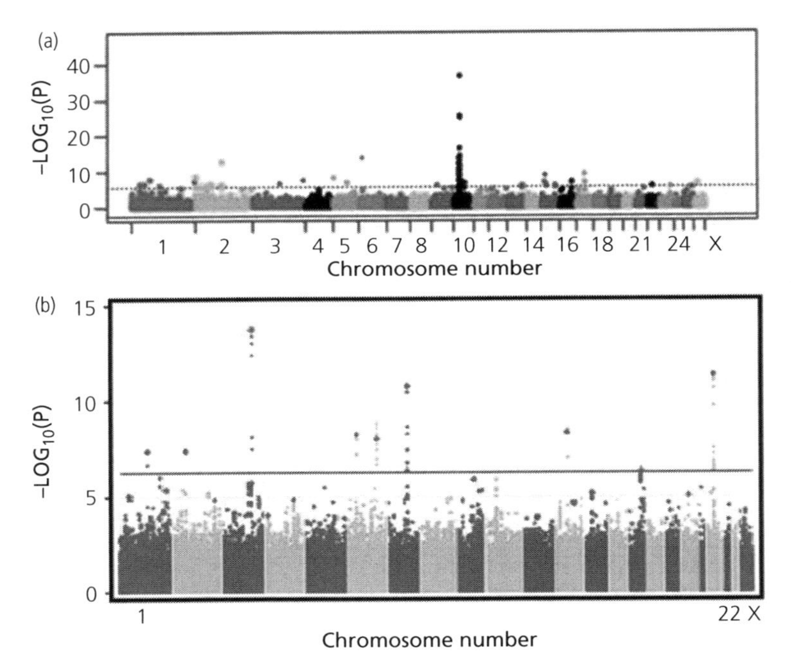

Figure 13.4 A Manhattan plot illustrating genome-wide associations between SNP markers (chromosomes coded by different shades of grey) and: (a) presence or absence of horn in a Soay sheep population. A strong association is particularly visible between horn type and a genomic region in chromosome 10. *p*-values were estimated using a chi-square statistic, and dotted lines indicate the significance threshold at $p = 1.859 \times 10^{-6}$, equivalent to an experiment-wide threshold of $p = 0.05$. An often more typical result for a QTL signal visualised on a Manhattan plot is to have several peaks of outlier SNPs, as seen in (b), where the influence of 400 000 SNPs was assessed against human adult height. Twenty SNP variants were associated with adult height at $p = 5 \times 10^{-7}$, and explained ~3% of height variation. Panel (a): From Johnston *et al*. (2011), modified. Panel (b): From Weedon *et al*. (2008), modified. Reproduced with permission from Nature Publishing Group.

unrelated individuals (or individuals of unknown relatedness) have been sampled can still be studied. The R-package GenABEL (Aulchenko *et al.* 2007) is particularly well suited to performing robust GWAS in wild populations even when samples have genetically structured data due to sampling different (sub)populations or because of the presence of related individuals. GWAS in wild populations remain relatively rare (Johnston *et al.* 2011; Santure *et al.* 2013), but this situation is surely set to change.

Attempts to perform a GWAS will only work if typed markers are in LD with (unknown) causal variants, and so it is important to know how far LD extends in the genome. LD is an important concept in population genetics and is defined as the non-random association of alleles at two or more loci (Hedrick 2005). That two loci are in LD does not mean that they need be physically linked. A lack of LD means that alleles at the two loci are not associated, either because they are located on different chromosomes or because recombination events are so frequent between the two loci that they appear to be unlinked. LD decreases at a rate that depends on the recombination frequency between the two loci, meaning that when other population genetic forces are not operating LD decreases by a factor of one half each generation for unlinked loci, and it decreases more slowly for closely linked loci (Hedrick 2005). Extensive LD will facilitate detection of associations between markers and genes underlying phenotypic traits or detection of signatures of selection. However, extensive LD also means that the physical distance between marker and gene may be considerable, making identification of the actual gene difficult. Low levels of LD will, on the other hand, complicate detection of an association (greater marker density will be required), but allow finer mapping of the location of a QTL or locus of selection when an association has been detected (Slate 2005). Consequently, knowing the level of LD across the genome will aid in the planning of GWAS studies because it will determine the number and density of markers necessary to detect causal variants of a given effect size at a given statistical power. Some ways to quantify and measure LD are described in Hedrick (2005) and Slatkin (2008).

The level of LD across the genome reflects a population's demographic history, and may affect its rate of response to selection (Hedrick 2005; Slatkin 2008). The genome-wide level of LD is affected by reductions in effective population size, population subdivision, non-random mating (e.g. inbreeding), which all are processes that result in increased LD across the genome. LD within particular genomic regions is on the other hand shaped by evolutionary forces such as selection and mutation, which both act to increase LD at loci close to the locus with the selectively advantageous or mutant allele. Thus local patterns of LD are informative with regard to identifying regions of the genome under selection (see Section 13.3.2).

In humans it has been found that LD varies considerably across populations and across different parts of the genome. Loci are organised in haplotype blocks with strong internal LD but weak LD with loci outside the haplotype block (Reich *et al.* 2001; Wall & Pritchard 2003). The haplotype blocks range in size from a few kilobases (kb) to more than 100 kb, which is very different from the theoretical prediction that little LD was expected beyond 3 kb (Wall & Pritchard 2003). LD in model organisms such as mouse and rat, and in domesticated animals like sheep, cattle and dogs generally extends much longer than in humans, probably as a result of the breeding history of these animals (Lindblad-Toh *et al.* 2005; Slatkin 2008). As a consequence, relatively fewer markers are needed to achieve the same statistical power when GWAS studies are carried out in these animals than in humans.

Studies of LD in wild vertebrate species are still quite rare (especially using SNPs), but patterns of LD seem to correspond with expectations from the demographic histories of the populations of the species that have been studied. Accordingly, LD in wild/commensal house mice (*Mus musculus*) decreased considerably faster than in laboratory strains (Laurie *et al.* 2007), and LD in species such as the pied flycatcher (*Ficedula hypoleuca*) and the collared flycatcher (*Ficedula albicollis*) (Backström *et al.* 2006; Ellegren *et al.* 2012), that are characterised by large population sizes, have very low LD levels and fast decline in LD across their genomes. On the other hand, studies of bighorn sheep (*Ovis canadensis*) and Soay sheep (*Ovis aries*) documented extensive LD similar to levels found in domestic animals, which was attributed to the populations being small

Box 13.1 Is there a missing heritability problem?

It has sometimes been suggested that expensive genome scans in humans have failed, because for traits that are known to be highly heritable (e.g. height), QTL mapped by linkage analysis or GWASs only explain a modest proportion of the additive genetic variation; for a review, see Manolio *et al.* (2009). The unexplained variation is frequently termed the 'missing heritability'. One proposed explanation for missing heritability is that epigenetic effects or other types of non-genetic inheritance are causing relatives to resemble each other, but are not detected in GWAS studies (Danchin *et al.* 2011). However, the problem with this explanation is that if epigenetic modifications contribute to a trait's heritability, they have to persist for a reasonably long number of generations and therefore are likely to be in LD with typed SNPs. That being the case, epigenetic effects will be detected in genome scans and contribute to the 'discovered' part of the heritability (Slatkin 2009). Therefore, more likely explanations for the missing heritability are that i) many causal variants have effect sizes too small to be detected in even the largest GWAS experiments and ii) typed markers are in imperfect LD with causal variants, possibly because they have different distributions of minor allele frequencies; so GWASs or molecular quantitative genetic approaches routinely underestimate the effects of causal variants detected by typed SNPs (Yang *et al.* 2010). Given this explanation, it seems premature to suggest that missing heritabilities are strong evidence of epigenetic effects contributing to a large proportion of phenotypic variation.

Papers by Yang *et al.* (2010, 2011) appear to have resolved the missing heritability problem. By developing a method based on genomic prediction, the authors showed that hundreds of thousands of SNPs jointly explain more than 50% of the phenotypic variation in human height. This is in contrast to results from large-scale GWASs that have identified approximately 180 QTL for height that together only explain approximately 10% of the observed variance (Lango Allen *et al.* 2010; Park *et al.* 2010), and suggests that variation in human height is due to many genes of small effect (see Section 13.2.2).

with polygynous mating systems (Miller *et al.* 2011; Kijas *et al.* 2013). Furthermore, a study of wild and domestic canids suggested a close correspondence with the demographic history of the different species and breeds of dogs and the extent of LD; in outbred grey wolf (*Canis lupus*) and coyote (*Canis latrans*) populations LD extended short distances (a few kb), whereas in dog breeds and wolf populations characterised by founder events and bottlenecks it extended for up to several mega base pairs (Mb; Gray *et al.* 2009). Similarly, Li and Merilä (2011) studied LD in three populations of Siberian jay (*Perisoreus infaustus*) and found a higher level of LD and slower reduction of LD with distance in an inbred population with strong population substructure compared to two outbred populations. The above studies clearly illustrate that careful examination of patterns of LD within the species and even the particular population of interest is needed. Understanding the causes of these patterns is important because it will enable sensible design of GWAS experiments, and will be of help in the interpretation of results from analyses of genetic signatures of selection (see Section 13.3.2).

13.3.2 Detecting genomic regions under selection

In this section, we briefly outline key features and limitations of analyses of signatures of selection in genetic data, and discuss how, when combined with relevant phenotypic information, genotype–phenotype associations can be used to infer patterns of individual variation in fitness. We start by examining how to detect selection that happened many (typically hundreds or thousands) generations ago and finish by discussing methods that can detect loci that have responded to selection in the last few generations (i.e. contemporary microevolution within the duration of a long-term study). The former approaches do not utilise phenotypic data, whilst the latter require individual fitness measurements.

13.3.2.1 Patterns of past selection

Different types of selection can be detected from DNA sequence data: positive selection (selection acting on advantageous mutations), negative (or purifying) selection (any type of selection acting against deleterious mutations) and balancing selection (selection that helps to maintain variability within a population, including frequency-dependent selection, overdominance and spatially/temporally fluctuating directional selection). Consequently, they generate contrasting patterns of genetic variability (Nielsen 2005), and some forms of selection (i.e. directional) are more easily detected than others (i.e. balancing selection) (Beaumont 2005). To detect signatures of selection in genetic data, statistical methods are developed in the form of neutrality tests (such as Tajima's *D* test), which usually take into account allelic frequency spectra (the allelic sample distribution in independent nucleotide sites) to identify changes in allelic frequencies relative to a null hypothesis of neutral evolution; for detailed reviews see Nielsen (2005) and Stinchcombe and Hoekstra (2008). There are important limitations regarding the robustness of conclusions made from neutrality tests, as they make important population genetic assumptions that are often difficult to validate. Indeed, neutrality tests are defined by restrictive demographic models assuming no population structure (e.g. random mating) and constant population size (Barrett & Hoekstra 2011; Nielsen *et al.* 2005). Simultaneous analysis of multiple genomic loci made available from next-generation sequencing may improve the estimation of parameters that are common among loci and gain independence from restrictive demographic assumptions (Akey *et al.* 2002; Vitalis *et al.* 2001), thereby increasing power and robustness of neutrality tests. Whilst whole genome sequencing is the gold standard for reducing ascertainment bias, it remains a rarely achievable goal in studies of wild populations (but see Ellegren *et al.* 2012). Given the recent dramatic (and continuing) drop in costs of genomic sequencing, it can be expected that many other wild populations will become the focus of whole genome sequence analyses.

The continued fall in the cost of the sequencing is important, because it means genomes can be sequenced at greater depth. Low-depth genome coverage results in an ascertainment bias towards the detection of SNPs with intermediate allele frequencies (Helyar *et al.* 2011). This in turn results in estimated allele frequency distributions that can be mistaken for evidence of balancing selection (Albrechtsen *et al.* 2010; Nielsen *et al.* 2005; Thornton & Jensen 2007). In the meantime, new methods for mitigating ascertainment bias are being developed, both by taking advantage of refinements in high-throughput sequencing experimental designs, and by improving statistical frameworks (Nielsen *et al.* 2011).

13.3.2.2 Population genomics

Another method for detecting signatures of selection in genetic data is to analyse patterns of population subdivision using F_{ST} statistics, relying on the assumption that high levels of F_{ST} at particular loci ('outlier' loci) between populations are indicative of selection that has acted on these loci, or on loci in LD with them (Akey *et al.* 2002, Beaumont & Balding 2004); but note the statistical concerns for this method highlighted by Nielsen (2001) and Przeworski (2002).

Since large-scale, genome-wide marker data have become increasingly available, genome scans to detect signatures of selection by identifying outlier markers are readily applied to detect signatures of selection in a large array of taxa (Beaumont 2005). The focus is often on evidence for positive directional selection leading to adaptive divergence of the population under selection (e.g. Hohenlohe *et al.* 2010), although these results are often tempered by theoreticians arguing that empirical genomic data from natural populations often use analytical frameworks resulting in high rates of false-positive and false-negative results (Bierne *et al.* 2011; Nei *et al.* 2010). It has also been suggested that loci displaying signatures of selection may not actually be within (or linked to) genomic regions causing adaptation to different environments, but instead are loci that diverged for other reasons such as pre-zygotic isolation or epistatic interactions (Bierne *et al.* 2011). That said, they are still of interest as loci that have been under some form of selection.

13.3.2.3 Detecting micro-evolutionary changes

By combining long-term genetic and phenotypic data, it might be possible to establish the respective roles of genetic drift and directional selection in shaping micro-evolutionary changes at individual loci, such as those that are shown to explain significant variation in GWAS. A recent report by Gratten *et al.* (2012) used gene-dropping simulations to show that the gene responsible for a coat pattern polymorphism in Soay sheep, *ASIP*, has undergone micro-evolutionary changes over the last twenty years that are probably too extreme to be attributable to genetic drift. Unfortunately, studies focusing on changes in allele frequency at loci associated with trait and/or fitness variation are still scarce. In the handful of studies that have examined selection-driven micro-evolution at specific loci, patterns are often much more complex than originally expected (see for example how *TYRP1* affects coat colour in Soay sheep (Gratten *et al.* 2008), or the role of *Eda* in armour development of three-spined sticklebacks (*Gasterosteus aculeatus*; Barrett *et al.* 2008)). These fairly complex relationships could not have been unravelled without experimental testing and/or population genetic approaches using molecular data on one hand, and demonstrating genetic correlations between different fitness-related phenotypic traits on the other (Barrett & Hoekstra 2011).

13.3.2.4 Combining selection scans with phenotypic data

It will not have escaped quantitative geneticists' attention that genetic signatures of selection revealed by genome scans make no reference to phenotypic or fitness information in contemporary samples. It is thus impossible to determine whether selection is currently acting at these loci, or to identify the traits that selection has acted upon. Taken alone, genome scans for loci under selection reveal little about current patterns of heritable quantitative variation segregating in the population. Therefore, it is advisable to use scans for loci under selection as part of a multi-step analysis, where population genomic and phenotype-focused quantitative genetic approaches are combined (Stinchcombe & Hoekstra 2008). Whilst knowledge of the biological system, its demographic history and ecological context is essential, ideally, a combination of genetic, phenotypic, fitness and environmental data should be used (Barrett & Hoekstra 2011) to understand how variation in natural populations is maintained. One question that urgently needs addressing is whether the loci that explain quantitative variation (and variation in fitness) today are the same loci that show signatures of past selection. We have little, or no, idea of the answer.

13.3.3 Whole genome (re)sequencing

The cost of genome sequencing is dropping rapidly, and as shown by the recent collared flycatcher genome assembly (Ellegren *et al.* 2012), whole genome sequencing is now a viable approach in non-model species living in the wild. Given the rapid progress in the use of next-generation sequencing techniques in non-model species during the last decade (Slate *et al.* 2010), it is certain that whole genome sequences of many non-model species and on many individuals within populations of these species will be available in the near future. There are (at least) two reasons why whole genome sequences data are useful beyond information given by high-density SNP chips or similar genotyping approaches. First, sequence data are preferable to SNP data for detecting signatures of selection or outlier loci, because orders of magnitude more variable sites are sampled, reflecting the true distribution of allele frequency data throughout the genome. SNP chips tend to be biased towards typing loci with relatively high variability (see Section 13.3.2). Second, due to recent progress in bioinformatic imputation methods, if only a subsample of individuals from a population are sequenced across their whole genome, other individuals, typed at SNP chips, can still have their genome sequences imputed, using the sequenced individuals as reference (Goddard *et al.* 2011). This approach relies on there being strong LD between the typed SNPs and other variable sites in the genome. A large number of individuals with phenotype data and imputed genome sequence will then be a powerful resource to, for example, discover causal mutations underlying quantitative traits (Meuwissen & Goddard 2010). Furthermore, this will also increase accuracy and applicability of GEBVs, because prediction equations will be improved by some of the imputed variable sites being in strong LD with causal variants.

13.4 Conclusions

In this chapter, we have pointed out ways in which high-throughput genomics approaches can complement and improve more 'traditional' quantitative genetic analyses. Power to identify particular genes underlying fitness-related quantitative traits, and quantify their effect sizes, may be low even in studies where extensive pedigree and phenotypic data have been collected over many generations. Certainly, low power will be likely if the growing evidence that many phenotypic traits really are determined by many genes of small effects is correct; i.e. that the central assumption of quantitative genetics holds. Of course, it could be argued that the loci of largest effect, even if rare in nature, are of greatest interest, and will also be the easiest ones to detect. Regardless, by combining long-term field data with high-density genomic data, it is now possible to carry out GWASs, chromosome partitioning and regional heritability mapping, and to estimate GEBVs. In tandem, these approaches provide important knowledge of the genetic architecture and evolutionary dynamics of fitness-related traits in natural systems. Finally, recent studies show that whole genome sequencing is now possible in non-model species, and increasingly sophisticated methods to identify micro-evolutionary change due to recent selection or historical signatures of selection can be employed on long-term ecological study systems. We believe that the application of high-throughput genomics tools in quantitative genetic studies of non-model species in the wild shows great promise to increase our understanding of ecological and evolutionary processes in natural populations.

Acknowledgements

Our research is funded by generous support from the Research Council of Norway (HJ), a Marie Curie Fellowship (MS) and the European Research Council (JS). Anne Charmantier, Dany Garant, Loeske Kruuk, an anonymous reviewer and Matt Robinson all made constructive comments on previous versions or excerpts of this chapter.

References

Aguilar, I., Misztal, I., Johnson, D.L., Legarra, A., Tsuruta, S., Lawlor, T.J. (2010) Hot topic: a unified approach to utilize phenotypic, full pedigree, and genomic information for genetic evaluation of Holstein final score. *Journal of Dairy Science*, **93**, 743–752.

Akey, J.M., Zhang, G., Zhang, K., Jin, L., Shriver, M.D. (2002) Interrogating a high-density SNP map for signatures of natural selection. *Genome Research*, **12**, 1805–1814.

Albrechtsen, A., Nielsen, F.C., Nielsen, R. (2010) Ascertainment biases in SNP chips affect measures of population divergence. *Molecular Biology and Evolution*, **27**, 2534–2547.

Altshuler, D., Daly, M.J., Lander, E.S. (2008) Genetic mapping in human disease. *Science*, **322**, 881–888.

Aulchenko, Y.S., Ripke, S., Isaacs, A., van Duijn, C.M. (2007) GenABEL: an R library for genome-wide association analysis. *Bioinformatics*, **23**, 1294–1296.

Backström, N., Brandström, M., Gustafsson, L., Qvarnström, A., Cheng, H., Ellegren, H. (2006) Genetic mapping in a natural population of collared flycatchers (*Ficedula albicollis*): conserved synteny but gene order rearrangements on the Avian Z chromosome. *Genetics*, **174**, 377–386.

Baird, N.A., Etter, P.D., Atwood, T.S., Currey, M.C., Shiver, A.L., Sachary, A.L., Selker, E.U., Cresko, W.A., Johnson, E.A. (2008) Rapid SNP discovery and genetic mapping using sequenced RAD markers. *PLoS ONE*, **3**, e3376.

Barrett, R.D.H., Hoekstra, H.E. (2011) Molecular spandrels: tests of adaptation at the genetic level. *Nature Reviews Genetics*, **12**, 767–780.

Barrett, R.D.H., Rogers, S.M., Schluter, D. (2008) Natural selection on a major armor gene in threespine stickleback. *Science*, **322**, 255–257.

Barton, N.H., Keightley, P.D. (2002) Understanding quantitative genetic variation. *Nature Reviews Genetics*, **3**, 11–21.

Beaumont, M.A. (2005) Adaptation and speciation: what can F_{st} tell us? *Trends in Ecology & Evolution*, **20**, 435–440.

Beaumont, M.A., Balding, D.J. (2004) Identifying adaptive genetic divergence among populations from genome scans. *Molecular Ecology*, **13**, 969–980.

Beavis, W.D. (1994) The power and deceit of QTL experiments: lessons from comparative QTL studies. In: *Proceedings of the forty-ninth annual corn and sorghum industry research conference*. American Seed Trade Association, Washington, D.C.

Beraldi, D., McRae, A.F., Gratten, J., Pilkington, J.G., Slate, J., Visscher, P.M., Pemberton, J.M. (2007a) Quantitative trait loci (QTL) mapping of resistance to strongyles and coccidia in the free-living Soay sheep (Ovis aries). *International Journal for Parasitology*, **37**, 121–129.

Beraldi, D., McRae, A.F., Gratten, J., Slate, J., Visscher, P.M., Pemberton JM (2007b) Mapping QTL underlying fitness-related traits in a free-living sheep population. *Evolution*, **61**, 1403–1416.

Bierne, N., Welch, J., Loire, E., Bonhomme, F., David, P. (2011) The coupling hypothesis: why genome scans may fail to map local adaptation genes. *Molecular Ecology*, **20**, 2044–2072.

Danchin, E., Charmantier, A., Champagne, F.A., Mesoudi, A., Pujol, B., Blanchet, S. (2011) Beyond DNA: integrating inclusive inheritance into an extended theory of evolution. *Nature Reviews Genetics*, **12**, 475–486.

Dominik, S., Henshall, J.M., Hayes, B.J. (2012) A single nucleotide polymorphism on chromosome 10 is highly predictive for the polled phenotype in Australian Merino sheep. *Animal Genetics*, **43**, 468–470.

Ekblom, R., Galindo, J. (2011) Applications of next generation sequencing in molecular ecology of non-model organisms. *Heredity*, **107**, 1–15.

Ellegren, E., Smeds, L., Burri, R., Olason, P.I., Backström, N., Kawakami, T., Künstner, A., Mäkinen, H., Ndachowska-Brzyska, K., Qvarnström, A., Uebbing, S., Wolf, J.B.W. (2012) The genomic landscape of species divergence in Ficedula flycatchers. *Nature*, **491**, 756–760.

Ellegren, H., Sheldon, B.C. (2008) Genetic basis of fitness differences in natural populations. *Nature*, **452**, 169–175.

Erickson, D.L., Fenster, C.B., Stenøien, H., Price, D. (2004) Quantitative trait locus analyses and the study of evolutionary process. *Molecular Ecology* 13, 2505–2522.

Falconer, D.S., Mackay, T.F.C. (1996) *Introduction to quantitative genetics*, 4th ed. Longman Group Ltd, Harlow.

Flint, J., Eskin, E. (2012) Genome-wide association studies in mice. *Nature Reviews Genetics*, **13**, 807–817.

Flint, J., Mackay, T.F.C. (2009) Genetic architecture of quantitative traits in mice, flies, and humans. *Genome Research*, **19**, 723–733.

George, A.W., Visscher, P.M., Haley, C.S. (2000) Mapping quantitative trait loci in complex pedigrees: a two-step variance component approach. *Genetics*, **156**, 2081–2092.

Goddard, M.E., Hayes, B.C. (2009) Mapping genes for complex traits in domestic animals and their use in breeding programmes. *Nature Reviews Genetics*, **10**, 381–391.

Goddard, M.E., Hayes, B.C., Meuwissen T.H. (2011) Using the genomic relationship matrix to predict the accuracy of genomic selection. *Journal of Animal Breeding and Genetics*, **128**, 409–421.

Gonzalez-Recio, O., Gianola, D., Rosa, G.J.M., Weigel, K.A., Kranis, A. (2009) Genome-assisted prediction of a quantitative trait measured in parents and progeny:

application to food conversion rate in chicken. *Genetics Selection Evolution*, **43**, 3.

Gratten, J., Pilkington, J.G., Brown, E.A., Clutton-Brock, T.H., Pemberton, J.M., Slate, J. (2012) Selection and microevolution of coat pattern are cryptic in a wild population of sheep. *Molecular Ecology*, **21**, 2977–2990.

Gratten, J., Wilson, A.J., McRae, A.F., Beraldi, D., Visscher, P.M., Pemberton, J.M., Slate, J. (2008) A localized negative genetic correlation constrains microevolution of coat color in wild sheep. *Science*, **319**, 318–320.

Gray, M.M., Granka, J.M., Bustamante, C.D., Sutter, N.B., Boyko, A.R., Zhu, L., Ostrander, E.A., Wayne, R.K. (2009) Linkage disequlibrium and demographic history of wild and domestic canids. *Genetics*, **181**, 1493–1505.

Hadfield, J.D. (2008) Estimating evolutionary parameters when viability selection is operating. *Proceedings of the Royal Society B-Biological Sciences*, **275**, 723–734.

Hadfield, J.D., Wilson, A.J., Garant, D., Sheldon, B.C., Kruuk, L.E.B. (2010) The misue of BLUP in ecology and evolution. *American Naturalist*, **175**, 116–125.

Hayes, B.J., Pryce, J., Chamberlain, A.J., Bowman, P.J., Goddard, M.E. (2010) Genetic architecture of complex traits and accuracy of genomic prediction: coat colour, milk-fat percentage, and type in Holstein cattle as contrasting model traits. *PLoS Genetics*, **6**, e1001139.

Hayes, B.J., Visscher, P.M., Goddard, M.E. (2009) Increased accuracy of artificial selection by using the realized relationship matrix. *Genetics Research*, **91**, 47–60.

Hedrick, P.W. (2005) *Genetics of populations*, 3rd edn. Jones and Batlett Publishers, Sudburry.

Helyar, S.J., Hemmer-Hansen, J., Bekkevold, D., Taylor, M.I., Ogden, R., Limborg, M.T., Cariani, A., Maes, G.E., Diopere, E., Carvalho, G.R., Nielsen, E.E. (2011) Application of SNPs for population genetics of non-model organisms: new opportunities and challenges. *Molecular Ecology Resources*, **11**, 123–136.

Hill, W.G. (2012) Quantitative genetics in the genomics era. *Current Genomics*, **13**, 196–206.

Hill, W.G., Kirkpatrick, M. (2010) What animal breeding has taught us about evolution. *Annual Review of Ecology, Evolution and Systematics*, **41**, 1–19.

Hill, W.G., Weir, B.S. (2011) Variation in actual relationship as a consequence of Mendelian sampling and linkage. *Genetics Research*, **93**, 47–64.

Hohenlohe, P.A., Bassham, S., Etter, P.D., Stiffler, N., Johnson, E.A., Cresko, W.A. (2010) Population genomics of parallel adaptation in threespine stickleback using sequenced RAD tags. *Plos Genetics*, **6**, e1000862.

Johnston, S.E., Beraldi, D., McRae, A.F., Pemberton, J.M., Slate, J. (2010) Horn type and horn length genes map to the same chromosomal region in Soay sheep. *Heredity*, **104**, 196–205.

Johnston, S.E., McEwan, J.C., Pickering, N.K., Kijas, J.W., Beraldi, D., Pilkington, J.G., Pemberton, J.M., Slate, J. (2011) Genome-wide association mapping identifies the genetic basis of discrete and quantitative variation in sexual weaponry in a wild sheep population. *Molecular Ecology*, **20**, 2555–2566.

Kijas, J.W., Lenstra, J.A., Hayes, B., Boitard, S., Neto, L.R.P., San Cristobal, M., Servin, B., McCulloch, R., Whan, V., Gietzen, K., Paiva, S., Barendse, W., Ciani, E., Raadsma, H., McEvan, J., Dalrymple, B., Int Sheep Genomics Consortium (2012) Genome-wide analysis of the world's sheep breeds reveals high levels of historic mixture and strong recent selection. *PLoS Biology*, **10**, e1001258.

Knief, U., Schielzeth, H., Kempenaers, B., Ellegren, H., Forstmeier, W. (2012) QTL and quantitative genetic analysis of beak morphology reveals patterns of standing genetic variation in an Estrildid finch. *Molecular Ecology*, **21**, 3704–3717.

Kruuk, L.E.B., Slate, J., Wilson, A.J. (2008) New answers for old questions: the evolutionary quantitative genetics of wild animal populations. *Annual Review of Ecology Evolution and Systematics*, **39**, 525–548.

Lango Allen, H., Estrada, K., Lettre, G., Berndt, S.I., Weedon, M.N., Rivadeneira, F., Willer, C.J., Jackson, A.U., Vedantam, S., Raychaudhuri, S., Ferreira, T., Wood, A.R., Weyant, R.J., Segrè, A.V., Speliotes, E.K., Wheeler, E., Soranzo, N., Park, J.H., Yang, J., Gudbjartsson, D., Heard-Costa, N.L., Randall, J.C., Qi, L., Vernon Smith, A., Mägi, R., Pastinen, T., Liang, L., Heid, I.M., Luan, J., Thorleifsson, G., Winkler, T.W., Goddard, M.E., *et al.* (2010) Hundreds of variants clustered in genomic loci and biological pathways affect human height. *Nature*, **467**, 832–838.

Lank, S.M., Wiseman, R.W., Dudley, D.M., O'Connor, D.H. (2010) A novel single cDNA amplicon pyrosequencing method for high-throughput, cost-effective sequence-based HLA class I genotyping. *Human Immunulogy*, **71**, 1011–1017.

Laurie, C.C., Nickerson, D.A., Anderson, A.D., Weir, B.S., Livingston, R.J., Dean, M.D., Smith, K.L., Schadt, E.E., Nachman, M.W. (2007) Linkage disequilibrium in mice. *PLoS Genetics*, **3**, e144.

Lee, S.H., Goddard, M.E., Visscher, P.M., van der Werf, J.H.J. (2010) Using the realized relationshipo matrix to disentangle confounding factors for the estimation of genetic variance components of complex traits. *Genetics Selection Evolution*, **42**, 22.

Lee, S.H., van der Werf, J.H.J., Hayes, B.J., Goddard, M.E., Visscher, P.M. (2008) Predicting unobserved phenotypes for complex traits from whole-genome SNP data. *PLoS Genetics*, **4**, e1000231.

Li, M.H., Merilä, J. (2011) Population differences in levels of linkage disequilibrium in the wild. *Molecular Ecology*, **20**, 2916–2928.

Lindblad-Toh, K., Wade, C.M., Mikkelsen, T.S., Karlsson, E.K., Jaffe, D.B., Kamal, M., Clamp, M., Chang, J.L., Kulbokas, E.J. 3rd, Zody, M.C,.Mauceli, E., Xie, X., Breen, M., Wayne, R.K., Ostrander, E.A., Ponting, C.P., Galibert, F., Smith, D.R., DeJong, P.J., Kirkness, E., Alvarez, P., Biagi, T., Brockman, W., Butler, J., Chin, C.W., Cook, A., Cuff, J., Daly, M.J., DeCaprio, D., Gnerre, S., Grabherr, M., Kellis, M. (2005) Genome sequence, comparative analysis and haplotype structure in the domestic dog. *Nature*, **438**, 803–819.

Lynch, M., Walsh, B. (1998) *Genetics and analysis of quantitative traits*. Sinauer Associates, Inc., Sunderland.

Mackay, T.F.C. (2001) The genetic architecture of quantitative traits. *Annual Review of Genetics*, **35**, 303–339.

Mackay, T.F.C., Stone, E.A., Ayroles, J.F. (2009) The genetics of quantitative traits: challenges and prospects. *Nature Reviews Genetics*, **10**, 565–577.

Makowsky, R., Pajewski, N.M., Klimentidis, Y.C., Vazquez, A.I., Duarte, C.W., Allison, D.B., de los Campos, G. (2011) Beyond missing heritability: prediction of complex traits. *PLoS Genetics*, **7**, e1002051.

Manolio, T.A., Collins, F.S., Cox, N.J., Goldstein, D.B., Hindorff, L.A., Hunter, D.J., McCarthy, M.I., Ramos, E.M., Cardon, L.R., Chakravarti, A., Cho, J.H., Guttmacher, A.E., Kong, A., Kruglyak, L., Mardis, E., Rotimi, C.N., Slatkin, M., Valle, D., Whittemore, A.S., Boehnke, M., Clark, A.G., Eichler, E.E., Gibson, G., Haines, J.L., Mackay, T.F.C., McCarroll, S.A., Visscher, P.M. (2009) Finding the missing heritability of complex diseases. *Nature*, **461**, 747–753.

Meuwissen, T., Goddard, M, (2010) Accurate prediction of genetic values for complex traits by whole-genome resequencing. *Genetics*, **185**, 623–631.

Meuwissen, T.H., Hayes, B.J., Goddard, M.E. (2001) Prediction of total genetic value using genome-wide dense marker maps. *Genetics*, **157**, 1819–1829.

Miller, J.M., Poissant, J., Kijas, J.W., Coltman, D.W. (2011) A genome-wide set of SNPs detects population substructure and long range linkage disequilibrium in wild sheep. *Molecular Ecology Resources*, **11**, 314–322.

Miller, M.R., Dunham, J.P., Amores, A., Cresko, W.A., Johson, E.A. (2007) Rapid and cost-effective polymorphism identification and genotyping using restriction site associated DNA (RAD) markers. *Genome Research*, **17**, 240–248.

Nagamine, Y., Pong-Wong, R., Navarro, P., Vitart, V., Hayward, C., Rudan, I., Campbell, H., Wilson, J., Wild, S., Hicks, A.A., Pramstaller, P.P., Hastie, N., Wright, A.F., Haley, C.S. (2012) Localising loci underlying complex trait variation using regional genomic relationship mapping. *PLoS ONE*, **7**: e46501.

Nei, M., Suzuki, Y., Nozawa, M. (2010) The neutral theory of molecular evolution in the genomic era. In: *Annual review of genomics and human genetics, Vol 11* (ed. A. Chakravarti & E. Green), pp. 265–289. Annual Reviews, Palo Alto.

Nielsen, R. (2001) Statistical tests of selective neutrality in the age of genomics. *Heredity*, **86**, 641–647.

Nielsen, R. (2005) Molecular signatures of natural selection. In: *Annual Review of Genetics*, pp. 197–218. Annual Reviews, Palo Alto.

Nielsen, R., Paul, J.S., Albrechtsen, A., Song, Y.S. (2011) Genotype and SNP calling from next-generation sequencing data. *Nature Reviews Genetics*, **12**, 443–451.

Nielsen, R., Williamson, S., Kim, Y., Hubisz, M.J., Clark, A.G., Bustamante, C. (2005) Genomic scans for selective sweeps using SNP data. *Genome Research*, **15**, 1566–1575.

Oliehoek, P.A., Windig, J.J., van Aarendonk, J.A.M., Bijma, P. (2006) Estimating relatedness between individuals in general populations with focus on their use in conservation programs. *Genetics*, **173**, 483–496.

Orr, H.A. (2005) The genetic theory of adaptation: a brief history. *Nature Reviews Genetics*, **6**, 119–127.

Park, J.-H., Wacholder, S., Gail, M.H., Peters, U., Jacobs, K.B., Chanock, S.J., Chatterjee, N. (2010) Estimation of effect size distribution from genome-wide association studies and implications for future discoveries. *Nature Genetics*, **7**, 570–575.

Peterson, B.K., Weber, J.N., Kay, E.H., Fisher, H.S., Hoekstra, H.E. (2012) Double digest RADseq: an inexpensive method for *de novo* SNP discovery and genotyping in model and non-model species. *PLoS One*, **7**, e37135.

Poissant J., Davis, C.S., Malenfant, R.M., Hogg, J.T., Coltman, D.W. (2012) QTL mapping for sexually dimorphic fitness-related traits in wild bighorn sheep. *Heredity*, **108**, 256–263.

Powell, J.E., Visscher, P.M., Goddard, M.E. (2010) Reconciling the analysis of IBD and IBS in complex trait studies. *Nature Reviews Genetics*, **11**, 800–805.

Przeworski, M. (2002) The signature of positive selection at randomly chosen loci. *Genetics*, **160**, 1179–1189.

Reich, D.E., Cargill, M., Bolk, S., Ireland, J., Sabeti, P.C., Richter, D.J., Lavery, T., Kouyoumjian, R., Farhadian, S.F., Ward, R., Lander, E.S. (2001) Linkage disequilibrium in the human genome. *Nature*, **411**, 199–204.

Riggio, V., Matika, O., Pong-Wong, R., Stear, M.J., Bishop, S.C. (2013) Genome-wide association and regional heritability mapping to identify loci underlying variation in nematode resistance and body weight in Scottish Blackface lambs. *Heredity*, **110**, 420–429.

Ritland, K. (1996) A marker-based method for inferences about quantitative inheritance in natural populations. *Evolution*, **50**, 1062–1073.

Ritland, K. (2000) Marker-inferred relatedness as a tool for detecting heritability in nature. *Molecular Ecology*, **9**, 1195–1204.

Robinson, M., Santure, A.W., De Cauwer, I., Sheldon, B.C., Slate, J. (2013) Partitioning of genetic variation across the genome using multi-marker methods in a wild bird population. *Molecular Ecology*, **22**, 3963–3980.

Santure, A.W., De Cauwer, I., Robinson, M.R., Poissant, J., Sheldon, B.C., Slate, J. (2013) Genomic dissection of variation in clutch size and egg mass in a wild great tit (*Parus major*) population. *Molecular Ecology*, **22**, 3949–3962.

Schielzeth, H., Forstmeier, W., Kempenaers, B., Ellegren, H. (2012a) QTL linkage mapping of wing length in zebra finch using genome-wide single nucleotide polymorphisms markers. *Molecular Ecology*, **21**, 329–339.

Schielzeth, H., Kempenaers, B., Ellegren, H., Forstmeier, W. (2012b) QTL linkage mapping of zebra finch beak color shows an oligogenic control of a sexually selected trait. *Evolution*, **66**, 18–30.

Slate, J., Santure, A.W., Feulner, P.G.D., Brown, E.A., Ball, A.D., Johnston, S.E., Gratten, J. (2010) Genome mapping in intensively studied wild vertebrate populations. *Trends in Genetics*, **26**, 275–284.

Slate, J., Visscher, P.M., MacGregor, S., Stevens, D., Tate, M.L., Pemberton, J.M. (2002) A genome scan for quantitative trait loci in a wild population of red deer (*Cervus elaphus*). *Genetics*, **162**, 1863–1873.

Slate, J. (2005) QTL mapping in natural populations: progress, caveats and future directions. *Molecular Ecology*, **14**, 363–379.

Slate, J. (2013) From Beavis to beak colour: a simulation study to examine how much QTL mapping can reveal about the genetic architecture of quantitative traits. *Evolution*, **67**, 1251–1262

Slatkin, M. (2008) Linkage disequilibrium—understanding the evolutionary past and mapping the medical future. *Nature Reviews Genetics*, **9**, 477–485.

Slatkin, M. (2009) Epigenetic inheritance and the missing heritability problem. *Genetics*, **182**, 845–850.

Steinsland, I., Thorrud Larsen, C., Roulin, A., Jensen, H. (2014) Quantitative genetic modeling and inference in the presence of non-ignorable missing data. *Evolution*, **68**.

Stinchcombe, J.R., Hoekstra, H.E. (2008) Combining population genomics and quantitative genetics: finding the genes underlying ecologically important traits. *Heredity*, **100**, 158–170.

Tarka, M., Åkesson, M., Beraldi, D., Hernandez-Sanchez, J., Hasselquist, D., Bensch, S., Hansson, B. (2010) A strong quantitative trait locus for wing length on chromosome 2 in a wild population of great reed warblers. *Proceedings Of The Royal Society B-Biological Sciences*, **277**, 2361–2369.

Thornton, K.R., Jensen, J.D. (2007) Controlling the false-positive rate in multilocus genome scans for selection. *Genetics*, **175**, 737–750.

Van Bers, N.E.M., Santure, A.W., van Oers, K., De Cauwer, I., Dibbits, B.W., Mateman, C., Crooijmans, R.P.M.A., Sheldon, B.C., Visser, M.E., Groenen, M.A.M., Slate, J. (2012) The design and cross-population amplification of a genome-wide SNP chip for the great tit *Parus major*. *Molecular Ecology Resources*, **12**, 753–770.

Van Bers, N.E.M., van Oers, K., Kerstens, H.H.D., Dibbits, B.W., Crooijmans, R.P.M.A., Visser, M.E., Groenen, M.A.M. (2010) Genome-wide SNP detection in the great tit *Parus major* using high throughput sequencing. *Molecular Ecology*, **19**, 89–99.

VanRaden, P.M. (2008) Efficient methods to compute genomic predictions. *Journal Of Dairy Science*, **91**, 4414–4423.

Visscher, P.M. (2009) Whole genome approaches to quantitative genetics. *Genetica*, **136**, 351–358.

Visscher, P.M., Hill, W.G., Wray, N.R. (2008) Heritability in the genomics era—concepts and misconceptions. *Nature Reviews Genetics*, **9**, 255–266.

Visscher, P.M., Macgregor, S., Benyamin, B., Zhu, G., Gordon, S., Medland, S., Hill, W.G., Hottenga, J.-J., Willemsen, G., Boomsma, D.I., Liu, Y.-Z., Deng, H.-W., Montgomery, G.W., Martin, N.G. (2007) Genome partitioning of genetic variation for height from 11,214 sibling pairs. *American Journal of Human Genetics*, **81**, 1104–1110.

Visscher, P.M., Medland, S.E., Ferreira, M.A.R., Morley, K.I., Zhu, G., Cornes, B.K., Montgomery, G.W., Martin, N.G. (2006) Assumption-free estimation of heritability from genome-wide identity-by-descent sharing between full siblings. *Plos Genetics*, **2**, 316–325.

Vitalis, R., Dawson, K., Boursot, P. (2001) Interpretation of variation across marker loci as evidence of selection. *Genetics*, **158**, 1811–1823.

Wall, J.D., Pritchard, J.K. (2003) Haplotype blocks and linkage disequilibrium in the human genome. *Nature Reviews Genetics*, **4**, 587–597.

Weedon, M.N., Lango, H., Lindgren, C.M., Wallace, C., Evans, D.M., Mangino, M., Freathy, R.M., Perry, J.R., Stevens, S., Hall, A.S., Samani, N.J., Shields, B., Prokopenko, I., Farrall, M., Dominiczak, A., Diabetes Genetics Initiative, Wellcome Trust Case Control Consortium, Johnson, T., Bergmann, S., Beckmann, J.S., Vollenweider, P., Waterworth, D.M., Mooser, V., Palmer, C.N.A., Morris, A.D., Ouwehand, W.H., Cambridge GEM Consortium, Zhao, J.H., Li, S., Loos, R.J., Barroso, I., Deloukas, P., *et al.* (2008) Genome-wide association analysis identifies 20 loci that influence adult height. *Nature Genetics*, **40**, 575–583.

Yang, J., Benyamin, B., McEvoy, B.P., Gordon, S., Henders, A.K., Nyholt, D.R., Madden, P.A., Heath, A.C., Martin, N.G., Montgomery, G.W., Goddard, M.E., Visscher, P.M. (2010) Common SNPs explain a large proportion of the heritability for human height. *Nature Genetics*, **42**, 565–569.

Yang, Z.H., Bielawski, J.P. (2000) Statistical methods for detecting molecular adaptation. *Trends in Ecology & Evolution*, **15**, 496–503.

Yang, J., Manolio, T.A., Pasquale, L.R., Boerwinkle, E., Caporaso, N., Cunningham, J.M., De Aandrade, M., Feenstra, B., Feingold, E., Hayes, M.G., Hill, W.G., Landi, M.T., Alonso, A., Lettre, G., Lin, P., Ling, H., Lowe, W., Mathias, R.A., Melbye, M., Pugh, E., Cornelis, M.C., Weir, B.S., Goddard, M.E., Visscher, P.M. (2011) Genome partitioning of genetic variation for complex traits using common SNPs. *Nature Genetics*, **43**, 519–525.

www.oup.co.uk/companion/charmantier

CHAPTER 14

Bayesian approaches to the quantitative genetic analysis of natural populations

Michael B. Morrissey, Pierre de Villemereuil, Blandine Doligez and Olivier Gimenez

14.1 Introduction

Evolutionary quantitative genetic analysis of natural populations is proving to be highly rewarding, but also comes with enormous challenges. Parameters that have always been regarded as difficult to estimate in the laboratory, for example genetic correlations, are even more difficult in data that contains the 'real' noise of nature. It is therefore very important to consider the best models that we can use for the study of data from natural populations, but also to consider the statistical uncertainty inherent to the estimates yielded by these models. Natural populations also present the quantitative geneticist with additional complications; in particular, it may become increasingly important to explicitly consider the process of observing data in conjunction with inferences about underlying biological processes. For example, in evaluating life histories when individuals are not perfectly observable, it will help to be cautious about genetic inferences. Bayesian techniques offer the empiricist some of the most promising ways of dealing with complicated and noisy data.

In this first section we take a very close look at estimation of heritability in a simple breeding experiment. This is not an analysis of a natural population, but provides a simple example that illustrates a range of non-trivial details with which any empiricist must familiarise him- or herself before putting Bayesian methods into practice. In short, the idea here is to turn the usual mode of presentation of a statistical method on its head. Rather than starting from a completely trivial model and building toward complex and scientifically interesting analyses, we are starting from an assumption that the reader is sufficiently familiar with the basic biological principle of inferring the genetic basis of traits from similarity among relatives (see Chapter 2, Postma). Given this biologist's view of the flow of phenotype data through a model to make genetic inferences based on similarity of relatives, we hope that the important statistical and Bayesian aspects of interpreting the models will be as intuitive as possible, and we will deal with these aspects as they arise. In doing so we hope to overcome the biggest obstacle to realising the potential benefits of Bayesian quantitative genetic analysis in the wild: getting off the ground.

Our biologist's view of a simple quantitative genetic analysis, and the subsequent more developed applications to natural populations in the subsequent section, do not provide any comprehensive guide to either Bayesian philosophies or methodologies. General practical texts include *Bayesian data analysis* (Gelman *et al.* 2004), *Data*

Quantitative Genetics in the Wild. Edited by Anne Charmantier, Dany Garant, and Loeske E. B. Kruuk

analysis using regression and multilevel/hierarchical models (Gelman & Hill 2007), and *Doing Bayesian data analysis* (Kruschke 2011). Those with an ecological background who are interested in incorporating Bayesian, 'modular' or hierarchical analyses, and potentially quantitative genetic methods into their work may find *Hierarchical modeling and inference in ecology* (Royle & Dorazio 2008) to be a useful introduction to Bayesian Markov Chain Monte Carlo (MCMC)-based analysis. An important general resource for MCMC analysis is the book *Handbook of Markov Chain Monte Carlo* (Brooks *et al.* 2011). A vastly more complete introduction to Bayesian quantitative genetic analysis, which includes many ideas that would greatly benefit the analysis of natural populations, is *Likelihood, Bayesian, and MCMC methods in quantitative genetics* (Sorensen & Gianola 2002). O'Hara *et al.* (2008) 'Bayesian approaches in evolutionary quantitative genetics' is a useful review that is less specifically focused on evolutionary problems in natural populations, but touches on similar and complimentary themes to ours.

In the second section, we explore some specific cases where extension of the approaches and thinking from the first section is currently allowing informative, cutting-edge quantitative genetic analyses of natural populations. In each example, the inherent flexibility of Bayesian approaches, and available Bayesian tools, is particularly important in allowing more direct inference of key evolutionary parameters than is often possible in frequentist frameworks. The goal of this section is to suggest several ways in which Bayesian analyses can potentially provide insight into current microevolutionary problems. This section is intended to be less didactic in terms of details of implementation. We describe several types of analyses that we expect to become increasingly common in the near future. Whilst we seek to explain the established aspects of their implementation, we do not intend that our treatment should in any way be regarded as a guide to 'best practices', because these are developing at a great pace. Rather, as before, we seek primarily to outline the utility and flexibility that Bayesian analysis can bring to the quantitative genetic analysis of natural populations. Each of the examples we discuss illustrates a specific way in which the Bayesian toolkit allows specific inference of evolutionary parameters that would be very difficult (but probably in no case impossible) to obtain otherwise.

14.2 Putting Bayesian methods into practice: a guided tour of a simple example

14.2.1 Heritability of morphological traits in crickets

We will be very explicit about i) how the maths in the example relate to quantitative genetic parameters and ii) how the specific maths can be implemented, using the BUGS programming language (Lunn *et al.* 2000), implemented with the software JAGS (Plummer 2003). This depth will subsequently prove valuable in the following subsection where we refit the model in several different ways to get a feel for some important details about prior specification, when we introduce the animal model in a Bayesian implementation, and more generally, as we move through the more interesting examples throughout the second section of the chapter.

We analyse phenotypic data from a quantitative genetic experiment on field crickets, *Teleogryllus oceanicus*, by Simmons and Garcia-Gonzalez (2007) who mated 30 males to a total of 84 females, and measured 378 female offspring for a range of traits, including pronotum length and ovary mass. The goal here is to estimate additive genetic variances, and heritabilities, and to control for and characterise any potentially confounding sources of variances, such as that arising from maternal effects. The main trick is to characterise the amount of variation due to sires. Four times the sire variance is the additive genetic variance, and the quotient of the additive genetic variance and the phenotypic variance is the heritability. The simplest mixed model with which to analyze this 'dams within sires' breeding experiment is

$$y_i = \mu + s_i + d_i + e_i \tag{14.1}$$

where y_i is the phenotype of individual i, μ is the population mean, s_i and d_i represent the effects of the dam and sire of individual i, and e are residual errors. Because we are interested in the variance among s and d values, not necessarily the effects of each dam and sire in isolation, we model them as random effects. What this means is that we assume that s_i, d_i, and similarly e_i values come from normal distributions, the variances of which are the parameters of interest, and which are estimated. This can be written $s_i \sim N(0, \sigma_s^2)$, $d_i \sim N(0, \sigma_d^2)$, and $e_i \sim N(0, \sigma_e^2)$, where σ_x^2 represents the variance among effects x. More fully, we could write the full likelihood as

$$\begin{aligned} \ell_i(\mu, s_i, d_i, \sigma_\epsilon^2) &= p(\mathbf{y}|\mu, s_i, d_i, \sigma_e^2) \\ &= \prod_{i=1}^{N_i} N(y_i|\mu + s_i + d_i, \sigma_e^2) \end{aligned} \quad (14.2a)$$

$$\ell(\sigma_d^2) = p(\mathbf{d}|\sigma_d^2) = \prod_{j=1}^{N_d} N(d_j|0, \sigma_d^2) \quad (14.2b)$$

$$\ell(\sigma_s^2) = p(\mathbf{s}|\sigma_s^2) = \prod_{k=1}^{N_s} N(s_k|0, \sigma_s^2) \quad (14.2c)$$

$$\ell(\mu, \sigma_{e(y)}^2, \sigma_d^2, \sigma_s^2) = p(\mathbf{y}|\mu, s_i, d_i, \sigma_e^2) \times p(\mathbf{d}|\sigma_d^2) \times p(\mathbf{s}|\sigma_s^2) \quad (14.2d)$$

where ℓ represents the likelihoods of parameters, p represents the probability of given observed data (i.e. y_i) or effects of dams and sires, i.e. the vectors **d** and **s**. The left-hand side of each expression in Eq. 14.2 represents the likelihood of parameter estimates, which is equated to the probability of some data, i.e. y in Eq. 14.2a or the unobservable sire and dam effects in Eqs. 14.2b,c, or all jointly in 14.2d. The right-most expressions in Eq. 14.2a,b,c are the core of the model. Here the probabilities are specified in terms of products of normal density functions associated with each data observation or random effect.

Eq. 14.2 may seem like an unnecessarily complex way of representing the mixed model that we managed with a single line in Eq. 14.1. However, this representation lies at the core of the problem, either in a frequentist likelihood, or in a Bayesian analysis. In order to analyse genetic parameters in the cricket dataset in a Bayesian framework, we have to do two things. First, we have to come up with priors for the parameters, and second, we have to think about implementation. To start with, we will try to apply the simplest priors we can think of. The parameter μ can in principle take any real value, so a very wide (high variance) normal density is simple and thorough. For the variance of each effect, i.e. the sire, dam, and residual sources of variation, any positive value is permissible. We will start by allowing all values greater than zero, and up to some large value (ideally, this should not be informed by the data in any way; in this case, making sure that the upper limit is substantially larger than the observed variance should be pragmatic) to be equally likely, i.e. we will apply uniform prior densities on the standard deviation (SD). Formally, we could express these prior choices as

$$\mu \sim N(0, 1000) \quad (14.3a)$$

$$\sigma_s \sim U(0, \sigma_{big}) \quad (14.3b)$$

$$\sigma_d \sim U(0, \sigma_{big}) \quad (14.3c)$$

$$\sigma_e \sim U(0, \sigma_{big}) \quad (14.3d)$$

where $N()$ is defined as above, and $U(x, y)$ represents a uniform density with minimum x and maximum y; σ_{big} is thus an arbitrarily large upper limit.

In BUGS code, the model is thus:

```
1 model{
2   #priors
3   mu~dnorm(0,0.001)  # bugs works with precision, i.e., 1/variance
4   sigma_s~dunif(0,sigma_big)
5   sigma_d~dunif(0,sigma_big)
6   sigma_e~dunif(0,sigma_big)
7
8   #random effects
9   for(j in 1:N_d){
10     d[j]~dnorm(0,1/sigma_d^2)
11   }
12   for(k in 1:N_s){
```

```
13      s[k]~dnorm(0,1/sigma_s^2)
14    }
15
16    #data
17    for(i in 1:N_i){
18      y[i]~dnorm(mu+s[sire[i]]+d[dam[i]],1/sigma_e^2)
19    }
20 }
```

Computer code can be as intimidating as math. However, having already taken a moment to write out the model in full (i.e. in Eq. 14.2), and to formally define some priors on the free parameters (i.e. in Eq. 14.3), all we have here is a statement of the same model in a different syntax. Lines 3 to 6 state that the priors have the normal and uniform distributions we stated. Lines 10 and 13 correspond to the expressions in Eq. 14.2b,c, and line 18 similarly corresponds to Eq. 14.2a. The information about the BUGS language that is needed to understand fully this code is that `x~a(b,c)` means 'x is sampled from distribution `a` with parameters `b` and `c`'. In the case of the uniform distribution (`dunif`), the parameters are the upper and lower bounds of the distribution, and for the normal distribution, the parameters are the mean and the precision, which is the inverse of the variance. Looping over each datum and random effect level is accomplished with 'for' loops, where `for(a in b:c){d}` means 'sequentially assign `a` to all inclusive integer values between `b` and `c`, and given these values of `a`, do `d`'. Indexing of elements of vectors is accomplished with square brackets. In the code above, several constants have to be provided to allow it to run. These are the numbers of dams, sires, and phenotypic observations (`N_d`, `N_s`, and `N_i`; note that indexes remain consistent with Eq. 14.2), a vector of the phenotypic observations (`y`) and vectors indicating which dam and sire is associated with each observation (`dam` and `sire`). Box 14.1 provides a brief overview of how models can be put to work, once coded in this way, using MCMC algorithms.

Box 14.1 Bayesianism

In statistical inference, Bayesianism is a paradigm in which (and contrary to frequentism), one consider parameters (hereafter θ) as random variables. In this sense, Bayesians are not interested in infering a point estimate (based on maximum likelihood for example) from the data Y, but a *posterior distribution* $P(\theta|Y)$. In order to do that, they use Bayes' inferential theorem:

$$P(\theta|Y) = \frac{P(Y|\theta)\,P(\theta)}{P(Y)}$$

where $P(Y|\theta)$ is the *likelihood* of the data given the model and the parameters, $P(\theta)$ is the *prior distribution* on the parameters and $P(Y)$ is a scaling constant. In essence, Bayesianism considers any inference as an *update* from your *prior belief* of what the parameters are to your *posterior belief* of what values are more probable now that you have analysed the data.

Different flavours of the MCMC algorithm represent the most popular Bayesian estimation algorithm, because they allow posterior distributions to be sampled for very arbitrary models. Thus, when models are becoming too complex for maximum likelihood estimation, Bayesianism and MCMC sampling are often used as alternative resources. MCMC algorithms yield samples of the parameters of a model in proportion to their posterior probability. For example, a well-conducted Bayesian analysis of a dataset wherein $V_A = 2$ would ideally yield many outputs of V_A in the vicinity of 2, and relatively fewer outputs of values substantially lower or greater than 2. The amount of evidence available will determine just how much those 'values substantially lower or greater than 2' actually differ from 2. Such samples are used to calculate posterior properties of the parameter (mean, median, variance, credible interval...), as well as to obtain posterior distributions of derived parameters of interest, such as h^2.

14.2.2 Posterior transformation to make inference of parameters of direct biological interest

Our main question was: what are the heritabilities of pronotum length and ovary mass? Given the posterior distributions of the sire models of the traits, we need only to apply the standard relationship between the sire variance and the additive genetic variance, and between the additive genetic and phenotypic variances, to obtain inference of the heritability. The most probable values of the genetic variance and heritability will be the modal values of their posterior distributions, and the posterior distributions in full provide inference of the probabilities, given the data, the model structure, and the priors, that the true values of the genetic variance and heritability take any other values. The posterior distribution of the additive genetic variance, obtained by applying $\sigma_A^2 = 4\sigma_s^2$ to each sample of the posterior distribution, is shown in Figure 14.1a,c, and the posterior distribution of the heritability, based on $\sigma_P^2 = \sigma_s^2 + \sigma_d^2 + \sigma_e^2$, is given in Figure 14.1b,d.

Thus, inference of genetic parameters can be very simple, given a fitted Bayesian model (Table 14.1). Parameter estimates are similarly easily obtained in a frequentist framework, but inference of the statistical support for estimates in a frequentist framework is much harder to obtain. For example, the implementation of a sire model with the software ASReml-R (Gilmour *et al.* 1999) is

```
asreml(fixed = pro ~ 1, random =
  ~sire + dam, data = crickets)
```

which is much easier than the implementation route we took above. Note however that Bayesian implementation with MCMCglmm (Hadfield 2010), an R-package for fitting Bayesian generalised linear mixed models by MCMC, would be similarly simple. However, we would see and understand less of what was happening under the hood, and some of the customising of the sire model in the next section would not be possible.

The restricted maximum likelihood (REML) solution of the model for pronotum length gives the

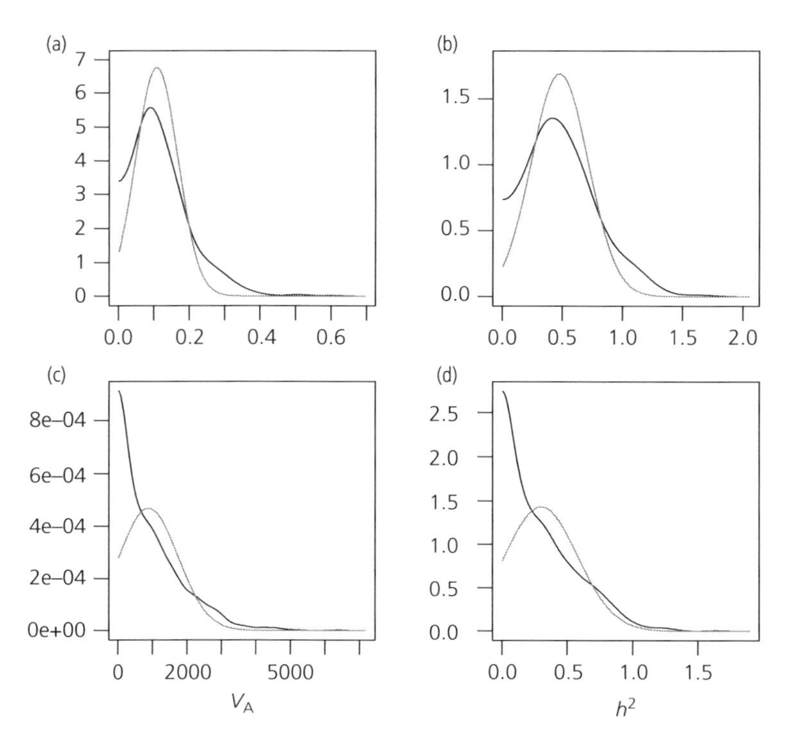

Figure 14.1 Additive genetic variance V_A (a, c) and heritability h^2 (b, d) of pronotum length (a, b) and ovary mass (c,d) in field crickets from Simons and Garcia-Gonzalez's (2007) breeding experiment. Black lines show posterior distributions of the parameters, and grey lines show frequentist approximations of the sampling error of their REML estimators, i.e. normal distributions with SD = standard error of each parameter.

Table 14.1 Bayesian and frequentist parameter values from mixed model (sire model) analysis of the cricket dataset. Directly modelled parameters are: the standard deviations associated with sire σ_s, dam σ_d, and the residual variation σ_e. Derived parameters are: the additive genetic variance σ^2_A, the maternal non-genetic variance σ^2_M, the total phenotypic variance σ^2_P, and the heritability h^2.

Parameter	Posterior mode	Posterior mean	SD of posterior	REML estimate	REML SE
(a) Directly modelled parameters; pronotum length					
σ^2_s	0.024	0.030	0.020	0.027	0.015
σ^2_d	<0.001	0.014	0.013	0.012	0.012
σ^2_e	0.182	0.191	0.016	0.189	0.016
(b) Derived parameters; pronotum length					
σ^2_A	0.095	0.121	0.078	0.108	0.059
σ^2_M	–0.022	–0.016	0.026	–0.015	0.021
σ^2_P	0.231	0.235	0.022	0.228	0.011
h^2	0.43	0.50	0.28	0.48	0.24
(c) Directly modelled parameters; ovary mass					
σ^2_s	2	251	235	222	213
σ^2_d	842	837	268	778	255
σ^2_e	1985	1989	183	1961	170
(d) Derived parameters; ovary mass					
σ^2_A	10	1004	942	886	853
σ^2_M	626	586	411	557	395
σ^2_P	3047	3078	310	2960	340
h^2	<0.01	0.32	0.276	0.30	0.28

SD = standard deviation; REML = restricted maximum likelihood; SE = standard error.

parameter estimates $\sigma^2_s = 0.027$, $\sigma^2_d = 0.012$, and $\sigma^2_e = 0.189$. This yields similar estimates of heritability to the parameter values we obtained from the Bayesian implementation (Table 14.1, Figure 14.1). However, inference of statistical uncertainty in the frequentist parameter estimates is much more challenging (see also Section 14.1); software may calculate *normal approximations of* standard errors (SEs) automatically, but normal approximations to the sampling distribution do not necessarily have sensible interpretations. The basic mechanics require that we obtain an estimate of the sampling variance–covariance matrix of the parameters (i.e. the variances about the estimates of σ^2_s, σ^2_d, and σ^2_e, and the covariances between each estimate). This is obtainable from the model fitted in ASReml-R (with some difficulty; ASReml-R returns the inverse of the average information matrix, which is a key parameter in the algorithm that ASReml uses to solve mixed models, for parameter estimates that are subject to internal scaling. We do not dwell here on the scaling, but we have done it). The sampling variances of the estimates of the variance components are σ^2_A: 0.0034, σ^2_M: 0.0004, and σ^2_P: 0.0001. We can obtain the sampling variance–covariance matrix of the derived parameters σ^2_A and σ^2_p by $\boldsymbol{A}\sigma^x\boldsymbol{A}^T$, where σ^x is the variance–covariance matrix of σ^2_s, σ^2_d, and σ^2_e, and $\boldsymbol{A}$ contains the coefficients describing the linear combination of the estimated parameters that is required for derivation of the parameter of interest. Thus, $\boldsymbol{A} = \left(\begin{smallmatrix} 4 & 0 & 0 \\ 1 & 1 & 1 \end{smallmatrix}\right)$, specifying the additive genetic variance as four times the sire variance plus zero times each of the other variances, and specifying the phenotypic variance as the sum of all variance components. The sampling variance of the estimated heritability is approximately $(\sqrt{\frac{\sigma^2[\sigma_a^2]}{\sigma_p^2}})^2 + (-\sigma_a^2\sqrt{\frac{\sigma^2[\sigma_p^2]}{(\sigma_p^2)^2}})^2 + 2\sigma_{a,p}\sqrt{\frac{\sigma^2[\sigma_a^2]}{\sigma_p^2}}(\frac{-\sigma_a^2\sqrt{\sigma^2[\sigma_p^2]}}{(\sigma_p^2)^2})$, where $\sigma^2[x]$ represents the sampling variance (SE^2) of x. As such, the SE (square root of the sampling variance) of σ_{A}^2 is 0.059 and the SE of the heritability is 0.24, for pronotum length (Table 14.1).

The tediously derived normal approximations to the sampling errors of the genetic variances and heritabilities of pronotum length and ovary mass are plotted with the posterior distributions of the Bayesian parameter estimates in Figure 14.1. Both approaches to making inferences about parameter values, and to obtaining information about how confident we can be about those inferences, tell us about the same thing. The data available suggest that the heritability of pronotum length is around 0.5, and that the heritability of ovary mass is probably lower. Also, both approaches suggest that our confidence that the real parameter values associated with both traits are close to our best inferences is weak. This is simply a result of the modest sample size.

The consideration of the full posterior distributions of parameters of biological interest that one is more inclined to do after application of a Bayesian analysis is quite useful. First, whilst it must be kept in mind that there are important ways in which a given posterior distribution may be influenced by model structure and prior specification (in addition to the data), a Bayesian posterior distribution is more interpretable as a complete representation of the statistical support for a given parameter having a given value. Where data are scarce, as is often the case for traits of ecological interest in wild populations, consideration of the full posterior distribution can lead to much more statistically sensible interpretations than might otherwise be made. For example, one might conclude that 'the most probable values of V_{A} and h^2 of cricket ovary mass are very low', but at the same time one should note that 'high values of V_{A} and h^2 of cricket ovary mass cannot be ruled out'. This latter component of the interpretation is key because there is appreciable density of the posterior distribution at high values of the genetic variances and heritabilities, even if the most probable values are lower.

There are also important differences between the two approaches to assessing uncertainty. First, the Bayesian posterior distribution is exact, given the priors, the data, and the model construction (although the MCMC implementation is an approximation, we expect it to yield the 'true' shape of the distribution). The frequentist approach we took to obtain SEs from the REML analysis is very fundamentally approximate. There is no need to dwell on the specific formulae applied, but the formula for the sampling variance is a very standard approximation, and more importantly, the whole process of obtaining the REML SEs assumes that sampling error of the directly estimated and the subsequently derived parameters, i.e. σ_s^2, σ_d^2, and σ_e^2, and then σ_{A}^2, σ_{P}^2 and h^2, are normally distributed. This assumption is of course fundamentally untrue, since normal distributions give non-zero density to all real values, whilst variance components are bounded at zero and heritability is bounded at zero and one. Furthermore, derivation of parameters such as σ_{A}^2, σ_{P}^2 and h^2 is very simple, and more complicated parameters that we might be interested in modelling will be totally intractable outside of a Bayesian context.

By roundabout means, we have obtained our first important message about the Bayesian quantitative genetic analysis of natural populations. Transformations of posterior distributions of directly modelled parameters yield valid posterior distributions of derived parameters. This is a powerful and practical feature of Bayesian analysis. However, it can also be dangerous. The details of how a model is specified, including prior specification, can have substantial and even dramatic influence both on directly modelled parameters and on derived parameters (for a discussion of the importance of priors, see Box 14.2).

14.2.3 Alternate model specifications: 'customising' the sire model, and enter the animal model

Inspection of the parameter estimates from the sire model reveals two incongruous results. First, there is non-zero posterior density (i.e. some statistical

Box 14.2 Prior distributions

One of the most problematic issues for new users of Bayesian inference is to deal with priors. This is for two reasons: first, one might think that we are not allowed to have prior knowledge before analysing the data; second, one can struggle to find which prior is the more sensible for one's analysis. Although a full Bayesian analysis includes constructing priors from previous experience, it is not the most popular practice because it requires that previous data exists, and requires objective rules to construct prior distributions from them.

Common practice is to seek analyses that have only weak prior influence, by seeking *non-informative prior distributions*, which in some cases can include *flat priors* assigning every event the same probability or *diffuse or vague priors* which are also very flat due to low precision (commonly: Gaussian prior with huge variance for fixed effects). The idea of *weakly informative priors* may be a more generally useful concept (Gelman *et al.* 2008). However, regardless of prior specification, it is good practice to check the *sensitivity to the prior distribution*, by running several concurrent priors and checking their influence on the posterior distributions. An example of prior sensitivity analysis applied to heritability is presented in de Villemereuil *et al.* (2013) Appendix B.

It is impossible to provide simple and general advice to beginner Bayesian analysts, but here are a couple of hints. Especially when using BUGS (or JAGS) one would be well advised to read Gelman (2006), as it is especially useful for understanding prior influence on variance components. If conducting quantitative genetic analyses of modest datasets using MCMCglmm (Hadfield 2010), it will be useful to consult the documentation for information about 'parameter-expanded priors' for variance components.

support) for values of heritability that exceed one (Figure 14.1) for both traits. Second, the dam variance is less than the sire variance (Table 14.1). Neither of these makes biological sense (in the latter case, assuming paternal effects are absent or are much smaller than maternal effects), and both occur despite the fact that the sire and dam sources of phenotypic variance have very simple biological interpretations. Being a proportion, heritability cannot exceed one; and the dam variance is the sum of genes and environments of mothers, and neither of these is likely to cause full sibs to resemble one another less than at random (assuming that unmodelled processes such as varying maternal allocation with birth order, or asymmetric sibling competition are not happening). However, the mixed model is blind to the biological interpretations, and since we have specified it in a way that the sire variance can exceed one quarter (which is not biologically meaningful as $V_{\text{sire}} = \frac{1}{4}V_{\text{A}}$, and $V_{\text{A}} \leq V_{\text{P}}$) of the phenotypic variance, and where the dam variance does not have a lower bound at the sire variance, the curious outcomes are not mathematically unsound.

If we have prior belief that we have conducted our analysis appropriately with regard to the interpretation of the sire and dam variances, we can reparameterise the sire model to reflect this belief. The statistical parameters that we directly modelled pertain to the variation associated with dam, sire and individuals. We can redefine these in terms of how the biologically interesting parameters about different sources of variation must be related to one another by

$$\sigma_{\text{P}} \sim U(0, \sigma_{\text{big}}) \tag{14.4a}$$

$$h^2 \sim U(0, 1) \tag{14.4b}$$

$$\omega_d \sim U(0, 1) \tag{14.4c}$$

$$\sigma_s = \sqrt{\frac{h^2\sigma_{\text{P}}^2}{4}} \tag{14.4d}$$

$$\sigma_d = \sqrt{\frac{h^2\sigma_{\text{P}}^2}{4} + \omega_d(1-h^2)\sigma_{\text{P}}^2} \tag{14.4e}$$

$$\sigma_e = \sqrt{\sigma_{\text{P}}^2 - \frac{h^2\sigma_{\text{P}}^2}{2} - \omega_d(1-h^2)\sigma_{\text{P}}^2} \tag{14.4f}$$

For lack of a standardised symbol to denote the proportion of the non-genetic phenotypic variance that is attributable to maternal identity, we denote this parameter ω_d. Expressions 14.4a,b,c simply define the biologically meaningful parameters as having reasonable ranges. The heritability is some fraction of the phenotypic variance, and the maternal variance is some fraction of the non-genetic variance. Eq. 14.4 simply translates these biologically more sensible parameters into the parameters of the sire model above. This small modification of the sire model to improve biological interpretability is straightforward given available Bayesian MCMC tools.

Another alternative parameterisation, and ultimately one that is very valuable and general, would be to directly model the process of inheritance of a multivariate trait, rather than modelling dam and sire effects, knowing that they represent some fraction of the genetic variation. This is the approach taken by the 'animal model'. In the animal model, variances associated with parental effects are replaced with a direct model of the effects of an individual's genes, i.e. a model of variation in *breeding values*. A breeding value is defined as twice the expected deviation of an individual's offspring's phenotype from the population mean. More intuitively, it can be thought of as the sum of the effects on an individual's phenotype of the genetic variants carried in that individual's genome. Specifically, an individual's breeding value is

$$a_i \sim N\left(\frac{a_{d_i} + a_{s_i}}{2}, \frac{\sigma_A^2}{2}\right) \quad (14.5)$$

i.e. an individual's expected breeding value is the mean of the breeding values of its parents, and its realized breeding value comes from a distribution with a variance that is half the additive genetic variance in the population. This additional variance is the segregational variance—it can be thought of as reflecting the fact that full siblings are not identical, i.e. full sibs differ in their particular proportional composition of grandparental alleles at different points in their genome, because of the segregation of these alleles into gametes that occurred in their parents. The animal model can still account for maternal effects: we include a dam effect as in the sire model (Eq. 14.1); but this effect is now directly interpretable as the non-(direct) genetic component of the maternal effect, because the genetic component is represented in Eq. 14.5.

Implementing the animal model in BUGS code can be relatively easy:

```
1 model{
2   #priors
3   mu~dnorm(0,0.001)
4   sigma_a~dunif(0,sigma_big)
5   sigma_m~dunif(0,sigma_big)
6   sigma_e~dunif(0,sigma_big)
7
8   #random effects
9   for(j in 1:N_d){
10     a_d[j]~dnorm(0,1/sigma_a^2)
11     m_d[j]~dnorm(0,1/sigma_m^2)
12   }
13   for(k in 1:N_s){
14     a_s[k]~dnorm(0,1/sigma_a^2)
15   }
16
17   #data
18   for(i in 1:N_i){
19     a_i[i]~dnorm((a_d[dam[i]]+a_s[sire[i]])/2,1/(0.5*sigma_a^2))
20     y[i]~dnorm(mu+a_i[i]+m_d[dam[i]],1/sigma_e^2)
21   }
22 }
```

The key lines are: 10, 11 and 14, where dams and sires are given genetic and non-genetic values sampled from normal distributions with estimated variances on lines 4 and 5; line 19, where offspring breeding values are modelled based on the mid-parent breeding value and the segregational variance following Eq. 14.5; and line 20, where offspring phenotypes are modelled based on the population mean, genetic and maternal effects, and the residual variance.

The posterior distributions of heritability and variance components for cricket pronotum length based on the sire model, the constrained sire model, and the animal model are shown in Figure 14.2. These distributions are all obtained from an identical dataset, with identical sets of assumptions for modelling the relationship between resemblance of relatives and genetic variation, i.e. the infinitesimal model (Chapter 2, Postma; Falconer & Mackay 1996). In analysis of data from single generation, the animal model uses no more information about the genetic basis of variation than does the sire model. This is because variation due to maternal genes and variation due to maternal identity are perfectly confounded in a single generation study. Thus, a closer look at the differences in the posterior distributions of the parameter estimates in Figure 14.2 provides an opportunity to better understand the models that we have implemented.

Two aspects of the posterior distribution of the heritability of pronotum length differ between the original sire model, and the constrained sire model and the animal model. In the first, there is non-zero posterior density at values greater than one, and

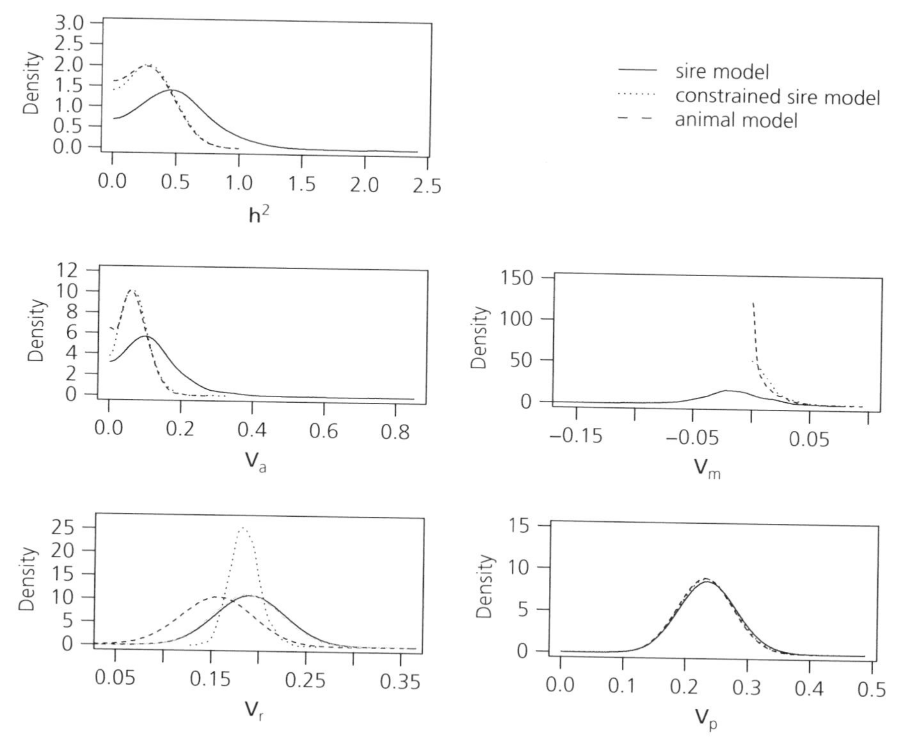

Figure 14.2 Posterior distributions of variance components for cricket pronotum heritability. Black lines represent estimates made with a sire model (as per Eqs. 14.1–3), dotted lines represent estimates made with a constrained sire model (reparameterisation following Eq. 14.4), and dashed lines represent estimates made with an animal model (additive genetic effects directly modelled following Eq. 14.5). The residual variance in a sire model includes segregational variance, whilst it does not in an animal model; in the panel with distributions of V_r, the dashed grey line shows the posterior distribution of the sum of the residual and the segregational variance.

the heritability estimates (and the estimates of the genetic variance) are lower in the latter two. The zero density for values of heritability greater than one is due to biologically reasonable structural constraints specifically built into the constrained sire model and inherent in the animal model. However, the lower estimates of σ_A^2 and h^2 do not necessarily reflect a shift in the posterior distributions to the left. Rather, because of the different structures of the models, more information is coming into play. Because the reparameterised models explicitly include information about maternal identity in the terms that we relate to genetic variation, and because maternal sibs covary less than paternal sibs (as discussed above, this does not have a simple biological interpretation, but even with the more biologically interpretable models, $\sigma_d < \sigma_s$ remains a property of the data), this lower covariance of maternal sibs results in inference of lower genetic variation.

The final very noticeable difference among the models is that the constrained sire model has a much narrower posterior distribution of the residual variance than the other models. This occurs despite the fact that the constrained sire model and the animal model have similar posterior distributions for all of the other parameters. The reason is simple: in the sire model, genetic variance is represented by the sire effect, but only in one quarter proportion—the rest is in the residual; whereas in the re-parameterisations, the genetic variance is fully attributed to the parents, and so in no part represented in the residual. In these relatively simple models, this does not complicate the implementation or interpretation of the models. However, in other analyses, sampling correlations (i.e. non-zero expected error in the estimate of one parameter, given error in another) can become important considerations. We will not dwell on this beyond pointing it out as one of the ways in which simple alternative model parameterisations to analyse the same data can have different properties.

A main take-home point from this section is that different model parameterisations can be important for allowing models to be constructed with structures that reflect either classical statistical models, or biological parameters, to a range of different degrees. Sometimes, this is essentially a matter of setting up the priors on the model in different ways. For example, deriving a sire variance from a phenotypic variance, a heritability, and a knowledge that variation due to sires represents one quarter of the additive genetic variance, can be seen as different specifications of the priors: essentially, the constrained sire model reflects the absolute prior belief, i.e. a very strong prior belief, that the heritability cannot exceed one. In other situations, different ways of setting up a model might be viewed more as a different model, rather than as re-parameterisations. In our simple example, analysing a quantitative genetic breeding experiment with a single generation of data, the animal model can be thought of either as a reparameterisation of the constrained sire model, or it can be seen as a different model. In a general pedigree with multiple generations, the animal model would fundamentally make use of more data than a sire model, and so would in fact be a different model beyond prior specification. From a practical perspective in our example analysis, though, the difference is unimportant. The message is the flexibility of Bayesian models, given the availability of tools such as the BUGS language.

14.3 Bayesian quantitative genetic analysis of natural populations: the present and beyond

14.3.1 Quantitative genetic analysis of non-normal quantitative traits

Of the current problems in evolutionary quantitative genetic analysis that will be most fruitfully pursued with the Bayesian toolkit, the analysis of non-normal traits is currently most developed. Traits of evolutionary and ecological significance can take a wide range of statistical distributions, such as binomial, exponential, Poisson (see for example Chapter 10, Kruuk *et al.*) or gamma distributions. Furthermore, trait distributions may be censored, truncated or zero-inflated. The analysis of variation in non-normal traits falls into the Bayesian realm largely by default. Non-Bayesian approaches for fitting generalised mixed models are not yet well developed, as they need to explicitly calculate or accurately approximate the likelihood, which can be extremely tricky for non-normal distributions of response variables (Bolker *et al.*, 2009). The Bayesian framework has the advantage of the MCMC algorithm (see Box 14.1), which is one of the most adaptable estimation algorithm of current statistics. In theory, as long as one can sample from a distribution (even 'exotic' ones like t-distributions, or censored or zero-inflated ones), MCMC can be used to sample a Bayesian model. As we saw in the first part of this chapter, the flexibility of Bayesian MCMC-based methods can be a great advantage, leading one to Bayesianism as much due to flexibility as due to philosophy. Here we highlight the use of Bayesian MCMC for quantitative genetic analysis of the types of non-normal data that are often of particular interest in realistic ecological scenarios (e.g. survival, fecundity).

The generalised animal model assumes a hypothetical latent trait θ, which is normally distributed and leads to the observable non-normal trait y through a 'link function' g. The model assumed for θ is identical to a 'classical' animal model

$$\theta = \mu + a + e \tag{14.6a}$$

$$a \sim N(0, \mathbf{A}\sigma_{\mathrm{A}}^2) \tag{14.6b}$$

$$e \sim N(0, \mathbf{I}\sigma_{\mathrm{R}}^2) \tag{14.6c}$$

where $\boldsymbol{A}$ is the relatedness matrix (generally derived from a pedigree). Letting y follow any distribution, noted π, which has parameters $g^{-1}(\theta)$ and φ, we can write

$$y \sim \pi(g^{-1}(\theta), \varphi) \tag{14.7}$$

Note that φ is often just a 'nuisance' parameter (for example the data dispersion parameter for a negative binomial distribution), in which we have little interest. Note that with this model, the data have a dispersion linked to the π distribution used, and that σ_{R} is in fact an 'overdispersion' parameter.

For example, for a Poisson distribution (hereafter noted $\mathcal{P}$), we will define Eq. 14.8 as

$$y \sim \mathcal{P}(\log^{-1}(\theta)) \quad \Longleftrightarrow \quad y \sim \mathcal{P}(\exp(\theta)) \tag{14.8}$$

using the canonical logarithm link function for the Poisson distribution. Note that no specific dispersion parameter φ (beyond overdispersion, or the variance in e in Eq. 14.6c) is typically applied in the case of the Poisson responses: the variance is equal to the mean for a Poisson distribution.

In addition to statistical convenience, a model such as defined above is very often justifiable, or even desirable, on biological grounds. The latent trait θ represents the additivity of many small sources of variance, including additive genetic effects, as assumed by the infinitesimal model and most of its quantitative genetic applications. Eq. 14.7, however, may be somewhat more arbitrary, in the sense that the link function may be merely a transformation that will be consistent with the distribution of our data, with often little attention to a biological justification. However, sometimes the link function will make good biological sense as well, for example the logarithmic link function for a Poisson trait implies multiplicative effects on the data scale, which may often be natural for skewed count variables.

A further advantage of the model defined above is that it is easily expanded to be multivariate. Indeed, since Eq. 14.6 is defined just as a non-generalised animal model, the multivariate version would be just as easily defined for each latent trait θ_k. Then, each θ_k can be independently transformed into the biological trait y_k via Eq. 14.7. Thus, a multivariate version of this model allows us to estimate the genetic correlation between several traits having different data distributions (for example a binary and a Gaussian trait). For the sake of simplicity and conciseness, we will continue with a univariate model.

When using the generalised animal model to study a quantitative trait, one question arises: on which 'scale' do you want to estimate the heritability? Indeed, regarding Eqs. 14.6 and 14.7, we have three 'traits' on which we can calculate heritability: the data scale y obviously, but also the latent (or linear predictor) scale θ and the link scale $g^{-1}(\theta)$. Each case can be justified: measuring the heritability on the data scale yields the most biologically sensible measure, but with the drawback that it will strongly depend on the distribution of the trait (it does not allow for comparison between Poisson and binomial traits, for example). Because one of the major interests in quantifying heritability lies in its ease of comparison between traits, it is rarely measured on this scale.

How does one calculate heritability on the different scales? On the data scale, there is actually no easy way to calculate heritability, since we need to have access to the additive genetic variance on this scale, which is not the same as the one on the other scales. The calculation of heritability on the latent scale is however pretty straightforward:

$$h^2_{\theta} = \frac{\sigma^2_A}{\sigma^2_A + \sigma^2_R} \tag{14.9}$$

In order to calculate the heritability on the link scale, we need to take into account the additional variance due to the link function:

$$h^2_{g^{-1}(\theta)} = \frac{\sigma^2_A}{\sigma^2_A + \sigma^2_R + \sigma^2_{\text{link}}} \tag{14.10}$$

Note that, in the case of the above formulated model, the term σ^2_R can be seen as an additive overdispersion parameter (the true residual dispersion being the link-specific variance). Neither scale of calculation of the heritability is either right or wrong. Heritability on the link scale is often sought (e.g. Nakagawa & Schielzeth, 2010; 2013; although the authors use the term 'latent scale' for the here-defined link scale), and may be most natural in many circumstances. To avoid confusion, explicit reporting of formulae used to calculate heritabilities in any particular study will be highly desirable.

We illustrate the study of non-normal traits with the special case of binary traits (such as presence/absence of a character or dichotomous behaviors like allo-/philopatry). We chose binary traits, because they are both a common and a quite difficult type of non-normal trait data, the main issue being the fact that a binary data point conveys little information (because it can only be 0 or 1). Dichotomous phenotypes do not always have a simple Mendelian genetic source, but may be (and most of the time are) the product of a large number of quantitative trait loci. In this case, they fall in the domain of quantitative genetics. The usual model for this kind of traits is the so-called 'threshold model', in which an underlying trait is normally distributed and a threshold is set to separate the dichotomous phenotypes (see Figure 14.3). In a

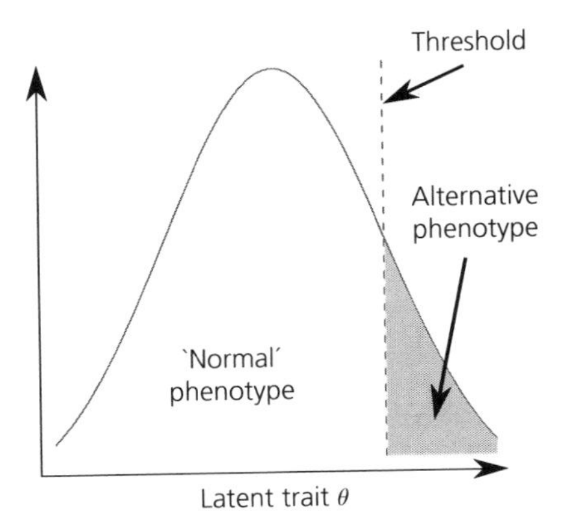

Figure 14.3 A threshold trait is assumed to be produced from a threshold effect depending on the value of the latent trait, depicted here as probability density θ: when θ is below the threshold, the expressed phenotype would be the 'normal' one; when θ is above the threshold, the expressed phenotype would the 'alternative' one.

generalised mixed model context, we assume for these traits a normally distributed latent trait θ and a binomial distribution for the actual binary trait:

$$y \sim \mathcal{B}(\text{probit}^{-1}(\theta)) \tag{14.11}$$

where $\mathcal{B}$ is the binomial distribution and the link function used is the probit function[1]. Note that, in this case, the transformation has a straightforward biological meaning. Indeed, when Wright (1934) first introduced the threshold model, he called the latent trait 'liability' with the idea that when this liability was too high (too many deleterious mutations), then the 'alternative' phenotype, which he considered as the 'sick' one, was expressed.

For the sake of comparison between models, the heritability is often estimated on the link scale:

$$h^2 = \frac{\sigma^2_A}{\sigma^2_A + \sigma^2_R + \sigma^2_{\text{link}}} = \frac{\sigma^2_A}{\sigma^2_A + 1 + 1} \tag{14.12}$$

When using binary data, the (overdispersion) residual variance σ^2_R in Eq. 14.6 is not identifiable.

[1] Defined as the cumulative density function of a standard normal distribution and, in practice, close to the canonical logit link. The choice of a probit link function is to ensure the continuity with the usual threshold model, which assumes a Gaussian distribution for the 'liability'.

Table 14.2 Comparison of σ_A^2 and h^2 estimates for a model with residual variance σ_R^2 fixed to 1 or 2. A probit link was used in the model; thus the heritability is calculated as $h^2 = \frac{\sigma_A^2}{\sigma_A^2+\sigma_R^2+1}$

Fixed σ_R^2	σ_A^2	h^2
1	2.00 (0.83)	0.53 (0.080)
2	2.96 (1.07)	0.497 (0.076)

Indeed, binary data cannot be 'overdispersed' in the sense that their variance is solely defined by the proportions p and $1-p$ of each phenotype: $V(y) = p(1-p)$. Thus, the total phenotypic variance is constrained for binary data. Therefore, we can estimate σ_A^2 for a given σ_R^2, but not *both* at the same time. Is this a problem? For a unique simulated data set ($h^2 = 0.5$ and 1000 individuals), we compared the MCMC outputs with residual variance fixed to one or two during the estimation process. The results (Table 14.2) show different estimated values for σ_A^2, but consistent results for h^2 (remember that both estimations are on the same data set). Indeed the actual estimated parameter in these models is more the intra-class coefficient, which is a class of coefficients the heritability belongs to. Thus the estimate of h^2 is invariant to the arbitrarily chosen fixed value for σ_R^2.

You do not necessarily need Bayesian tools to fit this kind of model. Approximate solutions can be obtained, for example, with the software ASReml (Gilmour *et al.*, 2006). However, when this kind of model is fitted outside of the Bayesian approach, the likelihood is too complex to be easily computed. Thus, two different workarounds are used: i) first consider the binary trait as normally distributed, estimate the heritability using REML, then use a correction (hereafter called corrected REML (REMLc); see Dempster & Lerner, 1950; Lynch & Walsh, 1998); or ii) approximate the likelihood, for example by using penalized quasi-likelihood (PQL; Breslow & Clayton, 1993). de Villemereuil *et al.* (2013) presented a study of the difference in distribution of estimates between these two frequentist and a Bayesian (MCMC) estimation method, based on simulated datasets. Simulations consisted of 1000 replicates of pedigrees and data, for each scenario. Each scenario consisted of one of three true levels of heritability (0.5, 0.3 or 0.1) and two levels of sample size ($n = 1000$ and $n = 200$). As illustrated by Figure 14.4, the PQL estimates are very biased (underestimation of the heritability). This is a known result for binary traits (Goldstein & Rasbash, 1996). The MCMC is a bit more biased than REMLc for small sample size, but is also more precise. The MCMC estimation method is here biased for small sample size because it becomes too sensitive to the particular form of prior used in this study (which was a bit informative; see Box 14.2 on prior sensitivity). Is the high imprecision of the REMLc an issue for estimating heritability? After all, if the standard error and the confidence interval are correctly calculated during the estimation process, we should prefer a non-biased estimator, such as REMLc. However when we calculate the coverage[2] associated with confidence intervals for REMLc and their Bayesian equivalents (credible intervals; CI) for MCMC, we see that MCMC has a better coverage (Figure 14.5, values should lie close to 95%). Thus, REMLc is altogether imprecise and very confident around its estimation, which is a bad combination.

Bayesian estimation methods (and especially MCMC) are very useful for the study of non-Gaussian traits for two main reasons: i) the MCMC algorithm is flexible enough to allow for almost any kind of distribution to be used as data distribution; and ii) the behaviour of the estimates is always 'correct' according to the model and the prior used (no asymptotic assumptions or approximations have to be made), which means that SEs and associated credible intervals are always a relevant estimation of the imprecision of the estimate[3]. However, as we saw with the binary example, the other side of the coin is that we should be very cautious with the choice of the prior, since for small sample

[2] The coverage is the proportion of time the confidence or credible interval contains the true value of the estimated parameter (here the heritability). The expected coverage corresponds to its nominal value, e.g. 95% for 95% confidence interval.

[3] Again, this is 'according to the prior', which means that a prior too informative might lead to overconfidence in the estimates; but for a Bayesian this is relevant: if you already are quite certain a priori of the results, then you are pretty confident of the estimates you get.

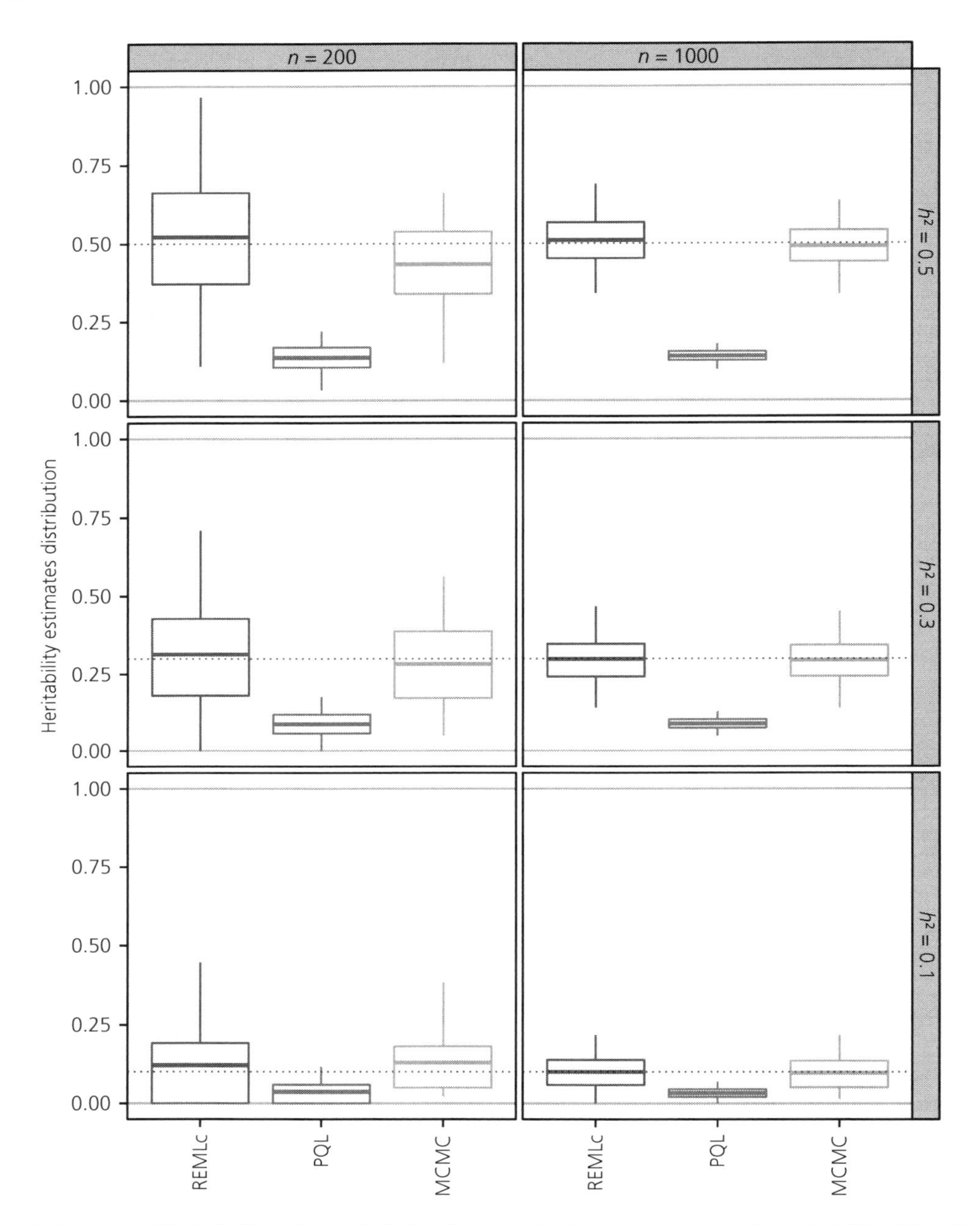

Figure 14.4 Comparison of the heritability estimates distributions for three estimation methods in a simulation study: REMLc, PQL and Bayesian MCMC. Lines show the average value for estimates, boxes show the interquartile interval and whiskers show 95% interquantile interval. Distributions are given for three levels of heritability (0.5, 0.3 and 0.1) and two sample sizes (1000 or 200 individuals).

size, it can have an effect on bias and precision of the estimate (more on this in de Villemereuil *et al.* (2013), Appendix B; Gelman (2006) also provides a useful discussion of the effects of priors for variance components).

14.3.2 Combining quantitative genetic and evolutionary inference

Evolutionary problems, as they play out under real ecological conditions, are inherently complex.

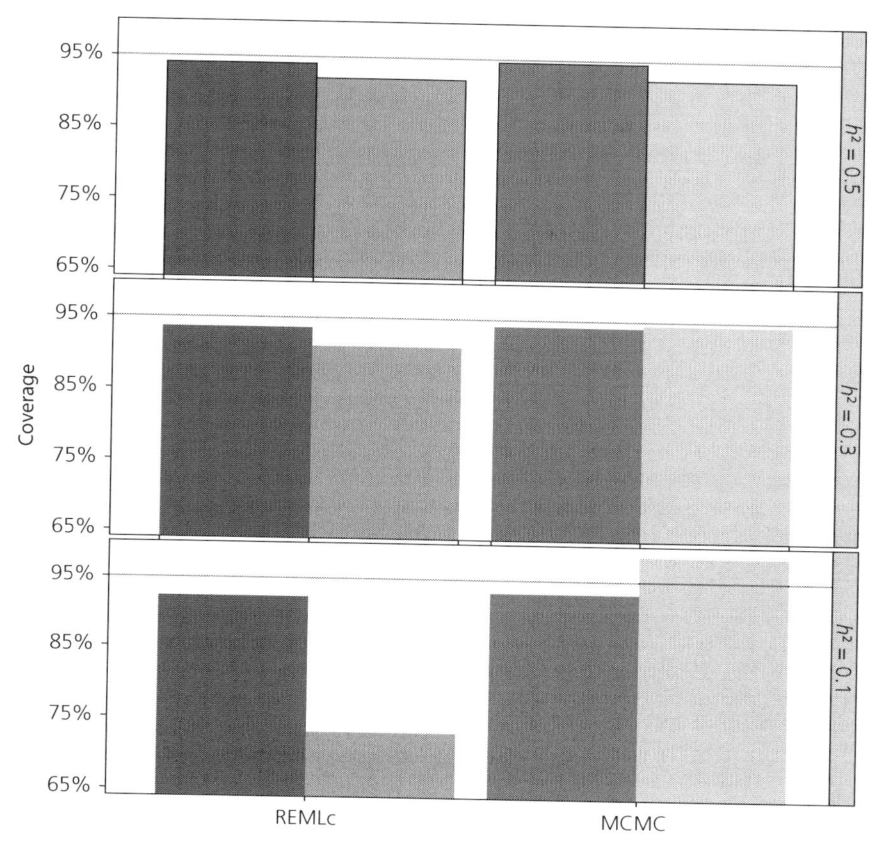

Figure 14.5 Coverage for frequentist REMLc and Bayesian MCMC methods in a simulation study. PQL is not shown, because its coverage is null. Heavy colored bars are for the large sample size (1000 individuals) and light colored bars are for the small sample size (200 individuals).

Consequently, even simplified empirical models will often have to account for more than one process in order to be useful. Bayesian analysis can greatly facilitate the simultaneous evaluation of multiple models, essentially allowing the researcher to break complex problems down into simpler hierarchical levels or 'modules'. In this section, we describe a hierarchical modelling (HM) approach to carry out inference in quantitative genetics. As an example, we illustrate how to calculate the heritability of survival for free–ranging populations in which individuals cannot exhaustively be seen or captured (Lebreton *et al.* 1992; Gimenez *et al.* 2008). Note that survival is intrinsically a non-normal trait and most of the material presented in the previous section applies here. This section is based on work by Papaix *et al.* (2010). In practice, we show how the HM framework allows combining a capture–recapture model to estimate survival whilst accounting for imperfect detection and an animal model to make the decomposition of variance in survival.

Hierarchical modelling (HM) is a powerful approach for analysing complex biological phenomena (Royle & Dorazio 2008; Cressie *et al.* 2009; Buoro *et al.* 2012). One of the most useful applications of HM is when observed data are influenced both by biological processes (e.g. survival probability) and and observation processes (e.g. capture probability). In such a case, a HM can be defined according to three levels: the data at hand Y, the underlying process of interest X and the parameters governing this process. The process X has some distribution governed by a set of parameters θ_X and is generally not directly or fully observable, e.g. due to issues of detectability or measurement error. Data Y have some distribution that depends on the process X and on a set of parameters θ_Y governing the relationship between Y and X. HM allows modelling the randomness both in the data and in the underlying process via the joint conditional distribution of Y and X given the set of associated parameters θ_Y and θ_X:

$$p(Y, X|\ \theta_X, \theta_Y) = p(Y|\ X, \theta_X) \times p(X|\ \theta_X) \quad (14.13)$$

where $p(A|B)$ stands for the probability of A given B. This HM formulation is generic and covers a wide variety of models, including so-called state-space models when the process of interest X has a temporal dynamic (e.g. Gimenez *et al.* 2012). HM offers a clear distinction between the biological process and its observation, and so it allows a focus on the former while accommodating uncertainties in the latter.

As an example of HM, let us consider the estimation of survival in capture–recapture models, with the aim of making inferences about the genetic control of variation in survival probability. In free-ranging populations, individual detectability is often less than one. This issue generates data generally collected in the form of 1s and 0s corresponding to a detection or not of I individuals over T sampling occasions. HM has been proposed as a flexible framework to deal with such capture–recapture data (Rivot & Prevost 2002; Gimenez *et al.* 2007; Royle 2008).

In this example, the process X is a binary random variable which represents the demography, with $X_{i,t} = 1$ if individual i is alive and available for detection at time t and 0 if it is dead. If individual i is alive at time $t-1$, it survives until time t with survival probability $\phi_{i,t}$ or dies with a probability $1 - \phi_{i,t}$; in other words,

$$p(X_{i,t}|\ X_{i,t-1}, \phi_{i,t-1}) = Bernoulli(X_{i,t-1} \times \phi_{i,t-1}) \quad (14.14)$$

Here, survival probability plays the role of θ_X in Eq. 14.13. Now the data Y is a binary random variable, with $Y_{i,t} = 1$ if the individual i is detected at time t and 0 otherwise. These observations are generated from the underlying demographic process, which is partially hidden from the observer, since when an individual is not detected, it is not possible to say whether it is alive or not. If individual i is alive at time t, then it has a probability $p_{i,t}$ of being encountered and a probability $1 - p_{i,t}$ otherwise; in other words, the link between survival and the detection of individuals is made through the observation equation:

$$\left[Y_{i,t}|\ X_{i,t}, p_t\right] = Bernoulli(X_{i,t} \times p_t) \quad (14.15)$$

Here, the detection probability (p_t) corresponds to the θ_Y in Eq. 14.13.

We have now developed one module: a Bayesian formulation of a simple version of the mark–recapture problem. We can combine this with a second module containing an animal model to bring in quantitative genetic inference based on similarity of relatives. To do so, we follow what was presented in Section 14.3.1. We assume that the random survival process X is related to a continuous underlying latent variable $l_{i,t}$, which, given $X_{i,t-1} = 1$, satisfies:

$$X_{i,t} = \begin{cases} 1 \text{ if } l_{i,t} > \kappa \\ 0 \text{ if } l_{i,t} \leq \kappa \end{cases} \quad (14.16)$$

for $t = f_i + 1, \ldots, T$, where f_i is the first time individual i is detected, κ is a threshold value, and T is the index of the last interval in time. We assume that the so-called liability $l_{i,t}$ is normally distributed with mean $\mu_{i,t}$ and variance σ_e. To ensure identifiability (because the residual variance of a Bernoulli variable is entirely determined by the mean), and without loss of generality, σ_e is set to 1 and κ to 0.

From this construction (see also Section 14.3.1), we have

$$\phi_{i,t-1} = \Pr(X_{i,t} = 1|X_{i,t-1} = 1) = F(\mu_{i,t}) \quad (14.17)$$

where F is the cumulative function of a normal distribution with mean 0 and variance 1. Noting that F^{-1} is the probit function often used to analyse binary data, we can specify an animal model on the mean of the liability:

$$\mu_{i,t} = \text{probit}(\phi_{i,t-1}) = \eta + b_t + e_i + a_i \quad (14.18)$$

where η is a constant term for the mean survival on the probit scale, b_t is a random yearly effect (i.e. year specific), e_i is an individual random effect which has no genetic basis and a_i is the genetic value for individual i. Note that the random effect e_i is random among individuals, but not among individuals at different time intervals, because this level of residual variance would be unobservable, i.e. totally confounded with the latent intercept for a Bernoulli response. Covariates can be incorporated as fixed effects possibly affecting survival, e.g. climate effects (Grosbois *et al.* 2008). We assume that the temporal effect b_t is normally distributed with mean zero and variance σ_t^2, e_i normally distributed with mean 0 and variance σ_e^2 whilst the distribution of **a**, the vector of the a_i's, is multivariate normal with mean 0 and variance–covariance matrix $\sigma_A^2 \boldsymbol{A}$, where σ_A^2 is the additive genetic variance and A the additive genetic relationship matrix (see previous sections). Heritability is calculated as the ratio of the additive genetic variance to the total variance:

$$h^2 = \frac{\sigma_A^2}{\sigma_t^2 + \sigma_e^2 + \sigma_A^2 + 1} \tag{14.19}$$

In order to completely specify the Bayesian model, we provide prior distributions for all parameters. All priors can be selected as sufficiently vague in order to induce little prior knowledge. For the purposes of demonstration, we chose $p \sim U[0,1]$ and $\eta \sim N(0,100)$. We assigned uniform distributions to the SD of the random effects, $\sigma_t \sim U[0,10]$, $\sigma_e \sim U[0,10]$ and $\sigma_A \sim U[0,10]$.

We illustrate this section by estimating the heritability of survival using data from a 29-year study of individually marked blue tits (*Cyanistes caeruleus*) monitored at Pirio, Corsica (see Papaix *et al.* 2010 for more details). The data comprises a total of 614 breeding individuals that were banded, released and recaptured in spring during breeding seasons between 1979 and 2007. The posterior distributions are displayed in Figure 14.6, and the resulting summary estimates are presented in Papaix *et al.* (2010). Detection probability p was high (p = 0.77, 95% CI: 0.71–0.82). Survival probability was in agreement with what we were expecting for a small

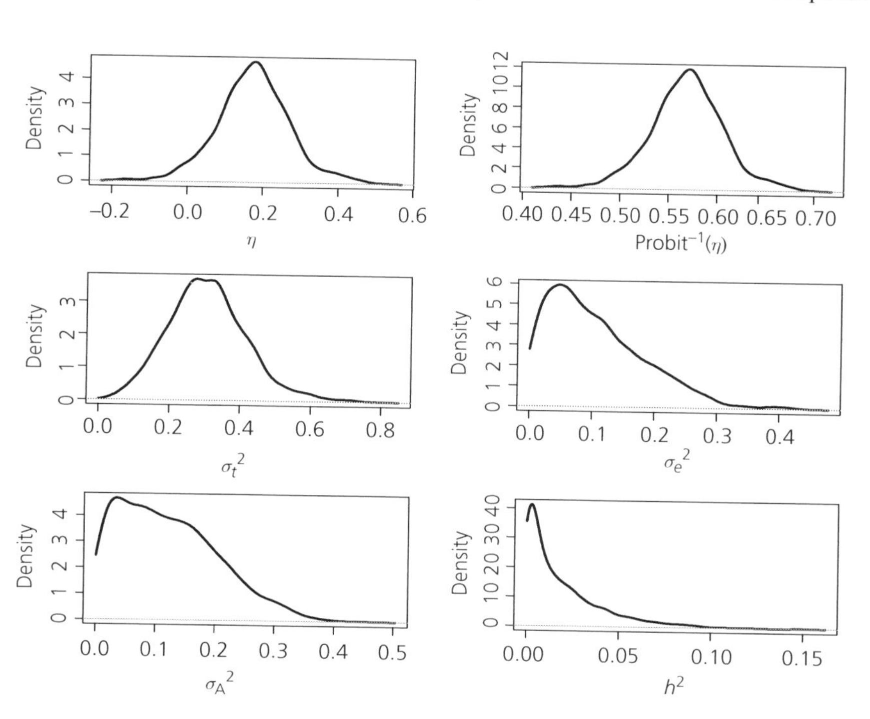

Figure 14.6 Posterior density distributions for parameters of the capture–recapture animal model used for the blue tit data. Notation: η is the mean survival on the probit scale, probit$^{-1}(\eta)$ is the mean survival after back-transformation, σ_t^2 is the variance of the yearly random effect, σ_e^2 is the variance of the non-genetic individual effect, σ_A^2 is the additive genetic variance and h^2 is the heritability.

passerine. The additive genetic variance σ_A^2 was low, resulting in a low heritability h^2 (Figure 14.6). The environmental variance σ_t^2 was moderate, suggesting that temporal variation in survival should not be neglected.

Overall, one can see the HM implementation of this capture–recapture animal model as simple 'modules' being run simultaneously: one module is for the demographic process, which is connected to an animal model to decompose variability in survival; the other module is for the observation process and is driven by the detection probabilities. Here, the flow of information is quite intuitive. It is easy to see how longitudinal individual-based data informs the inference of survival probability, and how variation in relatedness provides the basis for genetic inference. It is possible however, and indeed relatively easy, to code arbitrary models where the flow of information through the model structure is not so obvious. Dangerous situations arise easily if model complexity exceeds the intuitivity of the flow of information. Prior information, even if specified in a way that seems uninformative, can easily mask problems in multi-parameter models (Lele 2010, and Lavine 2010, generally); it is possible for innocuous priors to lead to apparent high precision (marked peaks) in posterior distributions of parameters for which there is actually no information. Whilst we generally avoid technical details here, it is worth noting that convergence (the desirable situation where an MCMC algorithm is collecting samples in the model's true region of highest probability density) can be difficult to diagnose as well in complex models. The first line of defence should be biological common sense: follow the flow of information from the data through the model to the parameters.

14.3.3 Towards comprehensive consideration of uncertainty in complex evolutionary analyses

In many empirical studies, multiple statistical procedures are applied to a given dataset. Often the outputs of some procedures serve as inputs for subsequent statistical tests or mathematical procedures. Consideration of statistical uncertainty, e.g. calculation of standard errors, and statistical hypothesis tests are then often conducted assuming that inputs to the very last statistical procedure represent error-free (i.e. statistical sampling error) observations, when in fact they are often themselves statistical parameters or summary statistics that may only be estimated with error. In general 'doing statistics on statistics' will not thoroughly account for statistical uncertainty in any but the last analytical procedures, and so will lead to anticonservative statistical inference. Furthermore, 'doing statistics on statistics', even if the 'statistics upon which statistics are done' are simple, i.e. means from multiple measures at some level of replication, can lead to very severe statistical biases, even when the first steps of statistical analysis seem to be very simple and pragmatic procedures. In this subsection, we present several illustrations of evolutionary quantitative genetic studies (not all exclusively about estimating genetic parameters) where Bayesian approaches have been demonstrated to provide robust analyses in complex problems.

The previous section lays out one potentially comprehensive way of avoiding 'doing statistics on statistics'. Bayesian methods will often facilitate explicit combination of two or more simple models into one more comprehensive model of critically related biological phenomena. More immediately though, samples of posterior distributions of fitted models obtained by MCMC methods have some very convenient properties. They can represent a complete description of (un)certainty of parameter values, and the ways in which uncertainty in one value correlates with uncertainty in another, given the data, the model structure, and the prior specifications. This is a very convenient feature of an applied Bayesian analysis.

A striking recent application in which consideration of the full uncertainty greatly changed the interpretation of a biological result was described by Hadfield *et al.* (2010). Given a fitted animal model, the predicted breeding values can be extracted. Several studies have extracted breeding values from animal model-based analysis of long-term studies, and used them to describe features of the genetics of those populations. A particularly interesting application is to conduct a test of whether or not mean breeding value has changed over time: simply, the regression of breeding values on time provides a test for microevolution.

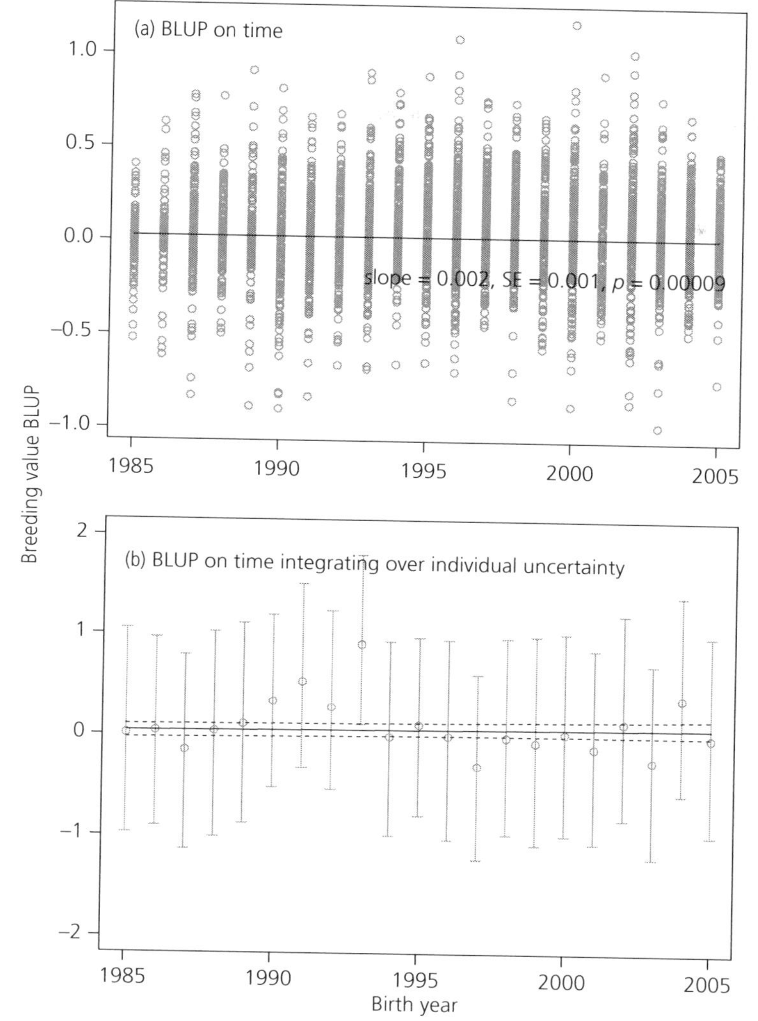

Figure 14.7 Animal model-based inference of microevolution in Soay sheep on St Kilda. (a) shows best linear unbiased predictions (BLUPs) of breeding values of individuals born 1985 to 2005, and their regression on birth year. (b) shows the same regression, but accounting for the fact that each BLUP, i.e. each point in (a), is not a known value, but rather an estimate with uncertainty. Grey points show the BLUP for one randomly chosen individual from each cohort, with 95% CI. The line (with 95% CI of the prediction, integrating over all uncertainty) is the regression corresponding to that in (a); note that this line represents a regression based on all individuals, as in (a), the selection of a single individual per cohort was conducted only for the purpose of plotting. In this example, 18.9% of the posterior distribution of the regression of breeding value on time has a negative slope, corresponding roughly to a two-tailed *P* value of 0.378. Note that Hadfield *et al.* 2010 considered the regression of mean cohort breeding value on time, and we consider the regression of individual breeding value on birth year. The analyses are similar but not identical; our alternative presentation is presented to highlight uncertainty in individual breeding values.

Figure 14.7 shows two regressions. Plot (a) shows the regression of the predicted breeding value of leg length of each Soay sheep (*Ovis aries*) from the ongoing study on the island of Hirta, St Kilda, Outer Hebrides, Scotland (Clutton-Brock & Pemberton 2004), born between 1985 and 2005. In plot (b), the breeding value of a randomly selected individual from each year is plotted with its associated 95% CI for illustration, along with the regression over all individuals in the population (the same line as in plot a). This prediction interval in (b) contains 95% of the density of predictions from regressions of breeding value on year, conducted for each of 1000 samples of the posterior distribution of the same (Bayesian) animal model that was used to get the breeding values in plot (a).

The difference between the two ascertainments of uncertainty in the regression of breeding value on year is very stark. In plot (a), the predicted breeding values are taken to be known values, but as the representative posterior distributions of breeding value in plot (b) show, they are anything but known. Furthermore, individuals that are alive in any given year tend to be closer relatives to other individuals alive at that time, or around that time, than to individuals that lived much earlier or later in the study. Therefore the very feature of the study that allows genetic parameters to be estimated, the pedigree, causes complex patterns of covariance both in similarity among individuals, but also causes complex patterns of uncertainty in the breeding values of individuals. Integrating over the full posterior distribution of the breeding values in the regression of breeding value on time acknowledges all of the complex patterns of uncertainty in breeding values.

More broadly, there is little tradition in quantitative genetics of considering and reporting uncertainty in estimates of many types of parameters. This is a little bit surprising, given that the field is necessarily so fundamentally statistical. However, it is also a natural result of the predominantly frequentist methods that have dominated the field, and the difficulties that can arise in even approximately describing statistical uncertainty in derived parameters (see Section 14.2.1). Parameters for which SEs, or any other assessment of statistical uncertainty, are rarely provided in quantitative genetic studies include predictions of evolutionary trajectories based on the breeder's or Lande equations (Lande 1979), and descriptions of the geometry of **G**-matrices (reviewed in Walsh & Blows 2009) such as the direction or length of $\mathbf{G}_{max}$ (the dominant eigenvector of the *G*-matrix), and the constraints that **G** may impose on adaptive evolution, as potentially described by evolvability and respondability (Hansen & Houle 2008), or metrics of constraint based on the effect of genetic correlations on the rate of adaptation (Stinchcombe and Agrawal 2009). As meta-analysis becomes more important in ecology, genetics, and evolutionary biology in general, there will be an increasing need for estimates of such parameters to be accompanied by metrics of their uncertainty.

Teplitsky *et al.* (2011) inferred the extent to which genetic correlations constrain the rate of adaptation of breeding traits in barn swallows (*Hirundo rustica*), and used Bayesian methods to evaluate the uncertainty in their estimate. One of their goals was to evaluate Agrawal and Stinchcombe 's (2009) metric of constraint due to genetic correlations, which is defined as

$$R = \frac{\Delta \bar{W}_{\mathrm{G}}}{\Delta \bar{W}_{\mathrm{I}}} \tag{14.20}$$

where $\Delta \bar{W}$ is the change in population absolute fitness due to one generation of response to selection, and subscripts G and I denote the change in mean fitness accounting for and discounting genetic correlations, respectively. $\Delta \bar{z}$ is obtained in the standard way according to the Lande (1979) equation $\Delta \bar{z} = \boldsymbol{G\beta}$, and the change in absolute fitness due to evolution is $\Delta W(\Delta \bar{z}) = \Delta \bar{z}^{\mathrm{T}} \boldsymbol{\beta} + \frac{1}{2} \Delta \bar{z}^{\mathrm{T}} \boldsymbol{\gamma} \Delta \bar{\mathbf{z}}$ where $\boldsymbol{\beta}$ and $\boldsymbol{\gamma}$ are vectors and matrices of directional and quadratic selection differentials, respectively.

Clearly, R is a very useful statistic (see also Chapter 12, Teplitsky *et al.*), as it boils down the influences and interactions of multiple aspects of genetic variation, covariation, and their complex relationship with fitness, into a single statistic with a straightforward evolutionary meaning. However, the equations involved in obtaining R represent a rather complex, if biologically highly interpretable, transformation of estimates of genetics and selection, and critically, these estimates are all made with error. Indeed, these estimates, i.e. genetic parameters and aspects of the phenotype–fitness map, are some of the most notoriously difficult parameters to characterise with precision in ecological studies. Teplitsky *et al.* (2011) estimated all of the parameters using Bayesian mixed and ordinary linear models, fitted using the R-package MCMCglmm (Hadfield 2010), and applied the calculation of R to many effectively independent samples of the posterior solution of mixed models characterising $\boldsymbol{G}$, $\boldsymbol{\beta}$, and $\boldsymbol{\gamma}$. The posterior distributions of each component parameter of genetics and selection are shown in Figures 14.8 and 14.9, respectively, and the posterior distribution of R is given in Figure 14.10a.

Morrissey *et al.* (2012) conducted a similar exercise to characterise the effect of genetic correlations on the rate of adaptation of female life-histories in red deer (*Cervus elaphus*). In their analysis, Bayesian techniques allowed demographic theory to be combined with quantitative genetic inferences in order to simultaneously evaluate genetic and selective parameters. As in Teplitsky *et al.* (2011), this allowed inference to be made of the full posterior distribution of R. The inference of R was remarkably similar to that in barn swallows (Figure 14.10), but somewhat less precise. Additionally, Morrissey *et al.* (2012) calculated R under the assumption of the phenotypic gambit, i.e. that phenotypic

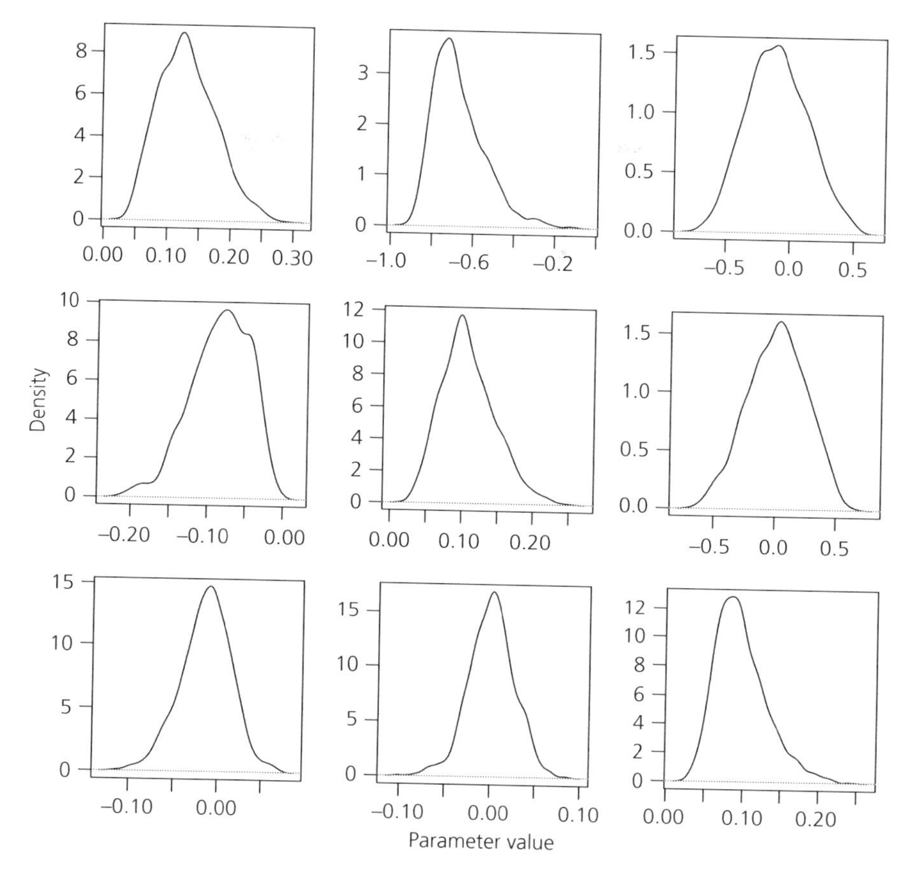

Figure 14.8 Posterior distributions of quantitative genetic parameters of breeding traits in Spanish barn swallows. Traits (top to bottom and left to right) are standardised arrival date, delay before breeding, and clutch size. Diagonal plots are genetic variances, below and above diagonal plots are genetic covariances and correlations, respectively. Posterior distributions are provided by C. Teplitsky from analyses reported in Teplitsky *et al.* (2011).

patterns are substitutable for genetic patterns, and showed that phenotypic correlations do not reveal the influence of genetics on evolutionary trajectories.

The evolutionary constraint metric *R*, and other such parameters, are difficult to characterise with certainty, and also are fundamentally parameters of particular populations at particular times and places, and in particular ecological conditions. Publishing the uncertainty in such metrics will ultimately be necessary to facilitate robust meta-analysis. In turn, meta-analysis will ultimately allow inferences about the general importance of different hypothesised processes, or the mean and range of values of parameters that are difficult to characterise, and that are specific to case studies. In support of meta-analysis, standardised reporting of features of posterior distributions will become highly worthwhile. At the very least, it would be useful if the SDs of the posterior distributions of important parameters were more generally reported (slightly turning a blind eye to philosophy, these may be used as SEs when implementing meta-analyses). Future developments in meta-analysis may allow the complexity of posterior distributions of parameters in individual studies to be accommodated, and so reporting of posterior means, modes, and quantiles (quartiles and 95% ranges may be generally useful) may also eventually prove beneficial.

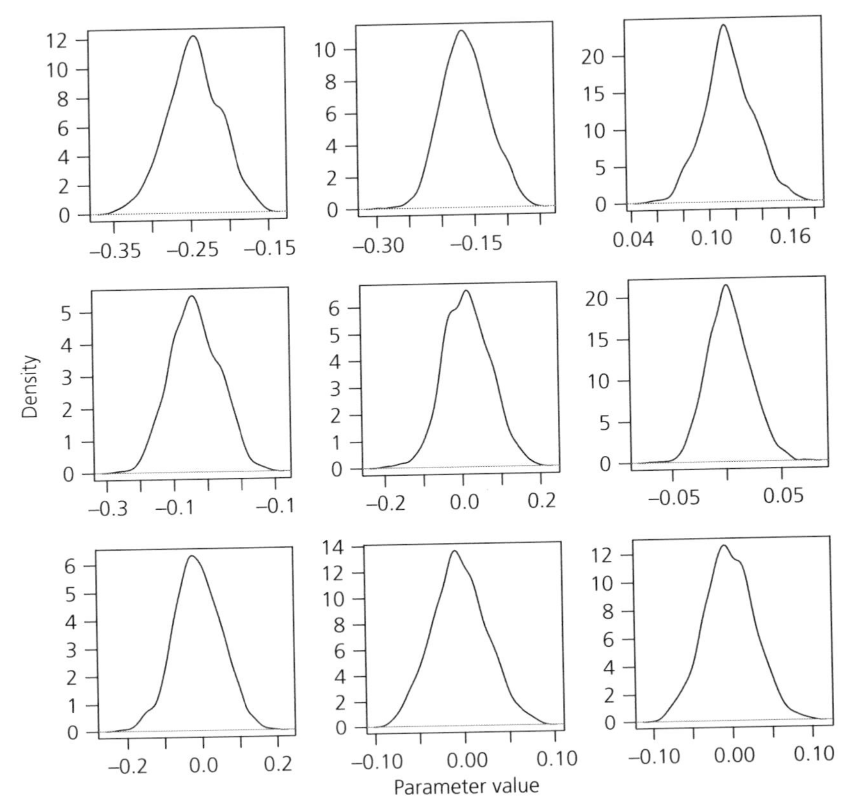

Figure 14.9 Posterior distributions of standardised selection gradients of breeding traits in Spanish barn swallows. Top and middle rows are (left to right) directional and quadratic gradients for arrival date, delay before breeding, and clutch size. Bottom row (left to right) are correlational selection gradients for arrival date and breeding delay, arrival date and clutch size, and breeding delay and clutch size. Posterior distributions are provided by C. Teplitsky from analyses reported in Teplitsky *et al*. (2011).

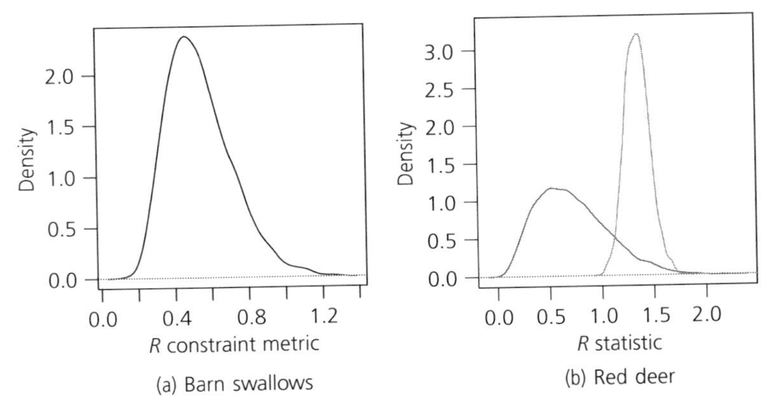

Figure 14.10 The influence of genetic correlations on the rate of adaptation of (a) breeding traits in Spanish barn swallows and (b) female life-history traits in red deer. In (b), the dark curve represents the posterior distribution of the constraint metric R, and the light curve represents R calculated under the assumption that $\mathbf{P}$ is substitutable for $\mathbf{G}$. The constraint metric R from Agrawal and Stinchcombe (2009) describes the proportion by which genetic correlations change the rate of adaptation (increase of population mean absolute fitness) relative to the rate of adaptation that would occur based on selection gradients and additive genetic variances if genetic correlations were zero. The posterior distribution in (a) is provided by C. Teplitsky from analyses reported in Teplitsky *et al*. (2011).

14.4 Conclusion

We hope that our first section's work through a simple problem highlighted some aspects of the flexibility of Bayesian approaches and currently available associated tools. Our sections on the current applications of Bayesian methodologies are not much more than scattered reports from the frontier of the quantitative genetic analysis of wild populations. For example, Bayesian methods are also contributing greatly to pedigree reconstruction, both in judging uncertainties in pedigrees, and allowing for complex patterns of missingness, whereby sibling relationships can be inferred despite missing the genetic information from the parents. Both of these desirable features of parentage analysis are obtainable without Bayesian methods, but particularly elegant Bayesian solutions have been provided in the software packages MasterBayes (Hadfield *et al.* 2006) and Colony2 (Wang 2004, Jones and Wang 2009). Our theme, throughout both the initial practical demonstration and the subsequent more cutting-edge examples, that Bayesian tools provide pragmatic ways forward in the complex realities of nature, is motivated by a belief that there is much more to come. Other areas where Bayesian methods may provide further benefits for wild quantitative genetics include:

(1) *Incorporation of spatial and temporal structure.* Relatives may generally vary in space for non-genetic reasons, and for reasons that may not be entirely explainable with available data (Stopher *et al.* 2012). Bayesian methods may facilitate much more widespread incorporation of spatial structure in quantitative genetic analyses of data from wild populations.

(2) *Robust modelling.* Data from natural populations tend to contain many outliers. Bayesian methods can generally allow incorporation of very general models of the distribution of any parameters, and this could include the use of thick-tailed distributions where outliers may exist. For example, t-distributions may generally be usable where normal distributions are currently more typical. Sorensen and Gianola (2002) discuss robust Bayesian quantitative genetic modelling in some detail.

(3) *More general models of the observation process.* Wild animals are notorious for moving around, and this creates problems for wild quantitative genetics beyond ascertainment of survival (as discussed in Section 14.2). Reproductive success is probably also generally underestimated in nature, and models of the process of observing reproduction (i.e. detectability of mating events and offspring; uncertainty in pedigrees; inference of survival on non-breeders when breeders are most easily censused) could be particularly beneficial. This will be beneficial both for understanding the quantitative genetics of reproduction, and also, considering quantitative genetics more broadly, for ascertainment of fitness in analyses of the selection of quantitative traits in nature.

References

Agrawal, A.F. & Stinchcombe, A.F. (2009) How much do genetic covariances alter the rate of adaptation? *Proceedings of the Royal Society B–Biological Sciences*, **364**, 1593–1605.

Bolker, B.M., Brooks, M.E., Clark, C.J., Geange, S.W., Poulsen, J.R., Stevens, M.H.H. & White, J.-S.S. (2009) Generalized linear mixed models: a practical guide for ecology and evolution. *Trends in Ecology and Evolution*, **24**, 127–135.

Breslow, N.E. & Clayton, D.G. (1993) Approximate inference in generalized linear mixed models. *Journal of the American Statistical Association*, **88**, 9–25.

Brooks, S., Gelman, A., Jones, G.L. & Meng, X.-L. (2011) *Handbook of Markov Chain Monte Carlo*. Chapman and Hall, Boca Raton.

Buoro, M., Prevost, E. & Gimenez, O. (2012) Digging through model complexity: using hierarchical models to uncover evolutionary processes in the wild. *Journal of Evolutionary Biology*, **25**, 2077–2090.

Clutton-Brock, T.H. & Pemberton, J.M., eds. (2004) *Soay sheep: dynamics and selection in an island population*. Cambridge University Press, Cambridge.

Cressie, N., Calder, C., Clark, J.S., Ver Hoef, J.M. & Wikle, C.K. (2009) Accounting for uncertainty in ecological analysis: the strengths and limitations of hierarchical statistical modeling. *Ecological Applications*, **19**, 553–70.

Dempster, E.R. & Lerner, I.M. (1950) Heritability of threshold characters. *Genetics*, **35**, 212–236.

Ellison, A.M. (2004) Bayesian inference in ecology. *Ecology Letters*, **7**, 509–520.

Falconer, D.S. & Mackay, T.F.C. (1989) *Introduction to quantitative genetics*. Longman Group, London.

Gelman, A. (2006) Prior distributions for variance parameters in hierarchical models. *Bayesian Analysis*, **1**, 515–533.

Gelman, A., Carlin, J.B., Stern, H.S. & Rubin, D.S (2004) *Bayesian data analysis*. Chapman and Hall, New York.

Gelman, A. & Hill, J.S. (2007) *Data analysis using regression and multilevel/hierarchical models*. Cambridge University Press, Cambridge.

Gelman, A., Jakulin, A., Pittau, M.A. & Su, Y.S. (2008) A weakly informative default prior distribution for logistic and other regression models. *The Annals of Applied Statistics*, **2**, 1360–1383.

Gilmour, A.R., Anderson, R.D. & Rae, A.L. (1985) The analysis of binomial data by a generalized linear mixed model. *Biometrika*, **72**, 593–599.

Gilmour, A.R., Cullis, B.R., Welham, S.J. & Thompson, R. (1999) *ASREML, reference manual*. Biometric bulletin, no 3, NSW Agriculture, Orange Agricultural Institute, Forrest Road, Orange, NSW Australia.

Gilmour, A.R., Gogel, B.J., Cullis, B.R. & Thompson, R. (2006) *ASReml user guide release 2.0*. VSN International Ltd. www.vsni.co.uk/software/asreml/.

Gimenez, O., Lebreton, J.-D., Gaillard, J.-M., Choquet, R. & Pradel, R. (2012) Estimating demographic parameters using hidden process dynamic models. *Theoretical Population Biology*, **82**, 307–316.

Gimenez, O., Rossi, V., Choquet, R., Dehais, C., Doris, B., Varella, H., Vila, J.-P. & Pradel, R. (2007) State-space modelling of data on marked individuals. *Ecological Modelling*, **206**, 431–438.

Gimenez, O., Viallefont, A., Charmantier, A., Pradel, R., Cam, E., Brown, C.R., Covas, R. & Gallard, J.M. (2008) The risk of flawed inference in evolutionary studies when detectability is less than one. *The American Naturalist*, **172**, 441–448.

Goldstein, H. & Rasbash, J. (1996) Improved approximations for multilevel models with binary responses. *Journal of the Royal Statistical Society Series A*, **159**, 505–513.

Grosbois, V., Gimenez, O., Gaillard, J.-M., Pradel, R., Barbraud, C., Clobert, J., Muller, A.P. & Weimerskirch, H. (2008) Assessing the impact of climate variation on survival in vertebrate populations. *Biological Reviews*, **83**, 357–399.

Hadfield, J.D. (2010) MCMC methods for multi-response generalized linear mixed models: the MCMCglmm R package. *Journal of Statistical Software*, **33**, 1–36.

Hadfield, J.D., Richardson, D.S. & Burke, T. (2006) Towards unbiased parentage assignment: combining genetic, behavioural and spatial data in a Bayesian framework. *Molecular Ecology*, **15**, 3715–3731.

Hadfield, J.D., Wilson, A.J., Garant, D., Sheldon, B.C., and Kruuk, L.E.B. (2010) The use and misuse of BLUP in ecology and evolution. *The American Naturalist*, **175**, 116–125.

Hansen, T.F. & Houle, D. (2008) Measuring and comparing evolvability and constraint in multivariate characters. *Journal of Evolutionary Biology*, **21**, 1201–1219.

Jones, O. & Wang, J. (2009) COLONY: a program for parentage and sibship inference from multilocus genotype data. *Molecular Ecology Resources*, **10**, 551–555.

Kruschke, J.K. (2011) *Doing Bayesian data analysis: a tutorial with R and BUGS*. Academic Press New York.

Lande, R. (1978) Quantitative genetic analysis of multivariate evolution applied to brain: body allometry. *Evolution*, **33**, 402–416.

Lavine, M. (2010) Living dangerously with big fancy models. *Ecology*, **91**, 3487.

Lebreton, J.-D., Burnham, K.P., Clobert, J. & Anderson, D.R. (1992) Modeling survival and testing biological hypotheses using marked animals: a unified approach with case studies. *Ecological Monographs*, **62**, 67–118.

Lele, S.R. (2010) Model complexity and information in the data: could it be a house built on sand? *Ecology*, **91**, 3493–3496.

Lunn, D.J., Thomas, A., Best, N. & Spiegelhalter, D. (2000) WinBUGS: a Bayesian modelling framework: concepts, structure and extensibility. *Statistics and Computing*, **10**, 325–337.

Lynch, M. & Walsh, B. (1998) *Genetics and analysis of quantitative traits*. Sinauer, New York.

Morrissey, M.B., Walling, C.A., Wilson, A.J., Pemberton, J.M., Clutton-Brock, T.H. & Kruuk, L.E. (2012) Genetic analysis of life-history constraint and evolution in a wild ungulate population. *American Naturalist*, **179**, E97–114.

Nakagawa, S. & Schielzeth, H. (2010) Repeatability for Gaussian and non Gaussian data: a practical guide for biologists. *Biological Reviews*, **85**, 935–956.

Nakagawa, S. & Schielzeth, H. (2013) A general and simple method for obtaining R2 from generalized linear mixed-effects models. *Methods in Ecology and Evolution*, **4**, 133–142.

O'Hara, R.B., Cano, J.M., Ovaskainen, O., Teplitsky, C. & Alho, J.S. (2008) Bayesian approaches in evolutionary quantitative genetics. *Journal of Evolutionary Biology*, **21**, 949–957.

Papaix, J., Cubaynes, S., Buoro, M., Charmantier, A., Perret, P. & Gimenez, O. (2010) Combining capture-recapture data and pedigree information to assess heritability of demographic parameters in the wild. *Journal of Evolutionary Biology*, **23**, 2176–2184.

Plummer, M. (2003) JAGS: A program for analysis of Bayesian graphical models using Gibbs sampling. In:

Proceedings of the 3rd international workshop on distributed statistical computing, March 20–22, Vienna, Austria.

R Core Development Team. (2005) *R: a language and environment for statistical computing*. R Foundation for Statistical Computing, Vienna, Austria.

Rivot, E. & Prevost, E. (2002) Hierarchical Bayesian analysis of capture-mark-recapture data. *Canadian Journal of Fisheries and Aquatic Sciences*, **1784**, 1768–1784.

Royle, J.A. (2008) Modeling individual effects in the Cormack-Jolly-Seber Model: a state-space formulation. *Biometrics*, **64**, 364–370.

Royle, J.A. & Dorazio, R.M. (2008) *Hierarchical modeling and inference in ecology: the analysis of data from populations, metapopulations, and communities*. Academic Press, San Diego.

Simmons, L.W. & Garcia-Gonzalez, F. (2007) Female crickets trade offspring viability for fecundity. *Journal of Evolutionary Biology*, **20**, 1617–1623.

Sorensen, D. & Gianola, D. (2002) *Likelihood, Bayesian and MCMC methods in quantitative genetics*. Springer, New York.

Stopher, K.V., Walling, C.A., Morris, A., Guinness, F.E., Clutton-Brock, T.H., Pemberton, J.M. & Nussey, D.H. (2012) Shared spatial effects on quantitative genetic parameters: accounting for spatial autocorrelation and home range overlap reduces estimates of heritability in wild red deer. *Evolution*, **66**, 2411–2426.

Teplitsky, C., Mouawad, N.G., Balbontin, J., de Lope, F. & Moller, A.P. (2011) Quantitative genetics of migration syndromes: a study of two barn swallow populations. *Journal of Evolutionary Biology*, **24**, 2025–2039.

de Villemereuil, P., Gimenez, O. & Doligez, B. (2013) Comparing parent–offspring regression with frequentist and Bayesian animal models to estimate heritability in wild populations: a simulation study for Gaussian and binary traits. *Methods in Ecology and Evolution*, **4**, 260–275.

Walsh, B. & Blows, M.W. (2009) Abundant genetic variation + strong selection = multivariate genetic constaints: a geometric view of adaptation. *Annual Review of Ecology, Evolution and Systemetics*, **40**, 41–59.

Wang, J. (2004) Sibship reconstruction from genetic data with typing errors. *Genetics*, **166**, 1963–1979.

Wright, S. (1934) An analysis of variability in number of digits in an inbred strain of guinea pigs. *Genetics*, **19**, 506–536.

CHAPTER 15

Evolutionary dynamics in response to climate change

Phillip Gienapp and Jon E. Brommer

15.1 Adaptation to environmental change

Human activities have drastically altered the environment of many species, with often negative consequences for population persistence. For example, increased human land use, introduced alien species, overexploitation and climate change have led to a loss of global biodiversity (Butchart *et al.* 2010). Warming temperatures, caused by human CO_2 emissions, have affected species and populations in various ways, e.g. shifted species' distributions and advanced their phenology, i.e. the timing of life-cycle events (Parmesan 2006). In principle, populations can respond to such environmental changes in three ways to avoid extinction: dispersal to unaffected areas, adaptation through phenotypic plasticity, or micro-evolution.

Dispersal to unaffected habitats mainly depends on the species' dispersal ability and the availability of unaffected habitat, neither of which may be sufficient to prevent extinction (Thomas *et al.* 2004). Even if unaffected habitat is available and can be reached with the species' dispersal ability, it is highly unlikely that dispersing individuals will encounter the exact same suite of biotic and abiotic conditions. For example, the species composition in the newly colonised areas will be different because not all species of a certain biome have similar dispersal abilities (e.g. Graham & Grimm 1990; Berteaux *et al.* 2004). Even if species would shift pole wards at the same rate as the climate warms and hence would not experience a different climate, they will still experience a different photoperiod, which has been shown to affect phenology (e.g. Lambrechts & Perret 2000; Bradshaw & Holzapfel 2001; Gienapp *et al.* 2010; Valtonen *et al.* 2011).

Phenotypic plasticity, the ability of genotypes to express different phenotypes depending on environmental conditions, enables individuals to cope with novel environmental conditions and thereby allows colonisation of novel environments (Yeh & Price 2004) or tracking of phenological changes in the environment (Charmantier *et al.* 2008). Reaction norms are however unlikely to be optimal outside the environmental conditions under which they evolved (Ghalambor *et al.* 2007). This limits the probability that adaptation to environmental change through phenotypic plasticity alone will be possible.

Whilst dispersal and phenotypic plasticity may enable individuals to cope with environmental change in the short term, in the long term, environmental change will lead to selection on ecologically important traits. A response to this selection obviously requires the affected trait(s) to be heritable. Quantifying heritabilities is hence important if we want to assess the possible extinction risk posed by environmental change. Any environmental change will initially move the optimal phenotype away from the population mean phenotype (assuming the population was well adapted to previous conditions); this will lead to selection but will also reduce population mean fitness (Lynch & Lande 1993; Bürger & Lynch 1995). Under ongoing environmental change the population mean

Quantitative Genetics in the Wild. Edited by Anne Charmantier, Dany Garant, and Loeske E. B. Kruuk

phenotype will consistently lag behind the optimal phenotype. If this 'evolutionary lag' becomes too large, due to fast environmental change or low genetic variance, the associated 'evolutionary load', i.e. the reduction in population mean fitness, can elevate extinction risk considerably (Lynch & Lande 1993; Bürger & Lynch 1995). Following environmental change, 'evolutionary rescue', i.e. the avoidance of population extinction by micro-evolutionary adaptation (Gonzalez *et al.* 2013), is hence only possible if sufficient genetic variation in the trait(s) that came under selection by environmental change exists. Consequently, we need reliable estimates of genetic (co)variances, among other things. Estimating genetic (co)variances is the key 'trade' of quantitative genetics, and it can thereby help us to understand adaptation to environmental change and its consequences. Another important aspect of understanding responses to environmental change is being able to disentangle changes in phenotypes due to phenotypic plasticity and micro-evolutionary change.

In this chapter we will not undertake to comprehensively review the literature about genetic adaptations to climate change. Instead we illustrate how quantitative genetics studies of natural populations can contribute to understanding adaptation to environmental change. First, we explore evidence for selection driven by climate change, as selection is obviously an important component of evolutionary change. Second, as pointed out above, genetic (co)variances of traits that are affected, or likely to be affected, by climate change are relevant for understanding potential evolutionary responses, and we give an overview of quantitative genetic studies of these traits. As phenotypic changes in response to environmental change can be due to phenotypic plasticity or evolutionary change, a thorough understanding of phenotypic plasticity can help to dissect whether an observed change is plastic or genetic (or a combination of both). Consequently, we devote two boxes and one section to this topic. After having explored selection, genetics, and phenotypic plasticity of traits affected by climate change, we 'sum up' our findings and also look beyond quantitative genetic approaches. We finish with general conclusions and an outlook on possible future directions in this field.

We will especially focus on phenological responses to climate change because, first, phenology, i.e. the seasonal timing of life-cycle events, as flowering, breeding, migration or hibernation, has important consequences for individual fitness; and, second, phenological changes are among the most frequently documented responses to climate change. About 90% of the studies that reported trait changes, and that were reviewed in Root *et al.* (2004), concerned phenological traits.

15.2 Climate change and selection

Adaptation through micro-evolution is necessary if environmental change has led to selection on a given trait because it has shifted the optimal phenotype away from the population mean phenotype. For most phenological events an 'optimal time window' exists which is often set by the phenology of other trophic levels (Visser & Both 2005). For example, optimal breeding time in birds and mammals depends strongly on the phenology of key food resources (Visser *et al.* 2006; Durant *et al.* 2007; Post & Forchhammer 2008). The phenology of lower trophic levels (producers and primary consumers) has advanced more strongly than the phenology of higher trophic levels (secondary consumers) due to climate change (Thackeray *et al.* 2010). The phenology of many secondary consumers is now too late in relation to the optimal time window, and climate change is hence likely to have led to selection on the phenology of higher trophic levels.

Despite the widespread changes in phenology attributed to climate change, direct evidence for selection on phenology induced by climate change is rare. This is likely because of a number of reasons. To demonstrate that climate change has led to selection, it is not sufficient to show that selection has changed in line with climate change but it is necessary to identify the ecological factor(s) that drive(s) selection and that they have been affected by climate change. One of the few good examples concerning phenology is timing of breeding in great tits (*Parus major*). In this species, the synchrony between the nestling phase and the peak abundance in caterpillars determines reproductive success (van Noordwijk *et al.* 1995; Visser *et al.* 2006).

Caterpillar phenology is determined by spring temperatures whose increase has led to an advancement of peak caterpillar abundance (Visser *et al.* 2006). This advanced caterpillar peak abundance has led to an increasing mismatch between birds and caterpillars, which increased selection on the birds breeding time (Visser *et al.* 1998; Visser *et al.* 2006). A number of other studies reported selection on phenology measured in the field but could or did not show that, e.g. selection has increased over time, that selection was causally related to an ecological factor or that this ecological factor has been affected by climate change (e.g. Réale *et al.* 2003a; Sheldon *et al.* 2003; Charmantier *et al.* 2006).

If we consider other traits besides phenology, the picture does not change much. Declines in body size of birds and mammals have frequently been reported (e.g. Yom-Tov 2001; Gardner *et al.* 2009; van Buskirk *et al.* 2010). These trends have often been interpreted as adaptive responses to increasing temperatures according to Bergmann's rule, although possible alternative explanations exist (Goodman *et al.* 2012). Actually, only very few studies estimated selection on body size, and those which did so typically found no selection or only weak selection even in the opposite direction (Teplitsky *et al.* 2008; Husby *et al.* 2011a, but see Ozgul *et al.* 2010). The brown colour-morph of tawny owls (*Strix aluco*) suffered greater mortality than the grey morph during the harsher winters of the 1980s in southern Finland, but as the winters became milder during the last two decades, the difference in survival between these morphs disappeared (Karell *et al.* 2011), suggesting that climate change has altered the survival selection acting on this species' plumage colouration.

Another possible reason that limits the direct evidence for climate change-induced selection is that quantifying individual fitness, especially reproductive success, in the field is still (very) difficult in certain taxa. Whilst linking offspring to their parents based on observations is straightforward (notwithstanding extra-pair paternities) in taxa with extensive brood care, such as birds and mammals, this approach is not possible in taxa without brood care, as e.g. most insects or fish. In these taxa, this is basically only possible using molecular genetic markers (e.g. DiBattista *et al.* 2007; Seamons *et al.* 2007, see also Chapter 9, Zajitschek & Bonduriansky).

Overall, the evidence for climate change-induced selection is (surprisingly) scarce, but this may be more a consequence of a lack of data rather than a true absence of climate change-induced selection. For example, given the facts that phenology has generally advanced at different rates on different trophic levels (Thackeray *et al.* 2010) and that the phenological synchrony between trophic levels is important for the reproductive success of the higher trophic level (Durant *et al.* 2005), it is probably safe to assume that climate change already has or will lead to selection on phenology.

15.3 Genetic (co)variances of climate change-affected traits

Genetic variances and covariances are key components of evolutionary change and reliable estimates are hence crucial to understand—and to some extent predict—responses to environmental change and possible demographic consequences (e.g. Lynch & Lande 1993; Bürger & Lynch 1995; Gienapp *et al.* 2013). Quantitative genetic analyses of relevant traits in wild populations using the animal model can make important contributions to this. As pointed out above, climate change has had strong effects on phenological traits which are expected to have led to selection on these traits. Consequently, it would be relevant to know how heritable such traits are. Several studies applying the animal model found avian breeding time and migration time, measured as arrival at the breeding grounds, to be heritable (see Charmantier & Gienapp, 2014 for a recent review). In mammals, four studies found significant heritability for a number of phenological traits, including hibernation date and parturition date, in red squirrels (*Tamiasciurus hudsonicus*), ground squirrels (*Urocitellus columbianus*), red deer (*Cervus elaphus*) and Soay sheep (*Ovis aries*; Réale *et al.* 2003b; Kruuk & Hadfield 2007; Clements *et al.* 2010; Lane *et al.* 2011). Studies on the quantitative genetics of phenological traits have of course not been limited to wild populations. Laboratory or experimental studies have provided further evidence in this field. For example, migratory behaviours in birds, that can be measured in the laboratory as migratory restlessness, have been found to be heritable (reviewed in Pulido 2007). Similarly, phenological traits have also been found to be moderately heritable in plants (reviewed in Geber & Griffen 2003) and salmonids (reviewed in Carlson & Seamons 2008).

The skew of quantitative genetic studies in wild populations towards avian phenology, and especially breeding time, reflects constraints and difficulties in obtaining the necessary high-quality data. Nest box-breeding birds are easily marked, and parent–offspring links are easily established, which has facilitated the many long-term studies with individual data. As pointed out at the end of the last section, establishing parent–offspring links is much more difficult in many species, which limits the application of quantitative genetic approaches. However, whilst our direct knowledge about the genetic architecture of traits that are relevant in the context of climate change may be limited, and certain traits relevant for adaptation to climate change showed low heritabilities (Hoffmann *et al.* 2003; Kelly *et al.* 2012), we would probably be safe to assume that most traits are heritable (Mousseau & Roff 1987; Houle 1992).

Genetic correlations can constrain evolutionary change but also speed it up depending on whether selection pressures are counteracting or working in the same 'direction' (see Chapter 10, Kruuk, *et al.* and Chapter 12, Teplitsky, *et al.*). For example, genetic correlations among life-history and morphological traits in an annual plant considerably reduced the expected response to selection predicted under climate change compared to models ignoring these correlations (Etterson & Shaw 2001). Hence, reliably measuring genetic correlations (and also multivariate selection) can be important for understanding evolutionary responses to climate change. There are by now many heritability estimates for phenological traits but the number of studies that estimated genetic correlations is still somewhat limited. Timing of emergence from hibernation and oestrus were highly genetically correlated in ground squirrels (Lane *et al.* 2011). In red deer, oestrus date, parturition date and timing of rut were genetically correlated, although oestrus date itself was not heritable (Clements *et al.* 2010). Timing of egg-laying and clutch size generally show a negative phenotypic correlation, and a corresponding genetic correlation has been found in one population of great tits (Sheldon *et al.* 2003; Husby *et al.* 2010) but not in another population of the same species (Gienapp *et al.* 2006; Husby *et al.* 2010). These examples illustrate that phenology can be linked with other life-history traits, and that such genetic correlations need to considered when investigating climate change evolutionary responses. It is however generally unclear, whether genetic correlations between traits that are already, or are likely to come, under selection through climate change will be constraining evolutionary responses (reviewed in Hellmann & Pineda-Krch 2007).

15.3.1 Stability of the *G*-Matrix

Another aspect that complicates fully understanding the evolutionary potential to adapt to environmental change is whether the additive genetic variances and additive genetic covariances of traits, making up the **G**-matrix, can be assumed to be stable when the environment changes or whether this **G**-matrix will, through genotype–environment interactions, be substantially altered. For example, Garant *et al.* (2008) estimated the **G**-matrix for laying date, clutch size and egg mass in Wytham Wood's great tits during two time periods (cold and warm) and found it to be stable. Robinson *et al.* (2009) used multivariate random regression to estimate how the genetic correlation between various traits measured in Soay sheep responded to changes in a (non-climatic) measure of environmental conditions. Some correlations increased, whereas other decreased in more favourable conditions. That the **G**-matrix can change rapidly in a wild population was recently demonstrated by Björklund *et al.* (2013), who considered **G**-matrices of several collared flycatcher morphological traits during a 25 year period. At present, there thus seems no coherent expectation for how the **G**-matrix would respond to climate change, and more work on this topic seems worthwhile. Nevertheless, given the obvious challenges in estimating $G \times E$ in a single trait, as documented in this section, extending approaches to multivariate trait relationships is likely to be prohibitively difficult in the majority of wild populations.

15.4 Climate change and phenotypic plasticity in wild populations

As pointed out above, a large fraction of traits affected by climate change are phenological traits (Root *et al.* 2004), and most, if not all, phenological traits are phenotypically plastic in response to weather conditions. For example, individual plants flower earlier, insects emerge earlier, or birds and mammals breed earlier in warmer years. Consequently, we need to consider a trait's phenotypic plasticity to get a more complete picture of potential climate change effects.

First, plasticity forms a 'null expectation' for observed changes in most traits. It is only when plasticity is absent or insufficiently strong to explain an observed change in trait means that one can start to conjecture that evolution has played a role in shifting the trait values. For labile traits, whose population mean values are commonly affected by a climatic variable, formally demonstrating plasticity with respect to climatic conditions requires showing that the trait–climate relationship also occurs on the individual level, which requires repeated measures on individuals under varying climatic conditions (e.g. Przybylo *et al.* 2000; see Box 15.1).

Second, there may be variation between individuals in the trait–climate relationship, which is termed $I \times E$ by Nussey *et al.* (2007), where the environmental variable E captures—in this case—an aspect of climate. Importantly, between-individual variation in plasticity ($I \times E$) may have a heritable component. Trait expression is then subject to genotype–environment interactions ($G \times E$). The interaction $G \times E$ implies that additive genetic variance of a trait changes when the climate changes. Climate change may hence, through $G \times E$, speed up or slow down the potential rate of evolution by increasing or decreasing the heritability of a trait. Hence, if we are to understand whether and how a wild population would respond to climate change, we need to specifically estimate and incorporate $G \times E$ in our models for evolutionary dynamics.

Third, climate change implies that the environment which a local population experiences is altered. This new environment is potentially characterised by 'new rules' which could lead to selection on plasticity itself rather than on the mean trait expression only. In that case, certain genotypic reaction norm slopes (Figure 15.1) may have higher fitness than others. One ecological pathway by which selection on plasticity could arise is if climate change alters the 'cue environment' (which drives trait expression) at a different rate than the 'selection environment' (which determines selection

Box 15.1 Phenotypic plasticity and its quantitative genetics

Broadly speaking, a trait is considered phenotypically plastic when its expression is not fixed at the level of population, individual or genotype (Figure 15.1). Within the context of phenological or life-history traits varying across a climate gradient, one typically deals with labile traits which are expressed repeatedly during an individual's lifetime. Examples are the seasonal timing of breeding and clutch or brood size. Plasticity associated with a labile trait can hence be assessed at three levels (Nussey *et al.* 2007): 1) the population level, where mean trait values vary with the environmental value (Figure 15.1a); and 2) the between-individual level where individuals may be plastic, but between-individual variation in plasticity may be absent (all individuals follow the population-level slope, Figure 15.1b). If among-individual variation in plasticity (Figure 15.1c, termed $I \times E$ by Nussey *et al.* (2007) exists, this can be further partitioned into 3) genotype-environment interaction ($G \times E$) and 4) the interaction with the environment of a non-genetic component called the permanent environment (PE; $PE \times E$). Phenotypic plasticity at the genetic level ($G \times E$) concerns the potential of a genotype to alter its expression as a function of the environment (Via & Lande 1985; Falconer & MacKay 1996). PE effects are any non-heritable effects that cause variation across individuals that is conserved across the repeated records. Examples include early ontogenetic effects due to rearing (maternal or natal effects), or differences in the local environmental quality experienced by the individuals throughout their lifetime (e.g. resource abundance in territory or home range).

continued

Box 15.1 *Continued*

The conceptual division between plasticity occurring at different levels, as described verbally above and as displayed in Figure 15.1, translates into a hierarchical modelling approach based on random regression (Henderson 1982). Random regression of trait z for individual i quantified at instance t under environmental condition E is, on the phenotypic level, described by:

$$z_{i,t,E} = \mu + bE_t + f_{ind}(i, x, E_t) + \varepsilon_{i,t,E} \quad \text{(B15.1.1a)}$$

which can be partitioned further into

$$z_{i,t,E} = \mu + bE_t + f_a(i, x, E_t) + f_{pe}(i, x, E_t) + \varepsilon_{i,t,E} \quad \text{(B15.1.1b)}$$

where μ denotes the fixed-effect mean and E_t the environmental value at instance t fitted as a fixed effect with the regression coefficient b denoting the overall, population-level change in mean z with E. The residual error ε is here allowed to be specific for each instance t and value of E (heterogeneous residuals). Random-regression functions $f_{ind}(i, x, E_t)$, $f_a(i, x, E_t)$ and $f_{pe}(i, x, E_t)$ describe an orthogonal polynomial of order x on the level of the phenotype, additive genetic effects and permanent environment, respectively. These are random effects in a linear mixed model quantifying the difference from the fixed-effect mean specific to each environmental value. For example, a first-order polynomial of the additive genetic effect f_a could for individual i represent the function $a_{0,i} + a_{1,i} \times E_t$. When order $x \geq 1$, the random-regression functions will estimate variances in ind_y in Eq. B15.1.1a, or a_y and pe_y in Eq. B15.1.1b (where $y = 0 \ldots x$) and all the covariances between these. A random-regression animal model (RRAM), described by Eq. B15.1.1b estimates the additive genetic (co)variances from information on the resemblance of population-wide relatives following animal model procedures (Lynch & Walsh 1998).

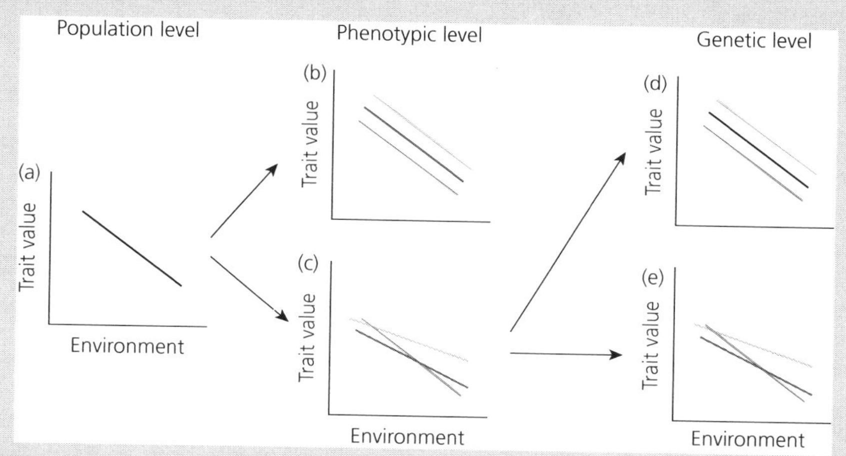

Figure 15.1 Levels of phenotypic plasticity. If a trait is plastic at the population level (a) all individuals may have the same reaction norm slope (but could differ in their reaction norm elevation, b) or vary in their reaction norms (c). If individual reaction norms vary, all variation seen at the phenotypic level (c) is only due to $PE \times E$ (d) or this variation may have a genetic basis ($G \times E$) (e) (this can be illustrated as 'family reaction norms').

on the trait). Although we do not know at present how common such a scenario is, the observed different rates of phenological shifts at different trophic levels (Thackeray *et al.* 2010) make it likely. In order for plasticity to evolve in response to such selection, variation in plasticity across individuals must be heritable, i.e. $G \times E$ needs to be present.

Despite their pivotal role for understanding the consequences of climate change, there are currently few published studies on genotype–climate interactions. Most studies have estimated only the between-individual variance in plasticity $I \times E$ (Table 15.1). A literature search revealed 19 estimates for $I \times E$ of a phenological or life-history

Table 15.1 Summary of studies testing for $I \times E$ for phenological or life-history traits across a climatic gradient in a wild population using random-regression mixed models (see Box 15.1). The reference (Ref) for each study is provided below the Table. Reported are for each species and, when relevant, population which trait was considered and which environmental covariate (Predictor), as well as 'Nind', the number of individuals included in the analysis, 'N/ind', the mean number of repeated records per individual, and 'Nyears', how many years of data were included. 'Pop. trend' indicates whether there was a significant population-level trend (Y) with its direction between brackets, or not (N). '$I \times E$' and '$G \times E$' indicate whether there was statistically significant variance in plasticity (Y) or not (N) on the between-individual ($I \times E$) and genetic ($G \times E$) levels respectively. '$G \times E/I \times E$' shows the proportion of $I \times E$ variance due to $G \times E$, irrespective of the statistical significance of the $G \times E$. The column 'R calculated?' indicates whether the random-regression estimates could be included (Y) or not (N) for calculation of the repeatability in two contrasting environmental conditions (as reported in Table 15.2), where the '(Y)' for one trait-climate relationship reported by Ref. 10 indicates the study was included by assuming that the variance in $I \times E$ equals null. Two estimates (Ref. 4) concern a reanalysis of a previous study using more data, and hence the original studies ((a) Nussey *et al.* 2005b; (b) Charmantier *et al.* 2008) are not included

Ref.	Species: population	Trait	Predictor	Nind	N/ind	Nyears	Pop. trend	$I \times E$	$G \times E$	$G \times E/I \times E$	R calculated?
1	Blue tit (*Cyanistes caeruleus*): D-Muro	Laying date	Temperature	540	1.4	18	Y (–)	N	–	–	Y
1	Blue tit (*Cyanistes caeruleus*): D-Rouviere	Laying date	Temperature	678	1.7	20	Y (–)	Y	–	–	Y
1	Blue tit (*Cyanistes caeruleus*): E-Muro	Laying date	Temperature	173	1.9	13	Y (–)	Y	–	–	Y
1	Blue tit (*Cyanistes caeruleus*): E-Pirio	Laying date	Temperature	640	2.0	22	N	N	–	–	Y
2	Collared flycatcher (*Ficedula albicollis*)	Laying date	Temperature	1126	2.4	23	Y (–)	Y	N	NA	Y
3	Common gull (*Larus canus*)	Laying date	Temperature	2262	5.1	37	Y (–)	Y	N	0.8	Y
4	Great tit (*Parus major*): Hoge Veluwe (a)	Laying date	Temperature	2243	1.6	34	Y (–)	Y	N	0.7	Y
4	Great tit (*Parus major*): Hoge Veluwe	Clutch size	Temperature	2243	1.6	34	N	N	N	1.0	Y
4	Great tit (*Parus major*): Wytham Woods (b)	Laying date	Temperature	4698	1.5	34	Y (–)	Y	N	0.6	Y
4	Great tit (*Parus major*): Wytham Woods	Clutch size	Temperature	4753	1.5	34	N	Y	Y	0.6	Y
5	Guillemot (*Uria aalge*)	Laying date	NAO	245	10.6	24	Y (–)	N	–	–	Y
6	Guillemot (*Uria aalge*)	Laying date	PC1 of various	89	6.6	20	Y (–)	Y	–	–	Y
7	House sparrow (*Passer domesticus*)	Provisoning rate	Rainfall	225	3.3	5	N	Y	–	–	Y
8	Painted turtle (*Chrysemys picta*)	Laying date	Temperature	268	4.9	12	Y (+)	Y	–	–	N
9	Red deer (*Cervus elaphus*)	Calving date: high density period	Rainfall	221	4.7	22	N	Y	–	–	N
9	Red deer (*Cervus elaphus*)	Calving date: low density period	Rainfall	185	6.0	10	Y (+)	Y	–	–	Y
10	Red deer (*Cervus elaphus*)	End date antler growth	Temperature	249	2.5	31	Y (–)	N	–	–	(Y)
10	Red deer (*Cervus elaphus*)	Start date antler growth	Temperature	412	3.8	39	Y	Y	–	–	Y

Ref: (1) Porlier *et al.* (2012); (2) Brommer *et al.* (2005); (3) Brommer *et al.* (2008); (4) Husby *et al.* (2010); (5) Reed *et al.* (2006); (6) Reed *et al.* (2009); (7) Westneat *et al.* (2011); (8) Schwanz & Janzen (2008); (9) Nussey *et al.* (2005a); (10) Clements *et al.* (2010).

traits affected by climatic variables, carried out in 10 wild populations (Table 15.1). In all these studies, the population-level plasticity was also found on the individual level, although it was not necessarily so that individuals differed in their plasticity ($I \times E$).

The variance due to $I \times E$ presents an upper limit on the $G \times E$ variance (under the assumption that permanent environment–climate $PE \times E$ variance is absent, see Box 15.1). As a consequence, absence of significant $I \times E$ implies there is no $G \times E$. In total, 14 out of 19 estimates revealed significant heterogeneity across individuals in plasticity, indicating that $I \times E$, and thus possibly $G \times E$, is a pervasive feature of natural variation. This compilation should be interpreted carefully as there only are a limited number of studies and because it may suffer from publication bias in that studies which tested but did not find $I \times E$ variation may have remained unpublished. Out of these 19 studies, only three studies have further partitioned the $I \times E$ variance into additive genetic and permanent environment variances (Table 15.1). Of these, Brommer *et al.* (2005) used an approach, which has since been shown to be inappropriate (Hadfield *et al.* 2010). Evidence of a statistically significant $G \times E$ was only found for the clutch size–temperature relationship in great tits breeding in Wytham Woods in the United Kingdom (Husby *et al.* 2010; Table 15.1). The available estimates of $G \times E$ variance, however, suggest that the proportion of $I \times E$ variance due to $G \times E$ ('$G \times E/I \times E$' column in Table 15.1) is high, where $\geq$60% of $I \times E$ variance typically is due to $G \times E$. Taken together, these studies therefore suggest that heterogeneity across individuals in plasticity is common (although not ubiquitous), but that establishing the additive genetic basis of plasticity is challenging in wild populations. In particular, the relatively large effect sizes of $G \times E$ in combination with the observation that the study with the highest sample size (in terms of number of individuals) was the only one to detect significant $G \times E$, suggests that lack of statistical power (rather than small effect size) is the primary reason for statistical non-significance of $G \times E$. Calculation of power to detect $G \times E$ is, however, complicated and unsolved at present (Stinchcombe *et al.* 2012), as it depends on both the available information on relatives and the number of repeated observations per relative.

15.4.1 What does *G* × *E* tell us about the potential for evolutionary responses to climate change?

$G \times E$ has the potential to alter the rate of evolutionary change as environmental conditions change. It is not straightforward, however, to a priori predict the direction in which environmental conditions will affect a trait's heritability because both higher and lower heritabilities under sub-optimal conditions have been reported (Hoffmann & Parsons 1991; Hoffmann & Merilä 1999; Charmantier & Garant 2005). Investigation of existing long-term data allows documentation of the pattern of $G \times E$ under the conditions prevailing thus far. This information can provide an indication of the change in heritability in the direction of the projected climate change across different study systems. Clearly, this exercise assumes that the climatic covariate considered in each study captures the main driver for phenotypic plasticity and hence $G \times E$ (see Box 15.2). In addition, it also assumes that the environmental conditions experienced in the future can be considered as a linear extension of the environmental conditions experienced by the population thus far.

We here attempt to outline some general knowledge and lessons gained from studies in the wild exploring changes in variance components across environmental conditions (Table 15.2). Because the number of studies on $G \times E$ in response to climatic variables in wild populations is still limited, we can consider here only the pattern of $I \times E$ documented (studies listed in Table 15.1) to provide us with an upper estimate of $G \times E$. We furthermore require a ranking of the environmental conditions, which we do by contrasting 'poor' vs 'good' environmental conditions. We define good environmental conditions as those environmental conditions in which the mean trait value changes in the direction favoured by selection. For example, warmer temperatures are associated with earlier laying (Table 15.1), and the overall selection gradient shows that earlier laying leads to increased fitness (the direction of selection is typically detailed in the studies listed in Table 15.1). Note that we simply adhere to this definition in order to facilitate cross-study comparisons and that we do not consider whether the

BOX 15.2 Challenges in modelling plasticity

The main difficulty in assessing climate-related plasticity is identifying the relevant environmental covariate. Finding the aspect of the climatic conditions which is causally determining trait variation is almost impossible in wild populations of animals and would require detailed laboratory studies. It feels ecologically intuitive that an apparent climatic driver (for example, temperature) is in reality a correlate of the causal factors underlying trait variation. This, in turn, implies that future forecasts (based on changes in the presumed climatic driver) need to consider how stable the relationship between the apparent climate driver and the actual causal driver(s) is under climate change, although this will in many cases be challenging. Furthermore, plasticity always occurs in response to particular environmental conditions, and the exact environmental driver for a plastic response is thus likely to be population specific (Scheiner 1993) as has been shown for the egg-laying date–temperature relationship in four blue tit (*Cyanistes caeruleus*) populations in southern France (Porlier *et al.* 2012). Hence, generalisation across populations may be limited.

Models are typically constructed on the assumption that there is an annual description of climate which determines phenotypes. Because local climatic features such as temperature and rainfall, and also large-scale descriptors such as the North Atlantic Oscillation (NAO), are essentially daily events, a choice of a relevant time window during which the annual climatic condition is measured is required. The time window during which a descriptor of climate was calculated (e.g. average temperature) can either be based on prior ecological knowledge or chosen as the period which correlated highest with the trait of interest. Importantly, however, the choice of environmental covariate affects the conclusion regarding the presence of $I \times E$ (and thus the potential of $G \times E$). For example, Brommer *et al.* (2005) contrasted $I \times E$ models with three alternative climatic descriptions (temperature, rainfall, NAO), which all explained variation in a phenological trait, breeding time, in collared flycatchers (*Ficedula hypoleuca*); but significant $I \times E$ variation was found for two of these three covariates. Husby *et al.* (2010) showed for the Wytham Wood great tit population that using average temperature led to detecting significant $I \times E$ in laying date, whereas all evidence of $I \times E$ variance disappeared when using the sum of daily temperatures ('warmth sum') instead. Interestingly, average temperature and 'warmth sum' were good correlates of annual mean laying date and were strongly correlated ($r = 0.96$). This finding implies that the choice of climatic description may radically change conclusions regarding the variation in plasticity within a population ($I \times E$, and likely also $G \times E$). This worrying observation potentially has important ramification for understanding plasticity in natural populations, but has received little attention thus far. Tackling this issue is clearly an important future challenge and alternative approaches than averages or temperature sums over fixed periods to describe the environment may be needed, such as techniques developed by Gienapp *et al.* (2005) and van de Pol and Cockburn (2011). These approaches are based on modelling daily probabilities that a phenological event, e.g. a females starts egg-laying, occurs and do not have to rely on environmental variables calculated over fixed periods.

environmental conditions alter the selection gradient, or whether prolonged exposure to 'good' environmental conditions may in fact be detrimental for population fitness.

When doing such an exercise based on random-regression estimates, one further detail to carefully consider is whether one univariate (homogeneous) residual variance was assumed or whether residual variances were allowed to be heterogeneous across the environmental gradient. This distinction is important, because we are primarily interested here in changes in the between-individual variances across the climatic gradient (provided by $I \times E$), but the residual variance is also likely to change across an environmental gradient (Hoffmann & Merilä 1999). When a trait's phenotypic variance changes over an environmental gradient, there are two non-mutually explanations: (1) the between-individual variance changes over the environmental gradient, that is, there is $I \times E$; and (2) the within-individual (residual) variance changes over the environmental gradient. Interestingly, many studies which test for $I \times E$, and thus specifically allow for between-individual variance to change over an environmental gradient, do not allow for residual variance to change over that gradient, because they assume homogeneous residuals (Table 15.2). Whenever compared directly, heterogeneous

Table 15.2 The change in variance components across climatic conditions based on the studies listed in Table 15.1 (except Ref. 8). The study (Ref.) is referred to as listed in Table 1, where 'Res' indicates whether residual variances were assumed to be heterogeneous (het) or univariate (uni) across the climatic gradient. 'Direction' indicates the direction of the population-level phenotypic plasticity with the environmental condition. The direction is interpreted (Int) such that higher values of the environmental covariate characterise either a better environment (trait changes in a direction for which the overall selection gradient indicates fitness is higher) or a worse environment (trait changes in the direction of lower fitness). Based on the details presented in the study, the residual variance (V_R) and between-individual variance (V_I) are given for poor, unfavourable (U) and good, favourable (F) environmental conditions, where favourable and unfavourable are as defined by 'Int'. When an additional variance component was estimated, it is reported under V_{other}. Variances are summed to give the phenotypic variance (V_P) and the repeatability R ($= V_I/V_P$) under unfavourable and favourable conditions. For one set of estimates provided by Ref. 10, quantitative estimation of variance components was not possible, and we made the conservative assumption that the non-significant $I \times E$ variance reported was zero. The direction of the change in V_P (ΔV_P) and change in R (ΔR) from unfavourable to favourable environmental conditions is summarised as either decreasing (–) or increasing (+) if the change in V_P or R exceeds 5% or considered as equal (0) if the change is smaller than 5%

Ref.	Res	Direction	Int	V_R(U)	V_R(F)	V_I(U)	V_I(F)	ΔV_I	V_{other}	V_P(U)	V_P(F)	ΔV_P	R(U)	R(F)	ΔR
3	het	–	better	13.9	13.9	2.0	10.0	+	5.9	21.8	29.8	+	0.09	0.34	+
6	het	–	better	33.0	19.0	45.0	40.0	–	11.5	89.5	70.5	–	0.50	0.57	+
4	het	–	better	16.0	13.9	5.8	20.9	+	8.3	30.1	43.0	+	0.19	0.49	+
4	het	–	better	16.8	14.0	11.3	14.3	+	16.0	44.1	44.2	0	0.26	0.32	+
4	het	0	better	1.8	1.8	1.4	1.4	0	0.6	3.8	3.8	0	0.37	0.37	0
4	het	0	better	1.1	1.2	1.4	1.5	+	0.4	3.0	3.2	+	0.48	0.48	0
1	het	–	better	13.9	15.5	22.7	27.3	+		36.5	42.8	+	0.62	0.64	0
2	uni	–	better	4.7	4.7	9.2	6.2	–		13.9	10.9	–	0.66	0.57	–
9	uni	+	worse	248.3	248.3	68.8	82.8	+	18.0	335.1	349.1	0	0.21	0.24	+
5	uni	–	better	10.3	10.3	8.6	9.0	0	10.6	29.5	29.9	0	0.29	0.30	0
1	uni	–	better	23.6	23.6	45.6	59.6	+		69.2	83.2	+	0.66	0.72	+
1	uni	–	better	20.2	20.2	18.4	33.5	+		38.6	53.7	+	0.48	0.62	+
1	uni	–	better	13.5	13.5	22.0	44.8	+		35.5	58.3	+	0.62	0.77	+
7	uni	0	worse	31.6	31.6	0.3	4.4	+		31.9	36.0	+	0.01	0.12	+
10	uni	–	better	44.0	44.0	56.3	64.2	+	8.5	108.8	116.7	+	0.52	0.55	+
10	uni	–	better					0				0			0

residuals typically fit the data better than univariate residuals (e.g. Reed *et al.* 2009), and it is possible that by a priori assuming homogeneous residuals, any existing heterogeneity in the residuals will inflate the estimate of $I \times E$ variance (Nicolaus *et al.* 2013).

We estimated between-individual variances and residual variances based on reported values or we used the random-regression results to estimate between-individual variance in poor and good environmental conditions following Meyer (1998). Values for the random-regression variance components are provided in the original publications, but some parameter estimates were obtained by contacting the authors, and these values are provided by Brommer (2013). We included all random-regression estimates (variance in elevation, slope and the covariance between these), irrespectively of whether they were statistically significantly. However, the study by Schwanz & Janzen (2008; Ref. 8 in Table 15.1) had to be omitted because it estimated negative variance for between-individual variance in plasticity ($I \times E$), which does not allow estimation of individual variance across the climatic gradient. Further, Nussey *et al.* (2005a, Ref. 9 in Table 15.1) estimated $I \times E$ for the same trait–climate relationship in the same species and population but in two differing time periods, and we here include only the latter time period (see Table 15.1) in order to avoid over-representation. Lastly, we could not include quantitative estimates for one trait–climate relationship estimated by Clements *et al.* (2010; Ref. 10 in Table 15.1) because the random-regression estimates of this non-significant $I \times E$ were not provided. We therefore conservatively in this case assumed that there was no change in between-individual variance across the environmental gradient. We chose the environmental values at which to evaluate the random-regression to lie close to the minimal and maximal values as based on information presented in the original studies. We arbitrarily set a threshold value for what constitutes a change in phenotypic variance and repeatability at 5% of the minimal value recorded for poor and good environmental conditions.

The above quantitative exercise provided 16 estimates of change in variance components and repeatability across an environmental gradient (Table 15.2). Most (9/16) estimates suggested that phenotypic trait variance V_P indeed increased from poor to good conditions by more than 5%, and only two estimates implied that V_P decreased (five estimates suggested no change >5%). In addition, most (11/16) estimates indicated that between-individual variance V_I increased under good environmental conditions, and only two estimates suggested a decrease in V_I (three estimates suggested no change >5%). Many studies were done on avian seasonal breeding timing (egg-laying date), which tends to be earlier in the season under favourable conditions (warmer temperature). The increase in phenotypic and between-individual variance was therefore not merely due to mean–variance scaling effects. Across studies, there was clear consistency on how between-individual variance changed in good environmental conditions independent of whether residuals were modelled as univariate (V_I increased in 5/8 studies) or heterogeneous (V_I increased in 6/8 studies). Thus, we find no strong evidence that fitting homogeneous errors would drive $I \times E$ variance in a certain direction, although we note that for a specific analysis, estimates of $I \times E$ variance are likely to change depending on the structure of the residual variance (cf. Nicolaus *et al.* 2013). Importantly, as a consequence of these changes in variance components, repeatability ($R = V_I/V_P$) increased under good environmental conditions in most (9/16) studies, whereas a decrease in repeatability was only suggested by one estimate (five estimates suggested repeatability did not change >5%). In summary, our estimates calculated on the basis of published literature studies on trait–climate plasticity suggest that it is likely that trait repeatability increases when climatic conditions change for the 'better' (Table 15.2). Because repeatability is the upper estimate of heritability (Falconer & MacKay 1996), this suggests that climate change could be associated with an increase in the potential of phenological and life-history traits to respond to selection in the wild.

A further worthwhile extension is to also consider what the available data tell us about how the selection gradient β varies across the environmental conditions. Two studies have linked not only additive genetic variance but also the strength of selection to an environmental covariate. Wilson *et al.*

(2006) found a negative correlation between maternal genetic variance and β, and Husby *et al.* (2011b) a positive correlation between additive genetic variance (V_A) and β. Although these findings point to opposite directions, $G \times E$ variance was itself not statistically significant in either of these studies, which may explain their inconsistency.

15.4.2 The challenge of evolving plasticity

Climate change may lead to selection on plasticity itself, i.e. reaction norm slopes, either because 'cue' and 'selection' environment change at different rates or because the variance in the environmental conditions experienced by the population increases (de Jong 1990). Selection on the $I \times E$ slope of the laying date-temperature relationship was documented in great tits (Nussey *et al.* 2005b), but the approach used was later recognised as anti-conservative (Hadfield *et al.* 2010). Approaches to calculate selection on plasticity exist (Nussey *et al.* 2008; Brommer *et al.* 2012), but are at present not properly explored within the context of climate-driven plasticity. An important technical aspect is that the degree by which plasticity itself is able to evolve independently from the trait mean is given by the genetic correlation between the $G \times E$ reaction norms' elevation and slope(s). A high genetic correlation indicates that selection on the mean trait value (which is a close correlate of elevation) will also change the reaction norm slope, but these properties can evolve more independently when their genetic correlation is low. Only few (seven) estimates of the genetic correlation between reaction norm elevation and slope have been published, but these estimates suggest that the magnitude of the genetic correlation between reaction norm slope and elevation tends to be high, and thus that the potential for evolution of plasticity independent from the trait mean may be limited (Brommer 2013). This does not imply that plasticity could not evolve in response to climate change, but it does suggest that separating selection on and evolutionary response of mean trait expression and trait plasticity is likely to be challenging. Nevertheless, one striking finding of Husby *et al.* (2010) is that the sign of the genetic correlation between elevation and slope can be reversed when comparing the same trait–climate relationship between populations of the same species. Another example of a change in sign of the genetic correlation between elevation and slope of plasticity is provided by Dingemanse *et al.* (2012). These findings imply that the genetic variance–covariance structure of plasticity is not fixed, but can respond to population-specific selection pressures or can differ between populations due to genetic drift.

15.5 Evolutionary responses to climate change

15.5.1 Where is the evidence?

Climate change has left a recognisable 'fingerprint' on a number of traits, especially phenology (Root *et al.* 2004). The question however is whether the observed changes in traits are an evolutionary response, i.e. a genetic change in response to selection, or products of phenotypic plasticity. If we want to confirm evolutionary change we need to demonstrate that the population in question has changed genetically (and preferably also that corresponding selection pressures exist). There are basically three ways to demonstrate genetic change in natural populations, and although only one makes use of quantitative genetic methods (the animal model), we briefly review all of them and the evidence they could provide.

First, the most obvious and direct way to test for genetic change is to study change at the molecular genetic level. Several studies have found changes in chromosomal arrangements or functional genes in *Drosophila*, which are consistent with climate change (e.g. Umina *et al.* 2005; Balanya *et al.* 2006; Rodríguez-Trelles *et al.* 2013). The main problem with this approach is, however, that in most cases the molecular genetic architecture of the traits that are likely to come under selection, mainly life-history and behaviour, is not well known (but see Hancock *et al.* 2011; Lasky *et al.* 2012). For example, candidate loci explained only little to no variation in avian migratory behaviour and breeding time, traits that are likely to come under climate change-induced selection (Mueller *et al.* 2011; Liedvogel *et al.* 2012).

Second, a more 'classical' way to test whether observed phenotypic changes have a genetic basis or are the product of phenotypic plasticity is the common garden experiment where individuals are kept under standardised conditions, thereby eliminating environmentally caused differences. By comparing sets of experiments carried out under identical conditions but at different points in time, three studies could show that traits in insects and birds have evolved in the direction expected under climate change (Bradshaw & Holzapfel 2001; Pulido & Berthold 2010; van Asch *et al.* 2013). Another study used a slightly different approach and compared individuals of an annual plant before and after a multiyear drought by 'resurrecting' pre-drought individuals from stored seed (Franks *et al.* 2007). However, although a common garden approach is powerful and versatile, its application is limited to organisms amenable for rearing under controlled conditions.

Thirdly, quantitative genetic analyses can disentangle phenotypic from genetic trends by assessing temporal changes in breeding values. In principle, this approach can be applied in wild populations provided the relatedness between individuals can be established. A number of studies used this feature of the animal model to disentangle genetic from phenotypic patterns, e.g. changes over time or differentiation among specific groups (reviewed in Hadfield *et al.* 2010). A few studies explicitly looked for genetic changes as a response to climate change: for example, in red-billed gulls (*Chroicocephalus scopulinus*), body mass declined as ambient temperatures increased, but there was no evidence for genetic change; maybe unsurprisingly, as there was no evidence for selection on body mass, either (Teplitsky *et al.* 2008). Breeding time is under strong climate change-induced selection in a Dutch great tit population (Visser *et al.* 1998), but predicted breeding values did not change (Gienapp *et al.* 2006). The advancement of parturition date in a Canadian red squirrel could be shown to be partly phenotypically plastic and partly genetic indicated by a change in breeding values (Réale *et al.* 2003b). There are, however, several technical problems with this approach. First, an absence of a change in breeding values may simply be a statistical power problem due to little pedigree information, which is the likely reason why Gienapp *et al.* (2006) found no trend in breeding values despite strong and consistent selection. Second, care has to be taken to entirely remove any phenotypic trend from the data before analysing temporal changes in breeding values (Postma 2006). Breeding values will always be biased towards the phenotypes, and hence a trend in breeding values could simply be a reflection of a phenotypic trend if the trend in phenotypes is not controlled for in the model (Postma 2006). Removing a phenotypic trend, by fitting year in the model, will also control for plastic responses to an unidentified environmental variable. Thirdly, temporal trends in breeding values could also be due to correlated prediction error among cohorts and genetic drift, both of which should be accounted for before drawing conclusions about evolutionary changes (Hadfield *et al.* 2010). When reanalysing two data sets, in which previously significant changes in breeding values were found, taking these issues into account, the statistical significance of the temporal changes was greatly reduced (Hadfield *et al.* 2010). A more recent study that used the approach outlined by Hadfield *et al.* (2010) and analysed temporal trends in avian body mass found phenotypic declines but no evidence for genetic change (Husby *et al.* 2011b).

15.5.2 To what extent are observed phenotypic changes due to plasticity?

One obvious shortcoming in many studies documenting putative evolutionary changes in response to climate change is a lack of critical evaluation of plasticity as a driving force of change (Gienapp *et al.* 2008). Here, careful analyses of within-individual plastic changes can be used to test whether phenotypic plasticity alone can explain the observed population trend. If suitable data are available to reliably estimate phenotypic plasticity, i.e. repeated measurements from individually marked individuals, one can test whether population-wide trends are in line with a trend as expected from phenotypic plasticity (Figure 15.2a) or whether alternative processes, such as a micro-evolutionary response to selection, are playing an (additional) role (Figure 15.2b). This is simply done by regressing individual phenotypes against a suitable environmental

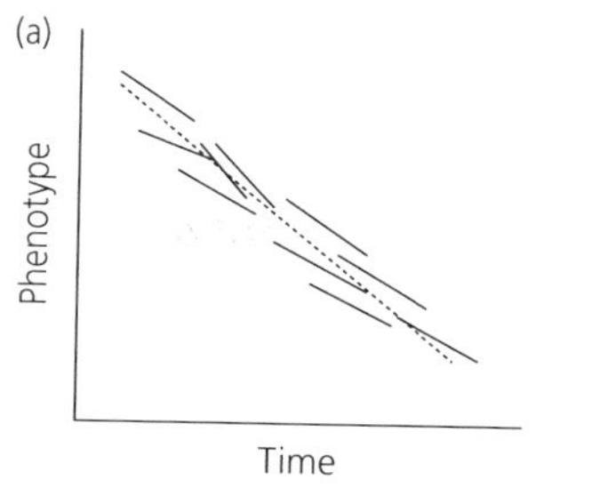

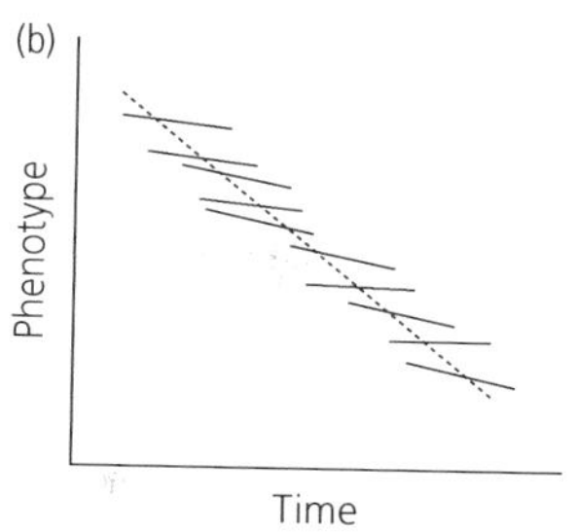

Figure 15.2 Disentangling phenotypic from genetic changes by analysing phenotypic plasticity. Any population mean change in a trait could be due to phenotypic plasticity or a genetic response to selection (or both). The dashed line indicates the temporal change at the population level (in a and b). The solid lines indicate temporal changes in individual phenotypes as predicted from a temporal change in the environment. In (a) the population mean trend can be explained by individual plasticity, but not in (b) because the individual reaction norms are shallower than population trend. Since the predicted temporal changes due to plasticity do not fully account for the observed population trend, other factors, as genetic change, have to drive it (if phenotypic plasticity has been reliably estimated).

variable using a random-regression framework (Box 15.1). To fully separate individual-level plasticity from population trends, which could possibly be genetic, within-individual centring of the environmental variable is often necessary (Kreft *et al.* 1995). In a second step, the mean of the individual reaction norms from the random-regression analysis and the observed values of the environmental variable are used to calculate the expected temporal trend due to phenotypic plasticity. For example, assume that a plasticity analysis reveals that individual birds start egg-laying earlier by 0.5 days per 1 °C and that temperatures have increased by 4 °C during the study period. These results lead to an expected advancement of two days during the study period, which is then compared to the observed population trend. If individual phenotypic plastic responses to environmental changes are the only driver of the observed trend, then the trend based on plasticity should agree with the observed trend. Any discrepancy between the two trends could be due to genetic changes but also to other processes. These other processes could be: a plastic response to a second environmental variable, plastic ontogenetic effects, or a change in the composition of immigrants. We discuss these in turn below.

A crucial assumption underlying any solid plasticity analysis is that all environmental factors responsible for the trait's plasticity have been identified (see Box 15.2). Otherwise, any phenotypic change could be due to plastic changes in response to changes in this, unidentified, environmental variable. A stronger or weaker advancement over time than expected from temperature-dependent phenotypic plasticity could for example be due to improved or worsened body condition. Second, a subtle form of plastic response to the environment is 'plastic ontogenetic effects'. If an individual's phenotype is affected by the environment only during its development (ontogenesis), but not later in life, a population trend could be a plastic response, whilst individuals would not show individual plasticity in relation to the environment. Finally, a changed immigration rate from a genetically differentiated population could also lead to a deviation between the observed trend and the trend predicted from mean individual plasticity.

Importantly, therefore, a discrepancy between observed trends with expectations from within-individual phenotypic plasticity will never be able to prove that a genetic change has occurred as there are alternative explanations. Nevertheless, this approach can offer useful first insights into the roles of plasticity and evolutionary change, which would not be possible to obtain otherwise. For example, if a trend predicted from individual plasticity matches the observed population trend, the most parsimonious explanation would be that the population trend is driven by phenotypic plasticity. A similar logic has, for example, been used to test for local adaptation in butterfly phenology and salmon run time to climate change (Crozier *et al.* 2011; Phillimore *et al.* 2012). We recommend this approach because there are ample published studies which have documented a strong phenotypic change and concluded it must be due to microevolution without a quantitative consideration of how substantial changes are possible assuming only phenotypic plasticity.

To illustrate how analyses of individual phenotypic plasticity can be used to test phenotypic vs genetic changes, we use datasets on colour morphs and body mass in birds. Colour-morph frequencies in tawny owls in southern Finland have changed, presumably in response to warmer winters (Karell *et al.* 2011). However, colour-morph does not show individual phenotypic plasticity in relation to winter climate (Figure 15.3), and the observed change in the highly heritable colour morphs may be due to a genetic change. Body weight of Siberian jays (*Perisoreus infaustus*) in southwestern Finland showed a marked decline over two decades followed by an equally strong increase over the following decade (Gienapp & Merilä, unpubl.). In this case no environmental variable driving these changes has (yet) been identified, so individual 'temporal reaction norms' were compared to population mean trends. This comparison showed that the observed trends could be explained by phenotypic plasticity (Figure 15.4). When this 'short-cut' is taken, care has to be taken that any analysis is restricted to periods during which the studied trait showed consistent temporal changes because otherwise individual phenotypic plasticity would be underestimated.

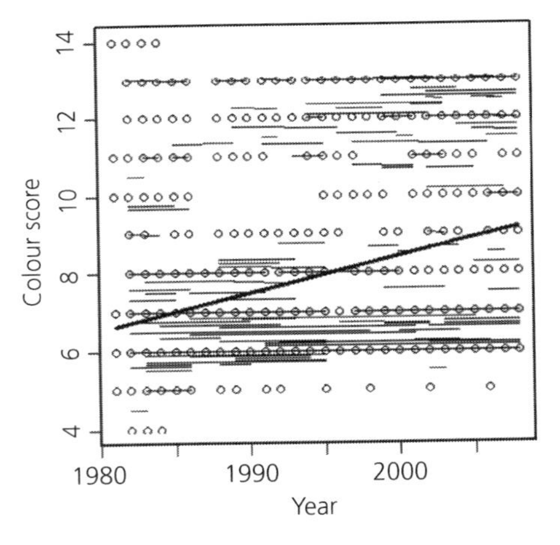

Figure 15.3 Population mean trends (thick line) and individual changes (thin lines) in colour morphs of tawny owls. Colour is scored from 4 (brown) to 14 (grey) based on the pigmentation of four different plumage parts (Karell *et al.* 2011). Colour-morph was regressed against a measure of winter climate, which was related to the population trend, and from these individual reaction norms temporal changes corresponding to the respective annual winter climates were calculated. Since the individual trends are shallower than the population mean trend this was some indication for genetic change, which was corroborated by other analyses (Karell *et al.* 2011).

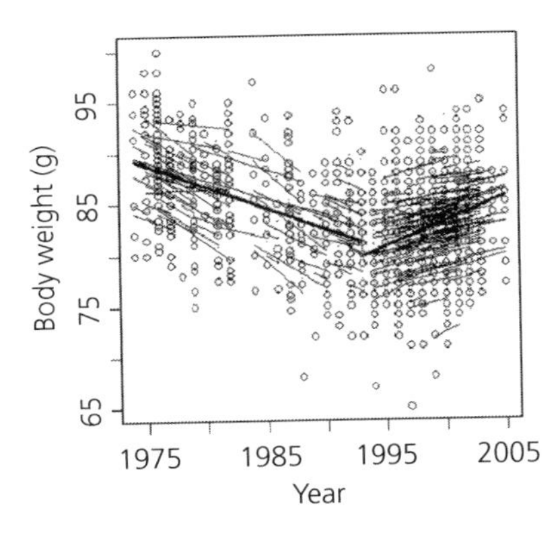

Figure 15.4 Population mean trends (thick line) and individual plasticity (thin lines) in body weight of Siberian jays. Since no environmental variable correlating with body weight could be identified, body weight was regressed against year, separately for the years 1974–1993 and 1994–2005, respectively, to estimate phenotypic plasticity. Reaction norms for this individual-level plasticity are plotted (thin lines). Since these reaction norms are similar to the population mean trends, phenotypic plasticity would be the most parsimonious explanation (Gienapp & Merilä, unpubl.).

15.6 Conclusions and outlook

The general evidence for genetic changes in response to climate change is still scarce and we may wonder whether this is a genuine lack of evolutionary adaptation or just a lack of suitable methods or the necessary high-quality data. Whilst Merilä (2012) states in his recent review that 'there appears to be little consensus—or even compelling general explanations—among evolutionary biologists as to why this is the case', we would however argue that evolutionary adaptation to environmental change is as common as evolutionary adaptation in general but that we 'simply' lack the appropriate methods for the existing data (see also Charmantier & Gienapp, 2014). As pointed out above, the use of molecular genetic or genomic approaches is constrained by our limited knowledge about suitable molecular markers or even functional genes for ecologically relevant traits in the majority of taxa. Quantitative genetic approaches could 'bridge this gap' but require data on individuals that are not always easily obtained, e.g. pedigrees are typically constructed by linking parents and offspring from behavioural information, which is difficult in species without parental care. In general, to address questions related to changes in response to climate change, long-term studies have to have been initiated and maintained and the relevant traits studied long before their relevance in the context of climate change was foreseeable. Consequently, most, if not all, available long-term data were not collected to study evolutionary effects of climate change, and this means that we have often extensive long-term datasets suitable for quantitative genetic analysis but on traits or populations unaffected by climate change.

Given these constraints imposed by existing data, we can ask whether quantitative genetics can still contribute to our understanding of evolutionary responses to climate change. A crucial requirement for 'classical' quantitative genetic studies in natural populations is the pedigree. However this may often be lacking, either simply due to uncollected data or due to the biology of the studied species. If DNA samples and (many) molecular markers are available, these markers can be used to reconstruct a pedigree or to directly estimate the relationship matrix (see Chapter 13, Jensen *et al.*). Early applications of these approaches gave unreliable and biased results due to the limited number of molecular markers used (Coltman 2005; Garant & Kruuk 2005). Increasing the number of markers is however now possible, since developing suitable molecular genetic markers and high-throughput genotyping, using e.g. SNPs, is now feasible also in molecular genetic 'non-model' species (Slate *et al.* 2010). Such approaches have been applied to natural populations in which a 'traditional' pedigree would have been impossible to obtain (e.g. Thériault *et al.* 2007; DiBattista *et al.* 2009; Koch *et al.* 2012, see also Chapter 9, Zajitschek & Bonduriansky). Clearly, these techniques also facilitate finding molecular markers underlying variation in functional traits and future quantitative genetic studies are likely to benefit from integrating this information to better model both the genetic architecture and the response of traits to selection.

To conclude, the evidence for evolutionary responses to climate change is still generally scarce, probably due to a lack of suitable data, but quantitative genetic studies in natural populations have made some contributions to this field and also have the potential to contribute further. Directly assessing genetic changes with the help of breeding values seems intuitively appealing but there are statistical issues which limit the usefulness of this approach in most study systems. Work on the genetic basis of phenotypic plasticity in wild populations is still a largely unexplored line of research. The statistical approach and concepts to study this issue are actively developed, and this avenue of research will hopefully become more important in the future with more studies explicitly considering $G \times E$ in a variety of systems. Overall, we point out that a better knowledge of the quantitative genetic architecture of ecologically relevant traits, e.g. whether low heritabilities or strong genetic correlations constrain evolutionary potential, is important for predicting whether populations are likely to adapt fast enough to climate change (e.g. Gienapp *et al.* 2013). By using molecular markers, it is now possible to carry out quantitative genetic analyses in species in which it has previously been impossible, which means that we can study species and populations that are affected by climate change rather than being constrained by available data. It is here that we envisage quantitative genetics contributing the most to studying evolutionary responses to climate change and its consequences.

References

Balanya, J., Oller, J.M., Huey, R.B., Gilchrist, G.W. & Serra, L. (2006) Global genetic change tracks global climate warming in *Drosophila subobscura*. *Science*, **313**, 1773–1775.

Berteaux, D., Réale, D., McAdam, A.G. & Boutin, S. (2004) Keeping pace with fast climate change: can arctic life count on evolution? *Integrative and Comparative Biology*, **44**, 140–151.

Björklund, M., Husby, A. & Gustafsson, L. (2013) Rapid and unpredictable changes of the G-matrix in a natural bird population over 25 years. *Journal of Evolutionary Biology*, **26**, 1–13.

Bradshaw, W.E. & Holzapfel, C.M. (2001) Genetic shift in photoperiodic response correlated with global warming. *Proceedings of the National Academy of Sciences of the United States of America*, **98**, 14509–14511.

Brommer, J.E. (2013) Phenotypic plasticity of labile traits in the wild. *Current Zoology*, **59**, 485–505.

Brommer, J.E., Kontiainen, P. & Pietiäinen, H. (2012) Selection on plasticity of seasonal life-history traits using random regression mixed model analysis. *Ecology and Evolution*, **2**, 695–704.

Brommer, J.E., Rattiste, K. & Wilson, A.J. (2008) Exploring plasticity in the wild: laying date-temperature reaction norms in the common gull *Larus canus*. *Proceedings of the Royal Society B-Biological Sciences*, **275**, 687–693.

Brommer, J.E., Sheldon, B.C., Gustafsson, L. & Merilä, J. (2005) Natural selection and genetic variation for reproductive reaction norms in a wild bird population. *Evolution*, **59**, 1362–1371.

Bürger, R. & Lynch, M. (1995) Evolution and extinction in a changing environment: a quantitative-genetic analysis. *Evolution*, **49**, 151–163.

Butchart, S.H.M., Walpole, M., Collen, B., van Strien, A., Scharlemann, J.P.W., Almond, R.E.A., Baillie, J.E.M., Bomhard, B., Brown, C., Bruno, J., Carpenter, K.E., Carr, G.M., Chanson, J., Chenery, A.M., Csirke, J., Davidson, N.C., Dentener, F., Foster, M., Galli, A., Galloway, J.N., Genovesi, P., Gregory, R.D., Hockings, M., Kapos, V., Lamarque, J.F., Leverington, F., Loh, J., McGeoch, M.A., McRae, L., Minasyan, A., Morcillo, M.H., Oldfield, T.E.E., *et al*. (2010) Global biodiversity: indicators of recent declines. *Science*, **328**, 1164–1168.

Carlson, S.M. & Seamons, T.R. (2008) A review of quantitative genetic components of fitness in salmonids: implications for adaptation to future change. *Evolutionary Applications*, **1**, 222–238.

Charmantier, A. & Garant, D. (2005) Environmental quality and evolutionary potential: lessons from wild populations. *Proceedings of the Royal Society B-Biological Sciences*, **272**, 1415–1425.

Charmantier, A. & Gienapp, P. (2014) Climate change and timing of avian breeding and migration: evolutionary *versus* plastic changes. *Evolutionary Applications*, **7**, 15–28.

Charmantier, A., McCleery, R.H., Cole, L.R., Perrins, C., Kruuk, L.E.B. & Sheldon, B.C. (2008) Adaptive phenotypic plasticity in response to climate change in a wild bird population *Science*, **320**, 800–803.

Charmantier, A., Perrins, C., McCleery, R.H. & Sheldon, B.C. (2006) Evolutionary response to selection on clutch size in a long-term study of the mute swan. *American Naturalist*, **167**, 453–465.

Clements, M.N., Clutton-Brock, T.H., Guinness, F.E., Pemberton, J.M. & Kruuk, L.E.B. (2010) Variances and covariances of phenological traits in a wild mammal population. *Evolution*, **65**, 788–801.

Coltman, D.W. (2005) Testing marker-based estimates of heritability in the wild. *Molecular Ecology*, **14**, 2593–2599.

Crozier, L.G., Scheuerell, M.D. & Zabel, R.W. (2011) Using time series analysis to characterize evolutionary and plastic responses to environmental change: A case study of a shift toward earlier migration date in sockeye salmon. *American Naturalist*, **178**, 755–773.

DiBattista, J.D., Feldheim, K.A., Garant, D., Gruber, S.H. & Hendry, A.P. (2009) Evolutionary potential of a large marine vertebrate: quantitative genetic parameters in a wild population. *Evolution*, **63**, 1051–1067.

DiBattista, J.D., Feldheim, K.A., Gruber, S.H. & Hendry, A.P. (2007) When bigger is not better: selection against large size, high condition, and fast growth in juvenile lemon sharks. *Journal of Evolutionary Biology*, **20**, 201–212.

Dingemanse, N.J., Barber, I., Wright, J. & Brommer, J.E. (2012) Quantitative genetics of behavioural reaction norms: genetic correlations between personality and behavioural plasticity vary across stickleback populations. *Journal of Evolutionary Biology*, **25**, 485–495.

Durant, J.M., Hjermann, D.Ø., Anker-Nilssen, T., Beaugrand, G., Mysterud, A., Pettorelli, N. & Stenseth, N.C. (2005) Timing and abundance as key mechanisms affecting trophic interactions in variable environments. *Ecology Letters*, **8**, 952–958.

Durant, J.M., Hjermann, D.O., Ottersen, G. & Stenseth, N. (2007) Climate and the match or mismatch between predator requirements and resource availability. *Climate Research*, **33**, 271–283.

Etterson, J.R. & Shaw, R.G. (2001) Constraint to adaptive evolution in response to global warming. *Science*, **294**, 151–154.

Falconer, D.S. & MacKay, T.F.C. (1996) *Introduction to quantitative genetics*. Longman, Harlow.

Franks, S.J., Sim, S. & Weis, A.E. (2007) Rapid evolution of flowering time by an annual plant in response to a climate fluctuation. *Proceedings of the National Academy of Sciences of the United States of America*, **104**, 1278–1282.

Garant, D., Hadfield, J.D., Kruuk, L.E.B. & Sheldon, B.C. (2008) Stability of genetic variance and covariance for reproductive characters in the face of climate change in a wild bird population. *Molecular Ecology*, **17**, 179–188.

Garant, D. & Kruuk, L.E.B. (2005) How to use molecular marker data to measure evolutionary parameters in wild populations. *Molecular Ecology*, **14**, 1843–1859.

Gardner, J.L., Heinsohn, R. & Joseph, L. (2009) Shifting latitudinal clines in avian body size correlate with global warming in Australian passerines. *Proceedings of the Royal Society B-Biological Sciences*, **276**, 3845–3852.

Geber, M.A. & Griffen, L.R. (2003) Inheritance and natural selection on functional traits. *International Journal of Plant Sciences*, **164**, S21–S42.

Ghalambor, C.K., McKay, J.K., Carroll, S.P. & Reznick, D.N. (2007) Adaptive versus non-adaptive phenotypic plasticity and the potential for contemporary adaptation in new environments. *Functional Ecology*, **21**, 394–407.

Gienapp, P., Hemerik, L. & Visser, M.E. (2005) A new statistical tool to predict phenology under climate change scenarios. *Global Change Biology*, **11**, 600–606.

Gienapp, P., Lof, M., Reed, T.E., McNamara, J., Verhulst, S. & Visser, M.E. (2013) Predicting demographically-sustainable rates of adaptation: can great tit breeding time keep pace with climate change? *Philosophical Transactions of the Royal Society of London Series B, Biological Sciences*, **368**, 20120289.

Gienapp, P., Postma, E. & Visser, M.E. (2006) Why breeding time has not responded to selection for earlier breeding in a songbird population. *Evolution*, **60**, 2381–2388.

Gienapp, P., Teplitsky, C., Alho, J.S., Mills, J.A. & Merilä, J. (2008) Climate change and evolution: disentangling environmental and genetic responses. *Molecular Ecology*, **17**, 167–178.

Gienapp, P., Väisänen, R.A. & Brommer, J.E. (2010) Latitudinal variation in breeding time reaction norms in a passerine bird. *Journal of Animal Ecology*, **79**, 836–842.

Gonzalez, A., Ronce, O., Ferriere, R. & Hochberg, M.E. (2013) Evolutionary rescue: an emerging focus at the intersection between ecology and evolution. *Philosophical Transactions of the Royal Society of London Series B, Biological Sciences*, **368**, 20120404.

Goodman, R.E., Lebuhn, G., Seavy, N.E., Gardali, T. & Bluso-Demers, J.D. (2012) Avian body size changes and climate change: warming or increasing variability? *Global Change Biology*, **18**, 63–73.

Graham, R.W. & Grimm, E.C. (1990) Effects of global climate change on the patterns of terrestrial biological communities. *Trends in Ecology & Evolution*, **5**, 289–292.

Hadfield, J.D., Wilson, A.J., Garant, D., Sheldon, B.C. & Kruuk, L.E.B. (2010) The misuse of BLUP in ecology and evolution. *American Naturalist*, **175**, 116–125.

Hancock, A.M., Brachi, B., Faure, N., Horton, M.W., Jarymowycz, L.B., Sperone, F.G., Toomajian, C., Roux, F. & Bergelson, J. (2011) Adaptation to climate across the *Arabidopsis thaliana* genome. *Science*, **334**, 83–86.

Hellmann, J.J. & Pineda-Krch, M. (2007) Constraints and reinforcement on adaptation under climate change: Selection of genetically correlated traits. *Biological Conservation*, **137**, 599–609.

Henderson, C.R. (1982) Analysis of covariance in the mixed model: higher-level, non-homogeneous, and random regressions. *Biometrics*, **38**, 623–640.

Hoffmann, A.A., Hallas, R.J., Dean, J.A. & Schiffer, M. (2003) Low potential for climatic stress adaptation in a rainforest *Drosophila* species. *Science*, **301**, 100–102.

Hoffmann, A.A. & Merilä, J. (1999) Heritable variation and evolution under favourable and unfavourable conditions. *Trends in Ecology & Evolution*, **14**, 96–101.

Hoffmann, A.A. & Parsons, P.A. (1991) *Evolutionary genetics and environmental stress*. Oxford University Press, Oxford.

Houle, D. (1992) Comparing evolvability of quantitative traits. *Genetics*, **130**, 195–204.

Husby, A., Hille, S.M. & Visser, M.E. (2011a) Testing mechanisms of Bergmann's rule: phenotypic decline but no genetic change in body size in three passerine bird populations. *American Naturalist*, **178**, 202–213.

Husby, A., Nussey, D.H., Visser, M.E., Wilson, A.J., Sheldon, B.C. & Kruuk, L.E.B. (2010) Contrasting patterns pf phenotypic plasticity in reproductive traits in two great tit (*Parus major*) populations. *Evolution*, **64**, 2221–2237.

Husby, A., Visser, M.E. & Kruuk, L.E.B. (2011b) Speeding up microevolution: the effects of increasing temperature on selection and genetic variance in a wild bird population. *PLoS Biology*, **9**, e1000585.

Karell, P., Ahola, K., Karstinen, T., Valkama, J. & Brommer, J.E. (2011) Climate change drives microevolution in a wild bird. *Nature Communications*, **2**, 208.

Kelly, M.W., Sanford, E. & Grosberg, R.K. (2012) Limited potential for adaptation to climate change in a broadly distributed marine crustacean. *Proceedings of the Royal Society B-Biological Sciences*, **279**, 349–356.

Koch, M., Wilson, A.J., Kerschbaumer, M., Wiedl, T. & Sturmbauer, C. (2012) Additive genetic variance of quantitative traits in natural and pond-bred populations of the Lake Tanganyika cichlid *Tropheus moorii*. *Hydrobiologia*, **682**, 131–141.

Kreft, I.G.G., Deleeuw, J. & Aiken, L.S. (1995) The effect of different forms of centering in hierarchical linear models. *Multivariate Behavioral Research*, **30**, 1–21.

Kruuk, L.E.B. & Hadfield, J.D. (2007) How to separate genetic and environmental causes of similarity between relatives. *Journal of Evolutionary Biology*, **20**, 1890–1903.

Lambrechts, M.M. & Perret, P. (2000) A long photoperiod overrides non-photoperiodic factors in blue tits' timing of reproduction. *Proceedings of the Royal Society B-Biological Sciences*, **267**, 585–588.

Lane, J.E., Kruuk, L.E.B., Charmantier, A., Murie, J.O., Coltman, D.W., Buoro, M., Raveh, S. & Dobson, F.S. (2011) A quantitative genetic analysis of hibernation emergence date in a wild population of Columbian ground squirrels. *Journal of Evolutionary Biology*, **24**, 1949–1959.

Lasky, J.R., Des Maraus, D.L., McKay, J.K., Richards, J.H., Juenger, T.E. & Keitt, T.H. (2012) Characterizing genomic variation of *Arabidopsis thaliana*: the roles of geography and climate. *Molecular Ecology*, **21**, 5521–5529.

Liedvogel, M., Cornwallis, C.K. & Sheldon, B.C. (2012) Integrating candidate gene and quantitative genetic approaches to understand variation in timing of breeding in wild tit populations. *Journal of Evolutionary Biology*, **25**, 813–823.

Lynch, M. & Lande, R. (1993) Evolution and Extinction in Response to Environmental Change. In: *Biotic Interactions and Global Change* (ed. P. M. Kareiva, J. G. Kingsolver & R. B. Huey), pp. 251–266. Sinauer Ass., Sunderland, Massachusetts.

Lynch, M. & Walsh, B. (1998) *Genetics and analysis of quantitative traits*. Sinauer, Sunderland.

Merilä, J. (2012) Evolution in response to climate change: in pursuit of the missing evidence. *Bioessays*, **34**, 811–818.

Mousseau, T.A. & Roff, D.A. (1987) Natural selection and the heritability of fitness components. *Heredity*, **59**, 181–198.

Mueller, J.C., Pulido, F. & Kempenaers, B. (2011) Identification of a gene associated with avian migratory behaviour. *Proceedings of the Royal Society B-Biological Sciences*, **278**, 2848–2856.

Nicolaus, M., Brommer, J.E., Ubels, R., Tinbergen, J.M. & Dingemanse, N.J. (2013) Cryptic patterns of variation in clutch size: density reaction norms in a wild passerine bird. *Journal of Evolutionary Biology*, **26**, 2031–2043.

Nussey, D.H., Clutton-Brock, T.H., Elston, D.A., Albon, S.D. & Kruuk, L.E.B. (2005a) Phenotypic plasticity in a maternal trait in red deer. *Journal of Animal Ecology*, **74**, 387–396.

Nussey, D.H., Postma, E., Gienapp, P. & Visser, M.E. (2005b) Selection on heritable phenotypic plasticity in a wild bird population. *Science*, **310**, 304–306.

Nussey, D.H., Wilson, A.J. & Brommer, J.E. (2007) The evolutionary ecology of individual phenotypic plasticity in wild populations. *Journal of Evolutionary Biology*, **20**, 831–844.

Nussey, D.H., Wilson, A.J., Morris, A., Pemberton, J., Clutton-Brock, T. & Kruuk, L.E.B. (2008) Testing for genetic trade-offs between early- and late-life reproduction in a wild red deer population. *Proceedings of the Royal Society B-Biological Sciences*, **275**, 745–750.

Ozgul, A., Childs, D.Z., Oli, M.K., Armitage, K.B., Blumstein, D.T., Olson, L.E., Tuljapurkar, S. & Coulson, T. (2010) Coupled dynamics of body mass and population growth in response to environmental change. *Nature*, **466**, 482–485.

Parmesan, C. (2006) Ecological and evolutionary responses to recent climate change. *Annual Review of Ecology, Evolution and Systematics*, **37**, 637–669.

Phillimore, A.B., Stålhandske, S., Smithers, R.J. & Bernard, R. (2012) Dissecting the contributions of plasticity and local adaptation to the phenology of a butterfly and its host plants. *American Naturalist*, **180**, 655–670.

Porlier, M., Charmantier, A., Bourgault, P., Perret, P., Blondel, J. & Garant, D. (2012) Variation in phenotypic plasticity and selection patterns in blue tit breeding time: between- and within-population comparisons. *Journal of Animal Ecology*, **81**, 1041–1051.

Post, E. & Forchhammer, M.C. (2008) Climate change reduces reproductive success of an Arctic herbivore through trophic mismatch. *Philosophical Transactions of the Royal Society of London Series B, Biological Sciences*, **363**, 2369–2375.

Postma, E. (2006) Implications of the difference between true and predicted breeding values for the study of natural selection and micro-evolution. *Journal of Evolutionary Biology*, **19**, 309–320.

Przybylo, R., Sheldon, B.C. & Merilä, J. (2000) Climatic effects on breeding and morphology: evidence for phenotypic plasticity. *Journal of Animal Ecology*, **69**, 395–403.

Pulido, F. (2007) Phenotypic changes in spring arrival: evolution, phenotypic plasticity, effects of weather and condition. *Climate Research*, **35**, 5–23.

Pulido, F. & Berthold, P. (2010) Current selection for lower migratory activity will drive the evolution of residency in a migratory bird population. *Proceedings of the National Academy of Sciences of the United States of America*, **107**, 7341–7346.

Réale, D., Berteaux, D., McAdam, A.G. & Boutin, S. (2003a) Lifetime selection on heritable life-history traits in a natural population of red squirrels. *Evolution*, **57**, 2416–2423.

Réale, D., McAdam, A.G., Boutin, S. & Berteaux, D. (2003b) Genetic and plastic responses of a northern mammal to climate change. *Proceedings of the Royal Society B-Biological Sciences*, **270**, 591–596.

Reed, T.E., Wanless, S., Harris, M.P., Frederiksen, M., Kruuk, L.E.B. & Cunningham, E.J.A. (2006) Responding to environmental change: plastic responses vary little in a synchronous breeder. *Proceedings of the Royal Society B-Biological Sciences*, **273**, 2713–2719.

Reed, T.E., Warzybok, P., Wilson, A.J., Bradley, R.W., Wanless, S. & Sydeman, W.J. (2009) Timing is everything: flexible phenology and shifting selection in a colonial seabird. *Journal of Animal Ecology*, **78**, 376–387.

Robinson, M.R., Wilson, A.B., Pilkington, J.G., Clutton-Brock, T., Pemberton, J. & Kruuk, L.E.B. (2009) The impact of environmental heterogeneity on genetic architecture in a wild population of Soay sheep. *Genetics*, **181**, 1639–1648.

Rodríguez-Trelles, F., Tarrío, R. & Santos, M. (2013) Genome-wide evolutionary response to a heat wave in *Drosophila*. *Biology Letters*, **9**, 20130228.

Root, T.L., Price, J.T., Hall, K.R., Schneider, S.H., Rosenzweigk, C. & Pounds, J.A. (2004) Fingerprints of global warming on wild animals and plants. *Nature*, **421**, 57–60.

Scheiner, S.M. (1993) Genetics and evolution of phenotypic plasticity. *Annual Review of Ecology and Systematics*, **24**, 35–68.

Schwanz, L.E. & Janzen, F.J. (2008) Climate change and temperature-dependent sex determination: can individual plasticity in nesting phenology prevent extreme sex ratios? *Physiological and Biochemical Zoology*, **81**, 826–834.

Seamons, T.R., Bentzen, P. & Quinn, T.P. (2007) DNA parentage analysis reveals inter-annual variation in selection: results from 19 consecutive brood years in steelhead trout. *Evolutionary Ecology Research*, **9**, 409–431.

Sheldon, B.C., Kruuk, L.E.B. & Merila, J. (2003) Natural selection and inheritance of breeding time and clutch size in the collared flycatcher. *Evolution*, **57**, 406–420.

Slate, L., Santure, A.E., Feulner, P.G.D., Brown, E.A., Ball, A.D., Johnston, S.E. & Gratten, J. (2010) Genome mapping in intensively studied wild vertebrate populations. *Trends in Genetics*, **26**, 275–284.

Stinchcombe, J.R., Function-valued Traits Working Group & Kirkpatrick, M. (2012) Genetics and evolution of function-valued traits: understanding environmentally responsive phenotypes. *Trends in Ecology & Evolution*, **27**, 637–647.

Teplitsky, C., Mills, J.A., Alho, J.S., Yarrall, J.W. & Merilä, J. (2008) Bergmann's rule and climate change revisited: disentangling environmental and genetic responses in a wild bird population. *Proceedings of the National Academy of Sciences of the United States of America*, **105**, 13492–13496.

Thackeray, S.J., Sparks, T.H., Frederiksen, M., Burthe, S., Bacon, P.J., Bell, J.R., Botham, M.S., Brereton, T.M., Bright, P.W., Carvalho, L., Clutton-Brock, T., Dawson, A., Edwards, M., Elliott, J.M., Harrington, R., Johns, D., Jones, I.D., Jones, J.T., Leech, D.I., Roy, D.B., Scott, W.A., Smith, M., Smithers, R.J., Winfield, I.J. & Wanless, S. (2010) Trophic level asynchrony in rates of phenological change for marine, freshwater and terrestrial environments. *Global Change Biology*, **16**, 3304–3313.

Thériault, V., Garant, D., Bernatchez, L. & Dodson, J.J. (2007) Heritability of life-history tactics and genetic correlation with body size in a natural population of brook charr (*Salvelinus fontinalis*). *Journal of Evolutionary Biology*, **20**, 2266–2277.

Thomas, C.D., Cameron, A., Green, R.E., Bakkenes, M., Beaumont, L.J., Collingham, Y.C., Erasmus, B.F.N., de Siqueira, M.F., Grainger, A., Hannah, L., Hughes, L., Huntley, B., van Jaarsveld, A.S., Midgley, G.F., Miles, L., Ortega-Huerta, M.A., Peterson, A.T., Phillips, O.L. & Williams, S.E. (2004) Extinction risk from climate change. *Nature*, **427**, 145–148.

Umina, P.A., Weeks, A.R., Kearney, M.R., McKechnie, S.W. & Hoffmann, A.A. (2005) A rapid shift in a classic clinal pattern in *Drosophila* reflecting climate change. *Science*, **308**, 691–693.

Valtonen, A., Ayres, M.P., Roininen, H., Poyry, J. & Leinonen, R. (2011) Environmental controls on the phenology of moths: predicting plasticity and constraint under climate change. *Oecologia*, **165**, 237–248.

van Asch, M., Salis, L., Holleman, L.J.M., van Lith, B. & Visser, M.E. (2013) Evolutionary response of the egg hatching date of a herbivorous insect under climate change. *Nature Climate Change*, **3**, 244–248.

van Buskirk, J., Mulvihill, R.S. & Leberman, R.C. (2010) Declining body sizes in North American birds associated with climate change. *Oikos*, **119**, 1047–1055.

van de Pol, M. & Cockburn, A. (2011) Identifying the critical climatic time window that affects trait expression. *American Naturalist*, **177**, 698–707.

van Noordwijk, A.J., McCleery, R. & Perrins, C. (1995) Selection for the timing of great tit breeding in relation to caterpillar growth and temperature. *Journal of Animal Ecology*, **64**, 451–458.

Via, S. & Lande, R. (1985) Genotype-environment interaction and the evolution of phenotypic plasticity. *Evolution*, **39**, 505–522.

Visser, M.E. & Both, C. (2005) Shifts in phenology due to global climate change: the need for a yardstick. *Proceedings of the Royal Society B-Biological Sciences*, **272**, 2561–2569.

Visser, M.E., Holleman, L.J.M. & Gienapp, P. (2006) Shifts in caterpillar biomass phenology due to climate change and its impact on the breeding biology of an insectivorous bird. *Oecologia*, **147**, 164–172.

Visser, M.E., van Noordwijk, A.J., Tinbergen, J.M. & Lessells, C.M. (1998) Warmer springs lead to mistimed reproduction in great tits (*Parus major*). *Proceedings of the Royal Society B-Biological Sciences*, **265**, 1867–1870.

Westneat, D.F., Hatch, M.I., Wetzel, D.P. & Ensminger, A.L. (2011) Individual variation in parental care reaction norms: integration of personality and plasticity. *American Naturalist*, **178**, 652–667.

Wilson, A.J., Pemberton, J.M., Pilkington, J.G., Coltman, D.W., Mifsud, D.V., Clutton-Brock, T.H. & Kruuk, L.E.B. (2006) Environmental coupling of selection and heritability limits evolution. *PLoS Biology*, **4**, e216.

Yeh, P.J. & Price, T.D. (2004) Adaptive phenotypic plasticity and the successful colonization of a novel environment. *The American Naturalist*, **164**, 531–542.

Yom-Tov, Y. (2001) Global warming and body mass decline in Israeli passerine birds. *Proceedings of the Royal Society B-Biological Sciences*, **268**, 947–952.

Index

Printed and bound by CPI Group (UK) Ltd, Croydon, CR0 4YY